Receptor Purification

Receptor Purification

Volume 1

Receptors for CNS Agents, Growth Factors, Hormones, and Related Substances

Edited by

Gerald Litwack

Fels Institute for Cancer Research and Molecular Biology, Philadelphia, PA

Humana Press
Clifton, New Jersey

Library of Congress Cataloging in Publication Data

Receptor purification / edited by Gerald Litwack
p. cm.
Includes index.
Contents: v. 1. Receptors for CNS agents, pituitary growth factors, hormones, and related substances.
ISBN 0-89603-167-5
1. Neurotransmitter receptors—Purification. 2. Hormone receptors—Purification. Cell receptors—Purification.
I. Litwack, Gerald.
[DNLM: 1. Receptors, Endogenous Substances—isolation & purification. QU 55 R2956]
QP364.7.R4278 1990
612.4—dc20
DLC
for Library of Congress 90-4689
CIP

Crescent Manor
PO Box 2148
Clifton, NJ 07015

Printed in the United States of America

Preface

The purpose of these volumes is to provide a reference work for the methods of purifying many of the receptors we know about. This becomes increasingly important as full-length receptors are overexpressed in bacteria or in insect cell systems. A major problem for abundantly expressed proteins will be their purification. In addition to purification protocols, many other details can be found concerning an individual receptor that may not be available in standard texts or monographs. No book of this type is available as a compendium of purification procedures.

Receptor Purification provides protocols for the purification of a wide variety of receptors. These include receptors that bind: neurotransmitters, polypeptide hormones, steroid hormones, and ligands for related members of the steroid supergene family and others, including receptors involved in bacterial motion. The text of this information is substantial, so as to require its publication in two volumes. Consequently, a division was made by grouping receptors by the nature of their ligands. Thus, in Volume One there are contributions on serotonin receptors, adrenergic receptors, the purification of GTP-binding proteins, opioid receptors, neurotensin receptor, luteinizing hormone receptor, human chorionic gonadotropin receptor, follicle stimulating hormone receptor, thyrotropin receptor, prolactin receptor, epidermal growth factor receptor, platelet derived growth factor receptor, colony stimulating factor receptor, insulin-like growth factor receptors, insulin receptor, fibronectin receptor, interferon receptor, and the cholecystokinin receptor. In some cases, for the better known receptors, more than one manuscript has been sought and included in this collection because of the manifold approaches to the purification of a given receptor. Representation of several approaches could be beneficial and outweigh the costs of increased numbers of pages. Also, for some receptors, purification is tailored for different forms of the receptor.

Many of the contributions to these volumes are from laboratories that lead the field not only in purification but in the phenomenology of receptors, both at the protein level and at the molecular biological level.

Any gaps in coverage of the specific receptors usually result from difficulties in obtaining manuscripts from representative laboratories or generate from contracts with laboratories that failed ultimately to provide a manuscript. Nevertheless, the coverage of various receptor types is broad and should be of considerable benefit to researchers in this overall area.

We now begin to witness the overexpression of full-length receptors in bacterial or baculovirus systems. Overexpression solves an important problem in receptor research in that it overcomes the difficulties of concentration typified by many receptors that occur in the nanomolar range. With the advent of overexpression systems it is now possible to generate receptors in the micromolar range advancing their status from trace proteins to abundant cellular proteins. These proteins need to be purified from the cells overexpressing them. Such purification processes must rely heavily on the information garnered for purification of the same receptor from various cells and tissues. This information is surveyed in this two volume work and appears at a time coinciding with the need to purify receptors from overexpression systems. The presentation of information about each receptor is complete enough so that all of the protocols required for the purification process are available without further pursuit of the literature for precise details. Moreover, in some cases, especially in Volume Two, there are contributions on specific agents useful for covalent binding of ligands to their receptors. In many cases information on cloning is presented that may be helpful to investigators considering overexpression systems. The material available in this two volume work should be of great use to biologists, biochemists, pharmacologists, molecular biologists, and endocrinologists, including graduate students as well as sophistocated workers.

I thank the Publisher for responslveness and rapid generation of these volumes in a diligent manner. I am indebted to the contributors for their conscientious preparation of manuscripts. I, and the contributors, are further indebted to a wide variety of journals and other publications that so freely allowed production of published data appearing in these volumes. Denise Valentine helped in the preparation of the index and other typing responsibilities.

Contents

Contributors

Takashi Akamizu • National Institutes of Health, Bethesda, MD
Om P. Bahl • State University of New York, Buffalo, NY
Robert C. Baxter • Department of Endocrinology, Royal Prince Alfred Hospital, Camperdown, Australia
Ricardo Brentani • Ludwig Institute for Cancer Research, Sao Paulo, Brazil
Joëlle Chabry • Centre de Biochimie du Centre National de la Recherche Scientifique, Nice, France
Deirdre Cooney • Department of Pharmacology, University College, Dublin, Ireland
Bosukonda Dattatreyamurty • Department of Chemistry, Albany Medical College, Albany, NY
Michele De Luca • Instituto Nazionale per la Recerca sul cancro, Genoa, Italy
Maria L. Dufau • National Institutes of Health, Bethesda, MD
Frances M. Finn • Department of Medicine, University of Pittsburgh, Pittsburgh, PA
Dina G. Fischer • Department of Virology, The Weizmann Institute of Science, Rehovot, Israel
Yoko Fujita-Yamaguchi • Department of Molecular Genetics, Beckman Research Institute of the City of Hope, Duarte, CA
Timothy J. Gallaher • Department of Biology, University of California, Santa Cruz, CA
Arieh Gertler • Department of Biochemistry and Human Nutrition, The Hebrew University of Jerusalem, Rehovot, Israel
Theresa L. Gioannini • Department of Psychiatry, New York University Medical Center, NY, NY

Henri J. Gozlan • INSERM, Paris, France
Michel Hamon • INSERM, Paris, France
Leonard C. Harrison • Walter and Eliza Hall Institute of Medical Research, Royal Melbourne Hospital, Parkville, Victoria, Australia
Carl-Henrik Heldin • Ludwig Institute for Cancer Research, Uppsala, Sweden
M. D. Hollenberg • Department of Pharmacology and Therapeutics, University of Calgary, Calgary, Alberta, Canada
Klaus Hofmann • Department of Medicine, University of Pittsburgh, Pittsburgh, PA
J. Justin Hsuan • Ludwig Institute for Cancer Research, London, UK
Alan K. Keenan • Department of Pharmacology, University College, Dublin, Ireland
Leonard D. Kohn • National Institutes of Health, Bethesda, MD
Thomas R. LeBon • Department of Molecular Genetics, Beckman Research Institute of the City of Hope, Duarte, CA
Peter J. Leedman • Walter and Eliza Hall Institute of Medical Research, Royal Melbourne Hospital, Parkville, Victoria, Australia
Craig C. Malbon • Department of Pharmacology, State University of New York, Stony Brook, NY
J. M. Maturo, III • C. W. Post College, Long Island University, Greenvale, NY
Jean Mazalla • Centre de Biochimie du Centre National de la Recherche Scientifique, Nice, France
Salah El Mestikawy • INSERM, Paris, France
Laurence J. Miller • Gastroenterology Unit, Mayo Medical School and Graduate School of Medicine, Rochester, MN
Julie D. Newman • Walter and Eliza Hall Institute of Medical Research, Royal Melbourne Hospital, Parkville, Victoria, Australia
Daniela Novick • Department of Virology, The Weizmann Institute of Science, Rehovot, Israel

George Panayotou • Ludwig Institute for Cancer Research, London, UK
Renata Pasqualini • Ludwig Institute for Cancer Research, Sao Paulo, Brazil
Leo E. Reichert, Jr. • Department of Chemistry, Albany Medical College, Albany, NY
Lars Rönstrand • Ludwig Institute for Cancer Research, Uppsala, Sweden
Menachem Rubinstein • Department of Virology, The Weizmann Institute of Science, Rehovot, Israel
Masamichi Satoh • Department of Pharmacology, Kyoto University, Kyoto, Japan
Carolyn D. Scott • Department of Endocrinology, Royal Prince Alfred Hospital, Camperdown, Australia
Eric D. Simon • Department of Psychiatry, New York University Medical Center, NY, NY
Hakimuddin T. Sojar • State University of New York, Buffalo, NY
E. Richard Stanley • Albert Einstein College of Medicine, Bronx, NY
Hiroshi Ueda • Department of Pharmacology, Kyoto University, Kyoto, Japan
Jean-Pierre Vincent • Centre de Biochimie du Centre National de la Recherche Scientifique, Nice, France
Howard H. Wang • Department of Biology, University of California, Santa Cruz, CA
Michael D. Waterfield • Ludwig Institute for Cancer Research, London, UK
Kenneth Y. Yamada • Laboratory of Molecular Biology, National Cancer Institute, Bethesda, MD
Susan S. Yamada • Laboratory of Molecular Biology, National Cancer Institute, Bethesda, MD
Yee-Guide Yeung • Albert Einstein College of Medicine, Bronx, NY
Nicole Zsurger • Centre de Biochimie du Centre National de la Recherche Scientifique, Nice, France

Serotonin Receptors

Timothy K. Gallaher
and Howard H. Wang

1. Background

1.1. Serotonin Receptors

Serotonin (5-hydroxytryptamine or 5-HT) is an indole neurotransmitter of the central and peripheral nervous systems. Pharmacological studies of serotonin receptors have revealed considerable heterogeneity of serotonin binding sites; as a result, a variety of receptor subtypes have been designated. Biochemical approaches have also revealed the involvement of second messengers in the serotonin-mediated receptor mechanism. Results obtained from autoradiographic methods combined with information from pharmacological studies have allowed the anatomical localization of the serotonin sensitive neurons in the central nervous system. The amino acid sequence of the receptor can be deduced from purified proteins and by molecular biological techniques. Further structural studies and studies of the functional receptor complex, however, require protein preparations isolated from brain tissues. This article will present a method for isolating serotonin receptors by affinity chromatography.

1.1.1. Pharmacological Definition of Serotonin Receptors

Peroutka and Snyder first demonstrated the heterogeneity of serotonin receptors (Peroutka and Snyder, 1979) based on a pharmacological analysis of binding to cortical membrane fragments by radiolabeled serotonin, lysergic acid diethylamide and spiperone. Serotonin was observed to bind to two distinct sites with a thousand-fold difference in affinity.

These sites were designated the 5-HT1 site, that binds serotonin at the nanomolar range, and the 5-HT2 site, that binds serotonin at the micromolar range, but binds spiperone at the nanomolar range. The hallucinogenic drug D-lysergic acid diethylamide (LSD) binds to both sites with high affinity. Thus, the concept of multiple serotonin receptors was introduced. Further, heterogeneity of 5-HT1 sites was demonstrated (Pedigo et al., 1981) by the ability of spiperone to compete, at low concentrations, with serotonin for some 5-HT1 sites. A subpopulation of 5-HT1a sites, that has a high affinity for spiperone and serotonin, was defined and designated 5HT1a. The 5-HT1a site has been further examined. The ligand (±)-8-hydroxy-2-(di-*N*-propylamino)tetralin (OH-DPAT) (Gozlan et al., 1983), was demonstrated to be the most selective ligand for labeling these sites. Since 5-HT1a sites were observed, and consist of a subpopulation of 5-HT1 sites, the 5-HT1b receptors were defined by exclusion. A specific ligand for the 5-HT1b site has yet to be developed (the most selective is RU 24969) (Hoyer et al., 1985). Since the definition of 5-HT1a and 1b receptors, two other classes of 5-HT1 receptors have been observed and designated 5-HT1c (Yagaloff and Hartig, 1985) and 5-HT1d (Heuring and Peroutka, 1987). The 5-HT2 site is labeled in the nanomolar range by spiperone, which also will label 5-HT1a sites and dopamine receptors. Since the earlier observation of the two main 5-HT subtypes, ketanserin (Leysen et al., 1982) and ritanserin (Korstanje et al., 1986) have been developed as the ligands of choice for labeling the 5-HT2 receptor owing to their high affinity and specificity for the site.

1.1.2. Serotonin Regulates Two Second Messenger Systems: Cyclic AMP and Inositol Triphosphate Formation

The ability of 5-HT to regulate second messenger systems through their effector enzymes via a GTP binding protein is now well established, but not completely understood. 5-HT was first shown to stimulate cAMP formation in newborn rat brain with a half-maximal response in the mid-nanomolar range (Von Hungen et al., 1975). Since then, 5-HT has also been shown to stimulate cAMP formation in adult rat brain (Barbaccia et al., 1983). Recent evidence suggests two sites may be involved in this process, one with a much lower activation concentration, in the Kd range of 5-HT1 type receptors (Shenkar et al., 1987). There is also evidence 5-HT inhibits cAMP formation by activating a 5-HT1a receptor (Dumuis et al., 1988). 5-HT inhibits cAMP formation in nonneural cells via 5HT1b receptors (Murphy and Bylund, 1988). Inositol triphosphate formation has been observed to be induced by 5-HT via the 5-HT1c (Conn et al., 1986) and the 5-HT2 receptors (Conn and Sanders-Bush, 1986). Although there is disagreement, the pharmacological evidence strongly attributes IP3 formation to 5-HT2 receptors (Godfrey et al., 1988). Nonneural cells and tissues have also shown response to serotonin with 5-HT2 pharmacology in the formation of IP3 (Schacter et al., 1985). The overall consensus is that cAMP is mediated by 5-HT1 receptors in either a positive or negative manner and 5-HT2 receptors and 5-HT1c receptors regulate IP3 formation. Biochemical studies support these observations with respect to the effect of guanine triphosphate nucleotides on ligand binding. It appears that for 5-HT to stimulate or attenuate a second messenger response, it must act via a guanine nucleotide binding protein. The 5-HT1a (Schlegel and Peroutka, 1986) receptor and 5-HT2 (Battaglia et al., 1984) receptors are modulated by guanine nucleotides in a way that is consistent with the mechanism of receptor mediated stimulation of second messengers. The binding affinity decreases in the presence of the nucleotide corresponding to an allosteric transition involved in the decoupling of the receptor-G-protein complex. No definite biochemical role has as yet been ascribed to the 5-HT1d receptor; however, it is known to be coupled to a G-protein (Herrick-Davis et al., 1988).

1.1.3. Physiological Effects of 5-HT

In hippocampus CA1 neurons, three different potassium conductance changes have been observed to be mediated by serotonin in sometimes opposite responses (Colino and Halliwell, 1987). One response is believed to be attributable to 5-HT1a receptors because the response was blocked by low concentrations of spiperone, or OH-DPAT acted as a partial agonist. Other studies have also reported OH-DPAT to be a partial agonist in this response (Andrade and Nicoll, 1987). This response consists of a Ca^{2+}-independent hyperpolarization where the increased K^+ conductance is mediated directly by a G-protein. Pertussis toxin blocks this hyperpolarization response. No second messenger system has been directly identified for the hyperpolarizing response although arachidonic acid metabolites have not been ruled out. This K^+ channel is shared with a $GABA_B$ receptor. Phorbol ester activation of protein kinase C destroys this hyperpolarization response. This may be because of phosphorylation of G-protein (Nicoll, 1988), resulting in the loss of the transduction function. This response is also seen in 26% of the pyrimidal neurons of the somatosensory cortex where OH-DPAT mimics the response (Davies et al., 1987).

A second observation is a depolarization attributed to 5-HT2 receptors in the cortex and 5-HT receptors of unknown subtype in the hippocampus (Nicoll, 1988; Davies, 1987). The cortical response was attributed to a decrease in K^+ conductance (Davies, 1987) owing to suppression of voltage-gated K^+ channels. Another observed response in the hippocampal neurons is suppression of the after hyperpolarization current (Colino and Halliwell, 1987; Andrade and Nicoll, 1987). This response is mediated by a third serotonin receptor which inhibits the Ca^{2+}-dependent K^+ conductance leading to an increase in neural discharge.

In summary, there are three serotonin-stimulated responses:

1. A hyperpolarization owing to direct activation of a K^+ channel via a G-protein;
2. A depolarization due to blockage of normal Ca^{2+}-dependent K^+ conductance; and
3. Inhibition of the intrinsic voltage-sensitive K^+ conductance.

These responses are most probably attributable to pharmacologically defined 5-HT receptors and a second messenger-mediated biochemical event not yet identified.

The biochemical event that mediates a physiological response is often a phosphorylation. The neuroblastoma NCB-20 has a 5-HT1-like receptor that stimulates adenylate cyclase, and another 5-HT receptor that stimulates release of acetylcholine (MacDermot et al., 1979). 5-HT activated adenylate cyclase has been seen to phosphorylate two proteins of 90 and 130 KD (Berry-Kravis et al., 1988). The incorporation of phosphate into the 90 KD band is increased in the presence of forskolin and isobutylmethylxanthine. The Ca^{2+} ionophore A23187 also increased phosphorylation of both proteins. The enkephalin DADLE decreased incorporation of the 90 KD protein predictably, because DADLE working through δ opiate receptors has been shown to decrease 5-HT stimulated adenylate cyclase (Berry-Kravis and Dawson, 1985). The phosphorylation is not reduced by phorbol esters, dibutrylcyclic GMP, or depolarization by high K^+. The 90 and 130 KD bands were phosphorylated via protein kinase A (a cAMP-dependent protein kinase) or PK-CM a Ca^{2+} concentration-dependent kinase), but not by protein kinase C.

5-HT activated Cl^- conductance has been extensively studied in *Xenopus* oocytes that have been injected with poly A mRNA. This Cl^- conductance is Ca^{2+}-activated and mediated by 5-HT1c receptors or 5-HT2 receptors (Ito et al., 1988). Calmodulin inhibitors are best at inhibiting the Cl^- response indicating the response was not owing to cAMP-dependent protein kinase nor protein kinase C because the isoquinoline sulfonamide kinase inhibitors were ineffective blockers at low concentration. These inhibitors at low concentrations are specific for the cAMP-activated kinase (PK-A) and for the protein kinase C activation, but at higher concentrations the isoquinoline sulfonamides act as a nonspecific kinase inhibitor and can abolish the 5-HT-stimulated Cl^- conductance. Therefore, although calmodulin regulates the response, other pathways can also modify the response. For example, cAMP or diacylglycerol suppresses the 5-HT elicited response when injected intracellularly. Injection of inositol triphosphate, though, did not suppress the response. On the other hand, the response is destroyed by pertussis toxin ADP-ribosylation of the receptor's G-protein. Thus studies on *Xenopus* have allowed identification of a Cl^- channel that is activated by a system of 5-HT-stimulated receptors, G-proteins, and second messengers.

Other cellular responses mediated by 5-HT receptors are also seen in neural and nonneural tissues. 5-HT stimulates smooth muscle cell mitogenesis via a serotonin type D receptor that is the closest equivalent to a neural 5-HT2 receptor (Nemecek et al., 1986). The 5-HT mitogenic response is as strong as that stimulated by the platelet derived growth factor; the two effects are synergistic. In rabbit brain, LSD induces formation of a heat shock protein that initiates a cessation of protein translation by disaggregation of polysomes (Fleming and Brown, 1986). This response is mediated through a 5-HT receptor and has been studied at a variety of different levels from the translational to transcriptional. The finding that LSD can induce a heat shock response resulting in complete abolition of protein synthesis at the ribosomal level is significant in terms of the molecular neurobiology of the brain.

1.1.4. Anatomical Distributions

The distribution of receptor subtypes to anatomical regions has been well documented by autoradiography for human (Hoyer et al., 1986a,b), rat (Pazos and Palacios, 1985a,b) and pig (Hoyer et al., 1985b) brain. Human and pig brain show the most similarity in their distribution. In human brain, the 5-HT1a receptor is most concentrated in the frontal cortex and the hippocampus. Layer II of the cortex is labeled about eight times more heavily than Layer III and about three times more than layer VI. The 5-HT1a binding, as defined by OH-DPAT, is 52, 11, and 26% of total $[^3H]$5-HT binding in layers II, III, and VI, respectively. In hippocampus, $[^3H]$OH-DPAT labels CA1 neurons more densely than the dentate gyrus and more than CA3 neurons. Respectively $[^3H]$OH-DPAT is 39, 66, and 47% of the total $[^3H]$ 5-HT binding in these regions. In contrast, 5-HT1b receptors were not directly observed in human brain nor pig brain, but are present in rat brain as labeled by (–)$[^{125}I]$iodocyanopindolol (Hoyer et al., 1985a). $[^3H]$5-HT binds with the highest density in the dorsal subiculum of the hippocampus, the substantia nigra, and the globus pallidus of the basal ganglia. The 5-HT1c receptor exists at a very high density in choroid plexus in human brain. It is also found in neural cells in layer II of the cortex where it represents about 13% of 5-HT1 sites. The 5-HT1c receptor is also seen in the hippocampus where it comprises 13, 77, and 20% of 5-HT1 sites in the CA1, CA3, and dentate gyrus, respectively. In the CA3 regions total [^{3}H] 5-HT binding is less than the sum of [^{3}H]OH-DPAT (5-HT1a) and [^{3}H] mesulergine (which labels 5-HT1c receptors), so there may be some recognition overlap for the subtypes in this region. The 5-HT1c receptor has been cloned (Julius et al., 1988) from chorid plexus tissue and the distribution of hybridizing mRNAs shows it to be expressed in not only chlorid plexus but also in basal ganglia, pons, medulla, hippocampus, and hypothalamus.

The 5-HT2 receptor is expressed in the frontal cortex. This is the case for human, pig, and rat. 5-HT2 sites are detected by using [^{3}H]ketanserin in autoradiographic studies of the hippocampus. These studies suggest a heterogeneity of 5-HT2 sites. Mesulergine does not label the 5-HT2 site in human brain, but does in pig brain. In human and pig brains, competition for [^{3}H]ketanserin binding is complex in contrast to the monophasic competition seen in rat brains. A ketanserin recognition site has been reported in rat brain (Roth et al., 1987) that may be involved in the observed multiphasic inhibition of [^{3}H]ketanserin binding. These findings indicate that the question of 5-HT2 receptor heterogeneity is still unanswered.

1.2. Serotonin Receptor as a Member of the Rhodopsin Family

Because serotonin receptors mediate their effects via a G-protein they may be related to the rhodopsin family of cell surface receptors. The rhodopsin proteins are structurally and functionally conserved from bacteria to mammals. The proteins are all transducers of extracellular signals. In rhodopsin, the signal is light and in the other family members it is a chemical signal. This chemical signal can be a simple molecule such as serotonin or epinephrine; it may also take the form of a small peptide such as substance K. The stimulated receptors activate a G-protein, which ultimately activates a cellular response. There are many responses elicited by these receptors. The structural conservation of the receptors indicates that a general signal transducing structure exists in which small changes in primary structure confer a number of different properties to the receptor. These differences make the individual receptor specific for a ligand and for responses mediated by a specific G protein. However, overall, the transmembrane structure is retained.

1.2.1. Structure of the Rhodopsin Family of Proteins

Rhodopsin and the β-adrenergic receptors are the first proteins of the rhodopsin family to be isolated and sequenced. They are transmembrane integral membrane proteins that span the bilayer seven times and have defined intracellular, extracellular, and transmembrane domains (Engelman and Zaccea, 1980). Mutagenesis studies have indicated the membrane "core" region of the proteins are essential for ligand binding to the β-adrenergic receptor (Dixon et al., 1987). Other mutagenesis studies have demonstrated the importance of the second and third transmembrane regions (Fraser et al., 1988; Chung et al., 1988), and in particular aspartic acid residues in these regions, for ligand binding and activation of the signal transduction. The ligand-binding domain and the G-protein-interacting domain have been shown to be distinct by research using chimeric receptors composed of part α-adrenergic receptors and part β-adrenergic receptors. If different domains of a receptor are substituted for the corresponding domain of another receptor that mediates a different response, the response is generated but has a pharmacological profile that normally corresponds to a different second messenger response. In this way, the cAMP stimulating β-adrenergic receptor can be made to stimulate cAMP with a pharmacological profile usually seen for the α-adrenergic receptor that normally inhibits cAMP formation (Kobilka et al., 1988). Mutagenesis studies such as these combined with monoclonal antibody studies (Weiss et al., 1988) are demonstrating the structural and functional domains of the receptors. Since the receptors are members of the rhodopsin family (Dohlman et al., 1987), the sequences of additional receptors will allow a better understanding of the structural and evolutionary relationships between these receptors. The more receptors that are sequenced, the more insight we can gain into their structure and function by comparative analysis, by mutagenesis studies, and by chimeric receptor studies.

1.2.2. Function of the Rhodopsin Family of Proteins

The function of the rhodopsin family receptors is to generate a transmembrane signal that activates a cellular response. These responses vary from receptor to receptor and are probably not all known. The most commonly identified responses are the stimulation or inhibition of cAMP, activation of phospholipase C, activation of cyclic GMP phosphodiesterase, and activation of ion channels. These responses are all mediated via a G-protein and the receptor-elicited response reflects the G-protein that is associated with the receptor. The structural basis for the receptor/G-protein interaction is also being studied in terms of functional domains of the receptor and the G-protein. The first protein in terms of structural and functional studies of this family is bacteriorhodopsin of the purple bacterium. This protein is a light-activated proton pump that exchanges sodium for protons. It has now been shown that the receptors coupled to Gi—the G-protein associated with receptor-mediated decreases in adenylate cyclase activity—are also proton pumps (Isom et al., 1987a,b) that alkalinize the intracellular environment upon ligand binding. The functional and physiological significance of this transient proton motive force is not known but it seems to be associated with synaptic transmission (Krishtal et al., 1987). An interesting speculation is that the Gi coupled receptors have retained their original proton pumping character, whereas the other receptors have lost it during evolution.

The discovery of the rhodopsin family is an important finding in terms of the molecular structure and evolution of the nervous system. The availability of primary sequences of the putative family members (such as the neural 5-HT receptors) will allow greater insight into the organization of the nervous system in that many related receptors can be compared and contrasted structurally and functionally. Understanding the individual receptors in terms of their pharmacology and function combined with their anatomical distribution will allow modeling of neural organization based on knowledge of the inherent properties of the receptors.

2. Methods

2.1. Preparation of Crude Membrane Fragments

Bovine brain is dissected and homogenized as a 10% wt/vol solution in a homogenization buffer (0.32*M* sucrose, 2.5 m*M* Tris pH 7.2–7.4). The initial homogenization can be carried out by a mortar/pestle method or by three 30-s homogenizations in a Waring™ blender. This homogenate is centrifuged for 10 min at 600 xg (2000 rpm in a GSA rotor). This supernatant is removed and centrifuged for 25 min at 50,000 xg (20,000 RPM in a Sorvall SS34 rotor) to sediment the membrane fragments. The supernatant is discarded. The pellet is resuspended in 2–3 pellet vol of the homogenization buffer and 5–10 vol of cold dH_2O with a teflon homogenizer. This suspension is centrifuged at 50,000 xg. The resuspension and centrifugation is repeated with sucrose. After the second 50,000 xg centrifugation with sucrose the pellet is resuspended in dH_2O only and centrifuged again at 50,000 xg. This is repeated once more and the final pellet is resuspended in a small vol of dH_2O and frozen in liquid nitrogen. It is stored in a high concentration, up to 50 mg/mL protein, in 5 mL or less aliquots. The last pellet can be used immediately or stored in liquid nitrogen (*see* flow diagram in Fig. 1). These membrane fragments are used either for crude membrane binding assay or for solubilization.

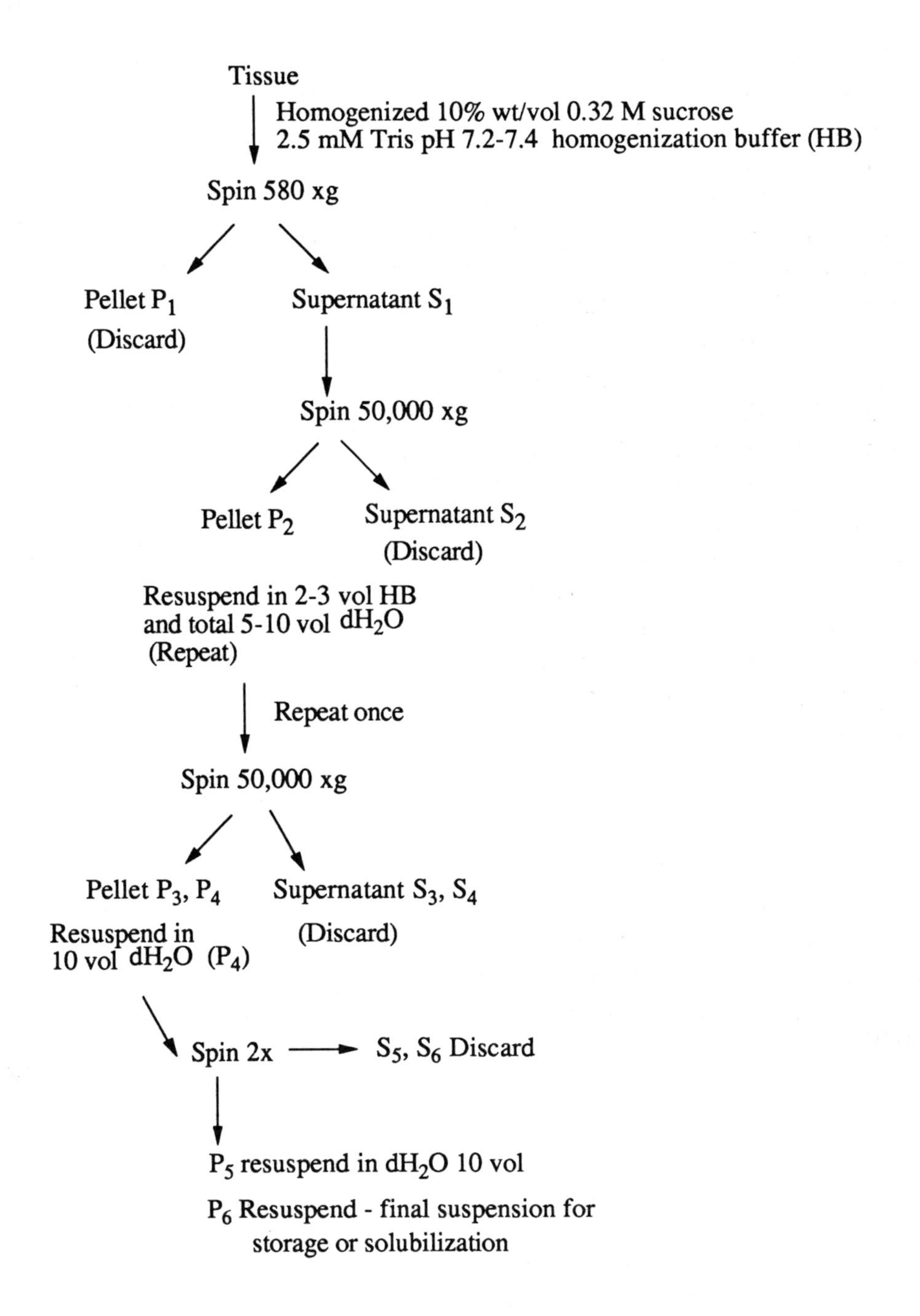

Fig. 1. Preparation of crude membrane fraction.

2.2. Solubilization of Crude Membrane Fragments

To isolate and purify membrane proteins the proteins must first be solubilized from the membranes by use of a detergent. The soluble proteins can then be subjected to chromatography and other biochemical techniques necessary to purify the protein. The protein's structure must be retained so that it exists in a functional state and can be assayed for ligand binding activity. Many detergents denature proteins. The detergent also must be readily removable if the proteins are to be reconstituted into artificial lipid bilayers. The detergent should have a relatively high critical micelle concentration so that it can be easily dialyzed. The zwitterionic detergent [3-(3-cholamidopropyl)dimethylammonio-*l*-propane sulfonate] (CHAPS) fulfills these criteria and has been shown to solubilize rhodopsin in functional form (Bennet and Brown, 1985).

Crude membrane fragments in dH_2O are centrifuged at 50,000 xg. The supernatant is discarded and the pellet is resuspended in the detergent solution [1.25% (~20 m*M*) CHAPS in a 50 m*M* Tris buffer, pH 7.2–7.4] by teflon homogenization. The resuspended pellet is stirred at 0–4°C for one hour with a magnetic stir bar, then centrifuged at 100,000 xg. The supernatant is the solubilized extract (*see* flow chart in Fig. 2). The protein to detergent solution ratio is 20 mg/mL; the overall protein yield is 20–25% after solubilization.

2.3. Radioligand Binding Assays

The crude membrane, solubilized and affinity-purified-reconstituted receptors are assayed for binding activity with [^{3}H]5-HT. The procedure involves incubating the sample with radioligand at a specific concentration and separating the bound from unbound ligand by filter adsorption of the membrane fragments or bound solubilized proteins to the filter. The solubilized proteins are not in a membrane and the filters must first be treated with polyethylenimine to adsorb the proteins.

Crude membrane and reconstituted membrane fractions are incubated in a total vol of 600 μL of binding buffer [50 m*M* Tris, 0.1% ascorbic acid, 6 m*M* $CaCl_2$, and 20 μ*M* pargyline (a monoamine oxidase inhibitor)]. This mixture is allowed to incubate at room temperature for 30 min, after which the bound and free ligand is separated by filtering with an Amicon vacuum manifold through a Whatman GF/B glass fiber filters. The filter is then washed with three 5 mL washes of ice-cold 5 m*M* Tris pH 7.2–7.4. The filters are then put in 10 mL scintillation cocktail and counted for radioactivity.

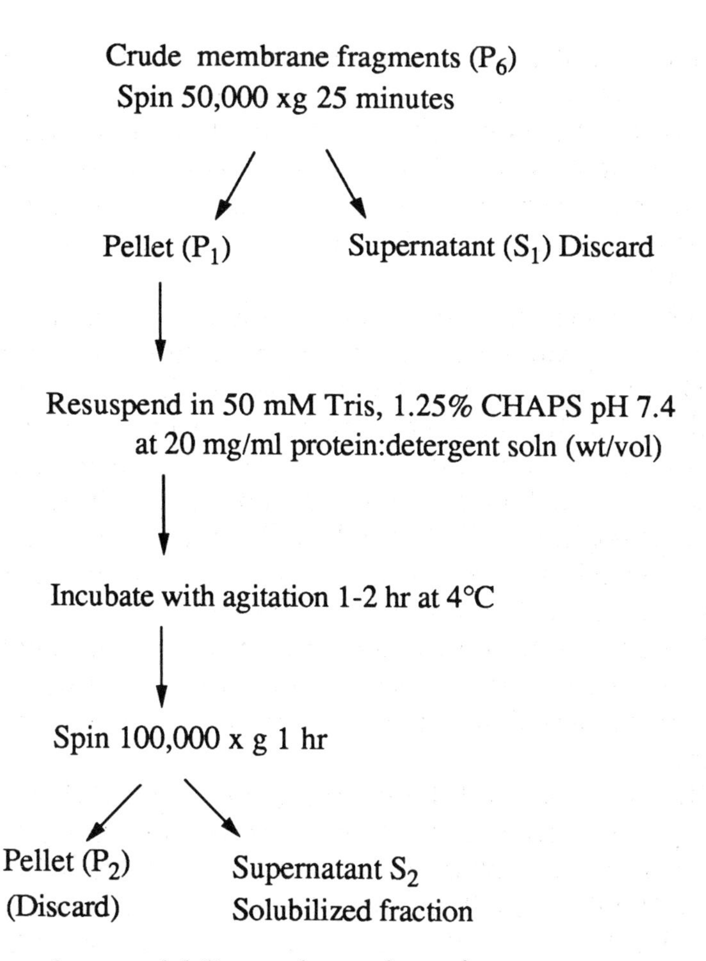

Fig. 2. Procedure to solubilize crude membrane fragments.

Solubilized extract will not be retained by a glass fiber filter. To retain the solubilized proteins, the GF/B filters must first be soaked in 0.3% polyethylenimine (Sigma) for at least three hours (Bruns et al., 1983). The binding assay is the same as for crude or reconstituted membrane except for this pretreatment of the filters.

3. Affinity Chromatography

3.1. Synthesis of Affinity Ligand

The synthesis of lysergic acid ethylamidoethylbromide (LAEB) utilizes a mixed anhydride reaction to form the intermediate complex that is then amidated. The reaction is two-step reaction and is based on the patent for LSD by Pioch at Eli Lilly Corporation (US Patent 2,736,728, 1954). The necessary reagents are lysergic acid, trifluoroacetic anhydride, 2-ethylaminoethylbromide, triethylamine. Acetonitrile (dry) and distilled chloroform are the needed solvents. All compounds and glassware must be completely dry and the reaction must take place within an air-tight flask and not come into contact with the atmosphere. The reactants can be transferred in solution by glass syringes. The reaction involves first forming the anhydride adduct by mixing the lysergic acid solution with the trifluoroacetic anhydride.

The first reaction occurs best at a 2.1:1 anhydride:acid mole ratio. The unwanted reaction products are evacuated with the solvent by a vacuum line that can be opened and closed to the reaction flask. A two-necked reaction vessel with a stoppered vacuum hose attachment on one neck and a serum stopper on the other neck is sufficient. The reactions are exothermic and must be carried out at low temperature, –40 –20°C.

The second reaction, which involves the anhydride adduct and the secondary amine, takes place at a 5.2:1 mole ratio of amine to anhydride. The 5.2:1 ratio of amine to anhydride can be maintained by a mixture of tertiary and secondary amines. Only the secondary amines will react and form the amide but tertiary amines are included. A 2:1 ratio of 2-ethylaminoethylebromide with a 3.2:1 ratio of triethylamine will result in the LAEB ligand.

A specific protocol is as follows: 0.5 g of lysergic acid is dissolved in 12.5 mL of dry acetonitrile. The solution is cooled to –40 –20°C. To this is added a suspension of 0.838 g trifluoroacetic anhydride (2.1:1 mole ratio TFAA:LA) in 7.5 mL acetonitrile, also at or below –20°C. This mixture is incubated while stirred at –20°C for 1.5 h. The mixed anhydride is then separated in the form of an oil by evaporating the solvent in vacuo below 0°C. The oil is resuspended in 20 mL acetonitrile at or below 0°C. To this solution is added a solution of 1.51 g ethylaminoethylbromide in 15 mL acetonitrile at room temperature. (This is a 5.2:1 amine:anhydride mole ratio. A mix of the secondary brominated amine and triethylamine can also be used as described.) This mixture is stirred in the dark for 2 h, after which the solvent is evaporated, leaving a mix of iso- and normal forms of the amide and unreacted reagents. The remaining residue is redissolved in a mixture of 15 mL chloroform and 2 mL dH_2O at 0°C. The chloroform layer is separated and the aqueous layer is extracted by 45 mL portions of chloroform. The chloroform extracts are combined and washed 4 times with 5 mL portions of dH_2O and 0°C to remove any residual amine salts. The chloroform extract is dried over anhydrous sodium sulfate and the chloroform is evaporated in vacuo. The residue is weighed and resuspended in a small volume of isopropyl alcohol and stored in the dark at –20°C. The reaction scheme is shown in Fig. 3.

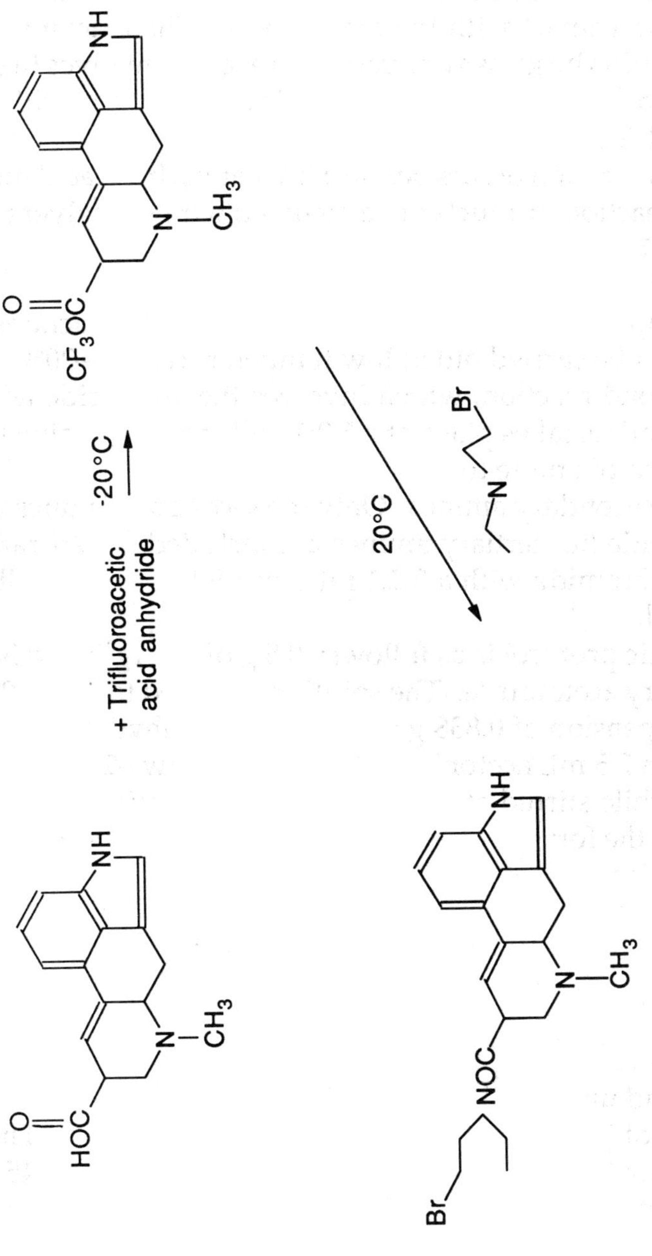

Fig. 3. The synthesis of lysergic acid ethylamidoethylbromide (LAEB).

3.2. Activation of Affinity Matrix

Affigel 401 is a suspension of agarose beads. A sulfhydryl group is coupled to each bead via a hydrophilic spacer arm. The gel is supplied in an aqueous solution in the oxidized disulfide form. The first step of the activation is to reduce the disulfides to provide free sulfhydryls for the bromine nucleophile to react with. The gel is suspended for 30 min in a solution of dithiothreitol (DTT) (20 m*M* DTT/200 m*M* Tris, pH 8.0), which acts as the reducing agent. The DTT is washed from the gel by pumping buffer through the column until a 50 μL sample of the eluate does not turn bright yellow when exposed to 5,5'-dithio-bis-(2-nitrobenzoic acid) (DTNB). At this time, the gel matrix should be resuspended and a 50 μL sample subjected to DTNB assay. A yellow color change indicates the presence of sulfhydryls. Because the LAEB affinity ligand is not readily soluble in water, the gel is equilibrated to the less polar environment of isopropyl alcohol. The gel is washed with 5 column vol each of 25% isopropyl alcohol in buffer, 50% isopropyl alcohol in buffer, and 100% isopropyl alcohol. When the matrix is completely equilibrated to isopropyl alcohol it is ready to be conjugated with the affinity ligand (*see* Fig. 4). This is done by resuspending the gel with LAEB at a 50:1 mole ratio of LAEB to the free sulfhydryls. The dextrorotatory enantiomer of LAEB is the stereospecific ligand for 5-HT receptors and will represent 25% of the crude LAEB mix. So the actual, important ratio is 12.5:1, but all isomers have bromine and will bind to the gel. After the excess LAEB is washed away with 5 column vol of isopropyl alcohol and after a 24-h incubation at 0–4°C, the gel matrix has been conjugated to the mixture of LAEB enantiomers. The matrix is reequilibrated to buffer by reversing the equilibration process described above. The degree of sulfhydryl group saturation by LAEB is determined by DTNB assay. DTNB in the presence of free sulfhydryl groups results in a yellow product (absorption at 467 nm). In the assay, a 50 μL sample is mixed with 1 mL DTNB (0.323 m*M*/100 m*M* Tris, pH 8.0). From the absorbance value, the sulfhydryl concentration is determined by comparison to a β-mercaptoethanol standard. LAEB typically saturates 80–90% of the sulfhydryl groups in Affigel-401 (*see* Table 1 below).

The unreacted sulfhydryl groups in the gel are inactivated by the following procedure. In buffer solution the gel is exposed to 0.25 g iodoacetamide to alkylate any unreacted sulfhydryls. After 1 h, the iodoacetamide is washed away with 10 column vol of buffer. The matrix is now fully activated. Before loading the solubilized sample the column is washed with 2 vol of 2 m*M* CHAPS/50 m*M* Tris, pH 7.2–7.4.

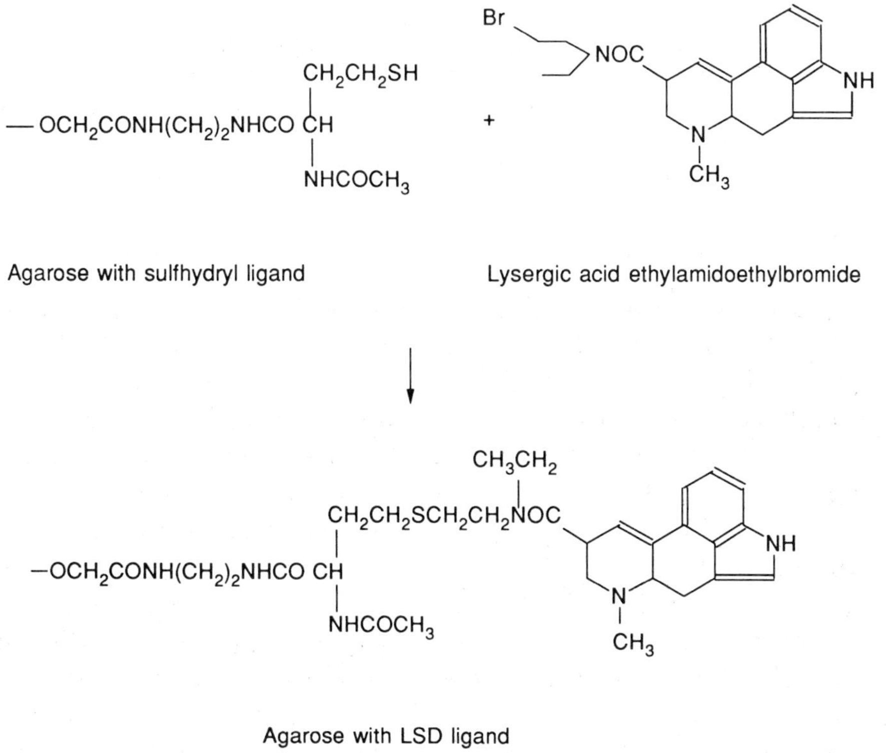

$-OCH_2CONH(CH_2)_2NHCO\,CH(CH_2CH_2SH)(NHCOCH_3)$

Agarose with sulfhydryl ligand

\+

Lysergic acid ethylamidoethylbromide

$-OCH_2CONH(CH_2)_2NHCO\,CH(NHCOCH_3)CH_2CH_2SCH_2CH_2N(CH_2CH_3)OC-$

Agarose with LSD ligand

Fig. 4. The reaction of lysergic acid ethylamide ethylbromide with the sulfhydryl ligand in Affigel-401 to give the LSD ligand.

3.3. *Affinity Purification and Reconstitution of 5-HT Receptors*

The affinity purification of the solubilized receptors involves first separating the 5-HT receptors from the bulk of the solubilized proteins by passing the solubilized extract through the LSD affinity column and eluting the receptors from the column in the presence of free 5-HT. The specific elution buffer (pH 7.2–7.4) contains 100 μ*M* 5-HT, 50 m*M* Tris, 4 m*M* CHAPS, 0.1% ascorbic acid, 6 m*M* $CaCl_2$, 20 μ*M* pargyline, and 3 mg/mL azolectin. The azolectin is a soybean lipid extract (from Associated Concentrate, Woodside, NY). The reconstitution of the receptors takes place when collected fractions are dialyzed against a Tris buffer. The detergent dialyzes out and lipid vesicles are formed. The dialysis is also necessary to remove the 5-HT from the eluate so that radioligand binding analysis can be done.

Twenty-five mL of activated LSD matrix is used and the frontal cortex of at least two bovine brains are loaded (675 mg solubilized protein) through the column. The solubilized extract is run through the column at a constant

Table 1
Saturation of Sulfhydryl Groups by LAEB

	A467	[-SH] m*M*	Percent saturation of sulfhydryl
Pre-LAEB	0.489	7.9	0
Post-LAEB	0.196	1.0	88

flowrate (usually 50 mL/h). The eluate is collected in 5 mL fractions and assayed for [^{3}H]5-HT binding and protein concentration. The [^{3}H]5-HT binding activity in the initial eluate (fractions 0–18 in Fig. 5) corresponds to receptors that did not bind the LSD column. This initial peak binding activity and peak protein concentration elute together. A new column retains about 90% of 5-HT receptors introduced, and a one-month-old column retains as little as 50% of receptors. The column is continuously flushed with buffer (2 m*M* CHAPS/50 m*M* Tris, pH 7.2–7.4) until no more proteins or activity are detectable. At this point the column has been washed and only LSD binding proteins are bound to the column. The specific elution buffer described earlier containing 5-HT and azolectin is then run through the column. Five mL fractions are collected and may be frozen in liquid nitrogen for storage. The fractions are extensively dialyzed (at least 4 × 4 L 50 m*M* Tris, pH 7.4) to remove the detergent. Fractions are then ready for radioligand assay. The dialyzed, reconstituted fractions can be assayed with glass fiber filters not treated with polyethylenimine. The maximum [^{3}H] 5-HT binding activity is detected in fractions 70–90 (Fig. 5) although the protein concentration in these fractions is below the sensitivity of our protein assay.

3.4. Preparation of Affinity-Purified Sample for SDS-PAGE

The affinity-purified sample can be analyzed by sodium dodecyl sulfate polyacrylamide gel electrophoresis (SDS-PAGE) and stained by either silver (Heukeshoven and Dernick, 1985) or Coomassie blue. The amount of protein recovered from the column is in the nanogram to microgram range for individual proteins. Loading more crude solubilized membranes through the column will achieve a higher net yield of proteins. To prepare the sample for electrophoresis the procedure is basically that of the affinity reconstitution described in Section 3.3, except that azolectin is omitted from the specific eluant and fractions need not be collected. One hundred and fifty mL of specific eluate is collected and is concentrated to 1–2 mL by reverse dialysis. This takes place by placing the eluate in dialysis tubing in a solu-

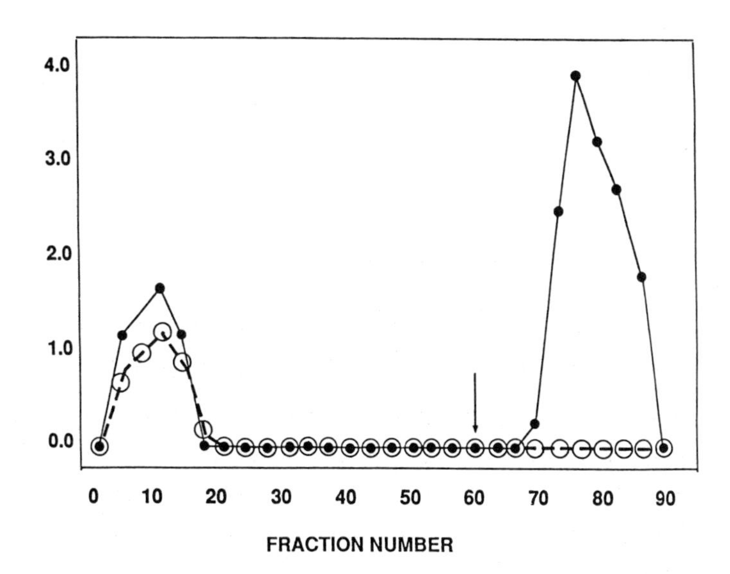

Fig. 5. Elution profile of solubilized extract through the LSD affinity column. The [^{3}H]5-HT binding activity (pmol/mL) is indicated in solid circles. The protein concentration as determined by Lowry assay (mg/mL) is represented by open circles. The arrow indicates the beginning of specific elution in the presence of serotonin.

tion of 15% polyethylene glycol (mol wt of 16,000–20,000) in the presence of 0.01% SDS. Overnight, the 150 mL will be concentrated to a few mL by osmosis. This 1–2 mL is then dialyzed against 4 L of distilled water to remove detergent or smaller mol wt PEG impurities. It is important to remove this detergent or PEG, since the presence of either will result in poorly resolved gels with the protein migrating as a broad smear rather than a distinct band. PEG is a protein precipitant and the presence of CHAPS will compete with SDS. If the proteins are not fully bound with SDS they will not migrate as distinct bands.

This dialyzed small sample is then frozen in either liquid nitrogen or acetone/dry ice and lyophilized. The lyophilized sample is then resuspended in the SDS sample application buffer. This resuspension should be in as small a volume as possible to load as much receptor as possible into an individual well. Failure to remove detergent or PEG is apparent after the lyophilization as a large amount of powder will be present. A resuspension into 50–100 µL is optimal for two lanes of purified sample. If the detergent or PEG is not suitably removed more than this amount of application buffer will be required to resuspend the sample, which will result in less net protein loaded into each well. The sample must be resuspended thoroughly or the proteins will smear.

Coomassie blue staining will stain in the microgram or high nanogram range and silver stain is approx 100 times more sensitive. We have used Coomassie blue G-250 and the silver stain method to visualize isolated proteins.

4. Discussion

4.1. [^{3}H]5-HT Binding in Solubilized and in Reconstituted Preparations

In the CHAPS solubilized preparations, dose-response analysis of [^{3}H]5-HT binding indicates three binding sites are present with dissociation constants of 1.5, 11.2, and 60.2 n*M*. Dose-response analysis in the presence of 100 µ*M* GTP reveals two binding components with K_ds of 4.3 and 60.2 n*M*. The presence of GTP has no effect on the maximum binding activity of 0.14 pmol/mg protein. Competition experiments for [^{3}H]5-HT binding at 5 n*M* by OH-DPAT and by 5-HT at 1 µ*M* indicate that 40% of the receptor population is 5-HT1a subtype. The results may be interpreted as demonstrating that a heterogeneous population of serotonin receptors including different G-protein-mediated allosteric states are present in the solubilized preparation.

In membranes reconstituted from affinity purified receptors, a different pattern of binding characteristics is displayed. An initial study was presented previously (Gallaher and Wang, 1988). Scatchard analysis indicates a single population is present with a K_d of 16.9 n*M* for 5-HT. The specific activity of 279 pmol/mg protein is an enrichment of almost 2000-fold over the CHAPS solubilized fraction. The results are summarized in Table 2 below.

4.2. Analysis of Affinity Purified Proteins by SDS-PAGE

A number of proteins in the mol wt range from 35,000 to 100,000 daltons are isolated, depending on the specific ligand used in the eluant and the method of staining used (unpublished observations). The frontal cortex of two or more bovine brains are needed to visualize the purified receptor proteins by silver stain and 5 or more for Coomassie blue visualization. Results from silver staining techniques were previously reported (Gallaher and Wang, 1988).

Table 2
HT Binding by Solubilized and Reconstituted Preparations

Fraction	$[^3H]$5-HT Bound per mg protein	K_d	Enrichment
Crude solubilized	0.14 pmol	1.5, 11.2, 60.2	1
Crude solubilized (+GTP)	0.14 pmol	4.3, 60.1	1
Reconstituted preparation	279 pmol	16.9	1993

The isolated proteins may be a mixture of receptor subtypes, different glycosylated forms of the same receptor, or G-protein components that have been copurified. Dopamine receptors have been shown to exist in multiple glycosylated forms (Jarvie et al., 1988) that migrate with mol wt from 137 KD for the fully glycosylated form to 44 KD for the nonglycosylated form. The higher mol wt proteins seen in our results may be glycosylated receptors. This question will be answered with the use of glycolytic enzymes and staining for sugar groups. The mol wt 55 (Saito and Shih, 1987) and 63 KD (Emerit et al., 1987) have all been identified as 5-HT1a receptors by photolabeling experiments. The amino acid sequence of a 5-HT1a receptor indicates a protein mol wt of approx 42 KD. The receptor migrates during SDS-PAGE as a 44 or 45 KD protein (Fargin et al., 1988). The presence of proteins of these mol wt after LSD affinity chromatography indicates the 5-HT1a receptor is present in a variety of glycosylated forms. The 5-HT1c receptor has a theoretical mol wt of 52 KD (Julius et al., 1988) and the presence of a 52–54 mol wt protein is consistent in all our specifically eluted samples. 5-HT2 receptors have been photolabeled and exist in various forms with mol wt 94, 68, and 52 KD (Wouters et al., 1987). In our studies, specific elution with ketanserin resulted in proteins with similar mol wt. The 94 KD protein was not seen with ketanserin but a faint band was seen at 70 KD and a heavy band was seen at 52 KD. These mol wt correlate with the photolabeled estimates.

4.3. Conclusions and Future Directions

The LSD affinity column combined with CHAPS solubilization is a simple technique for purifying serotonin receptors from a receptor-rich tissue. The isolated proteins are still being analyzed in terms of their pharmacological and structural identities. By scaling up the procedures described in this paper, it will be possible to obtain purified receptors in amounts adequate for protein mapping studies. Even when the primary structures are clearly established, such mapping studies are still a powerful means of determining the three-dimensional structure of this very important protein.

Abbreviations:

DDT dithiothreitol
DTNB 5, 5, -dithiobis-(2-nitrobenzoic acid)
LAEB lysergic acid ethylamidoethylbromide
LSD lysergic acid diethylamide
5-HT serotonin or 5-hydroxytryptamine
CHAPS 3-[3-cholamidopropyl)-dimethylammonio]-*1*-propane sulfonate

References

Andrade, R. and Nicoll, R. A. (1987) *J. Physiology* **394,** 99–124.

Barbaccia, M. L., Brunello, N., Chuang, D.-M., and Costa, E. (1983) *J. Neurochem.* **40,** 1671–1679.

Battaglia, G., Shannon, M., and Titeler, M. (1984) *J. Neurochem.* **43,** 1213–1219.

Bennet, R. and Brown, P. (1985) *Vision Research* **25,** 1771–1781.

Berry-Kravis, E. and Dawson, G. (1985) *J. Neurochem.* **45,** 1739–1747.

Berry-Kravis, E., Kazmierczak, B. K., and Dawson, G. (1988) *J. Neurochem.* **50,** 1287–1296.

Bruns, R. F., Lawson-Wendling, K., and Pugsley, T. A. (1983) *Anal. Biochem.* **132,** 74–81.

Chung, F.-Z., Wang, C.-D., Potter, P. C., Venter, J. C., and Fraser, C. M. (1988) *J. Biol. Chem.* **263,** 4052–4055.

Colino, A. and Halliwell, J. V. (1987) *Nature* **328,** 73–77.

Conn, P. J. and Sanders-Bush, E. (1986) *J. Neurosci.* **6,** 3669–3675.

Conn, P. J., Sanders-Bush, E., Hoffman, B. J., and Hartig, P. R. (1986) *PNAS* **83,** 4086–4088.

Davies, M. F., Deisz, R. A., Prince, D. A., and Peroutka, S. J. (1987) *Brain Res.* **423,** 347–352.

Dixon, R. A. F., Sigal, I. S., Rands, E., Register, R. B., Candelore, M. R., Blake, A. D., and Strader, C.D. (1987) *Nature* **326,** 73–77.

Dohlman, H. G., Caron, M. G., and Lefkowitz, R. J. (1987) *Biochem.* **26,** 2657–2664.

Dumuis, A., Sebben, M., and Bockaert, J. (1988) *Mol. Pharmacol.* **33,** 178–186.

Emerit, M. B., El Mestikaway, S., Gozlan, H., Cossery, J. M., Besselievre, R., Marquet, A., and Hamon, M. (1987) *J. Neurochem.* **49,** 373–380.

Engelman, D. M. and Zacca, G. (1980) *PNAS* **77,** 5894–5898.

Fargin, A., Raymond, J. R., Lohse, M. J., Kobilka, B. K., Caron, M. G., and Leffkowitz, R. J. (1988) *Nature* **335,** 358–360.

Fleming, S. W. and Brown, I. R. (1986) *J. Neurochem.* **46,** 1436–1443.

Fraser, C. M., Chung, F.-Z., Wang, C.-D., and Venter, J. C. (1988) *PNAS* **85,** 5478–5482.

Gallaher, T. K. and Wang, H. H. (1988) *PNAS* **85,** 2378–2382.

Godfrey, P. P., McClue, S. J., Young, M. M., and Heal, D. J. (1988) *J. Neurochem.* **50,** 730–738.

Gozlan, H., El Mestikaway, S., Pichat, L., Glowinski, J., and Hamon, M. (1983) *Nature* **305,** 140–142.

Herrick-Davis, K., Titeler, M., Leonhardt, S., Struble, E., and Price, D. (1988) *J. Neurochem.* **51,** 1906–1912.

Heukeshoven, J. and Dernick, R. (1985) *Electrophoresis* **6,** 103–112.

Heuring, R. E. and Peroutka, S. J. (1987) *J. Neurosci.* **7,** 894–903.
Hoyer, D., Pazos, A., Probst, A., and Palacios, J. M. (1986a) *Brain Res.* **376,** 85–96.
Hoyer, D., Pazos, A., Probst, A., and Palacios, J. M. (1986b) *Brain Res.* **376,** 97–107.
Hoyer, D., Engel, G., and Kalkman, H. O. (1985a) *Eur. J. Pharmacol.* **118,** 1–12.
Hoyer, D., Engel, G., and Kalkman, H. O. (1985b) *Eur. J. Pharmacol.* **118,** 13–23.
Isom, L. L., Cragoe, E. J., and Limbird, L. E. (1987a) *J. Biol. Chem.* **262,** 6750–6757.
Isom, L. L., Cragoe, E. J. ,and Limbird, L. E. (1987b) *J. Biol. Chem.* **262,** 17504–17509.
Ito, I., Hirono, C., Yamagishi, S., Nomura, Y., Kaneko, S., and Sugiyama, H. (1988) *J. Cell. Physiol.* **134,** 155–160.
Jarvie, K. R., Niznik, H. B., and Seeman, P. (1989) *Mol. Pharmacol.* **34,** 91–97.
Julius, D., MacDermot, A. B., Axel, R., and Jessel, T. M. (1988) *Science* **241,** 558–564.
Kobilka, B. K., Kobilka, T. S., Daniel, K., Regan, J. W., Caron, M. G., and Lefkowitz, R. J. (1988) *Science* **240,** 1310–1316.
Korstanje, C., Sprenkels, R., Doods, H. N., Hutgenburg, J. G., Boddekee, E., Batink, H. D., Thoolen, M. J., and Van Zweiten, P. A. (1986) *J. Pharm. Pharmacol.* **38,** 374–379.
Krishtal, O. A., Osipchuk, Y. V., Shelest, T. N., and Smirnoff, S. W. (1987) *Brain Res.* **436,** 352–356.
Leysen, J. E., Niemegeers, C. J. E., Van Neuten, J. M., and Laduron, P. M. (1982) *Mol. Pharmacol.* **21,** 301–314.
MacDermot, J., Higaashida, H., Wilson, S. P., and Matsuzawa, H. (1979) *PNAS* **76,** 1135–1139.
Murphy, T. J. and Bylund, D. B. (1988) *Mol. Pharmacol.* **34,** 1–7.
Nemecek, G. M., Coughlan, S. R., Handley, D. A. , and Moskowitz, M. A. (1986) *PNAS* **83,** 674–678.
Nicoll, R. A. (1988) *Science* **241,** 545–551.
Pazos, A. and Palacios, M. (1985a) *Brain Res.* **346,** 205–230.
Pazos, A. and Palacios, M. (1985b) *Brain Res.* **346,** 231–249.
Pedigo, N. W., Yamamura, H. I., and Nelson, D. L. (1981) *J. Neurochem.* **36,** 220–226.
Peroutka, S. J. and Snyder, S. H. (1979) *Mol. Pharmacol.* **16,** 687–699.
Roth, B. L., McClean, S., Zhu, X. Z., and Chuang, D. M. (1987) *J. Neurochem.* **49,** 1833–1838.
Saito, T. and Shih, J. C. (1987) *J. Neurochem.* **49,** 1361–1366.
Schachter, M., Godfrey, P. P., Minchin, M. C., McClue, S. J., and Young, M. M. (1985) *Life Sciences* **37,** 1641–1647.
Schlegel, J. R. and Peroutka, S. J. (1986) *Biochem. Pharmacol.* **36,** 1943–1949.
Shenkar, A., Maayani, S., Roberts, S., and Hill, D. (1987) *Mol. Pharmacol.* **31,** 357–367.
Von Hungen, K., Roberts, S., and Hill, D. (1975) *Brain Res.* **84,** 257–267.
Weiss, E. R., Kelleher, D. J., and Johnson, G. L. (1988) *J. Biol. Chem.* **263,** 6150–6154.
Wouters, W., Van Dun, J., and Laduron, P. M. (1987) *FEBS Letters* **213,** 359–364.
Yagaloff, K. A. and Hartig, P. R. (1985) *J. Neurosci.* **5,** 3178–3183.

Purification and Cloning of Central Serotonin Receptors

Henri Gozlan, Salah El Mestikawy, and Michel Hamon

1. Introduction

Serotonin (5-hydroxytryptamine, 5-HT) is a neurotransmitter involved in the control of various physiological functions and brain disorders, and its effects are mediated through several classes of specific 5-HT receptors (Osborne and Hamon, 1988). Indeed the demonstration of the heterogeneity of 5-HT receptors was reported as far back as 1957 (Gaddum and Picarelli, 1957), and became more evident during the past five years. To date, the development of radioligand binding techniques has established that 5-HT receptors are divided into three main classes, designated 5-HT_1, 5-HT_2 and 5-HT_3, with different pharmacological properties, regional distributions, and functions in the central nervous system (CNS) (Bradley et al., 1986; Tricklebank, 1985).

The 5-HT_1 class, characterized by a high (nanomolar) affinity for $[^3H]$-5-HT, has been recently divided into at least four different subtypes: 5-HT_{1A}, 5-HT_{1B}, 5-HT_{1C}, and 5-HT_{1D} (Pedigo et al., 1981; Hoyer et al., 1985; Vergé et al., 1986; Heuring and Peroutka, 1987). The availability of a specific and selective agonist of 5-HT_{1A} receptors, 8-hydroxy-2-(di-*n*-propylamino) tetralin (8-OH-DPAT), has led to considerable progress in the knowledge of the pharmacological and biochemical properties of this particular subtype (Hamon et al., 1987). 5-HT_{1A} receptors are concentrated in the limbic system (Pazos and Palacios, 1985) and are probably involved in various physiological functions (Dourish et al., 1987). Evi-

Receptor Purification, vol. 1

dence has also been reported for their implication in the mechanism of action of new anxiolytics drugs, such as buspirone, gepirone and ipsapirone (Hamon et al., 1988).

5-HT_{1B} sites exist in the CNS of rats and mice, but are apparently absent in other species, such as, human, bovine, and pig. Instead, 5-HT_{1D} sites with different pharmacological properties are present in the latter, in the same brain regions (notably the extrapyramidal areas) as 5-HT_{1B} sites in small rodents. No specific ligand (except [^{3}H]5-HT) is available yet for the labeling of 5-HT_{1B} and 5-HT_{1D} receptors and their physiological roles remain to be clearly established.

5-HT_{1C} sites are abundant in the choroid plexus of all species (Pazos et al., 1985; Pazos and Palacios, 1985). In contrast to other 5-HT_1 subtypes, which exhibit rather low affinity (millimolar) for antagonists, the 5-HT_{1C} receptor is recognized by selected antagonists in the ergot series. Thus [^{3}H]mesulergine and [^{125}I]iodo-LSD have been successfully used for the selective labeling of 5-HT_{1C} receptors in the choroid plexus.

All 5-HT_1 receptors are coupled to a transduction mechanism (Hamon, 1987): 5-HT_{1A} (Devivo and Maayani, 1986), 5-HT_{1B} (Bouhelal et al., 1988), and 5-HT_{1D} (Hoyer and Schoeffter, 1988), receptors are negatively coupled to adenylate cyclase, and 5-HT_{1C} receptors are linked to phospholipase C (Conn et al., 1986).

5-HT_2 receptors are characterized by a low (micromolar) affinity for serotonin and are usually labeled with antagonists, such as [^{3}H]ketanserin (Leysen et al., 1982). These receptors are highly concentrated in the frontal cortex (Pazos et al., 1985a), and are coupled to phospholipase C (Conn and Sanders-Bush, 1985, Kendall and Nahorski, 1985).

5-HT_3 receptors originally described in the periphery have been found in the rat CNS (Kilpatrick et al., 1987), and in neuroblastoma-glioma cell lines (Hoyer and Neijt, 1987). They are characterized by a low affinity for agonists and a high affinity for selective antagonists (Richardson and Engel, 1986). 5-HT_3 antagonists possess potent antiemetic properties, but the antipsychotic and anxiolytic properties that have been claimed are not yet firmly established.

Thus, it appears that there are at least 6 different pharmacologically characterized 5-HT receptors. Are these receptors chemically different proteins or are they structurally related? One way to answer this question would be to purify and establish the sequence of each 5-HT receptor. For this purpose, two approaches can be used: the purification of the protein receptor by standard biochemical techniques, and cloning of the receptor gene. During the past five years, the second way was initiated, which led to the knowledge of the sequences of several neurotransmitter receptors: α- and β-adrenergic, muscarinic, GABA , and so on. Very recently, the application of the cloning strategy also appeared to be a very fruitful approach for a better knowledge of central 5-HT receptors.

To date, biochemical as well as cloning data support that 5-HT_{1A}, 5-HT_{1C} and 5-HT_2 receptors, at least, are different proteins, but all belong to the same family of receptors coupled to G-proteins. Both sets of data are discussed in the present review.

2. Biochemical Purification of 5-HT Receptors

5-HT receptors are membrane proteins, the concentration of which is rather low in the central nervous system. Therefore, one approach to purify a given 5-HT receptors would be to irreversibly label the corresponding protein with a radioactive ligand and then follow this incorporated radioactivity within the different purification steps. Several photoaffinity probes have been designed for this purpose.

2.1. Photoaffinity Labeling

2.1.1. 5-HT_1 Binding Sites

Originally, the molecular structure of serotonin was modified in order to introduce a photosensitive moiety. Thus, NAP-5-HT (Cheng and Shih, 1979), and ANPA-5-HT (Saitoh and Shih, 1987) have been synthesized (Fig. 1), but their respective affinities for 5-HT_1 receptors were in the micromolar range or less. Therefore, it was not surprising that they were inefficient as ligands for labeling 5-HT_1 receptors. [^{3}H]5-HT itself has been irradiated in the presence of horse synaptosomal membranes, and SDS-PAGE analysis of the resulting solubilized membranes indicated that the radioactivity was incorporated into a 60 kDa band (Rousselle et al., 1985). Agonists, and, to a lower extent antagonists prevented the covalent binding of [^{3}H]5-HT. However, these pharmacological investigations were too limited to assign this band to a particular 5-HT_1 receptor (Rousselle et al., 1985).

2.1.2. 5-HT_{1A} Receptors

Modifications of more specific 5-HT_{1A} ligands, 8-OH-DPAT (Gozlan et al., 1983) and PAPP (Ransom et al., 1985), have led to 2 azido derivatives: [^{3}H]8-MeO-NAP-amino-PAT (Emerit et al., 1986; 1987) and [^{3}H]*p*-azido-PAPP (Ransom et al., 1986) (Fig. 1). In the dark, both compounds possess a high affinity for the 5-HT_{1A} receptor, and, upon irradiation, an irreversible labeling of this receptor was demonstrated. In both cases,

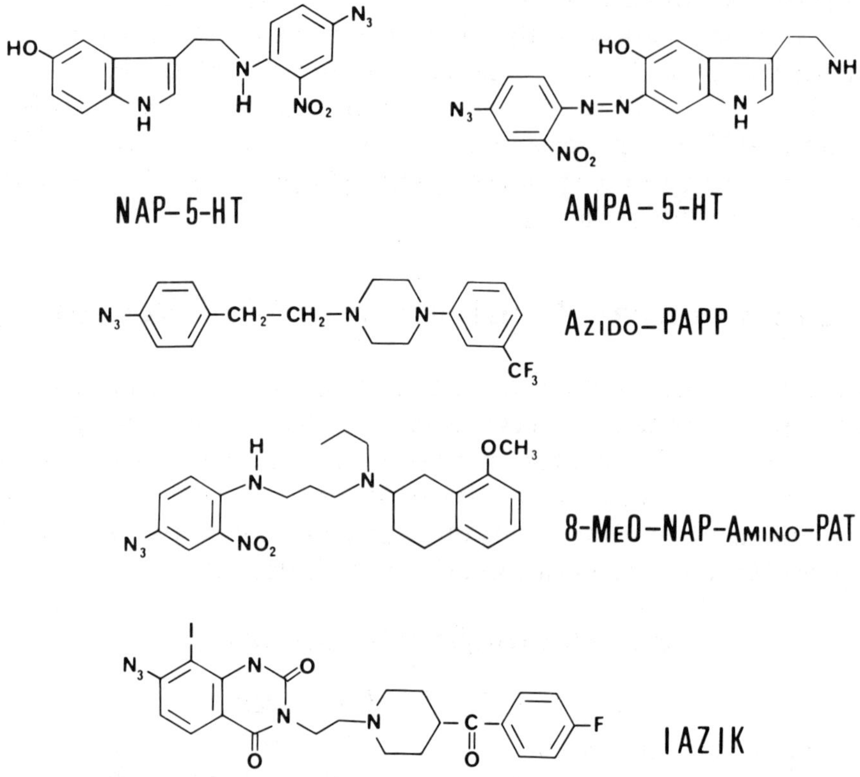

NAP-5-HT

ANPA-5-HT

AZIDO-PAPP

8-MEO-NAP-AMINO-PAT

IAZIK

Fig. 1. Chemical structures of photoaffinity ligands of central 5-HT receptors.

the covalent binding was prevented by 5-HT_{1A} agonists and antagonists but not by 5-HT_{1B} and 5-HT_2 ligands, confirming the selectivity of the photolabeling. SDS-PAGE of radioactive material in brain membranes indicated a major band at 63kDa (Emerit et al., 1987) or 55kDa (Ransom et al., 1986). These tritiated probes have not been used to further purify the 5-HT_{1A} receptor, since the specific radioactivity of tritium was not enough for such a purpose. Iodinated derivatives of PAPP or 8-OH-DPAT would be more valuable tools in this respect.

The determination of the mol wt of the 5-HT_{1A} receptor protein subunit has also been performed using target size analysis allowing comparison with that derived from photoaffinity labeling. First estimates by the radiation-inactivation technique gave a value of 58.4kDa for the 5-HT_{1A} receptor labeled with [^{3}H]8-OH-DPAT in the rat brain (Gozlan et al., 1986), in close agreement with those obtained using photosensitives probes (Ransom et al., 1986; Emerit et al., 1987). However, a recent study, employing the same methodology, but with a different ligand [^{3}H]ipsapirone, indicated a mol wt of 37.3kDa for 5-HT_{1A} sites in the rat hippocampus (Ferry et al., 1988). Even if there are some data showing that mol wts determined by SDS-PAGE are overestimated by up to 30% (Glossmann and Neville, 1971) and even if the glycosylated part of a membrane protein does not contribute to its mol wt estimated by the radiation-inactivation technique (Lowe and Kempner, 1982), the discrepancy between the low value found by Ferry et al. (1988), and that reported earlier (Gozlan et al., 1986) is rather puzzling. The most probable explanation should deal with the different conditions used by the two groups for the radiation-inactivation technique, since Nielsen and Braestrup (1988) recently pointed out how critical this procedure is for the accurate evaluation of the mol wt of proteins.

2.1.3. 5-HT$_2$ Receptors

IAZIK, a ^{125}I-azido-derivative of ketanserin has been recently developed by the Janssen group (Fig. 1). This probe bound irreversibly to a 67.5kDa protein in rat brain membranes (Wouters et al., 1987). A comparison of this value with those obtained using the radiation-inactivation technique reveals obvious discrepancies. Indeed, Gozlan et al. (1986) did find a mol wt for 5-HT_2 binding sites in the range of 60kDa, but higher values: 145–152kDa (Nishino and Tanaka, 1985), and 200–209kDa (Brann, 1985) have been also reported. Conversely, Ferry et al. (1988) found a much lower value: 35kDa for the 5-HT_2 receptor from rat cortical tissue. Again, these discrepancies emphasized the limits of the radiation-inactivation technique for the determination of the mol wt of membrane receptors.

2.1.4. Other 5-HT Receptors

No other photosensitive probes have been designed to specifically label 5-HT_{1B}, 5-HT_{1C}, 5-HT_{1D} and 5-HT_3 receptors. Interestingly, these receptor types are also those for which biochemical investigation has been so far limited.

2.2. Solubilization of 5-HT Receptors

In addition to using radioactive photosensitive probes for monitoring the labeled 5-HT receptor proteins throughout various chromatographic steps, another approach can be selected for purifying such membrane receptors. The method consists of solubilizing the receptor protein with a detergent and to follow this protein throughout the different purification steps by its binding property for selective radioactive ligands. The problems associated with this method result from the reduction or the loss of the binding capacity of the protein during the purification procedure. However, successful solubilization has been described for several 5-HT receptor subtypes.

The solubilization procedure involves the treatment of membranes or synaptosomal preparations with a detergent in buffer, generally 50 m*M* Tris-HCl (pH 7.4), which may contain protease inhibitors and/or MAO inhibitors, cation chelators, and salts. Ionic and nonionic detergents have been used. Notably, a zwitterionic detergent, 3-(3-cholamidolapropyl)dimethylammonio-*1*-propane sulfonate) (CHAPS), has provided the best results. After a brief sonication, or simple agitation, the mixture of membranes plus detergent was left for about 60 min at 4°C, and then centrifuged at 100,000 g. In some cases, the supernatant was also passed through a 0.22 µm filter before its use as a source of solubilized receptors.

The concentration of the initial membrane preparation has been quite variable from one group to another, between 1–40 mg prot/mL, for the solubilization step (Asarch and Shih, 1987; Gallaher and Wang, 1988). The yield, estimated by the binding capacity of the solubilized receptors compared to membrane receptors, ranges between 30–40% in the best cases, but much lower values have been also reported.

Since the soluble receptor becomes associated with the detergent, its binding characteristics are highly dependent on the concentration and the nature of the detergent. A reduction in the concentration of detergent can be achieved by dialysis, gel filtration, or polystyrene bead adsorption. In addition, binding conditions have to be modified for measuring soluble receptors. This includes, for instance, the precipitation of the receptor with polyethylene glycol (PEG), or its retention on glass fiber filters coated with a basic polymer such as polyethylenimine (Bruns et al., 1983).

2.2.1. 5-HT_1 Binding Sites

Vandenberg et al. (1983) first reported the solubilization of 5-HT_1 sites from the bovine cerebral cortex using a mixture of 4% octyl β-D-glucopyranoside, 3% Triton X-100, and 1% Tween 80. This mixture of nonionic detergents was found to be more efficient than the use of any one alone. Digitonin, lysolecithin, Brij-58, and SDS were also less efficient.

[^{3}H]5-HT binding assays were conducted on the solubilized fraction before and after elimination of the detergents by dialysis and treatment with Bio-Beads SM-2. Separation of the receptor protein from the free radioligand was obtained using the precipitation method with PEG (25%

w/v) plus bovine gamma globulin (0.1% w/v). Under these conditions, a specific binding of 80–90% was reported for [^{3}H]5-HT. In the presence of detergents, a monophasic saturation curve of solubilized 5-HT_1 binding sites was observed and a Kd of 50–100 n*M* was estimated. When detergents were eliminated, the saturation curve indicated the presence of two classes of binding sites (Kd1 = 1.3 n*M* and Kd2 = 22 n*M*), as found in membranes. Two years after the publication of this first paper, further data were reported by the same group (Allgren et al., 1985) with slight modifications of the solubilization procedure. This time, bovine membranes were treated with 1% Tween 80 and 3% Triton X-100 at 20°C for 60 min and incubated at 4°C for 45 min. Again, detergents were eliminated from the solubilized fraction by dialysis and Bio-Beads treatment. Partial pharmacological characterization indicated that the solubilized site did not correspond to a 5-HT_2 site but rather to a 5-HT_{1B} site. However, the solubilized receptor was probably altered during the treatment since the stereospecificity for LSD was lost. In addition, it could not be of the 5-HT_{1B} subtype since Heuring and Peroutka (1987) have clearly demonstrated the absence of this particular subtype in the bovine brain. Instead, it might correspond to the 5-HT_{1D} subtype of 5-HT_1 receptors (Heuring and Peroutka, 1987).

Another attempt to solubilize 5-HT_1 sites from rat and horse membranes has been made using either 1% digitonin or 1% sodium cholate (Rousselle et al., 1985), in the presence of EDTA (2 m*M*), the protease inhibitor, PMSF (1 m*M*), and aprotinin (5 μ/L). In the presence of detergent [^{3}H]5-HT bound to the solubilized material with two different affinities, indicating the heterogeneity of the solubilized binding sites. Furthermore, the modulation of agonist binding by guanine nucleotides, regularly observed with brain membranes, was no longer detected in the soluble fraction. However, this effect could be restablished after the addition of phospholipids (0.2% asolectin). Whether phospholipids are essential for the coupling of the receptor with the G-protein cannot be inferred from this study, since, at the same time as they added asolectin, detergents were also eliminated by gel filtration.

More recently, Gallaher and Wang (1988) also described the solubilization of 5-HT_1 sites from bovine membranes using the zwitterionic detergent, CHAPS (1.9 m*M*). Binding assays performed using the PEG precipitation technique indicated that even in the presence of the detergent, 5-HT_1 binding sites can be detected. However, no pharmacological characterization of these sites has been reported.

In all these studies [^{3}H]5-HT has been used to label the solubilized sites. However, this ligand possesses approximately the same affinity for the different 5-HT_1 subsites, and, since whole brain or areas containing more than one site have been generally studied, it appears that no information regarding a given solubilized subtype can be derived from these reports. For instance, Allgren et al. (1985) have concluded that 5-HT_{1B} sites have been preferentially solubilized under their own conditions, but satisfactory pharmacological characterization of the solubilized site was not really achieved. In addition, experiments using [^{3}H]5-HT are difficult to interpret when they are conducted in the absence of reducing agents since the most part of the resulting "specific" binding does not involve 5-HT_1 sites (Asarch and Shih, 1987). Indeed the same group was unable to reproduce previous finding (Vandenberg et al., 1983) in more recent experiments with ascorbic acid included in the buffer (Hamblin et al., 1987).

2.2.2. 5-HT_{1A} Receptors

Two groups have successfully solubilized and fully characterized the soluble 5-HT_{1A} receptor protein from the hippocampus of the rat (Gozlan et al., 1987; El Mestikawy et al., 1988) and of the bovine (Asarch and Shih, 1987; Takeuchi et al., 1988). These investigators optimized the nature and the concentration of the detergent(s) as well as the ratio of detergent/protein. Using 10 m*M* CHAPS, 35% of the rat 5-HT_{1A} receptor could be solubilized from rat hippocampal membranes. In case of bovine membranes, a yield of 25% was obtained with a mixture of 0.3% digitonin and 0.1% Nonidet NP-40. When rat or bovine cerebellum membranes were extracted under the same conditions, no 5-HT_{1A} receptors could be detected in the soluble fraction. In addition, solubilization of the bovine frontal cortex with digitonin and Nonidet did not reveal the presence of 5-HT_2 receptors in the soluble fraction (Asarch and Shih, 1987). El Mestikawy et al. (1988) added PMSF in the solubilizing mixture in order to prevent possible proteolysis, but Asarch and Shih (1987) did not use protease inhibitors since the yield of the solubilization procedure was apparently not improved in their presence. Nethertheless, the latter authors and Takeuchi et al. (1988) reported that the solubilized 5-HT_{1A} receptor was highly sensitive to calcium-dependent proteases.

Both groups reported an increased thermolability of the 5-HT_{1A} under the soluble form. Thus, a rapid inactivation occurs when the temperature increases. In addition, congelation at low temperatures and subsequent thawing, completely suppresses [^{3}H]5-HT or [^{3}H]8-OH-DPAT binding. 5-HT_{1A} receptors solubilized from the bovine brain seem however, more sensitive to temperature changes than those from rat hippocampus. Notably [^{3}H]8-OH-DPAT binding could be found after 48 h at 4°C in the rat soluble fraction, whereas it was markedly reduced after 24 h at 4°C in the bovine soluble fraction. Therefore, binding assays were conducted either at 0°C (Asarch and Shih, 1987) or at 15°C (El Mestikawy et al., 1988). In both cases, the filtration technique with glass fiber filters presoaked with 0.3% PEI was used since other methods tested were less efficient (El Mestikawy et al., 1988). Interestingly, the elimination of detergents also contributes to some reduction or even the suppression of 5-HT_{1A} binding activity in the soluble fractions (Asarch and Shih, 1987). This indicates that detergents contribute to the stabilization of the protein structure.

Clearly, pharmacological experiments conducted on both soluble preparations have shown that the solubilized site has a 5-HT_{1A} profile:

1. Binding of either [^{3}H]5HT or [^{3}H]8-OH-DPAT was saturable andtheir affinity for the solubilized sites was in agreement with the corresponding values using membrane preparations. Scatchard plot indicated only one site, i.e., the 5-HT_{1A} receptor, could be evidenced with [^{3}H]8-OH-DPAT in the soluble fraction from the rat hippocampus.
2. This site possesses high affinity for 5-HT_{1a} drugs including 5-HT, 8-OH-DPAT, PAPP, Ipsapirone, Spiroxatrine, and a good correlation was found between the respective IC_{50} values of 13 drugs against[^{3}H]-8-OH-DPAT binding to the soluble and the membrane receptors (El Mestikawy et al., 1988).
3. The solubilized receptor can be modulated by guanine nucleotides. However, the bovine 5-HT_{1A} receptor seems to be less sensitive than that from the rat. This indicates that both preparations have cosolubilized the receptor protein with the regulatory G-protein, but in the bovine preparation one component may have been slightly altered during the purification procedure.
4. Modulations by mono and divalent cations of [^{3}H]8-OH-DPAT binding to the rat soluble 5-HT_{1A} receptor were also observed, although hey were less striking than in membranes (El Mestikawy et al., 1988).
5. The affinity of [^{3}H]8-OH-DPAT for the soluble site was markedly dependent on the pH, as previously reported for membrane-bound 5-HT_{1A} receptors (Hall et al., 1986).

2.2.3. 5-HT_{1B} and 5-HT_{1D} Receptors

No data are available yet about the solubilization of 5-HT_{1B} receptors from the rat brain. However, the work of Asarch and Shih (1987) has provided some information concerning the solubilization of 5-HT_{1D} receptors from the bovine brain. These investigators claimed that the solubilization conditions described for 5-HT_{1A} receptors were also efficient for 5-HT_{1D} sites (called 5-HT_{1B} in this work), and, as expected, the displacement by 8-OH-DPAT of [^{3}H]5-HT specifically bound to the soluble fraction of the bovine cerebral cortex was clearly biphasic: the high affinity component corresponding to 5-HT_{1A} sites, and the low affinity component to 5-HT_{1D} sites. Furthermore, the proportions of the two sites were the same as in membrane preparations, indicating that the 5-HT_{1A} and 5-HT_{1D} receptor binding proteins have the same solubilization requirement (regarding, for instance, the detergent/protein ratio, phospholipids). This is not the case for 5-HT_2 receptors, since under the same conditions no [^{3}H]spiperone specific binding could be detected in soluble cortical extracts (Asarch and Shih, 1987).

2.2.4. 5-HT_{1C} Receptors

The choroid plexus represents a unique structure where only one class of 5-HT_1 receptor is found. In this context, Yagaloff and Hartig (1986) have successfully solubilized the 5-HT_{1C} receptor from pig choroid plexus, using CHAPS as a detergent. *N-1*-methyl-2-[^{125}I]lysergic acid diethylamide ([^{125}I]MIL) was used for labelling the soluble 5-HT_{1C} sites that were subsequently separated from the free ligand (23% PEG 8000 or Sephadex G 50 gel filtration). A yield of 45% was obtained with 10 m*M* CHAPS but could not be improved by higher concentrations of the detergent.

As reported for the 5-HT_{1A} site, the solubilized 5-HT_{1C} site was thermolabile. Its pharmacological properties were nearly identical to those of membrane-bound receptors, with a high affinity for mianserin and a marked stereoselectivity toward LSD. However, the affinity for 5-HT was 20 times higher in the soluble fraction than in membranes. A similar increase of the affinity of agonists for solubilized sites has been also reported about catecholaminergic receptors and might reflect an alteration of the interaction between the receptor protein and the G-protein. However, the loss of another modulatory component of the 5-HT_{1C} site or a structural change of the receptor protein during solubilization could not be excluded (*see* Yagaloff and Hartig, 1986).

2.2.5. 5-HT_2 Receptors

The first reported solubilization of 5-HT_2 receptors from rat and dog brain was achieved using nonionic detergents such as lysolecithin (0.25%) (Ilien et al., 1982,1982a) and digitonin (1%) (Chan and Madras, 1982). Recently, using CHAPS in the presence of a high concentration (1.4 *M*) of NaCl, Wouters et al. (1985) markedly improved the solubilization procedure, with up to 40% of [^{3}H]ketanserin binding sites of rat frontal cortex being recovered in the soluble fraction. The optimum concentration of the detergent was 6.8 m*M*, and lower yields were obtained with higher or lower concentration. This contrasts with 5-HT$_1$ receptors since no modification of the solubilization characteristics can be observed with concentrations of CHAPS higher than 10 m*M*.

The binding of [^{3}H]ketanserin was saturable and possesses the characteristics of a 5-HT$_2$ binding site, being displaced by nanomolar concentrations of 5-HT$_2$ antagonists and by micro- to millimolar concentrations of agonists. However, as already reported for 5-HT$_{1C}$ receptors, agonist affinity was 10 times higher than in corresponding membranes. The high concentration of salt in the buffer may explain the observed differences.

2.3. Analysis of Solubilized 5-HT Receptors

Following solubilization of 5-HT receptors, the evaluation of some physicochemical parameters of the binding proteins and their purification have been undertaken using different approaches: exclusion chromatography, gradient sedimentation, and affinity chromatography.

2.3.1. 5-HT$_1$ Binding Sites

Vandenberg and colleagues reported that high affinity [^{3}H]5-HT binding activity was preserved after chromatography on Sephadex S-300 of 5-HT$_1$ receptors solubilized from bovine brain (Vandenberg et al., 1983). Elution of 5-HT$_1$ sites from the column was performed with the solubilizing buffer supplemented with 0.2*M* NaCl and 0.2% (w/v) octyl-β-D-glucogalactoside. Under these conditions, a single peak of [^{3}H]5-HT binding activity was found in the eluted fractions that corresponded to an apparent mol wt of 98 ± 18 kDa. The same group was also able to observe [^{3}H]5-HT binding activity in fractions collected after the sedimentation of solubilized material through a 10–30% linear glycerol gradient. The maximal binding activity was found in a fraction corresponding to a sedimentation coefficient of 3.5 S (equivalent to a mol wt of 58kDa). The addition of the protease inhibitor PMSF did not change the result, indicating that this low mol wt is unlikely to be a result of cleavage of the native protein receptor. However, when sedimentation was performed through the same gradient but without detergent, a sedimentation coefficient of 11.5 S was obtained, corresponding to a calculated mol wt of 150kDa (Allgren et al., 1985).

Different values were reported using AcA 22 chromatography of sodium cholate-solubilized horse brain extracts (Rousselle et al., 1985). Thus, [^{3}H]-5-HT binding activity was found in two broad peaks eluted with the solubilizing buffer, corresponding to apparent mol wt of 235 ±

33 and 438 ± 24kDa. An enrichment of 30–40-fold in [^{3}H]5-HT specific binding activity (per mg protein) was noted at the level of each peak compared to the starting soluble extract.

Usually, affinity chromatography is one of the best techniques for the purification of receptors, and attempts have also been made with this approach in case of unidenfied 5-HT$_1$ subtypes. Gallaher and Wang (1988) coupled a LSD derivative to an agarose matrix (Fig. 2C) on which the solubilized 5-HT$_1$ sites from bovine cortex could absorb. Elution with a buffer containing 120 μ*M* 5-HT, 4 m*M* CHAPS, and lipids (asolectin) and dialysis to eliminate the detergent allowed the reconstitution of the eluted 5-HT receptors into lipid vesicles. A specific [^{3}H]5-HT binding activity (kDa = 16.9 n*M*) could be measured in these vesicles. 5-Methoxytryptamine, but not ketanserin, displaced [^{3}H]5-HT from this vesicular binding site. An enrichment of 1000-fold was evaluated in the reconsti-tuted vesicles compared to the starting extract. SDS-PAGE of the richest fraction revealed 4 broad bands in the 92–95, 78–84, 68–73, and 61–66kDa ranges. The possibility that these bands correspond to various 5-HT receptors types was suggested. Indeed, several 5-HT$_1$ subtypes do exist in the bovine cerebral cortex and the selected affinity ligand derived from LSD bound to all of them, and also probably to 5-HT$_2$ receptors. The fact that no fragment corresponding to a G-protein subunit can be observed in SDS-PAGE led Gallaher and Wang (1988) to speculate that they have purified a "naked" receptor, but further studies are needed before reaching a definitive conclusion on this point.

2.3.2. 5-HT$_{1A}$ Receptors

Extensive analysis of the 5-HT$_{1A}$ receptor solubilized from the rat hippocampus has been described (El Mestikawy et al., 1988,1989). Thus, when the CHAPS-solubilized fractions were passed through a Sephacryl S-400 column, two peaks with [^{3}H]8-OH-DPAT binding activity were eluted. The first one corresponded to the void vol of the column, and the second allowed the calculation of a mol wt of 155kDa for the eluted material. Since the mol wt of the 5-HT$_{1A}$ binding subunit determined by affinity chromatography (Emerit et al., 1987; Ransom et al., 1986) or by radiation-inactivation (Gozlan et al., 1986) was in the 60kDa range, and because the affinity of [^{3}H]8-OH-DPAT for the second peak could be modulated (decrease) by GppNHp, it has been assumed that this value of 155kDa was that of a complex formed by the binding subunit still attached to a G protein.

Since the solubilized receptor was sensitive to temperature, various experimental conditions were investigated in order to improve its

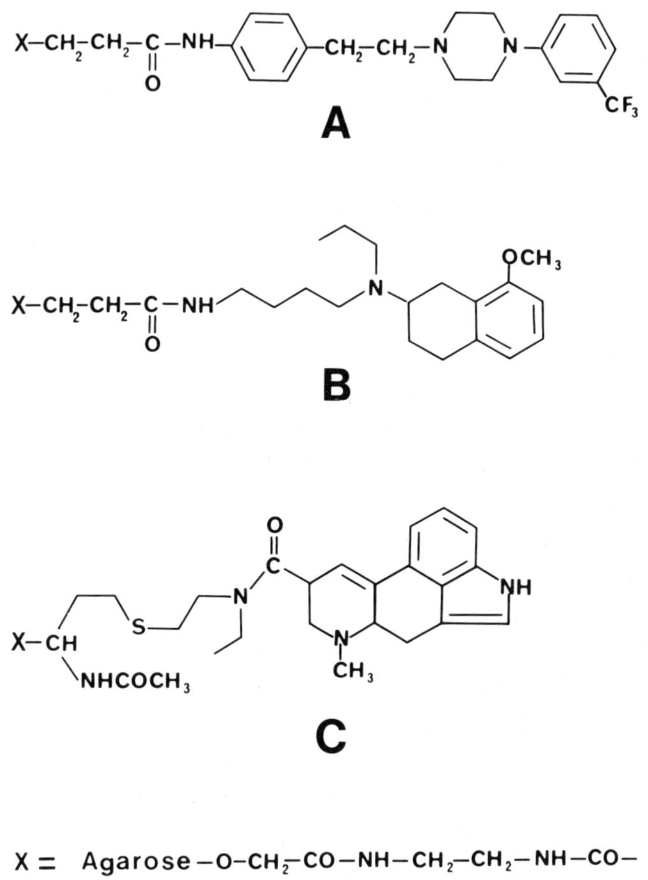

A

B

C

X = Agarose–O–CH_2–CO–NH–CH_2–CH_2–NH–CO–

Fig. 2. Affinity columns for the purification of 5-HT receptors. **A** PAPP was coupled to affi-gel 10 (Takeuchi et al., 1988). **B** 8-MeO-2-[(*N*-Propyl, *N*-butylamino) amino]tetralin was coupled to affi-gel 202 (El Mestikawy et al., 1989). **C** Lysergic acid ethylamidoethyl bromide was coupled to affi-gel 401 (Gallaher and Wang, 1988).

thermostability (El Mestikawy et al., 1989). Thus, a 50 m*M* Tris-HCl buffer containing 0.06% CHAPS, 10% glycerol, 50 μ*M* DTT, 0.1 m*M* $MnCl_2$, and 50 mg/L of cholesterol hemisuccinate at pH 7.4 proved to ensure a high degree of stability of the solubilized sites during chromatographic analysis.

Purification of the 5-HT_{1A} receptor was then attempted through two consecutive affinity chromatography steps. Indeed, a derivative of 8-OH-DPAT was coupled to an agarose matrix (Fig. 2B) on which the solubilized receptor was retained, then eluted with 1 m*M* 5-HT, directly on a wheat-germ agglutinin (WGA) column. Elution was performed with *N*-acetyl glucosamine, and [^{3}H]8-OH-DPAT binding activity was detected in the eluate. An enrichment of 400-fold was calculated in the best fraction compared to the starting extract, but the actual factor of purification was probably much higher, since no attempt was made to improve the assays conditions for the measurement of [^{3}H]8-OH-DPAT binding to the eluted material (El Mestikawy et al., 1989). SDS-PAGE of the richest fractions confirmed the presence of the expected 60kDa band, probably the 5-HT_{1A} binding subunit. Two other bands with M_r of 50 and 41kDa were also detected. They might correspond to proteolytic fragments of the 5-HT_{1A} subunit, or to proteins associated with the receptor. In fact, evidence has been reported that the solubilized receptor was still attached to a G-protein even after its partial purification on a WGA column (El Mestikawy et al., 1988). Other data from the same study led to the conclusion that the 5-HT_{1A} receptor binding subunit is a 60kDa glycosylated protein, negatively charged at pH 7.4, containing SH groups of some importance for the binding of agonists (El Mestikawy et al., 1989).

Most of these properties are in agreement with those recently described for the bovine brain 5-HT_{1A} receptor by Takeuchi et al. (1988). Using a PAPP-affinity column (Fig. 2A) these authors were able to purify the 5-HT_{1A} receptor but the enrichment was not evaluated since no binding activity could be found after the affinity chromatography step. The mol wts of the two proteins specifically eluted from this column by 5-HT were 55 and 38kDa. The first value is again in agreement with the previous evaluation of the 5-HT_{1A} receptor mol wt by the same group. This was confirmed by photoaffinity labeling of the eluted fraction using [^{3}H]azido-PAPP since SDS-PAGE indicated also the presence of two labeled proteins of 55 and 38kDa. The isoelectric points of these two proteins were in the same range: 6.0 and 6.5. Whether the 38kDa protein was a proteolytic fragment of the larger binding subunit has not specially been addressed, but remains an interesting possibility.

2.3.3. 5-HT_{1C} Receptors

Yagaloff and Hartig (1986) have shown that the 5-HT_{1C} receptor solubilized by CHAPS from the pig choroid plexus has a high mol wt. Thus, a value of 800kDa was deduced from gel filtration (AcA 22) and equilibrium sedimentation experiments. This high value may indicate an aggregation of the receptor protein.

2.3.4. 5-HT_2 Receptors

The purification of the 5-HT_2 receptor has not been achieved yet, but some information regarding the physicochemical properties of 5-HT_2

binding subunit have been reported, mainly by the Janssen group. Thus, the sedimentation of solubilized 5-HT_2 through a 20–30% linear sucrose gradient, allowed the recovery of [^{3}H]ketanserin binding in a 5 S fraction (Wouters et al., 1985), corresponding to a M_r of 80 Kda (Wouters et al., 1987). A previous evaluation by the same technique has led to a higher value (Ilien et al., 1982), but the former seems more likely, since photo affinity labeling (Wouters et al., 1987), and radiation-inactivation (Gozlan et al., 1986), gave a value in the same range.

3. Cloning of 5-HT Receptors

Beside the purification of 5-HT receptors using various biochemical techniques, another approach can be used for their extensive characterization: the cloning of cDNAs from specific messenger RNAs coding for the receptors. Considerable progress has been made in the knowledge of 5-HT receptors thanks to the application of this technique during the present year. Thus, three of them have been cloned and their respective sequence established.

3.1. 5-HT_{1A} Receptors

Using a human β_2-adrenergic cDNA probe, in order to isolate the gene encoding for the related β_1 receptor, an intronless clone termed G-21 was selected that did not correspond to any adrenergic receptor (Kobilka et al., 1987). Further investigations have shown that this genomic clone encodes in fact, the 5-HT_{1A} receptor (Fargin et al., 1988).

This 421 amino acid protein (Fig. 3A) belongs to the class of G-protein-coupled receptors and possesses a high degree of analogy with the human and the hamster β_2-adrenergic receptor proteins. The highest homology is found in the hydrophobic transmembrane domains (up to 76% in the 6th membrane-spanning domain). In contrast, the third cytoplasmic loop (Fig. 3A) which is assumed to be involved in the coupling with G-proteins (Kobilka et al., 1988), is markedly different in the 5-HT_{1A} receptor and in the (β_2-adrenergic receptor (<5% homology). In the 5-HT_{1A} receptor, this loop contains a greater number of amino acid residues than in the β_2 receptor (128 instead of 30), and its length is comparable with that of the recently cloned α_2 receptors (150 residues, Regan et al., 1988; Kobilka et al., 1987a), which are also negatively coupled to adenylate cyclase.

The short C-terminal chain contains no phosphorylation sites, but these sites are found on the third cytoplasmic loop. Antibodies against

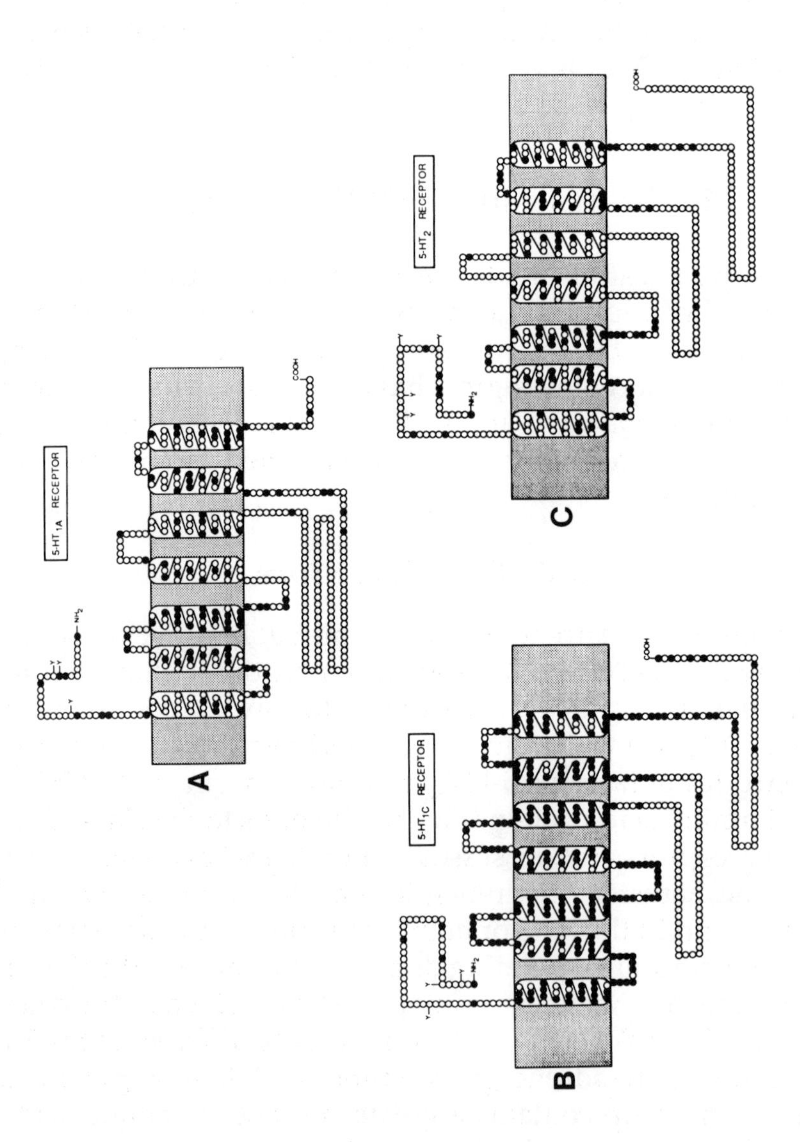

Fig. 3. Representation of transmembrane topology of **A** 5-HT_{1A} receptor; **B** 5-HT_{1C} receptor, and **C** 5-HT_2 receptor. Each receptor contains seven hydrophobic domains. The NH_2 terminus is located on the extracellular face of the lipid bilayer and the COOH terminus is on the cytoplasmic side. Solid circles indicate common amino acid residues in: **A** 5-HT_{1A} and 5-HT_{1C} receptors, **B** 5-HT_{1C} and 5-HT_2 receptors, and **C** 5-HT_2 and 5-HT_{1A} receptors.

a part of the third cytoplasmic loop have been obtained and they should constitute useful tools for isolating the receptor protein.

From its 421 amino acids, a mol wt of 45,763kDa was calculated, but some asparagine residues also exist in the amino terminal chain, the glycosylation of which could increase this value. Therefore, the actual mol wt fits perfectly with that established previously by radiation-inactivation (Gozlan et al., 1986) and irreversible photolabeling (Ransom et al., 1987; Gozlan et al., 1987).

Recently, the 5-HT_{1A} receptor from the rat brain has also been cloned. Its sequence possesses a high degree of analogy with the human receptor (Civelli, personal communication).

3.2. 5-HT_{1C} Receptor

This receptor was first cloned from a mouse choroid plexus papilloma, and was pharmacologically characterized, but its sequence was not established in these studies (Lübbert et al., 1987,1987a). Recently, the molecular characterization of a functional cDNA encoding the 5-HT_{1C} receptor was published (Julius et al., 1988), clearly demonstrating that this receptor also belongs to the family of G-protein-coupled receptors (Fig. 3B). This 460 amino acid protein shares only 25% identity with the β_2-adrenergic receptor. Even in the transmembrane domains, the 5-HT_{1C} receptor is quite different from the 5-HT_{1A} receptor (Fig. 3A). Furthermore, the length of the third cytoplasmic loop is shorter in the 5-HT_{1C} receptor (77 residues) than in the 5-HT_{1A} receptor (128 residues), but in the same order of that in the α_1 loop (63 residues, Cotecchia et al., 1988). Interestingly, both α_1 and 5-HT_{1C} receptors are coupled to phospholipase C. A mol wt of 51,899kDa was calculated for the 460 amino acids of the 5-HT_{1C} receptor.

3.3. 5-HT_2 Receptor

A cDNA encoding the 5-HT_2 receptor has been cloned and sequenced recently using two oligonucleotides directed against amino acid residues 88-104 and 134-149 of the 5-HT_{1C} receptor (Pritchett et al., 1988). These residues are mainly located in the second and the third transmembrane domains, where analogy clearly exist in their respective sequences of 5-HT_{1C} (Fig. 3B) and 5-HT_2 receptors (Fig. 3C). Indeed, 60–100% analogy was found between these two receptors at the level of the seven transmembrane hydrophobic domains. Both receptors are coupled to phospholipase C, and, as expected, the third loop is of the same length in the 5-HT_2 (67 residues), as in the 5-HT_{1C} (77 residues) and α_1- (63 residues) receptors. In addition, the C-terminal chain of the 5-HT_2 receptor has just one more residue than that of the 5-HT_{1C}. However, less analogy between 5-HT_{1C} and 5-HT_2 receptors was found in the last two cytoplasmic domains, which indicates that the coupling with the G-protein is not only related to the length of the third loop but involves other parameters. In fact, for all G-protein-coupled receptors that have been sequenced so far, the third loop contains a great number of acidic and basic amino acids whose role(s) remain(s) to be established.

Finally, as expected from pharmacological data, the 5-HT_2 receptor (Fig. 3C) is clearly different from 5-HT_{1A} receptor (Fig. 3A). The calculated mol wt of the 5-HT_2 receptor, 58,420kDa, is again in the same range as that determined by photoaffinity labeling (Wouters et al., 1987) and by radiation inactivation (Gozlan et al., 1986).

4. Conclusion

The purification of 5-HT receptors has been attempted by conventional biochemical methods, but, clearly, no receptor protein has been yet isolated and completely purified. In contrast, cloning of 5-HT receptor genes has led to the determination of the complete sequence and primary structure of receptor proteins. The models proposed for the receptor structures are in agreement with their known biological and pharmacological properties. However, the complete purification of native receptors has to be achieved for a definitive demonstration of the relevance of such models to actual status of receptors in brain membranes. Other major problems are still pending: for instance the purification and the cloning of 5-HT_{1B} and 5-HT_{1D} receptors. However, on the account of the considerable contribution of molecular biology for the present knowledge of other 5-HT_1 receptors, it can be reasonably predicted that these problems will be solved within the next few months. Finally, a formidable challenge remains, the purification of 5-HT_3 receptors, but further pharmacological characterization of these receptors in the CNS will have to be achieved first.

Acknowledgments

We thank P. H. Seeburg for the kind communication of his results on 5-HT_2 cloning prior to publication, A. Carayon for preparing the MacIntosch manuscript, and J.-J. Benoliel for drawing the receptor proteins.

References

Allgren, R. L., Kyncl, M. M. A., and Ciaranello, R. D. (1985) *Brain Res.* **348,** 77–85.

Asarch, K. B. and Shih, J. C. (1987) *J. Neurochem.* **48,** 1494–1501.

Bouhelal, R., Smounya, L., and Bockaert, J. (1988) *Eur. J. Pharmacol.* **151,** 189–196.

Bradley, P. B., Engel, G., Feniuk, W., Fozard, J. R., Humphrey, P. P. A., Middlemiss, D. N., Mylecharane, E. J., Richardson, B. P., and Saxena, P. R. (1986) *Neuropharmacology* **25,** 563–576.

Brann, M. R. (1985) *Biochem. Biophys. Res. Commun.* **133,** 1181–1186.

Bruns, R. F., Lawson-Wendling, K., and Pugsley, T. A. (1983) *Anal. Biochem.* **132,** 74–81.

Chan, B. and Madras, B. K. (1982) *Eur. J. Pharmacol.* **83,** 1–10.

Cheng, S. H. and Shih, J. C. (1979) *Life Sci.* **25,** 2197–2203.

Conn, P. J. and Sanders-Bush, E. (1985) *J. Pharmacol. Exp. Ther.* **234,** 195–203.

Conn, P. J., Sanders-Bush, E. Hoffman, B. J., and Hartig, P. R. (1986) *Proc. Natl. Acad. Sci. USA* **83,** 4086–4088.

Cotecchia, S., Schwinn, D. A., Randall, R. R., Lefkowitz, R., Caron, M. G., and Kobilka, B. K. (1988) *Proc. Natl. Acad. Sci. USA* **85,** 7159–7163.

Devivo, M. and Maayani, S. (1986) *J. Pharmacol. Exp. Ther.* **238,** 248–253.

Dourish, C. T., Ahlenius, S., and Hutson, P. H. (1987) *"Brain 5-HT_{1A} Receptors,"* Ellis Horwood, Chichester, p. 306.

El Mestikawy, S., Cognard, C., Gozlan, H., and Hamon, M. (1988) *J. Neurochem.* **51,** 1031–1040.

El Mestikawy, S., Taussig, D., Gozlan, H., Emerit, M. B., Ponchant, M., and Hamon, M. (1989) *J. Neurochem.* **53,** 1555–1566.

Emerit, M. B., Gozlan, H., Marquet, A., and Hamon, M. (1986) *Eur. J. Pharmacol.* **127,** 67–81.

Emerit, M. B., El Mestikawy, S., Gozlan, H., Cossery, J. M., Besselievre, R., Marquet, A., and Hamon, M. (1987) *J. Neurochem.* **49,** 373–380.

Fargin, A., Raymond, J. R., Lohse, M. J., Kobilka, B. K., Caron, M. G., and Lefkowitz, R. J. (1988) *Nature* **335,** 358–360.

Ferry, D. R., Goll, A., Glossmann, H., Glaser, T., and Traber, T. (1988) *Neurosci. Res. Commun.* **2,** 9–17.

Gaddum, J. H. and Picarelli, Z. P. (1957) *Brit. J. Pharmacol. Chemother.* **12,** 323–328.

Gallaher, T. K. and Wang, H. H. (1988) *Proc. Natl. Acad. Sci. USA* **85,** 2378–2382.

Glossman, H. and Neville, D. M. (1971) *J. Biol. Chem.* **246,** 6339–6346.

Gozlan, H., El Mestikawy, S., Pichat, L., Glowinski, J., and Hamon, M. (1983) *Nature* **305,** 140–142.

Gozlan, H., Emerit, M. B., Hall, M. D., Nielsen, M., and Hamon, M. (1986) *Biochem. Pharmacol.* **35,** 1891–1897.

Gozlan, H., Emerit, M. B., El Mestikawy, S., Cossery, J. M., Marquet, A., Besselievre, R., and Hamon, M. (1987) *J. Recept. Res.* **7,** 195–221.

Hall, M. D., Gozlan, H., Emerit, M. B., El Mestikawy, S., Pichat, L., and Hamon, M. (1986) *Neurochem. Res.* **11,** 891–912.

Hamblin, M. W., Ariani, K., Adriaenssens, P. I., and Ciaranello, R. D. (1987) *J. Pharmacol. Exp. Ther.* **243,** 989–1001.

Hamon, M., Fattaccini, C. M., Adrien, J., Gallissot, M.-C., Martin, P., and Gozlan, H. (1988) *J. Pharmacol. Exp. Ther.* **246,** 745–752.

Hamon, M. (1987) in *"Pharmacology"* (Rand, M. J. and Raper, C., eds.), Elsevier, Amsterdam, pp. 281–284.
Hamon, M., Emerit, M. B., El Mestikawy, S., Vergé, D., Daval, G., Marquet, A., and Gozlan, H. (1987) in *"Brain 5-HT$_{1A}$ Receptors"* (Dourish, C. T., Alhenius, S., and Hutson, P. H., eds.),Horwood, Chichester, pp. 34–51.
Heuring, R. E. and Peroutka, S. J. (1987) *J. Neurosci.* **7,** 894– 903.
Hoyer, D., Engel, G., and Kalkman, H. O. (1985) *Eur. J. Pharmacol.* **118,** 13–23.
Hoyer, D. and Neijt, H. C. (1987) *Eur. J. Pharmacol.* **143,** 291, 292.
Hoyer, D. and Schoeffter, P. (1988) *Eur. J. Pharmacol.* **147,** 145–147.
Ilien, B., Gorissen, H., and Laduron, P. M. (1982) *Mol. Pharmacol.* **22,** 243–249.
Ilien, B., Schotte, A., and Laduron, P. M. (1982a) *FEBS Lett.* **138,** 311–315.
Julius, D., MacDermott, A. B., Axel, R., and Jessell, T. M. (1988) *Science* **241,** 558– 564.
Kendall, D. A. and Nahorski, S. R. (1985) *J. Pharmacol. Exp. Ther.* **233,** 473–479.
Kilpatrick, G. J., Jones, B. J., and Tyers, M. B. (1987) *Nature* **330,** 746–748.
Kobilka, B. K., Frielle, T., Collins, S., Yang-Feng, T. L., Kobilka, T. S., Francke, U., Lefkowitz, R. J., and Caron, M. G. (1987) *Nature* **329,** 75–79.
Kobilka, B. K., Matsui, H., Kobilka, T. S., Yang-Feng, T. L., Francke, U., Caron, M. G., Lefkowitz, R. J., and Regan, J. W. (1987a) *Science* **238,** 650-656.
Kobilka, B. K., Kobilka, T. S., Daniel, K., Regan, J. W., Caron, M. G., and Lefkowitz, R. J. (1988) *Science* **240,** 1310–1316.
Leysen, J. E., Niemegeers, C. J. E., Van Nueten, J. M., and Laduron, P. M. (1982) ^{3}H-Ketanserin (R 41468) *Mol. Pharmacol.* **21,** 301–314.
Lowe, M. E. and Kempner, E. S. (1982) *J. Biol. Chem.* **257,** 12478–12480.
Lübbert, H., Hoffman, B. J., Snutch, T. P., Van Dyke, T., Levine, A. J., Hartig, P. R., Lester, H. A., and Davidson, N. (1987) *Proc. Natl. Acad. Sci. USA* **84,** 4332–4336.
Lübbert, H., Snutch, T. P., Lester, H. A., and Davidson, N. (1987) *J. Neuroscience* **7,** 1159–1165.
Nielsen, M. and Braestrup, C. (1988) *J. Biol. Chem.* **263,** 11900–11906.
Nishino, N. and Tanaka, C. (1985) *Life Sci.* **37,** 1167–1174.
Osborne, N. N. and Hamon, M. (1988) *"Neuronal Serotonin."* J. Wiley and Sons, Chichester, Amsterdam, p. 555.
Pazos, A. and Palacios, J. M. (1985) *Brain Res.* **346,** 205–230.
Pazos, A., Hoyer, D., and Palacios, J. M. (1985) *Eur. J. Pharmacol.* **106,** 539–546.
Pazos, A., Cortes, R., and Palacios, J. M. (1985a) *Brain Res.* **346,** 231–249.
Pedigo, N. W., Yamamura, H. I., and Nelson, D. L. (1981) *J. Neurochem.* **36,** 220–226.
Pritchett, D. B., Bach, A. W. J., Wozny, M., Taleb, O., Dal Toso, R., Shih, J. C., and Seeburg, P. H. (1988) *EMBO J.* , **7,** 4135–4140.
Ransom, R. W., Asarch, K. B., and Shih, J. C. (1985) *J. Neurochem.* **46,** 68–75.
Ransom, R. W., Asarch, K. B., and Shih, J. C. (1986) *J. Neurochem.* **47,** 1066–1072.
Regan, J. W., Kobilka, T. S., Yang-Feng, T. L., Caron, M. G., Lefkowitz, R. J., and Kobilka, B. K. (1988) *Proc. Natl. Acad. Sci. USA* **85,** 6301–6305.
Richardson, B. P. and Engel, G. (1986) *TINS* 424–428.
Rousselle, J. C., Gillet, G., and Fillon, G. (1985) *J. Pharmacol. (Paris)* **16,** 421– 438.
Saitoh, T. and Shih, J. C. (1987) *J. Neurochem.* **49,** 1361–1366.
Takeuchi, Y., Yang, W., and Shih, J. C. (1988) *J. Neurochem.* **51,** 1343–1349.
Tricklebank, M. D. (1985) *Trends Pharmacol. Sci.* **6,** 403–407.
Vandenberg, S. R., Allgren, R. L., Todd, R. D., and Ciaranello, R. D. (1983) *Proc. Natl. Acad. Sci. USA* **80,** 3508–3512.

Vergé, D., Daval, G., Marcinkiewicz, M., Patey, A., El Mestikawy, S., Gozlan, H., and Hamon, M. (1986) *J. Neurosci.* **6,** 3474–3482.
Wouters, W., Van Dun, J., Leysen, J. E., and Laduron, P. M. (1985) *Eur. J. Pharmacol.* **115,** 1–9.
Wouters, W., Van Dun, J., Leysen, J. E., and Laduron, P. M. (1987) *FEBS Lett.* **213,** 359–364.
Yagaloff, K. E. and Hartig, P. R. (1986) *Mol. Pharmacol.* **29,** 120–125.

Purification of β-Adrenergic Receptors

Isolation of Mammalian β_1- and β_2-Subtypes

Craig C. Malbon

1. Introduction

Information transduction across biological membranes is a fundamental process in nature, providing the means by which cells communicate with neighboring cells, distant cells, and the changing, often challenging melieu in which they exist. Hormone receptors and the photopigments like rhodopsin represent one class of cell-surface membrane proteins that transduce information from ligands (or photons in phototransduction) to their intracellular effector units via GTP-binding regulatory proteins, or "G-proteins" (Gilman, 1987; Stiles et al., 1984; Graziano and Gilman, 1987). The light-sensitive cGMP phosphodiesterase of vertebrate retinal rods (Stryer and Bourne, 1986) and the hormone-sensitive adenylate cyclase system (Gilman, 1987) represent two well-characterized systems in which activation of the receptor is transduced into the enhanced activity of a membrane-bound enzyme through the mediation of specific G-proteins, G_t and G_s, respectively. Of the many hormones that operate via G-proteins to control effector systems such as adenylate cyclase (Gilman, 1987), phospholipase C (Cockcroft, 1987), and ion channels (Neer and Clapham, 1988), the catecholamines have been the most extensively studied. The receptors that mediate β-catecholamine responses, the β-adrenergic receptors, will be the focus of this chapter. These very-low-abundance receptors, often present at about 1000 copies/cell, have been the focus of intense interest by biochemists for more than a decade. Isolation of these receptors has been a formidable task for reasons that will be illuminated. Although first isolated from amphibian (Shorr et al., 1981) and avian (Vauquelin et al., 1979; Shorr et al., 1982b) erythrocyte membranes, β-adrenergic receptors of mammalian origin have been the targets for much of the large-scale purification. Isolation of mammalian β-adrenergic receptors will be highlighted in this chapter. The most recent work on the isolation of the human receptors will be described. The methodologies developed in the pursuit of the purification of the β_1 and β_2-receptor peptides can be generally applied to other sources of starting material. For the sake of brevity and clarity, details of the evolution of the technologies not pertinent to the general topic of receptor isolation will be omitted purposely.

2. Experimental Parameters

This section provides the basic strategy that has been developed in this laboratory for the successful isolation of mammalian b-adrenergic receptors from a variety of tissues, isolated cells, cells in culture, and, currently, from cells transfected with an expression vector that directs high expression of these receptors. The protocols are designed for large-scale purification in which kilogram quantities of tissues or hundreds of liters of cells grown to high density in spinner culture are the starting sources. The dynamics of the separations are such that the protocols are not easily scaled down without incurring considerable losses in the overall yield. Although it is possible to modify aspects of the procedures to accommodate the processing of much smaller quantities of starting material, the reader is encouraged to experiment with this parameter only after having successfully isolated receptor on a large scale from one of the sources described below.

2.1. Source of Receptor for Purification

More often than not, economics dictate the selection for the starting source of a large-scale isolation of receptor. Having successfully isolated receptor from a variety of tissues and cells, we found that the most economical and readily available source is the human placenta. The human placenta is a rich source of receptor (nearly 2 nmol of receptor per placenta), and this tissue displays both pharmacological subtypes, β_1 and β_2-adrenergic receptors (Moore and Whitsett, 1981). This latter feature, the presence of both receptor subtypes, is of no small consequence, since the only other ready source for the β_1-receptor peptides has been the isolated rat fat cell (Cubero and Malbon, 1984). In addition, the specific activity of crude and purified basal membranes are nearly an order of magnitude greater than those for similar preparations derived from other tissues (Kelley et al., 1983). These four features—economy, availability of fresh tissue, rich abundance of β_1- and β_2-receptors, and ready preparation of membranes with specific activities of 3000–5000 fmol/mg of membrane protein—provide a strong case for selecting human placenta as the starting material. Sources of homogenous mammalian β_1-receptor that have been successfully employed for large-scale isolation are confined to the rat fat cell (Cubero and Malbon, 1984), whereas sources of β_2-receptors include canine lung (Homcy et al., 1983), bovine lung (Insoft and Homcy, 1984), hamster lung (Benovic, 1984), rat liver (Graziano et al., 1985), as well as several cells in culture, including S49 mouse lymphoma cells (George and Malbon, 1985), and human A431 epidermoid carcinoma cells (Guillet et al., 1985). For details of tissue handling and membrane preparation from these sources, the reader should consult the specific references provided.

The acquisition of fresh placenta generally poses little problem. In our experience University-affiliated or local community hospitals are willing to provide ready access to placentas from full-term births following Ceasarian section or vaginal delivery. It is important to note, howver, that the tissue must be fresh and chilled in an isotonic saline buffer immediately. For our studies, the placentas were obtained, chilled, and then processed within 1 h of delivery. As is true for any situation in which fresh human tissue is processed, proper precautions to eliminate possible infection (including infection with HIV-I and hepatitis B) from the tissue must be taken. Each of the steps includes a protocol followed by a discussion of important aspects of that step and application to the isolation of human β–adrenergic receptors.

2.2. Membrane Isolation

2.2.1. Protocol

Basal plasma membranes are prepared as described elsewhere (Bahouth et al., 1986). Briefly, 20 g portions of villous tissue are sonicated at 0–1°C in 100 mL of 50 m*M* Tris-HCl buffer (pH 7.4, 4°C), 10 µg/mL leupeptin, 10 μg/mL aprotinin, 100 μ*M* phenylmethylsulfonyl fluoride (PMSF). PMSF must be omitted from preparations used for adenylate cyclase assays, bacterial toxin-catalyzed labeling, and reconstitution experiments with membranes from mouse lymphoma S49 *cyc*-cells (devoid of G_s). The sonicated tissue is filtered and washed on a nylon mesh, stirred in 5 m*M* Tris-HCl (pH 7.4) for 30 min, and again washed to remove cellular debris. The tissue is resuspended in 10 m*M* EDTA, 42 m*M* Tris-HCl (pH 7.4), 208 m*M* sucrose, 10 μg/mL leupeptin, 10 μg/mL aprotinin, and 100 μ*M* PMSF, and the temperature raised to 20°C just prior to the sonication step. The sonication, performed with a Branson Sonic Probe (Model L) for 20 s at 0°C, releases the basal plasma membrane from collagenous substrata. The tissue fragments are removed with a gauze filter and the filtrate is centrifuged for 10 min at $3300g_{av}$. The resulting supernatant is centrifuged at $81,000g_{av}$ for 40 min. The resuspended pellet, a crude preparation of basal plasma membranes, displays a specific activity of β-adrenergic receptor binding approximately 30-fold higher than that of the tissue homogenate. Membranes at this stage should be characterized with respect to β–receptor content and subtype, adenylate cyclase activity, and bacterial toxin-catalyzed ADP-ribosylation (if evaluation of G-protein levels is required).

For further purification, the pellets are resuspended in a buffer (TSE) composed of 50 m*M* Tris-HCl (pH 7.4, 4°C), 250 m*M* sucrose, 1 m*M* EDTA, and applied to a discontinuous gradient of Ficoll (mol wt = 400,000) in TSE buffer with steps of 4 and 10% Ficoll, having densities of 1.05 and 1.07 g/mL, respectively. The gradients tubes are centrifuged at $90,000g_{av}$ for 2 h. Material at the 4–10% interface is collected, diluted approximately five fold with TSE and centrifuged at $81,000g_{av}$ for 40 min. Membrane pellets are resuspended in TSE, and used immediately. Membranes at this stage display a specific activity of β–adrenergic receptors ranging from 3000–5000 fmol/mg protein, a 45-fold increase over that of the initial tissue homogenate. These purified membranes should be assayed for β–adrenergic receptor content, subtype determination, and should be analyzed by photoaffinity labeling to ascertain receptor integrity, if at all possible.

2.2.2. Comments

In order to preserve the intergrity of β-adrenergic receptors it is imperative to keep tissue (or cells) well-buffered from proteolysis and thermal denaturation. These receptors can be proteolyzed to smaller fragments that often retain some ligand binding capacity. For this reason, measurements of the specific activity of the membranes are often misleading and can greatly overestimate the probable yield of intact receptor. Used in tandem with SDS-polyacrylamide gel electrophoresis, photoaffinity labeling and immunoblotting (*see below*) provide two strategies with which the molecular status of the receptor can be ascertained throughout the purification. Our experience dictates that the inclusion of leupeptin and high concentration of freshly-prepared PMSF are indispensable in minimizing the proteolytic degradation of the β-receptor.

2.3. Solubilization of Membrane Receptors

2.3.1. Protocol

The basal plasma membranes are resuspended at a protein concentration of 5–6.25 mg/mL in a buffer composed of 250 m*M* NaCl, 5 m*M* EDTA, 1% digitonin, 10 m*M* Tris-HCl (pH 7.4, 4°C), 10 μg/mL aprotinin, 10 μg/mL leupeptin, 100 μ*M* PMSF, and then homogenized with 10 strokes of a motor-driven Potter-Elvejhem tissue grinder operated at maximum speed. The digitonin extract is stirred at 4°C for 60 min and then centrifuged at 100,000g_{av} at 4°C for 1 h. This high-speed supernatant is aspirated from the pellet using a large plastic Pasteur pipet. The extract should be maintained in an ice slurry. Specific activity measurements of the diluted digitonin extract are critical at this point of the purification (Table 1). Under the conditions detailed above, digitonin (1%, w/v) effectively solubilized at least 40% of the β-adrenergic receptors of human placental basal membranes. For the reasons described below, the yield of solubilized receptors in the extract actually may be much higher. We have succeeded in solubilizing more than 60% of the receptor complement in rat liver membranes (β_2-subtype) and more than 70% of the receptor complement of purified rat fat cell membranes (β_1-subtype) (Graziano et al., 1985; Cubero and Malbon, 1984). In these specific cases the solubilization step provides an apparent increase in specific activity. Rarely does 1% digitonin extract more than 50% of the membrane protein, so that high yield of solubilized receptor often leads to a modest purification.

2.3.2. Comments

In maintaining proper "bookkeeping" of the progress of the purification, it is important to recognize that the analysis of the receptor at various steps must be performed under differing conditions. Whereas radioligand binding analysis of the homogenate, crude membranes, and purified basal membranes can be performed using either tritiated dihydroalprenolol (DHA) or radioiodinated cyanopindolol (ICYP), analysis

Table 1
Large Scale Isolation of Human β-Adrenergic Receptors[a, b]

Fraction	Protein (ug)	β-Adrenergic receptors Total (pmol)	Specific activity (nmol/mg)	β_1-receptors %	Purification (fold)
Human placenta basal membranes	233,000 (21)	280 ± 15	0.001	65 ± 3 (7)	0
Digitonin extract	91,600 (21)	110 ± 9	0.001 ± 0.0002	65 ± 5 (7)	0
Affinity chromatography	230 (13)	17 ± 1	0.08 ± 0.03	62 ± 4	80
C_7—Sepharose chromatography	22 ± 4 (7)	11 ± 2	0.5 ± 0.03	60 ± 2 (4)	500
DEAE-Sephacel chromatography	9 ± 2 (5)	9 ± 1	1.08 ± 0.1	50 ± 5 (3)	990
Steric-exclusion h.p.l.c.	1.2 ± 0.2 (4)	3.7 ± 0.5	3.08 ± 0.2	50 ± 5 (4)	3,080

[a]The β-adrenergic receptors in digitonin extracts of basal membranes of human placenta were purified sequentially by affinity chromatography, C7-Sepharose chromatography, DEAE-Sephacel chromatography, followed by HPLC on steric-exclusion columns. Experimental details are outlined in Section 2. The quantity and subtype distribution of β-adrenergic receptors at each step of the isolation was determined. Data are expressed as the mean ± SE from the indicated number (■) of large scale preparations (from Bahouth et al., 1987).

[b]The fractions obtained at each step of the purification are listed. The specific activities and total receptor yields are provided along with the protein content at each purification step. The extent of purification (-fold over starting fraction) is provided also.

of the digitonin extract is best performed with ICYP under conditions were the final concentration of detergent is <0.05%. Higher concentrations of digitonin prohibit accurate assessment of receptor content using ICYP. It is best to measure receptor content using ICYP and extracts diluted to 0.05, 0.01, and 0.005% digitonin. An accurate "limit" specific activity will be obtained using this approach. The advantage of using ICYP at this point is to minimize the amount of receptor consumed in the analysis of purification.

2.4. Purification Strategies

2.4.1. Affinity Chromatography—Protocol

The synthesis of the affinity matrix Sepharose support, to which the β-adrenergic antagonist alprenolol has been covalently coupled, follows the procedure developed by Caron et al. (1979). The details of this procedure are described in Section 3. The digitonin extract (75 mL) is maintained on ice and loaded at a flowrate of 40 mL/h onto a 75 mL (1.6 × 40 cm) Sepharose-alprenolol column equilibrated at ambient room temperature in a buffer (TEN) composed of 10 m*M*-Tris-HCl (pH 7.4), 2 m*M* EDTA, 100 m*M* NaCl, and supplemented with 0.05% digitonin, 1 m*M* PMSF, 10 μg/mL leupeptin, and 10 μg/mL aprotinin. The column is maintained at 4°C and washed overnight with 300 mL TEN buffer at 25 mL/h. After washing, the column is thermally equilibrated to 22°C. The adsorbed receptor is eluted with a 180-mL linear gradient of 0–100 μ*M* (-)alprenolol prepared in TEN buffer and loaded at a rate of 37 mL/h (Fig. 1). Bound receptors elute from the alprenolol-Sepharose matrix in a symmetrical peak at a concentration of competing ligand of 35–40 μ*M*. The immobilization of the alprenolol results in an apparent affinity of receptor for matrix that requires these concentrations for sharp, clean elution. Aliquots (0.5 mL) of each fraction (8–10 mL) are dialyzed at 4°C in TEN buffer without digitonin to reduce the concentration of free (-)alprenolol and assayed for β-adrenergic receptor activity by radioligand binding.

2.4.2. Comments

One of the major obstacles encountered in this step of the purification is the efficient reduction of the (-)alprenolol concentration to levels that do not interfere with the assay of receptor content by radioligand binding. In addition, when much larger volume columns of affinity matrix are employed, the volume of the eluted receptor may become unwieldy. Although we have employed ion-exchange chromatography at this point in several previous strategies (Cubera and Malbon, 1984; Graziano et al., 1985; George and Malbon, 1985), hydrophobic chromatography offers several real advantages over ion-exchange chromotography for this purpose.

2.4.3. Hydrophobic Chromatography—Protocol

Heptylamine-Sepharose (C7-Sepharose) is conveniently synthesized by the method of Shaltiel (1974). For our studies, we use Sepharose 4BCL as the insoluble support. Fractions eluted from the Sepharose-alprenolol matrix are maintained on ice, pooled, and applied two consecutive times to a 5 mL column (1 × 6 cm) of C7-Sepharose that had been equilibrated at 4°C with TEN buffer. The column is washed successively with 20 mL of TEN buffer, and then with 20 mL of 500 m*M* NaCl, 2 m*M* EDTA, 10 m*M* Tris (pH 7.4), 0.05% digitonin, 1 m*M* PMSF. β–Adrenergic receptors are eluted with a 50-mL linear gradient between 250 m*M* NaCl,

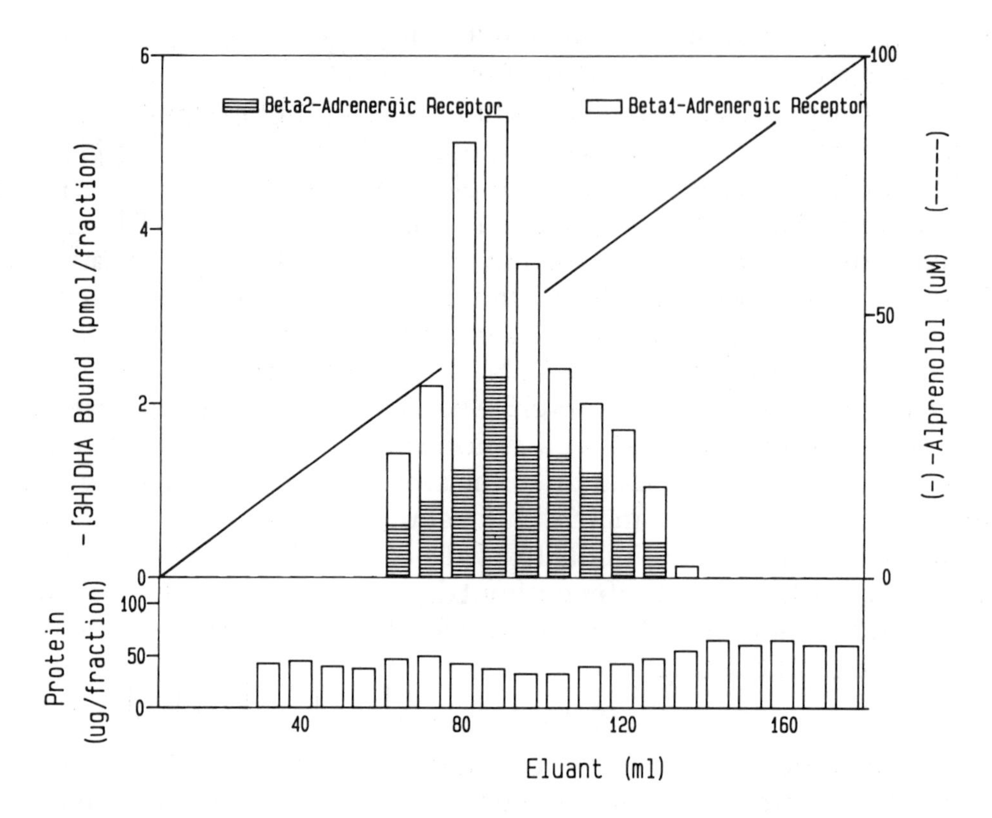

Fig. 1. Affinity chromatography of human β-adrenergic receptors: *Upper panel* β-adrenergic receptors (0.1 nmol) were applied (40 mL/h) to a 75 mL Sepharose 4B-alprenolol column. The column was washed at 4°C with 5 column volume. Matrix-adsorbed receptors were eluted using a gradient of (-)alprenolol (0–100 μM). Aliquots of each fraction (8 mL) were dialyzed at 4°C to reduce the concentration of alprenolol. Competition of [^{3}H]DHA binding by 1 μM CGP-20712A was used to define receptor subtype distribution. *Lower panel* Protein content in the fractions eluted by alprenolol. These data are taken from Bahouth et al. (1987).

2 m*M* EDTA, 10 m*M* Tris (pH 7.4), 1 m*M* PMSF, 0.1% digitonin (initial buffer) and 50 m*M* NaCl, 2 m*M* EDTA, 10 m*M* Tris (pH 7.4), 1 m*M* PMSF, 1% digitonin (final buffer). The increasing digitonin concentration and decreasing ionic strength provide an unusually high yield step in which the receptor can be concentrated and freed of alprenolol (Fig. 2). Approximately 50% of the protein and little β–adrenergic receptor are found in the "flow through" of the C7-Sepharose matrix in the hydrophobic chromatography of human receptors from basal placental membranes. Using the gradient described above, elution of the receptors is achieved at a concentration of 0.5% digitonin. A sharp peak of receptor and a 7–10-fold increase in specific activity was obtained (Fig. 2, Table 1). Peak fractions from the affinity and hydrophobic chromatographs are analyzed

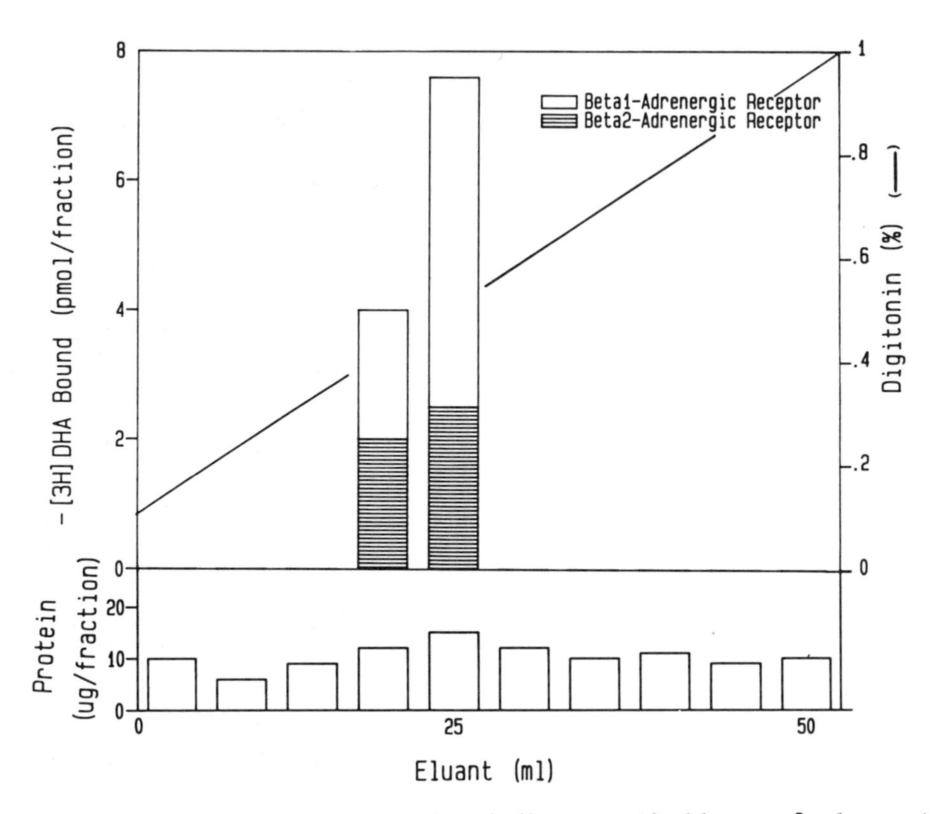

Fig. 2. Hydrophobic chromatography of affinity-purified human β-adrenergic receptors. *Upper panel* β-Adrenergic receptors eluted by alprenolol from the Sepharose 4B-alprenolol matrix (16 pmol) were pooled (80 mL) and applied twice over a 5 mL heptylamine-Sepharose column at 4°C. The column was washed over 2 h, then eluted with a gradient (50 mL) at 10 mL/hr as described in the text. Ten fractions of 5 mL each were collected and a 0.5 mL aliquot from each fraction was diluted with 100 m*M* NaCl, 2 m*M* EDTA, 10 m*M* Tris (pH 7.4) and 1 m*M* PMSF to a final digitonin concentration of 0.1% prior to quantification of the amount and subtype distribution of receptors. Receptor subtype composition was defined by competition of radioligand binding by 1 μ*M* CGP-20712A (Bahouth et al., 1986). *Lower panel* Total protein content of each fraction. Data are expressed as the mean of triplicate determinations from a representative experiment. These data are taken from Bahouth et al. (1987).

by SDS-polyacrylamide gel electrophoresis (Fig. 3). A prominent peptide with M_r = 67,000 and a second minor peptide with M_r = 45,000 are observed in the peak fractions from the hydrophobic chromatography.

2.4.4. Comments

During the development of this step in the procedure, a number of hydrophobic matrices of differing hydrophobicity were evaluated. In general, matrices of higher hydrophobic character (C10, C12, C18) bound a large portion of the receptor in an apparently irreversible manner. The

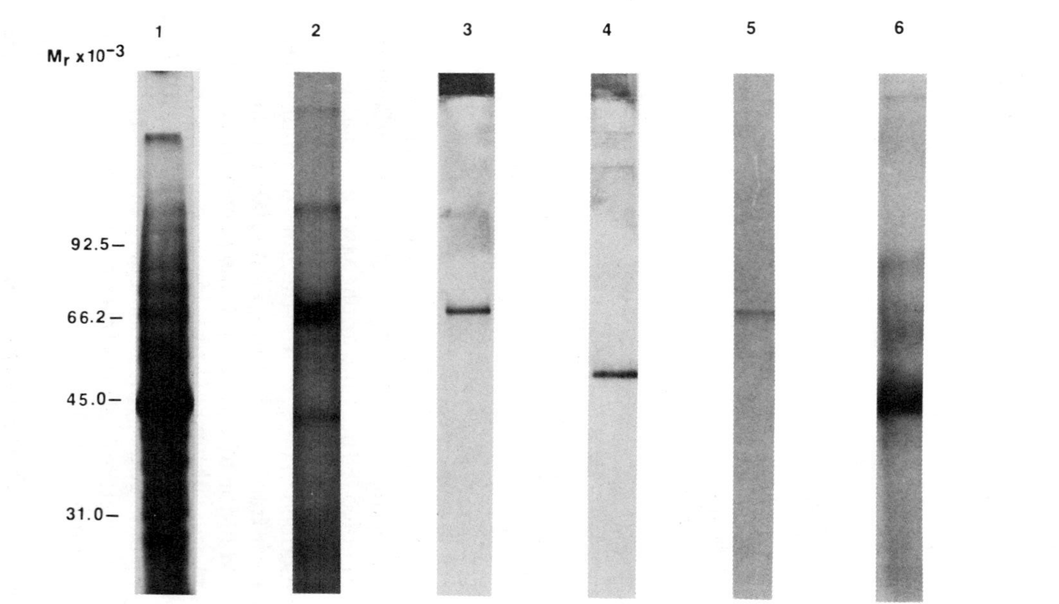

Fig. 3. Large-scale isolation of human β-adrenergic receptors: Analysis by SDS-polyacrylamide gel electrophoresis. Lanes **1** and **2** An aliquoit (0.05 mL) of Sepharose-alprenolol (lane **1**) and C7-Sepharose (lane 2) eluates displaying β-adrenergic receptor binding activites of 0.1 and 0.5 nmol receptor / mg protein, respectively, were radioiodinated and subjected to SDS-polyacrylamide gel electrophoresis. The autoradiograms are 2-d exposures using XO-Mat film and a Cronex plus intensifying screen. Lanes **3** and **4** A0.2 mL aliquot (200 ng protein) of Peak 1 of HPLLC-purified receptors was lyophilized and subjected to SDS-polyacrylamide gel electrophoresis in the presence of 10% 2-mercaptoethanol (lane **3**) or in the absence of 2-mercaptoethanol (lane **4**). The proteins were made visible by silver staining of the gel. Lane **5** A0.1 mL aliquot (100ng protein) of Peak 1 of the HPLC-purified receptors was lyophilized and subjected to SDS-polyacrylamide gel electrophoresis in the presence of 2-mercaptoethanol and the proteins made visible by the periodic acid-silver staining of the gel. Lane **6** an aliquot of purified β_2-adrenergic receptor isolated from S49 mouse lymphoma cells (George et al., 1985) was alkylated without chemical reduction of disulfide bonds (Moxham and Malbon, 1985) and then subjected to SDS-polyacrylamide gel electrophoresis. This lane provides a point of reference for the comparison of the nonreduced forms of the mouse compared to the human β-adrenergic receptors. Portions of these data are taken from Bahouth et al. (1987).

C3 and C4 matrices displayed greater flow-through of receptor and offered little advantage over the heptylamine-Sepharose. The C7-Sepharose chromatography was considerably more rapid than the ion-exchange and dialysis methods used previously. For the isolation of the human β–receptors, this advance improved the overall yield and integrity of the final product. It is at this point that the receptors are further concentrated by ion-exchange chromatography in preparation for steric-exclusion HPLC.

2.4.5. Ion-Exchange Chromatography—Protocol

Ion-exchange chromatography is routinely performed on DEAE-Sephacel. Fractions containing receptor are pooled, diluted with 10 m*M* Tris (pH = 8.0), 1 m*M* EDTA, 0.05% digitonin, 1 m*M* PMSF to a final NaCl concentration of 10 m*M*, and applied twice to a 1 mL DEAE-Sephacel column (1.1 × 1.1 cm). The column is washed with the dilution buffer (4 vol, two times) and the receptors are eluted from the column with 1–1.5 mL of 0.2*M* NaCl, 0.1*M* Na_2HPO_4/NaH_2PO_4 (pH 7.0), 1 m*M* EDTA, 0.1% digitonin, and 0.1 m*M* PMSF. Chromatography of DEAE-Sephacel under these conditions increases the specific activity 1–2-fold (Table 1). Manipulation of the conditions for elution of the bound receptor by salt can improve the level of purification at this step.

2.4.6. Comments

Previously, we used ultrafiltration on Amicon devices to concentrate the receptor in preparation for steric-exclusion HPLC (Cubero and Malbon, 1984). Although this method is still acceptable, ion-exchange chromatography is superior. We have experimented with a number of ion-exchange matrices, including small, high-capacity systems designed for use with HPLC systems (BioRad MA7P cartridges). The HPLC systems offer some advantages in that complex elution profiles can be created either to minimize the elution time or to sharpen the profile of the eluted receptor. However, loading times on the HPLC are frustratingly slow and recoveries often suffer when these alternative matrices or hardware configurations are employed. The system described above is optimal for rapid, large-scale preparations.

2.4.7. Steric-Exclusion HPLC—Protocol

The DEAE-Sephacel eluate was subjected next to steric-exclusion HPLC, as described (Cubero and Malbon, 1984). HPLC was performed on macroporous, spherical silica supports using a Beckman/Altex 342 system following the approach of Shorr et al. (1982a). In general, samples (0.1–0.5 mL) were chromatographed on two Beckman/Altex TSK-3000

(7.5 mm × 30 cm) and one Beckman/Altex TSK-2000 (7.5 mm × 30 cm) columns, tandem-linked. A short (7.5 mm × 10 cm) guard column packed with TSK-3000 should precede the analytical columns. Precipitated digitonin and other insoluble materials can be collected in this column and the column replaced as needed at significantly lower cost than replacing an analytical column. This configuration is optimal for most routine purification (Fig. 4A). β–Adrenergic receptors subjected to steric-exclusion HPLC (Fig. 4B) displayed M_r = 180,000 (peak I) and 90,000 (peak II). In this chromatograph, peak II- receptor accounted for about 15%. In many preparations, little or no peak II-receptor material is observed. The specific activity of receptor at this point is usually at least 3 nmol ICYP binding/mg protein (Table 1). SDS-polyacrylamide gel analysis of the peak I material from the HPLC that was chemically reduced and then alkylated revealed a prominent single peptide with M_r = 67,000 (Fig. 3, lane 3). Under nonreducing conditions, the receptor displayed M_r = 55,000 (Fig. 3, lane 4), as first noted for the β–receptor isolated from rat fat cells (Moxham and Malbon, 1985; Moxham et al., 1986). The glycoprotein nature of the β-adrenergic receptor (George et al., 1986) permits identification of the receptor by specific staining with the periodic acid-silver reagent (Dubray and Bezard, 1982).

Inserting additional columns into the system can often improve (sharpen) the peaks of eluting receptor protein. In our experience, the purity of the receptors that are typically loaded onto the HPLC system is sufficiently great that separation of the lower and higher molecular weight non-receptor contaminants from the peak of receptor does not require any greater resolution by the system. If the receptor is not highly purified (less than 0.5–2.0 nmol ICYP binding/mg protein) after affinity and hydrophobic chromatography, it may be advantageous to repeat these two steps prior to the steric-exclusion HPLC (Graziano et al., 1985).

The mobile phase is composed of 0.2*M* NaCl, 0.1*M* Na_2HPO_4/ NaH_2PO_4 (pH 7.0), 1 m*M* EDTA, 0.1 m*M* PMSF, and 0.1% digitonin. The mobile phase must be subjected to filtration through 0.22 μm filters (Millipore) just prior to use in the HPLC. The HPLC columns must be equilibrated with the mobile phase. Equilibration can be determined by spectroscopic analysis of the effluent mobile phase. At equilibration, the concentration of digitonin stabilizes. Chromatography of the receptor should not be attempted until the digitonin concentration in the entire system is constant. Leupeptin and aprotinin are not included in the final stages of the purification in order to permit accurate assessment of the protein content. Freshly prepared PMSF is included throughout these final steps and is indispensible. Chromatography is best performed at

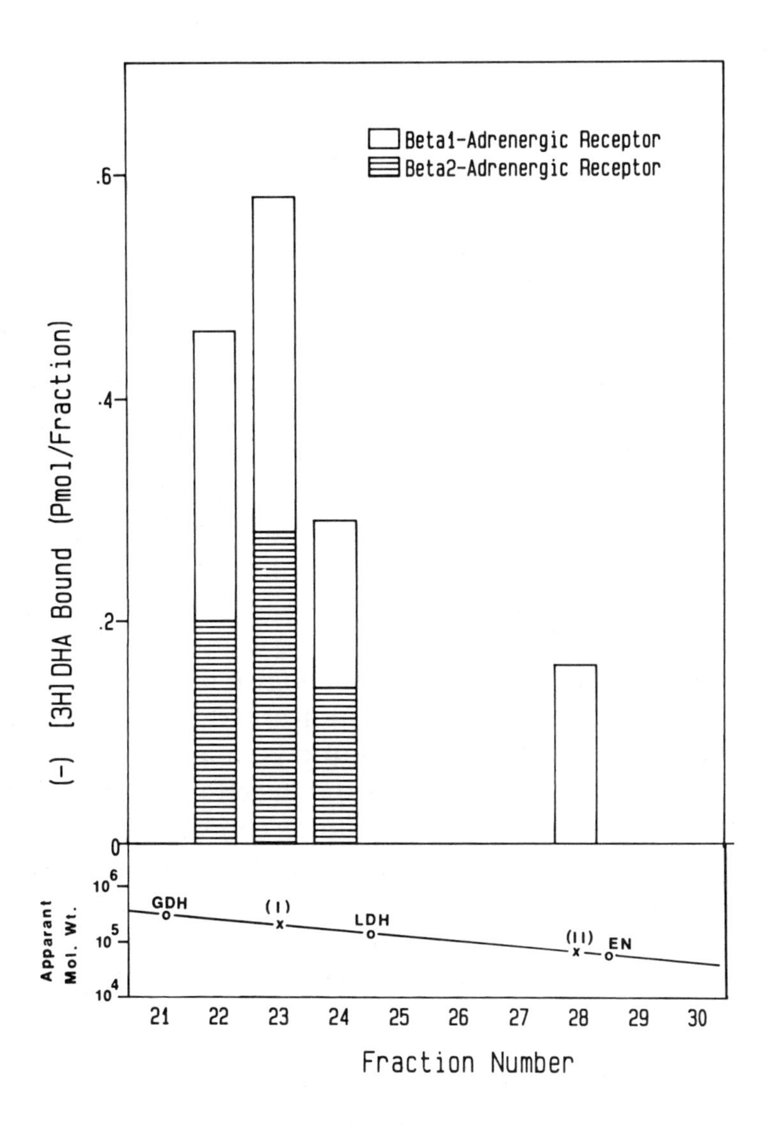

Fig. 4. Steric-exclusion HPLC profile of human β-adrenergic receptors. *Upper panel* 0.5-mL aliquot of the DEAE-Sephacel eluate containing 2 pmol of receptor binding activity was chromatographed on two TSK-3000 (7.5 mm × 30 cm) and one TSK-2000 (7.5 mm × 30 cm) steric-exclusion columns, tandem-linked. Aliquots of each fraction eluted from the HPLC were assayed by radioligand binding for content of β-adrenergic receptors. β_1-Adrenergic receptors were measured in the fractions eluted in peak 1 only (fractions 22–24). *Lower panel* Molecular weight marker proteins in a total vol of 0.25 mL were subjected to HPLC under identical conditions used to purify β-adrenergic receptors. Elution of the mol wt marker proteins in parallel chromatographs was followed by monitoring their absorbance at 280 nm. GDH, glutamate dehydrogenase (M_r = 290,000; LDH, lactate dehydrogenase) M_r = 140,000; EN, enolase (M_r = 67,000). These data are taken from Bahouth et al. (1987).

22°C. Through the use of water-jackets and a cooling core it is possible to perform the chromatography at lower temperatures. However, precipitation of the digitonin in the mobile phase and of the concentrated receptor may occur. To obviate this problem, the HPLC is routinely performed at ambient room temperature (22°C).

2.4.8. Comments

The macroporous, spherical silica matrix, like the affinity, hydrophobic, and ion-exchange matrices, "seasons" with continued use. Yields from the columns improve with repetitive use, a reflection of the saturation of sites in the matrix that may irreversibly bind protein. Unfortunately, as the seasoning of the HPLC matrix proceeds, so increases the back-pressure of the column. The increase in column back-pressure is predictable and can be ignored until the back-pressure approaches the maximum permissible pressure recommended by the manufacturer. Although it is possible to "strip" the column with mixtures of methanol:water and high-salt buffers, the success of these efforts is usually less than satisfactory. Replacement of the column(s) is the best solution.

A flowrate of 0.2–0.5 mL/min at room temperature provides the best compromise between peak resolution and development time for the chromatograph. The fraction size should be maintained at 0.2–0.5 mL. Increasing the fraction size collected from the HPLC above 1.0 mL greatly reduces the yield and purity by virtue of collection of contaminating proteins in the shoulders of the receptor peak. It is advisable to maintain a small fraction size until the receptor content and protein profile can be established in an aliquot each of the fractions. Fractions can then be pooled with confidence, weighing concerns of purity against those of overall yield.

2.5. Analysis of Receptor During Isolation

2.5.1. Radioligand Binding Analysis and Photoaffinity Labeling

2.5.1.1. Radioligand Binding Assays. β–Adrenergic receptors are measured in the basal plasma membranes as described by Bahouth et al. (1986). Purifed basal plasma membranes (60–100 μg) and [^{3}H]DHA (concentration range of 10^{-10}– 10^7*M*) are incubated for 30 min at 20°C in a vol of 0.2 mL in a buffer (TM) composed of 50 m*M* Tris-HCl (pH 7.4) and 10 m*M* MgC_{12}. A time-course with 2 n*M* [^{3}H]DHA demonstrates that binding reaches equilibrium at 10 min (data not shown). The incubation is stopped by the addition of 6 mL ice cold TM buffer. The mixture is immediately filtered onto glass-fiber filters (Whatman, GF/C). The filters are washed 2 times with 6 mL vol of the stopping solution and the radioactivity retained on the filters can be measured either by liquid scintillation spectrometry (for tritium-labeled compounds) or by gamma counting (for radioiodinated ligands). For determination of β-adrenergic receptor binding to digitonin extracts, ICYP is the preferred radioligand. The high speed supernatant from the extraction must be diluted ten-fold in incubation buffer (100 m*M* NaCl, 2 m*M* EDTA, 10 m*M* Tris-HCl, pH 7.4) to reduce the final digitonin concentration in the assay to 0.05%. It was necessary to reduce the digitonin concentration to 0.1% to achieve optimal binding of the ICYP to the detergent-dispersed receptor. The solubilized basal membranes (12–20 µg) are incubated with 10–1000 p*M* ICYP for 1 h at 20°C in a total vol of 0.1 mL. Bound radioactivity of detergent-dispersed extracts is determined using the polyethylene glycol technique of Vauquelin et al. (1979).

Specific binding is defined as total radioligand binding minus the binding that was insensitive to competition by either 1 µ*M* (-)alprenolol or 30 µ*M* (-)isoproterenol. Total binding of the ligand in the assay must be maintained at less than 10% (Scatchard, 1949). At either concentrations of DHA of 0.2–5 n*M* or concentrations of ICYP of 10–100 p*M*, specific binding accounts for more than 90% of the bound radioactivity. In studies employing highly purified membranes, the nonspecific binding component can be assessed with the use of the "LIGAND" program and radioligand concentrations up to 50-fold greater than the K_d (Munson and Rodbard, 1980). All receptor binding parameters are analyzed by computer using nonlinear, least-squares, curve-fitting procedure and the modified LIGAND program (Biosoft-Elsevier, Cambridge, UK). Data are presented as means ± SEM from the pooled data of at least three experiments that are analyzed simultaneously. The data generated from this analysis are then plotted using the EnerGraphics plotting program (Enertronics Research Inc., St. Louis, MO). The pharmacology of human β-adrenergic receptors is shown in Table 2.

2.5.1.2. Photoaffinity Labeling of β-Adrenergic Receptors. Photoaffinity labeling of purified basal plasma membranes is routinely performed according to Rashidbaigi and Ruoho (1981), using the β-adrenergic photolabel iodoazidobenzylpindolol [^{125}I]IABP. Membranes (1 mg/mL protein) in 0.3 mL lysis buffer (5 m*M* $MgCl_2$, 10 m*M* Tris-HCl pH 7.4) are incubated either with or without 1 µ*M* (-)alprenolol or the desired concentration of (±)CGP-20712A and [^{125}I]IABP ($0.3–0.4 \times 10^{-9}M$) in the dark for 30 min at 30°C in 8 mL Pyrex® glass tubes. After incubation, the contents of the tubes are cooled quickly in ice cold water to 1°C, diluted to 5 mL with ice cold lysis buffer, and photolyzed immediately with a 1-

kW high-pressure mercury lamp for 4 s at a distance of 10 cm from the lamp (Rashidbaigi et al., 1983). After photolysis, the contents are centrifuged and the pellet dissolved in 0.15 mL of a buffer containing 75 m*M* Tris-HCl (pH 6.8), 2% SDS, 5% 2-mercaptoethanol, 10% glycerine, and incubated at room temperature for 10–15 min. The entire sample is then subjected to electrophoresis on 9% SDS-polyacrylamide gels (*see below*). The relative molecular weight of the samples separated by gel electrophoresis are determined with the use of unlabeled or radioiodinated markers: myosin (M_r = 200,000), phosphorylase *b* (M_r = 92,500), bovine serum albumin (BSA) (M_r = 66,200), ovalbumin (M_r = 45,000), carbonic anhydrase (M_r = 31,000), soybean trypsin inhibitor (M_r = 21,500), lysozyme (M_r = 14,400). An example of photoaffinity labeling of human placental basal membranes with IABP is shown in Fig. 5. Note that in addition to the two major receptor peptide species with M_r = 65,000 and 54,000, is a 76,000-M_r receptor peptide first reported by this laboratory in 1986 (Bahouth et al.). The existence of this higher molecular weight species of human b-adrenergic receptor has since been confirmed in several laboratories. Our attempts to isolate pmol quantities of this novel receptor peptide have been unsuccessful to date.

2.5.2. SDS-Polyacrylamide Gel Electrophoresis—Protocol

Fractions are treated with or without agents that chemically reduce disulfide bonds of proteins, alkylated, and then subjected to electrophoresis on homogeneous 10% polyacrylamide gels in the presence of 0.1% SDS (Laemmli, 1970). The separating gels are polymerized overnight at room temperature, the stacking gels are polymerized for 4 h prior to use. Proteins were made visible by silver staining using the Bio-Rad kit. Glycoprotein staining was performed by the periodic acid-silver stain method (Dubray and Bezard, 1982). Radioiodination of β–adrenergic receptors is performed to minimize the amount of receptor consumed for analysis during purification. The proteins of fractions from the various steps of purification are radioiodinated with Na[^{125}I] by the chloramine-T method (Greenwood et al., 1963), then desalted on Sephadex G-50 (Graziano et al., 1985) or precipitated in 20% ice cold trichloroacetic acid. Protein content of membranes and digitonin-extracts can be measured by the method of Lowry et al. (1951). In highly purified preparations, protein is best determined by the Amido Schwarz assay of Schaffner and Weissmann (1973), using bovine serum albumin as a standard.

Table 2
Affinities of Membrane Bound, Solubilized, and Purified Human β-Adrenergic Receptors for β-Adrenergic Ligands[a]

	β-Adrenergic receptors		
Ligand	Membrane bound	Digitonin-extract	HPLC-purified
	K_d, μM		
(-)isoproterenol	0.4	0.02	0.06
(+)isoproterenol	34	1.2	3.0
(-)norepinephrine	30	3.0	5.0
(-)epinephrine	20	1.2	2.0
(-)propranolol	0.004	0.004	ND
(+)propranolol	0.9	0.2	ND

[a]The inhibition of [^{3}H]DHA binding by the ligands was performed using 2 n*M* [^{3}H]DHA (in membranes and digitonin extracts) and 10 n*M* (in purified preparations of β-adrenergic receptors). The inhibition of [^{3}H]DHA binding to basal membranes of human placenta by agonists was performed in a buffer composed of 50 m*M* Tris (pH 7.4), 10 m*M* $MgC1_2$ and in the presence of 0.1 m*M* GTP. The affinity constants K_d for (-)[^{3}H]DHA in basal membranes of human placenta and digitonin extracts of these membranes was determined by equilibrium binding. For calculating the K_d of agonists for purified receptors, the K_d for [^{3}H]DHA was 1.5 n*M*. The concentration of each ligand that inhibited the specific [^{3}H]DHA binding by 50% (IC_{50}) was determined and used to calculate the K_d of this ligand by the equation of Cheng and Prusoff (1973). Data are expressed as the mean of triplicate determinations from representative experiments. ND, not determined. These data are taken from Bahouth et al. (1987).

2.5.3. Immunoblotting and Immunoprecipitation– Immunoblotting

Purified β–adrenergic receptors are subjected to gel electrophoresis in the presence or absence of 20 m*M* dithiothreitol, alkylated with a threefold molar excess of *N*-ethymaleimide, and then electrophoretically transferred from the resolved gel to nitrocellulose by the method of Towbin et al. (1979), as modified by Erikson et al. (1982). After transfer, the nitrocellulose is washed in distilled water and allowed to air dry. The nitrocellulose is rehydrated, blocked for 1 h with phosphate-buffered saline (PBS) containing 100 mg/mL bovine serum albumin for 1 h, rinsed once with distilled water, and then incubated for 2 h at 22°C with antireceptor antibodies (1:100-fold dilution, 10 mL) in 0.3% (v/v) Tween 20/PBS. The blot is washed three times with Tween 20/PBS for 10 min each and then incubated for 2 h at 37°C with goat antirabbit antibody conjugated to calf alkaline phosphatase (Kirkegard and Perry Laborato-

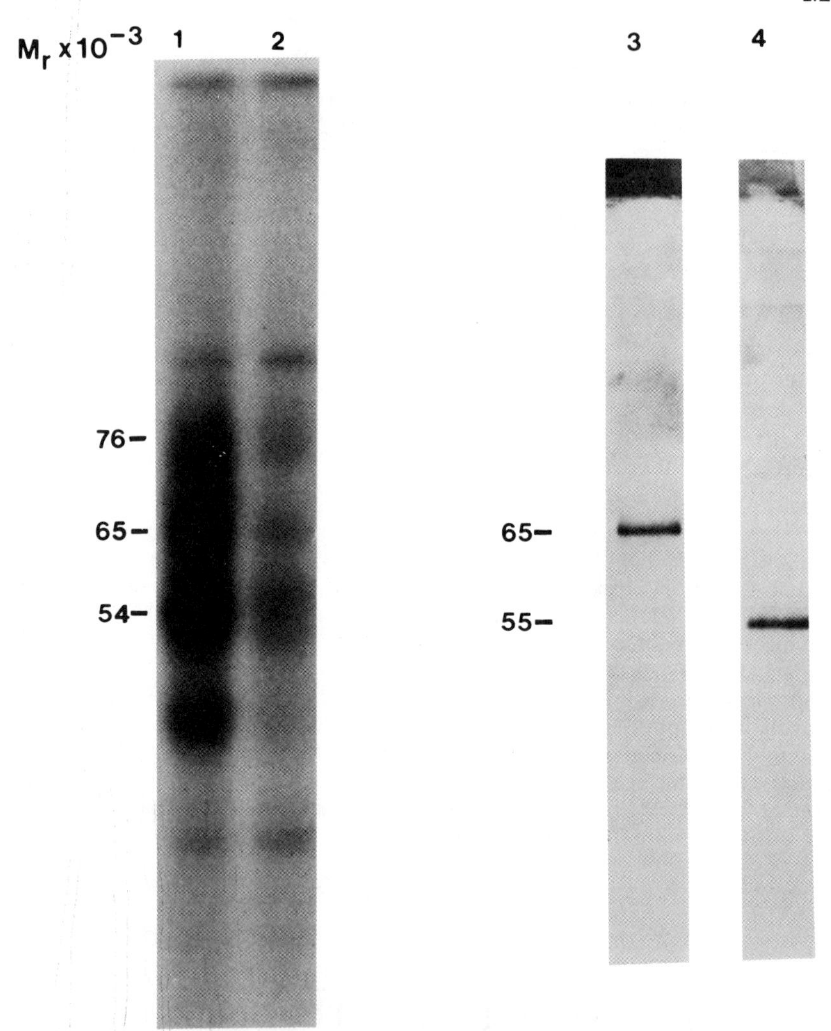

Fig. 5. Photoaffinity radiolabeling of human β–adrenergic receptors. Basal membranes isolated from human placenta (0.2 mg) possessing 4.0 pmol of β–adrenergic receptor/mg protein were incubated with 0.3 n*M* radioiodinated iodoazidobenzylpindolol alone (lane 1) or in the presence of 1 μ*M* (-)alprenolol (lane 2). After incubation for 30 min at 30°C in the dark, the membranes were photolyzed, washed, solubilized, and subjected to SDS-polyacrylamide gel electrophoresis. The gels were stained, destained, dried, and exposed Kodak™ XAR-5 film for 24 h. Purified β–adrenergic receptors isolated from S49 mouse cells (George and Malbon, 1985) were subjected to SDS-polyacrylamide gel electrophoresis with (lane 3) and without (lane 4) chemical reduction of disulfide bonds. The gel was stained with silver and provides a reference for comparison of the mobility of the photoaffinity-labeled human vs stained β–adrenergic receptors.

ries, Gaithersburg, MD) diluted 1:1000 in Tween 20/PBS. The blot is washed with 50 m*M* glycine buffer (pH 9.6) for 5 min and then incubated at 22°C with substrate solution (5.0 mL of 50 m*M* glycine (pH 9.6), 0.17 mg/mL *p*-nitro blue tetrazolium chloride, 7 m*M* $MgCl_2$, and 0.08 mg/mL

5-bromo-4-chloro-3-indoyl phosphate) until immune complexes begin to appear. The reaction is stopped by removing the substrate solution and rinsing the blot in distilled water.

2.5.4. Immunoprecipitation

Radioiodinated receptor preparations are incubated overnight at 4°C with antireceptor antibodies (1:400 final dilution) in a vol of 0.125 mL containing 100 m*M* NaCl, 10 m*M* Tris (pH 7.4), 2 m*M* EDTA, 0.01% digitonin (NTED buffer). A 0.025-mL aliquot of undiluted, goat antimouse or goat antirabbit second antibody (Cooper Biomedical) is added and incubated for 4 h at 4°C. For immunoprecipitation with mouse antibodies, control mouse ascites fluid (0.025 mL) is added as the carrier. The samples are collected by centrifugation and the pellets washed three times with each 1 mL of NTED buffer. The final immunoprecipitate was incubated for 60 min at 37°C in 0.1 mL of solubilizing solution containing 0.125*M* Tris (pH 6.8), 4% SDS, 20% sucrose, and 20% 2-mercaptoethanol, alkylated with molar excess of *N*-ethylmaleimide, and then subjected to gel electrophoresis.

2.5.5. Comments

Both of these techniques capitalize on the use of high quality antireceptor antibodies as probes of the quantity (by quantitative immunoblotting) and molecular nature (by analysis on SDS-polyacrylamide gel electrophoresis) of the receptor at various stages of purification. Used in tandem with radioligand binding analysis, these techniques provide accurate assessment of receptor inactivation (loss of ligand binding capacity) and receptor integrity (reduction in apparent M_r on SDS-polyacrylamide gel electrophoresis). As a result of the amplification offered by the use of a second antibody to which an enzyme has been immobilized, it is possible to minimize the amount of receptor consumed in the analysis of specific activity and molecular nature. Additionally, we have used the Pharmacia Phast system to run "micro" gels that can resolve and stain extremely low amounts of protein (0.5–5.0 ng of receptor). These approaches are indispensible for investigators seriously considering routine purification of β-adrenergic receptors, or many other, low-abundance G-protein-linked receptors.

2.5.6. Co-Reconstitution of β-Adrenergic Receptors with G_s in Defined Liposomes—Protocol

β-Adrenergic receptors purified from S49 mouse lymphoma cells (George and Malbon, 1985) and G_s purified from rabbit liver (Sternweis et al., 1981) can be functionally co-reconstituted into unilamellar vesicles composed of phosphatidylethanolamine, phosphatidylserine, and cholesterol (3:2:1) (Brandt et al., 1983; Brandt and Ross, 1986). Functional activation of G_s is measured according to [^{35}S] GTPγS binding, exactly as described earlier (Pedersen and Ross, 1985; Asano et al., 1984).

2.5.7. Comments

The critical demonstration of true receptor purification relies on achieving two goals—demonstrating that the specific activity of the receptor is sufficiently close to theoretical to support the protein staining data, and demonstrating that this same receptor preparation is functionally active in a defined co-reconstitution system with its G-protein, G_s. Advances made by Elliott Ross (Dept. Pharmacology, Univ. Texas, Dallas) and coworkers have provided the technical basis for the routine co-reconstitution of β–adrenergic receptors and G_s in liposomes. Our collaborations with Dr. Ross have enabled us to evaluate several features of the activation of the mammalian receptor by agonist and by thiols (Moxham et al., 1988) that would have been virtually impossible by any other means.

3. Special Considerations

3.1. Synthesis of Alprenolol-Sepharose— "A Major Stumbling Block"

One of the most commonly encountered problems for most laboratories seeking to isolate β–adrenergic receptors is the successful synthesis of the Sepharose-alprenolol. Two procedures have been reported for the immobilization of alprenolol to an insoluble matrix (Vauquelin et al., 1979; Caron et al., 1979). We have synthesized affinity matrices using both of these procedures (Cubero and Malbon, 1984). We routinely obtained greater recovery with the matrix prepared according to Caron et al. (1979). Successful isolations of mammalian β_1-adrenergic receptor (Cubero and Malbon, 1984) and then the β_2-receptor (Benovic et al., 1984; Graziano et al., 1985; George and Malbon, 1985) have been achieved using the Sepharose-alprenolol matrix. Regarding the actual synthesis, it is important to pay strict attention to the details of the protocol and prepare the matrix using standard techniques employed in organic synthesis. All reagents must be of "synthetic" grade, many must be redistilled just prior to use. In our experience, it has been impossible to scale down the synthesis for a few hundred mL of starting matrix. Routinely, we prepare 1–2 L of matrix. Not all of the matrices display the same quality with regard to efficient adsorption and release of receptor. Therefore, we often prepare several liters and evaluate each lot for suitability.

3.2. Selection of Digitonin

Digitonin is routinely prepared as a 3% solution in a 10 m*M* Tris-HCl (pH 7.4), 90 m*M* NaCl buffer that is then heated at 95–100°C for 30 min, and allowed to stand for 48 h at 25°C. The solution is centrifuged, and the resultant supernatant filtered through a 0.22-μm cartridge filter (Millipore). The filtered solution is diluted to 1% digitonin with the Tris-NaCl buffer just prior to use for solubilizing the membranes. For use in HPLC buffers, a 5% solution of digitonin is prepared in HPLC-grade water following the procedure outlined above. The filtered solution is aliquoted, lyophilized and stored at –80°C. The lyophilized digitonin is dissolved in the mobile phase buffer and filtered again just prior to use in HPLC separations. The ability of various lots and grades of digitonin to solubilize receptor binding activity from purified cell membranes (including basal membranes from human placenta) has been found to be quite variable. All of the lots of digitonin, both purified (95–98% pure, from Sigma) and highly purified (>98% pure, from Gallard-Schlessinger, Carle Place, NY) grades, solubilize 35–40% of the protein from cell membranes when the extraction is performed with 1% digitonin. The amount of β–receptor binding activity solubilized by these various lots of digitonin often ranges, however, from 0–40%, with the average being 15%. Over the years, two lots of purified digitonin (lots 62F-0135 and 122F-0135 from Sigma), were identified that permitted solubilization of 50–70% of the receptor binding activity from several membrane sources. For large-scale purification of β–adrenergic receptors, it would seem most profitable to screen as wide a variety of lots and grades of digitonin as can be obtained in order to maximize the yield at this crucial first step. Our recent experience suggests that for purposes of solubilizing β–adrenergic receptors, digitonin obtained from Gallard-Schlessinger is of higher uniform quality. This vendor screens large lots of digitonin for its ability to solubilize receptor prior to release to the general public. If resources do not permit analysis of multiple lots, the Gallard-Schlessinger digitonin should be evaluated first. Once a lot has been proven to be useful for a particular isolation it is best to acquire a "stockpile" for future routine use.

3.3. Simultaneous Purification of β_1– and β_2–Adrenergic Receptor Peptides

3.3.1. Protocol

HPLC-purified β-adrenergic receptors (1 mL) are pooled, diluted fourfold with 10 m*M* Tris-HCl (pH 7.4), 1 m*M* EDTA, 0.05% digitonin, 1 m*M* PMSF, and applied to a 1 mL (0.6 cm × 1.4 cm) Sepharose-alprenolol column. The column is washed at 4°C with 20 mL of TEN buffer at 2 mL/h, then returned to room temperature and eluted sequentially with 2.5 mL of 1 μ*M* CGP-20712A followed by 2.5 mL of 100 μ*M* (-)alprenolol in TEN buffer. The eluates are dialyzed in TEN buffer without digitonin to permit measurement of β–adrenergic receptor content by [^{125}]iodocyanopindolol binding.

3.3.2. Comments

The purified preparations composed of both receptor subtypes migrates as a single, prominent species when subjected to analytical SDS-polyacrylamide gel electrophoresis (Fig. 3). Thus, these human, pharmacologically-distinct receptor peptides are structurally similar, as previously reported for the rat β_1- and β_2-receptors (Graziano et al., 1985; Moxham et al., 1986). The selectivity of the interaction of CGP-20712A for β_1-adrenergic receptors (Dooley and Bittiger, 1987, Bahouth et al., 1986, 1987) can be exploited to separate the β_1-subtype from the mixture of subtypes observed in the purified receptor preparations. For this purpose, the HPLC-purified receptor (Peak I) was subjected to a second pass of affinity chromatography on Sepharose-alprenolol. The bound receptor was then eluted sequentially by 1 μ*M* CGP-20712A followed by 100 μ*M* (-)alprenolol.

The concentration of CGP-20712A that was selected for elution competes solely for the β_1-subpopulation of adrenergic receptors (*see* Fig. 6). The pharmacology of the β_1-adrenergic receptors eluted by CGP-20712A has been examined by equilibrium binding techniques utilizing the high β_1-affinity adrenergic antagonist [^{125}I]iodocyanopindolol (Bahouth et al., 1987). Metoprolol is a β–adrenergic antagonist selective for receptors of β_1-subtype (Minneman et al., 1979). Competition of [^{125}I] ICYP binding by metoprolol was performed in order to quantify the amount of receptors present in the CGP-20712A-eluted material that are truly β_1-subtype. Analysis of the equilibrium binding data as an Eadie-Hofstee plot yielded a straight line (Bahouth et al., 1987). These data suggest that metoprolol was binding to a homogeneous population of binding sites. Under the conditions employed, the IC_{50} for metoprolol inhibition of [^{125}I]ICYP binding was $4 \times 10^{-7}M$, indicating a K_d for metoprolol of 80 n*M* (Bahouth et al., 1987). These results agree well with the reported affinity of metoprolol to β_1-adrenergic receptors in other tissues (Minneman et al., 1979).

The β–adrenergic agonists (-)epinephrine and (-)norepinephrine were equipotent competitors of [^{125}I]ICYP binding to the purified receptors (Bahouth et al., 1987). This rank order of potencies of β–adrenergic agonists indicates again that the receptor eluted by CGP-20712A was β_1 in nature (Lands et al., 1967). Similar analysis was performed on the receptors eluted by alprenolol from the CGP-20712A-stripped, Sepharose-alprenolol matrix. The competition of [^{125}I]ICYP binding to the alprenolol-eluted receptors by ICI-118,551 was shallow, with a Hill slope of 0.8. Hofstee analysis of the competition data (Bahouth et al., 1987) indicates that ICI-118,551 is binding, (K_d = 3 n*M*) to a population of β_2-adrenergic receptors that accounts for 75% of the receptor population, and to a smaller population of β_1-adrenergic receptors (25% of total) with a K_d = 0.1 μ*M*. Unlike the situation observed for the receptor eluted with CGP-20712A, these results demonstrate that the receptors eluted with alprenolol after CGP-20712 A were a mixed population of β–adrenergic subtypes, predominantly β_2- in character.

The fractions eluted from the affinity matrix by CGP-20712A (as compared to alprenolol), when radioiodinated and analyzed by SDS-polyacrylamide gel electrophoresis, demonstrate the similar nature of the two pharmacologically-distinct receptor peptides (Fig. 6). The radioiodinated peptides eluted by either CGP-20712A or alprenolol alone migrate, with an M_r = 67,000 under reducing conditions, and with M_r = 54,000 under nonreducing conditions.

It is important to note that when the isolation of either receptor subtype is performed without sufficient protease inhibitors, one prominent 45,000-M_r species of receptor will be observed. This peptide appears to be a "limit" peptide with respect to ability to bind β–adrenergic ligands. Proteolytic activity is the single major problem confronting investigators seeking to isolate β–adrenergic receptors from the basal membranes of human placenta. Careful analysis of the receptor purified by affinity chromatography (Fig. 3, lane 1) reveals a prominent 45,000-M_r peptide species. Although also appearing in fractions from the hydrophobic chromatography step, this limit peptide does not copurify in the HPLC (owing to its smaller size or behavior in aggregates) steps. A partial proteolytic peptide map (Cleveland et al., 1977) of the 67,000- vs

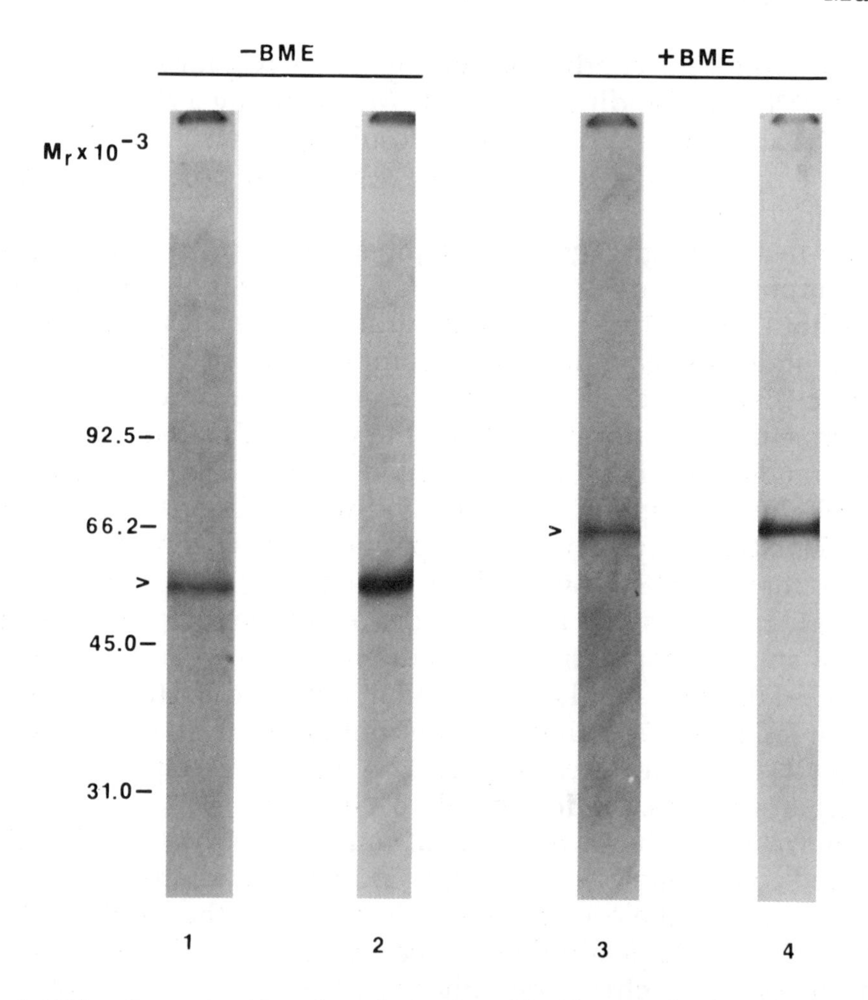

Fig. 6. SDS-polyacrylamide gel analysis of purified human β-adrenergic receptors isolated by selective elution from affinity matrix. An aliquot of HPLC-purified receptors was subjected to affinity chromatography on Sepharose- alprenolol and eluted sequentially first with 1 μ*M* CGP-20712A and then with 100 μ*M* (-)alprenolol. Aliquots of each eluant (0.5 mL) were concentrated by using a Centricon-30 microconcentrator (Amicon). The concentrate was radioiodinated and desalted on Sephadex G-50. Aliquots (0.2 mL) were lyophilized, solubilized, and subjected to SDS-polyacrylamide gel electrophoresis without (lanes 1 and 2) or with (lanes 3 and 4) prior chemical reduction of disulfide bonds with β-mercaptoethanol (BME). All samples were alkylated with *N*-ethylmaleimide prior to gel electrophoresis (Moxham and Malbon, 1985). Lanes 1 and 3 are 3-d autoradiograms of receptor eluted by 1 μ*M* CGP-20712A. Lanes 2 and 4 are 1-d autoradiograms of receptor eluted by 100 μ*M* (-)alprenolol. These data are taken from Bahouth et al. (1987).

45,000-M_r peptides reveals striking homology (Fig. 7). The 45,000-M_r peptide is also observed in the photoaffinity labeling studies (Fig. 5, lane 1). Thus the presence of this 45,000-M_r species by protein staining (Fig.

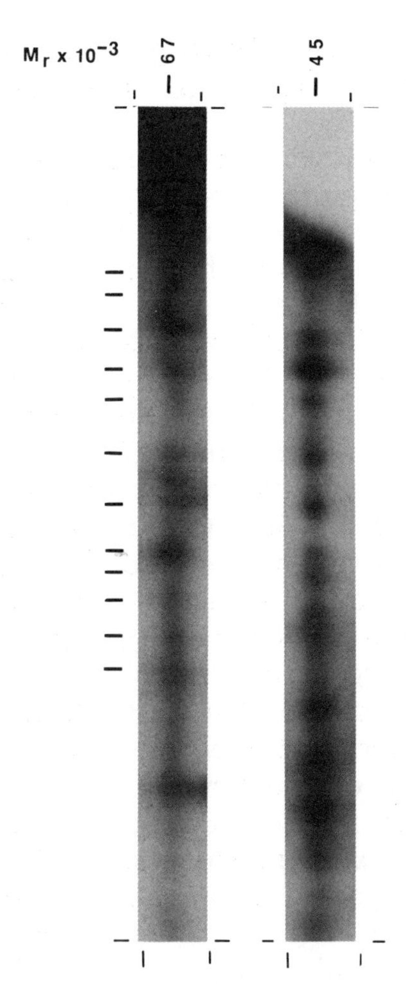

Fig. 7. Partial proteolytic peptide maps of human β–adrenergic receptor peptides of M_r = 67,000 and 45,000. Radioiodinated receptor species were chemically reduced, alkylated, and subjected to SDS-polyacrylamide gel electrophoresis in the first dimension (Graziano et al., 1985), partially digested with protease, and then subjected to SDS-polyacrylamide gel electrophoresis again in the second dimension. The gel was stained, destained, dried, and subjected to autoradiography. The patterns of radiolabeled fragments derived from the 67,000-M_r species (67) and from the 45,000-M_r species (45) are striking in their similarity. Tick marks on left-hand margin denote peptide fragments common to both species.

3, lanes 1 and 2), photoaffinity labeling (Fig. 5, lane 1), or by immunoblotting (data not shown) dictates that more rigorous attention be paid to solving the problem of proteolytic degradation. More liberal use of protease inhibitors boosts the yield that we obtain currently from human placental basal membranes (Bahouth et al., 1987).

3.4. Molecular Biology and Creating "High Expressor" Cell Lines

Identification of tissues or cells with relatively high levels of expression of β-adrenergic receptors has been a primary effort directed toward the goal of large-scale isolation of these receptors. With respect to tissues, lung (Benovic et al., 1984) and human placenta (Bahouth et al., 1986) appear to lead the list of high expressors. The successful isolation of β_1–receptors from rat fat cells (Cubero and Malbon, 1984) and β_2-receptors from rat liver (Graziano et al., 1985) was primarily undertaken to establish the biochemical nature of the two subtypes. Later, these receptor subtypes would be compared by immunological criteria (Moxham et al., 1986) and then isolated simultaneously from the human placenta basal membranes (Bahouth et al., 1987). Cells in culture, too, provide a useful starting point for the isolation of receptor (George and Malbon, 1985). However, the necessity to culture cells in suspension and in large vol (100 L) precludes the routine adoption of this strategy for most laboratories. For these reasons, we targeted our efforts to applying recombinant DNA technology to creating a cell line that constitutively expresses a high level of β–adrenergic receptor. Such a cell line may offer many advantages in the large-scale isolation of receptor.

The plasmids used in the construction of an expression vector may be obtained from the following sources: pUC13B2AR containing the hamster β_2-adrenergic gene (Dixon et al., 1986) from R. A. F. Dixon at Merck Sharp and Dohme Research Laboratories (West Point, PA); and pSP70 from Promega (Madison, WI); pSV2-dhfr from ATCC (Rockville, MD). The construction of this vector is summarized in Fig. 8, and is discussed in detail elsewhere (George et al., 1988). The resultant vector pSV2BAR provides for the expression of the hamster cDNA under the control of the SV40- early promoter and utilizes, in addition to the promoter, the intron of the SV40 small-T antigen and the polyadenylation site from the SV40 early region. Chinese hamster ovary (CHO) cells are cotransfected with pSV2BAR and pHOMER (containing the selectable marker for neomycin-resistance) following coprecipitation of the DNA with calcium phosphate. Cells are selected in RPMI 1640 media containing 10% fetal calf serum, penicillin (100 U/mL), streptomycin (100 μg/mL), and 300 μg/mL of the neomycin analog, G418 (Gibco). More than 500 clones have been selected and characterized. Stable clones that constitutively express 2–3 million receptors/cell have been isolated and characterized (George et al., 1988). The receptors of the CHO transfectant cells are fully functional and identical to the native receptors that are expressed at very low levels in wild-type CHO cells.

Only recently have we initiated the scaling up of growth of these cells for the large-scale isolation of receptor. The specific activity of the crude cell membranes isolated from these cells is often 6–12 pmol of receptor/mg of membrane protein, nearly an order of magnitude higher than most highly purified membranes. Although these mutant cells display slower growth characteristics, it seems likely that CHO high-expressor cell lines (or other transfected cells) may develop into the ideal starting source for the large-scale isolation of the mammalian β–adrenergic receptors.

Two new approaches to receptor expression have appeared recently. Marullo et al. (1988, 1989) reported the expression of human β_2- and β_1-adrenergic receptors in *E. coli.* George et al. (1989) demonstrated a high-efficiency expresssion of mammalian β–adrenergic receptors in baculovirus-infected insect cells. In the former case, the coding region of the gene for the human β_2-adrenergic receptor was fused to the β-galactosidase gene of the lambda gt11 expression vector. The expressed fusion protein displayed the pharmacological properties expected for a β_2-adrenergic receptor (Marullo et al., 1988). The level of expression of active receptor was very low (25 copies/cell) contrasting with the high expression of total fusion protein expected for a gene product under the control of the *lac* promoter. Our own experience with the expression of G-protein-linked receptor in *E. coli* has been very similar to that reported by Marullo et al. (1988). Fusion proteins designed with the leader sequence of the OmpA protein to direct integration in the outer membrane of *E. coli*, too, displayed very low expression when constructs contained the entire coding region of the receptor (Rapiejko and Malbon, unpublished observations). More recent work by Strosberg and coworkers reports the expression of both the human β_1- and β_2-adrenergic receptors as a fusion protein with the *lam*B protein (Marullo et al., 1989). The improved expresion in *E. coli* of bacterio-opsin, a membrane protein with a topology similar to that of the β–adrenergic receptor (Wang, et al., 1989), demonstrates the feasibility of this approach (Karnik et al., 1987).

We adapted the baculovirus expression system (Smith and Summers, 1982) to the problem and have succeeded in obtaining expression of 10–20 million β_2-adrenergic receptors/cell (George et al., 1989). A transfer vector was constructed (pVL941-BAR) and the receptor gene was transferred into the viral genome AcMNPV. Infection of a clonal isolate of *Spodoptera frugiperda* cells (Sf9) with this baculovirus expres-

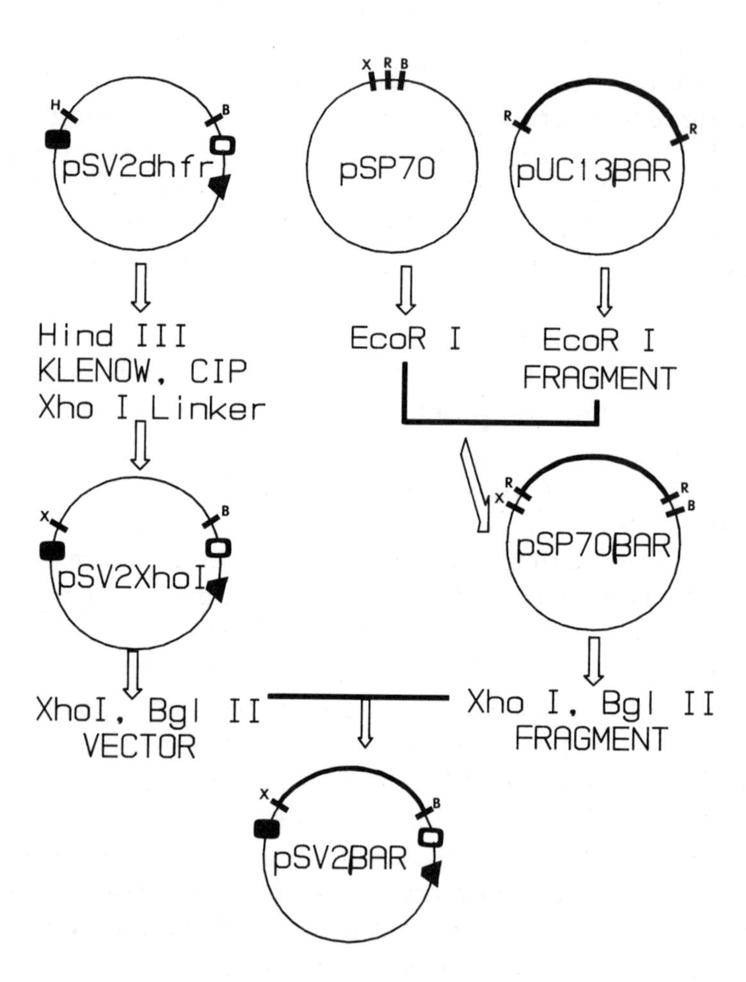

Fig. 8. Construction of pSV2BAR, an expression vector for mammalian cell transfection. The mammalian expression vector was constructed by standard recombinant DNA techniques. The hamster cDNA encoding the β_2-adrenergic receptor was excised as an Eco RI fragment from pUC13BAR and inserted into pSP70, creating pSP70BAR in which the cDNA is now flanked by Xho I and Bgl II restriction sites. The pSV2dhfr expression vector harbors the SV40 early promoter (●), the gene for dihydrofolate reductase flanked by Hind III and Bgl II restriction sites, the intron to the small-T antigen (O), and the polyadenylation site from the SV40 early region (trapezoid,▲). The Hind III site of pSV2dhfr plasmid was converted to an Xho I site. The dhfr gene was excised from the expression vector by restriction with Xho I and Bgl II, and the Xho I-Bgl II fragment of pSP70BAR was inserted into this site. The resultant expression vector pSV2BAR was used to transfect mammalian cells (George et al., 1988).

sion vector harboring the cDNA encoding the β–adrenergic receptor resulted in high-efficiency expression. By the second day of infection, the level of expression was 12 million receptors/cell. Specifc activities of the crude lysates of infected Sf9 cells were >30 pmol/mg of protein. One liter of infected Sf9 cells displays the capacity to express 20–40 nmol or receptor. Photoaffinity labeling of the receptor revealed two major species of receptor with M_r 46,000 (presumably unglycosylated) and 48,000, compared to 65,000 for the Chinese hamster ovary cells stably transfected with a mammalian expression vector (George et al., 1988). The baculovirus expression system provides the high-efficiency expression of receptor required for physicochemical analyses of the purified protein.

4. Closing Remarks

The task of isolating low-abundance, G-protein-linked receptors on a large scale is a formidable one. Based on a decade of experience in several laboratories, one might expect the purification of the β–adrenergic receptor to be more facile than it is. The fact that the molecular cloning of the human β–adrenergic receptor using rodent probes (Dixon et al., 1986) was reported simultaneously (Kobilka et al., 1987) with the isolation of the receptors from basal membranes of human placenta (Bahouth and Malbon, 1987) is a stark reminder of the technical problems encountered when applying routine protein biochemical techniques to the purification of these membrane proteins. Through the combined use of molecular biology and protein chemistry, the isolation of mg quantities of receptors produced by recombinant DNA technology would appear to be likely. The protocols and strategies outlined in the present work provide the tools with which this final goal will be realized.

Acknowledgments

This work was supported by US Public Health Service Grants DK 30111 and DK25410 from the National Institutes of Health. C. C. M. is the recipient of a Career Development Award (K04 AM00786) from the NIH. Foremost, my thanks to Suleiman Bahouth (Department of Pharmacology, Health Sciences Center, Univ. of Tennessee, Memphis) for his diligent effort in the development of the protocol for the successful isolation of the human $β_1$- and $β_2$-adrenergic receptors. I would like to express my thanks also to David Baker and Cynthia Kaplan (Dept. of Pathology, University Hospital, SUNY at Stony Brook) for their assistance in procuring the placentas.

References

Asano, T., Pedersen, C. W., Scott, C. W., and Ross, E. M. (1984) *Biochemistry* **23,** 5460–5467.

Bahouth, S. W., Kelley, L. K., Smith, C. H., Arbabian , M., Ruoho, A. E., and Malbon, C. C. (1986) *Biochem. Biophys. Res. Commun.* **141,** 411–417.

Bahouth, S. W. and Malbon, C. C. (1987) *Biochem. J.* **248,** 557–566.

Benovic, J. L., Shorr, R. G. L., Caron, M. G., and Lefkowitz, R. J. (1984) *Biochemistry* **23,** 4510–4518.

Brandt, D. R., Asano, T., Pedersen, S. E., and Ross, E. M. (1983) *Biochemistry* **22,** 4357–4362.

Brandt, D. R. and Ross, E. M. (1986) *J. Biol. Chem.* **261,** 1656–1664.

Caron, M. G., Srinivasan, Y., Pitha, J., Kociolek, K., and Lefkowitz, R. J. (1979) *J. Biol. Chem.* **254,** 2923–2927.

Cheng, Y.-C. and Prusoff, W. H. (1973) *Biochem. Pharmacol.* **22,** 3099–3108.

Cleveland, D. W., Fischer, S. G., Kirschner, M. W., and Laemmli, U. K. (1977) *J. Biol. Chem.* **252,** 1102–1106.

Cockcroft, S. (1987) *Trends Biochem. Sci.* **12,** 75–78.

Cubero, A. and Malbon, C. C. (1984) *J. Biol. Chem.* **259,** 1344–1359.

Dixon, R. A. F., Kobilka, B. K., Strader, D. J., Benovic, J. L., Dohlman, H. K., Frielle, T., Bolanowski, M. A., Bennett, C. D., Rands, E., Diehl, R. E., Mumford, R. A., Slater, E. E., Sigal, I. S., Caron, M. G., Lefkowitz, R. J., and Strader, C. D. (1986) *Nature* **321,** 75–79.

Dooley, D. J. and Bittiger, H. (1987) *J. Pharmacol. Methods* **18,** 131–136.

Dubray, G. and Bezard, G. (1982) *Anal. Biochem.* **119,** 325–329.

Erikson, P. F. I., Minier, L. N., and Lasher, R. S. (1982) *J. Immunol. Methods* **51,** 241–249.

George, S. T. and Malbon, C. C. (1985) *Biochem.* **15,** 349–366.

George, S. T., Ruoho, A. E., and Malbon, C. C. (1986) *J. Biol. Chem.* **261,** 16559–16570.

George, S. T., Arbabian, M. A., Ruoho, A. E., Kiley, J. and Malbon, C. C. (1989) *Biochem. Biophys. Res. Commun.* **163,** 1265–1269.

George, S. T., Berrios, M., Haadcock, J. R., Wang, H. Y., and Malbon, C. C. (1988) *Biochem. Biophys. Res. Commun.* **150,** 65–672.

Gilman, A. G. (1987) *Annu. Rev. Biochem.* **56,** 615–649.

Graziano, M. P. and Gilman, A. G. (1987) *Trends in Pharmacol. Sci.* **8,** 478–481.

Graziano, M. P., Moxham, C. P., and Malbon, C. C. (1985) *J. Biol. Chem.* **260,** 7665–7674.

Greenwood, F. C., Hunter, W. M., and Glover, J. S. (1963) *Biochem. J.* **89,** 114–123.

Guillet, J. G., Kaveri, S. V., Durieu, O., Delavier, C., Hoebeke, J., and Strossberg, A. D. (1985) *Proc. Natl. Acad. Sci. USA* **82,** 1781–1784.

Homcy, C. J., Rockson, S. G., Countaway, J., and Egan, D. A. (1983) *Biochemistry* **22,** 660–668.

Insoft, R. and Homcy, C. (1984) *Fed. Proc.* **43,** 1580.

Karnik, S. S., Nassal, M., Doi, T., Jay, E., Sgaramella, V., and Khorana, H. G. (1987) in *Escherichia coli. J. Biol. Chem.* **262,** 9255–9263.

Kelley, L. K., Smith, C. H., and King, B. F. (1983) *Biochim. Biophys. Acta* **734,** 91–98.

Kobilka, B. K, Dixon, R. A. F., Frielle, T., Dohlman, H. G., Bolanowski, M. A., Sigal, I. S., Yang-Feng, T., Francke, U., Caron, M. G., and Lefkowitz, R. J. (1987) *Proc. Natl. Acad. Sci. USA* **84,** 46–50.

Laemmli, U. K (1970) *Nature* **227,** 680–685.

Lands, A. M., Arnold, A., McAauliff, J. P., Ludueno, F. P., and Brown, T. G. (1967) *Nature* **214,** 597, 598.
Lowry, O. H., Rosebrough, N. J., Farr, A. L., and Randall, R. J. (1951) *J. Biol. Chem.* **193,** 265–275.
Malbon, C. C., Rapiejko, P. J., and Watkins, D. C. (1988) *Trends in Pharm. Sci.* **9,** 33–36.
Marullo, S., Delavier-Klutchko, C., Eshdat, Y., Strosberg, A. D., and Emorine, L. (1988) *Proc. Natl. Acad. Sci. USA* **85,** 7551–7555.
Marullo, S., Delavier-Klutchko, C., Guillet, J.-G., Charbit, A., Strosberg, A. D., and Emorine, L. J. (1989) *Biotechnology* **7,** 923–927.
Minneman, K. P., Hegstrand, L. R., and Molinoff, P. B. (1979) *Mol. Pharmacol.* **16,** 34–46.
Moore, J. J. and Whitsett, J. A. (1981) in *Placenta: Receptors, Pathology, and Toxicology,* Univ. of Rochester, NY., pp. 103–114.
Moxham, C. P. and Malbon, C. C. (1985) *Biochemistry* **24,** 6072–6077.
Moxham, C. P., George, S. T., Graziano, M. P., Brandwein, H., and Malbon, C. C. (1986) *J. Biol. Chem.* **261,** 14562–14570.
Moxham, C. P., Ross, E. M., George, S. J., and Malbon, C. C. (1988) *Mol. Pharm.* **33,** 486–492.
Munson, P. J. and Rodbard, D. (1980) *Anal. Biochem.* **107,** 220–239.
Neer, E. J. and Clapham, E. (1988) *Nature* **333,** 129–134.
Pedersen, S. E. and Ross, E. M. (1985) *J. Biol. Chem.* **260,** 14150–14157.
Rashidbaigi, A. and Ruoho, A. E. (1981) *Proc. Natl. Acad. Sci. USA* **78,** 1609–1613.
Rashidbaigi, A., Ruoho, A. E., Green, D. A., and Clark, R. B. (1983) *Proc. Natl. Acad. Sci. USA* **80,** 2849–2853.
Scatchard, G. (1949) *Ann. NY Acad. Sci.* **51,** 660–672.
Schaffner, W. and Weissmann, C. (1973) *Anal. Biochem.* **34,** 502–514.
Shaltiel, S. (1974) *Methods Enzymol.* **34,** 126–140.
Shorr, R. G. L., Lefkowitz, R. J., and Caron, M. (1981) *J. Biol. Chem.* **256,** 5820– 5826.
Shorr, R. G. L., Heald, S. L., Jeffs, P. W., Lavin, T. N., Strohsaker, M. W., Lefkowitz, R. J., and Caron, M. (1982a) *Proc. Natl. Acad. Sci. USA* **79,** 2778– 2782.
Shorr, R. G. L., Strohsacker, M. W., Lavin, T. N., Lefkowitz, R. J., and Caron, M. G. (1982b) *J. Biol. Chem.* **257,** 12341–12350.
Smith, G. E. and Summers, M. D. (1982) *Virology* **123,** 393–406.
Sternweis, P. C., Northup, J. K., Smigel, M. D., and Gilman, A. G. (1981) *J. Biol. Chem.* **256,** 11517–11526.
Stiles, G. L., Caron, M. G., and Lefkowitz, R. J. (1984) *Physiol. Rev.* **64,** 661–743.
Stryer, L. and Bourne, H. R. (1986) *Annu. Rev. Cell Biol.* **2,** 391–419.
Towbin, H., Staehlin, T., and Gordon, J. (1979) *Proc. Natl. Acad. Sci. USA* **76,** 4350–4354.
Vauquelin, G., Geynet, P., Hanoune, J., and Strosberg, A. D. (1979) *Eur. J. Biochem.* **98,** 543–556.
Wang, H.-Y., Lipfert, L., Malbon, C. C., and Bahouth, S. B. (1989) *J. Biol. Chem.* **264,** 14424–14431.

Purification of β,γ Subunits of Guanine Nucleotide Binding Proteins

Deirdre Cooney and Alan K. Keenan

1. Introduction

Guanine nucleotide binding proteins (G proteins) are now known to constitute a family of regulatory or transducer molecules that link external membrane-bound receptors to corresponding intracellular effector signals in a wide variety of cell types. These established/putative effector systems are summarized in Table 1 for G proteins that have been purified to homogeneity and whose functions have been directly tested to date. It is important to realize that the fact that a particular purified G protein can be shown to stimulate a given effector in a reconstitution system does not prove this to be the function normally served by that G protein *in situ*.

The heterotrimeric nature of G proteins (α,β,γ subunits of 39–45, 35–36 and 8–10kDa, respectively), which was first established for transducin (G_T) by Fung et al. (1981), has since been demonstrated for all members of the family so far identified. We will not be concerned with the so-called monomeric G proteins recently identified in mammalian cells ($p21^{ras}$), in *Saccharomyces cervisiae* (RAS 2), and purified from human placenta (G_p), since it has not been proven that these are normally associated with β,γ subunits. α Subunits of different G proteins are highly homologous, however, they are the products of separate genes (Neer and Clapham, 1988, and references therein), and G proteins generally are

Receptor Purification, vol. 1

Table 1
Functions of Purified G Proteins

G protein	Effector system	Reference
G_S	a) Adenylate cyclase catalytic unit	Codina et al.,1984a,b; Strittmatter and Neer, 1980; May et al., 1985; Neer et al., 1987
	b) Cardiac calcium channels	Yatani et al., 1987
G_i	a) Adenylate cyclase catalytic unit	Hildebrandt et al.,1984
	b) f-MLP-stimulated PI turnover	Kikuchi et al., 1986
G_o	a) f-MLP-stimulated PI turnover	Kikuchi et al., 1986
	b) Neuronal calcium channels	Hescheler et al., 1987
G_T	a) Retinal cGMP phosphodiesterase	Fung et al.,1981
	b) G_S-stimulated adenylate-cyclase catalytic unit	Cerione et al., 1985

distinguished on the basis of their α subunits. Although originally only one type of a subunit was identified for a given G protein, molecular cloning of these subunits has revealed in some cases two or more closely related proteins. For example, distinct G_T α subunits have been identified in retinal rods and cones by cDNA sequencing (Lerea et al., 1986). In addition, a number of similar "G_i" proteins have been found using cDNA for a G_i α subunit (Jones and Reed, 1987). On the other hand, two forms of G_S α have been identified electrophoretically, although they appear to arise from a single gene by differential splicing of mRNA (Robishaw et al., 1986).

The β and γ subunits of G proteins occur in the cell membrane as a complex that can only be disrupted under denaturing conditions. Furthermore, in contrast to the diversity among α subunits, β,γ subunits show greater similarities in different G proteins, and, in some cases, appear to be shared between different α subunits. The β subunits generally occur as a doublet of mol wts 36 and 35kDa (also referred to as β1 and β2 forms, respectively) of varying relative abundance, and although these two forms can be immunologically distinguished (Roof et al., 1985), no functional differences have yet been identified. The γ subunits appear to be identical in nonretinal cells (with one known exception, *see* Evans et al., 1987), whereas the γ subunit of G_T is immunologically (Gierschik et al., 1985) and structurally (Hildebrandt et al., 1985) distinct.

The α subunits of G proteins have been regarded as the "active species" since G α expresses GTPase activity and contains the receptor recognition site and the site of ADP-ribosylation by cholera toxin (G_S, G_T) or pertussis toxin (G_i, G_o,G_T). There is, however, a recent report of βγ subunits binding to β-adrenergic receptors (Im et al., 1988), although the role of this process in G protein activation remains to be clarified. The observed ability of β,γ subunits to dissociate from the heterotrimer has led to the dissociation-activation model of G protein action proposed by Katada et al. (1984b) for G_i. According to this model, the α subunits liberated on activation of G interact directly with the effector enzyme (e.g., the catalytic unit of adenylate cyclase), whereas β,γ subunits serve as switchoff signals, by scavenging free α subunits and thus causing reassociation of the heterotrimer. This may be the mechanism by which β,γ inhibits the activity of G_Sα, but it is not clear whether G_i-mediated effects such as inhibition of adenylate cyclase, are a result of liberated β,γ subunits combining with G_Sα, or of a direct interaction of G_i α with the catalyst, C.

It has also been proposed by Levitzki's group (1988, and references therein) that G_S or, at least, G_S α is permanently associated with C. Thus, β,γ could dissociate from G_Sα•C on activation, and the role of β,γ may be confined to that of anchoring the G α subunit to the membrane. However, the fact that more "active" functions have been recently assigned to the β,γ complex (*see* Table 2) would seem to argue against their merely serving a silent anchoring role.

2. Purification of G Proteins

Advances in our knowledge of the diversity of G proteins in recent years should not necessarily affect purification protocols for β,γ subunits, but would have consequences for α subunit purification. For example, when the purity of "$α_i$" was monitored on SDS PAGE after ADP-ribosylation with radioactive NAD^+, the expected single band was observed, but we now know that up to three $α_i$ subspecies of almost identical mol wt exist in certain cell types and, therefore, the use of more highly specific probes is necessary when characterizing α subunits.

The purification of β,γ subunits from specific sources (described below in detail) includes references to the corresponding G protein purification protocols. This Section will outline the general principles that are applicable to the purification of a number of G proteins after first considering the "atypical" member of the family, G_T. The latter, which was

Table 2
Functions of Isolated β,γ Subunits

Effector system	Nature of effect	Reference
Muscarinic receptor-gated cardiac potassium channel	Activation	Logothetis et al., 1987
Retinal phospholipase A_2	Activation	Jelsema and Axelrod, 1987
Brain cAMP phosphodiesterase	Inhibition	Asano et al., 1986
Calmodulin-stimulated brain adenylate cyclase	Inhibition	Katada et al., 1987a
G_S-stimulated adenylate cyclases	Inhibition	Northup et al., 1983b; Neer et al., 1987; Cerione et al., 1987

the first to be purified, has the characteristics of a peripheral protein and can be solubilized readily from disk membranes of rod outer segments (ROS) simply by eluting with hypotonic buffer containing GTP or nonhydrolyzable analogs such as Gpp(NH)p or GTPγS. All other G proteins, however, can be fully extracted from membranes only in the presence of detergent (usually cholate). The first of these to be successfully purified to homogeneity was the stimulatory protein G_S, which has been obtained from a number of sources such as rabbit liver (Northup et al., 1980; Sternweis et al., 1981), turkey erythrocytes (Hanski et al., 1981), and human erythrocytes (Codina et al., 1984a,b). Subsequently, other tissues have served as sources of pure G_S and other G proteins, e.g., G_i, G_o, from bovine brain (Sternweis and Robishaw, 1984; Neer et al., 1984) and rat brain (Katada et al., 1986); G_i, G_p from human placenta (Evans et al., 1986) and blood platelets (Evans et al., 1987). Most recently, a novel pertussis toxin substrate "G_n," thought to be linked to stimulation of phospholipase C activity, has been purified from bovine neutrophils by Gierschik et al. (1987) and by Dickey et al. (1987) from rabbit neutrophils.

A general purification protocol is outlined in Table 3, adapted from that of Bokoch et al. (1984) for purification of G_i and G_S from rabbit liver and applicable to a number of tissues containing more than one G protein. Solubilization of washed membranes in cholate is followed by ion exchange (DEAE-Sephacel) and gel filtration (Ultrogel AcA-34) chromatography, leading to copurification of G_S with G_i (rabbit liver), or $G_i + G_o$ (bovine brain), or with G_i (human placenta). This step, however, leads to resolution of G_p from $G_S + G_i$ in placenta. G_S can be resolved from the other proteins by hydrophobic ion-exchange chromatography on heptylamine-Sepharose. G_i/G_o mixtures can finally be resolved on

Table 3
Summary Scheme for G Protein Purification

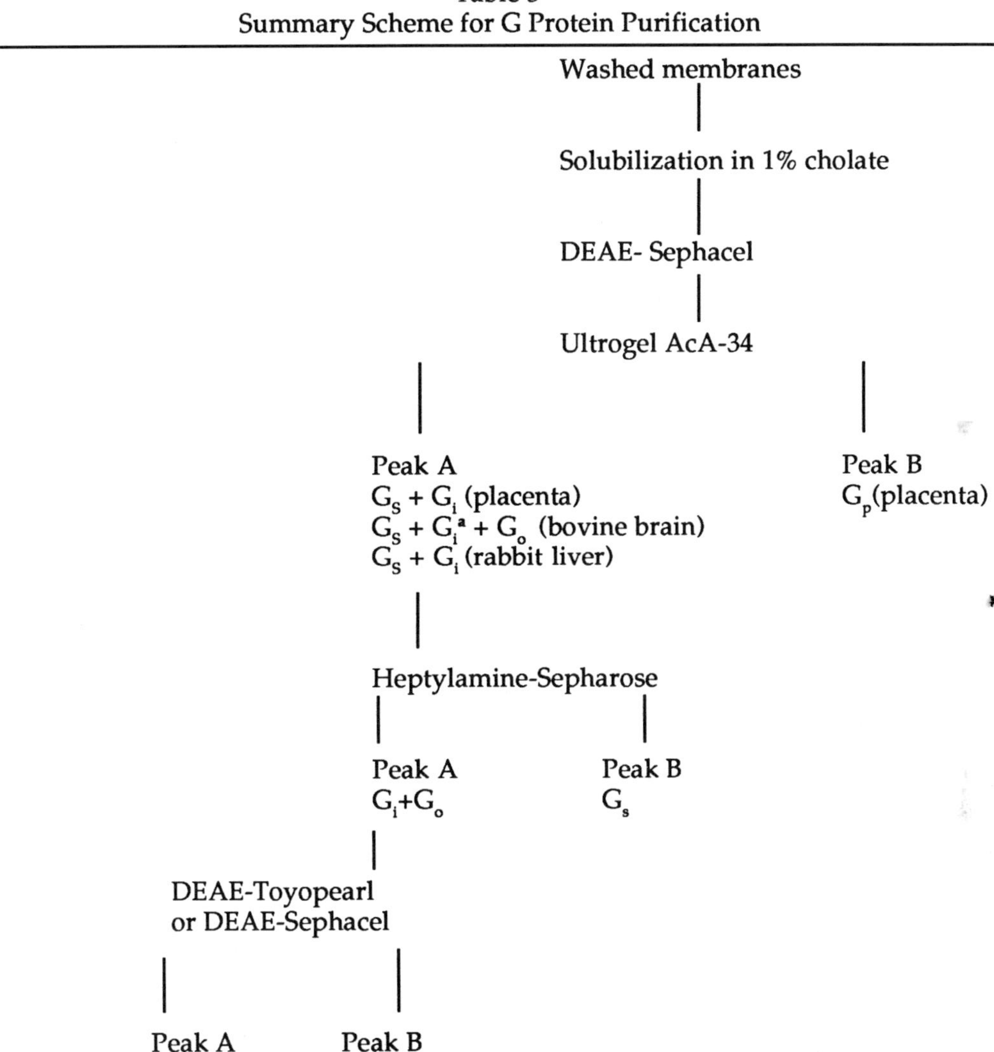

[a]G_i may be one or more G proteins.

DEAE-Toyopearl or DEAE-Sephacel followed by treatment with hydroxyapatite to remove activating ligands if required (*see* β,γ purification from rat brain, below).

2.1. The Use of Activating Ligands

The earliest reported purifications of turkey erythrocyte G_S and rabbit liver G_S were carried out under activating conditions (with AMF: 1μ*M* ATP or 20 μ*M* $AlCl_3$, 6 m*M* $MgCl_2$,10 m*M* NaF), but later protocols

(Bokoch et al., 1984) have not used such activators, sometimes because of the desire to preserve the heterotrimeric structure of the G protein throughout the course of the purification. The use of activating ligands is routine when resolution of α from β,γ subunits is required and is also a necessary condition for successful purification of functional α subunits from G_n (Gierschik et al., 1987) though Dickey et al. (1987) found that stabilization could be achieved with ethylene glycol.

3. Functionality of Purified G Proteins

Reconstitution assays have played a major role in assessing the functionality of purified G_S and G_i. Hanski et al. (1981) originally showed that when fluoride-activated pure G_S was reconstituted with membranes prepared from the cyc⁻ mutant of S49 cells, the cyclase responsiveness to β-adrenergic agonists (absent in this mutant) was restored. Since there is no corresponding mutant cell line deficient in G_i, an alternative strategy was necessary in this case. When pertussis toxin-treated human platelet membranes (which express G_i-linked α_2-adrenoceptors) were treated with pure G_i, the ability of adrenaline to inhibit adenylate cyclase was restored (Katada et al., 1984a). Other functional assays for G proteins are described in the references listed in Table 1 and will not be further expanded here, since many of the functions of G proteins are carried out by their α subunits. Those functions served by β,γ subunits will be expanded in the following Sections.

4. Purification of β,γ Subunits

Following the proposal (Kuhn, 1980) and establishment of a subunit structure (Fung et al., 1981) for G_T, it became clear that striking similarities existed between G_T and the other G proteins. Northup et al. (1980) had noted the presence of an apparent excess of a 35 kDa polypeptide in fractions obtained during purification of rabbit liver G_S that, although it could be resolved from G_S on hydroxyapatite, was subsequently shown to be identical to the 35 kDa component of purified G_S. This group later isolated β,γ subunits from both turkey erythrocyte (Hanski et al., 1981) and rabbit liver G_S (Northup et al., 1983a). Sternweis and Robishaw (1984) also purified this 35 kDa band from bovine brain as a byproduct of G_S, which is less abundant in this tissue. In all cases, during the early work, the γ subunit associated with β was not identified (except for G_T), since it ran with the dye front on the gels used, nor was the doublet nature of β (35/36kDa) recognized. In the purification schemes to follow, the progress of preparative subunit separations by chromatographic techniques is routinely monitored by analytical means on SDS PAGE, which originally served as the means of confirming the subunit composition of purified G proteins.

4.1. β,γ from Transducin

Bovine ROS have been widely used as a source of β,γ and the purification strategy employed by Chabre's group (Deterre et al., 1984) uses G_T isolated by the method of Kuhn (1980). The overall scheme is outlined in Table 4. A bleached ROS suspension is pelleted and the membrane-bound G_T washed with 5 m*M* Tris-HCl to remove soluble proteins and with 100 m*M* Tris-HCl to release other membrane-bound proteins (1 m*M* DTT is present in all buffers). G_T is then extracted by treatment with 10 μ*M* GTPγS in 5 m*M* Tris-HCl, which also causes subunit dissociation. Subsequent ion exchange chromatography of the activated G_T by FPLC using a Polyanion SI column yields two peaks corresponding to α and β,γ subunits. The order of elution of the two peaks is determined by the nature of the salt gradient used for the elution step. As can be seen in Fig. 1, irrespective of the eluting salt, a good separation of subunits is achieved. Amphibian G_T can also serve as a source of β,γ subunits and the purification scheme of Yamazaki et al. (1987), starting with *Bufo marinus* ROS, is shown in Table 5. The washed pellet, obtained by Kuhn's method (1980) is treatedwith 100 m*M* Tris-HCl containing 0.1 m*M* Gpp(NH)p, centrifuged and washed repeatedly (again DTT, 5 m*M*, is present in all buffers). The pooled supernatants contain α subunits. Further washing of membranes with 5 m*M* Tris-HCl removes a fraction enriched in β,γ subunits that is applied to a Blue Sepharose CL-6B column equilibrated with the same buffer. The eluate from this column contains, in addition to pure β,γ, a second peak of β, free of other G_T subunits, though somewhat contaminated with the γ subunit of phosphodiesterase. The elution pattern of subunits is shown in Fig. 2 and their purities as assessed on SDS gels are given in Fig. 3.

4.2. β,γ from Rabbit Liver

The purification of G_i from rabbit liver described above (Bokoch et al., 1984) under nonactivating conditions, can serve as the starting point for resolution of G_i α and β,γ subunits, as described by Katada et al. (1984a). Purified G_i is activated in HED buffer (50 m*M* Hepes, pH 8, 1 m*M* EDTA, 1 m*M* DTT) containing 0.1% lubrol with either (i) GTPγS +

Table 4
Purification of β,γ Subunits from Bovine ROS (*see* Deterre et al., 1984)

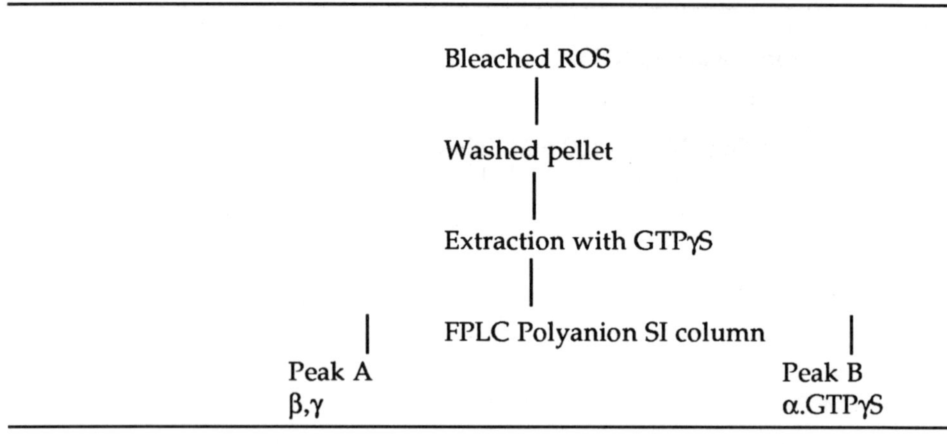

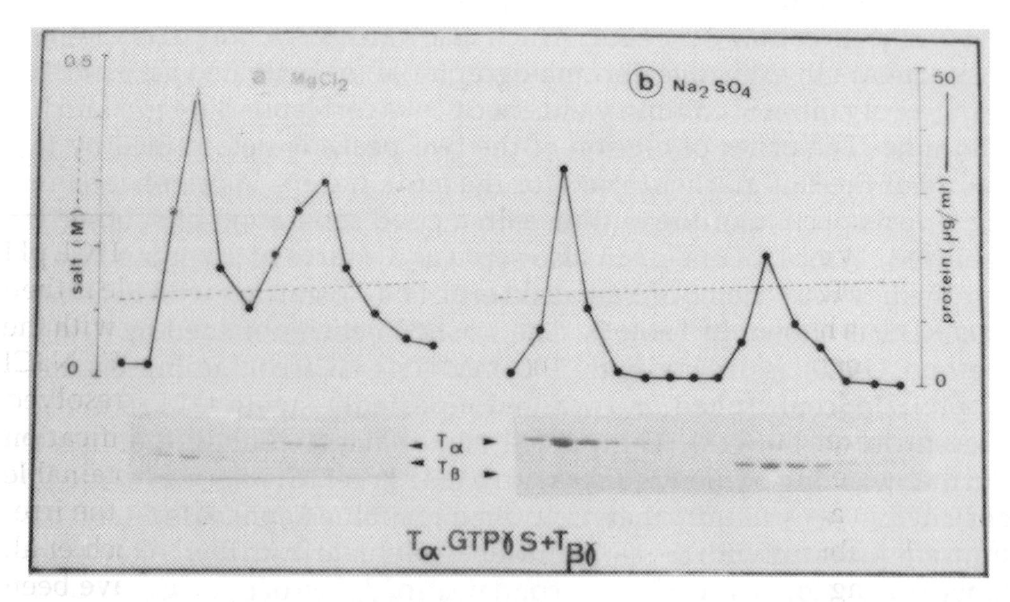

Fig. 1. Elution profile and gel electrophoresis patterns of transducin subunits after FPLC, as a function of the nature of the eluting salt; a, $MgCl_2$; b, Na_2SO_4. Protein (–) and salt (---) concentrations are indicated. γ Subunits are not shown on gels (reproduced from Deterre et al., 1984, with permission).

$MgCl_2$ (10 μ*M* and 10 m*M*, respectively) or (ii) AMF. In case (i), protein is then filtered through Sephadex G-25 to remove unbound GTPγS and applied to a hydroxyapatite column equilibrated in buffer containing 1% sodium cholate but no lubrol. After elution of protein with buffer containing 300 m*M* phosphate, pH 8, the material is applied to two TSK-125 HPLC columns and after elution the output is recycled through the columns for 90 min and fractions are collected. In case (ii), activation is

Table 5
Purification of β,γ Subunits from Amphibian ROS (*see* Yamazaki et al., 1987)

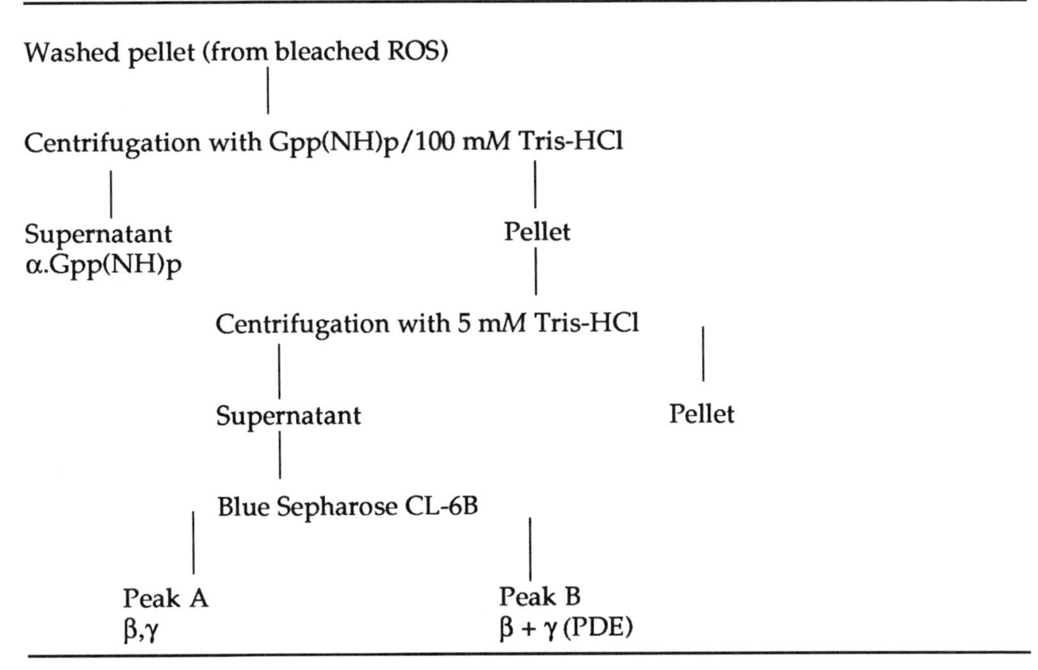

followed by dilution of the sample with TED buffer (20 m*M* Tris-HCl, pH 8,1 m*M* EDTA,1 m*M* DTT) in 0.3% cholate and AMF. The sample is then applied to a heptylamine-Sepharose column, and, after washing with the same buffer, is eluted with a reverse gradient (cholate 0.3–1.2%, NaCl 250–50 m*M*, in TED containing AMF). Fractions containing resolved subunits are filtered through G-25 to remove AMF. The purification schemes are outlined in Table 6. A better resolution of β,γ is obtainable when G_i is activated with GTPγS/Mg, which may be owing to the irreversible subunit dissociation caused by this nucleotide. Bokoch et al. (1984) using modified SDS gels containing 15% acrylamide, have been able to visualize the ~10kDa γ subunit which copurifies with the α and β subunits of G_i.

4.3. β,γ *from Rat Brain*

Ui's group has recently isolated β,γ subunits in high yield from the "IAP (pertussis toxin) substrates" fraction obtained during purification of rat brain G proteins (Katada et al., 1986). This fraction is now known to contain at least three distinct α subunits whose associated β,γ subunits are indistinguishable. The overall scheme is based on the rabbit liver G_i purification discussed above and the protocol using the two most abundant IAP substrates (α_{41} β,γ and α_{39} β,γ) as a starting point is given in Table

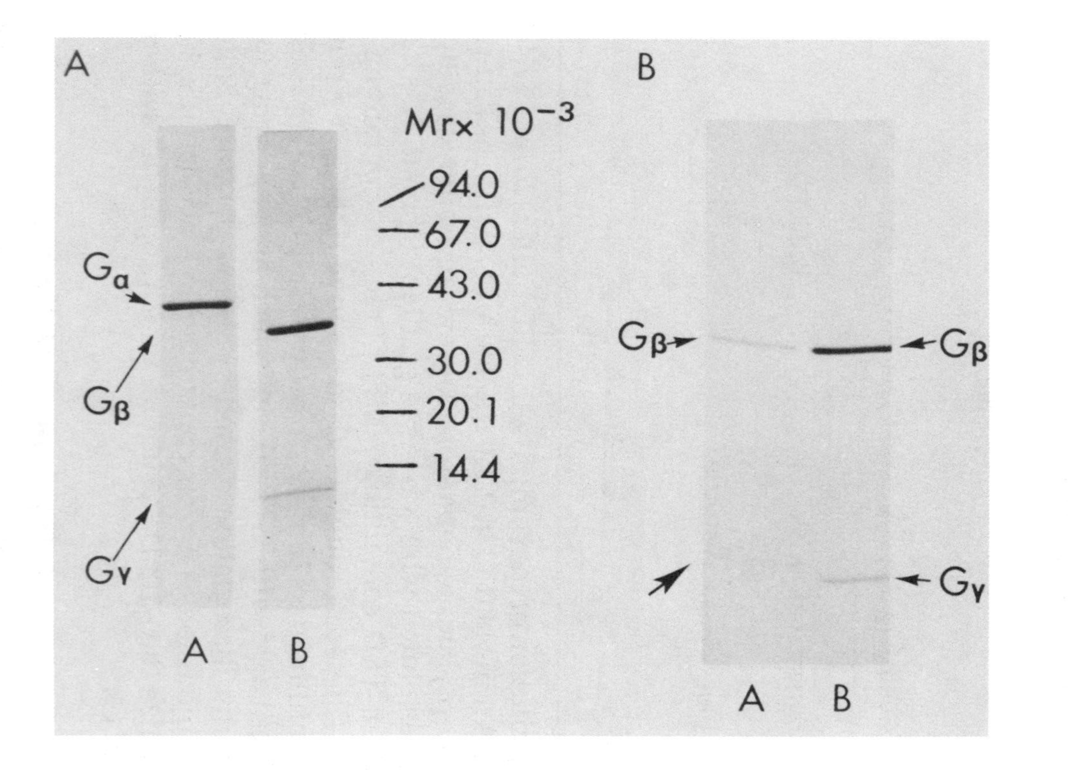

Fig. 2. Elution pattern of transducin subunits after Blue Sepharose CL-6B chromatography. **A** α Subunits are assayed in terms of their capacity to bind Gpp(NH)p in the presence of purified β,γ + urea-treated ROS membranes. **B** β,γ Subunits are assayed in terms of their capacity to stimulate Gpp(NH)p binding to α subunits in the presence of urea-treated ROS membranes (reproduced from Yamazaki et al., 1988, with permission).

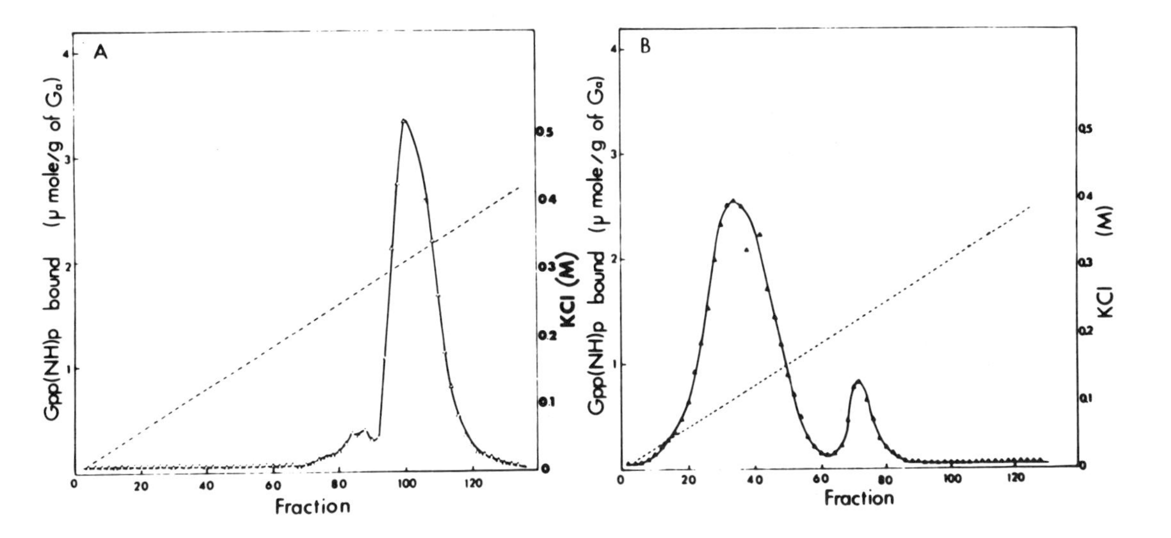

Fig. 3. Purity of transducin obtained as described in Fig. 2. In **A**, the purity of α is compared with β,γ. In **B**, the purity of β,γ is compared with β and the arrow in lane **A** indicates the contaminant protein (M_r = 13kDa). In each case 10 μg of protein was applied to the gel (reproduced from Yamazaki et al., 1987, with permission).

7. The yields of resolved purified heterotrimers obtained after the DEAE-Toyopearl and hydroxyapatite steps are given in Table 8 and are expressed as a percentage of the amounts present in the rat brain cholate extract. Table 7 shows the resolution of each αβ,γ complex by one activation method only, but both methods have been used for each hetero-trimer. The resolution of α and β,γ subunits is shown in Fig. 4 after HPLC on TSK G 3,000 SW columns of GTPγS/Mg-treated material. Similar resolution is shown for both sets of subunits after AMF/heptylamine-Sepharose chromatography, using a reverse gradient as for rabbit liver (Fig. 5). In each case, following the chromatographic step β,γ subunits were concentrated on hydroxyapatite columns and filtered through Sephadex G-25 to remove GTPγS or AMF. Although it is not clear from the accompanying gels in Figs. 4 and 5, the band corresponding to the β-subunits occurs as a 35/36kDa doublet in each case. A third brain pertussis toxin substrate, α_{40} βγ has also been recently resolved (Katada et al., 1987b) on a Mono Q HR5/5 column by FPLC, and the corresponding β,γ subunits again resolved on TSK G 3,000 SW columns.

Table 6
Purification of β,γ Subunits from Rabbit Liver
(*see* Katada et al., 1984a, and references therein)

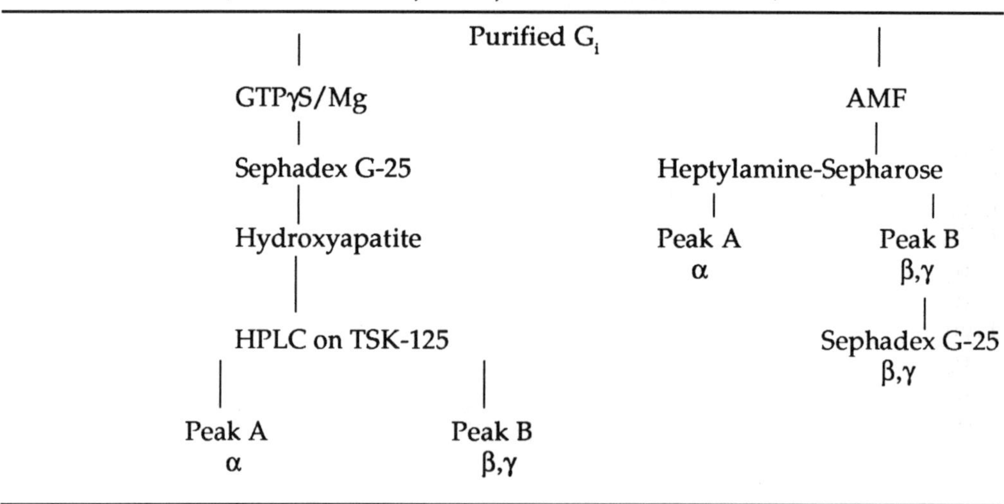

Table 7
Purification of β,γ Subunits from Rat Brain IAP Substrates
(*see* Katada et al., 1986)

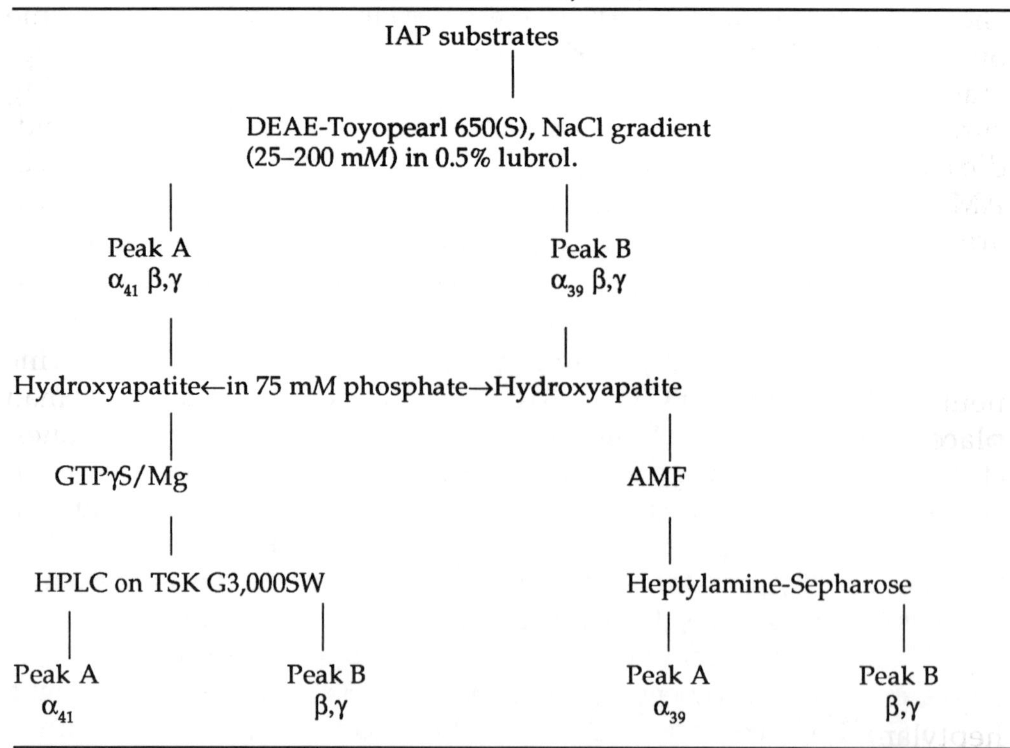

Table 8
Purification of IAP Substrates from Rat Brain Membranes
(adapted from Katada et al., 1986 with permission)

	ADP-ribosylation			GTPγS binding		
	Amount	Specific act.	Recovery[a]	Amount	Specific act.	Recovery[a]
Purification step	nmol	nmol/mg	%	nmol	nmol/mg	%
Cholate extract	308	0.16	100	642	0.32	100
DEAE-Toyopearl						
α_{41} βγ	30	8.1	10	28	7.6	4.4
α_{39} βγ	46	7.9	15	48	8.3	7.5
Hydroxyapatite						
α_{41} βγ	25	8.9	8.1	24	8.6	3.7
α_{39} βγ	37	8.8	12	38	9.1	5.9

[a]Measured in terms of α subunit properties, i.e., the ability to undergo ADP-ribosylation in the presence of β,γ and to bind GTPγS.

4.4. β,γ *from Bovine Neutrophils*

The G protein purified in high yield from bovine neutrophils (G_n) by Gierschik's group can also serve as a source of β,γ subunits. The scheme outlined in Table 9 involves the use of activating ligands. The partial separation of α_n and β,γ on DEAE-Sephacel is shown in Fig. 6. Following chromatography on heptylamine-Sepharose, α_n and β,γ elute in the middle and at the end of the gradient (0.3–1.5% cholate, 250-0 m*M* NaCl, in AMF), respectively. Recoveries of β,γ are equivalent to those of α_n, which are indicated in Table 10.

4.5. β,γ *from Human Placenta*

In a protocol essentially similar to that described above for bovine neutrophils, Evans et al. (1987) purified the β_{35} form of β,γ from human placenta in high yield. After cholate extraction of the membranes, chromatography on DEAE-Sephacel (without AMF) resolves oligomeric G proteins from β_{35} that elutes at high salt. The overall yields of β_{35} during purification are shown in Table 11. The 33-fold enrichment of β subunit activity contrasts with the virtual absence of any change in GTPγS binding to minor contaminating α subunits (*see* Evans et al., 1987, for further details). The $\beta_{35/36}$ doublet can also be purified from placenta by hydroxyapatite chromatography of G_i/G_s-containing fractions eluted from heptylamine-Sepharose.

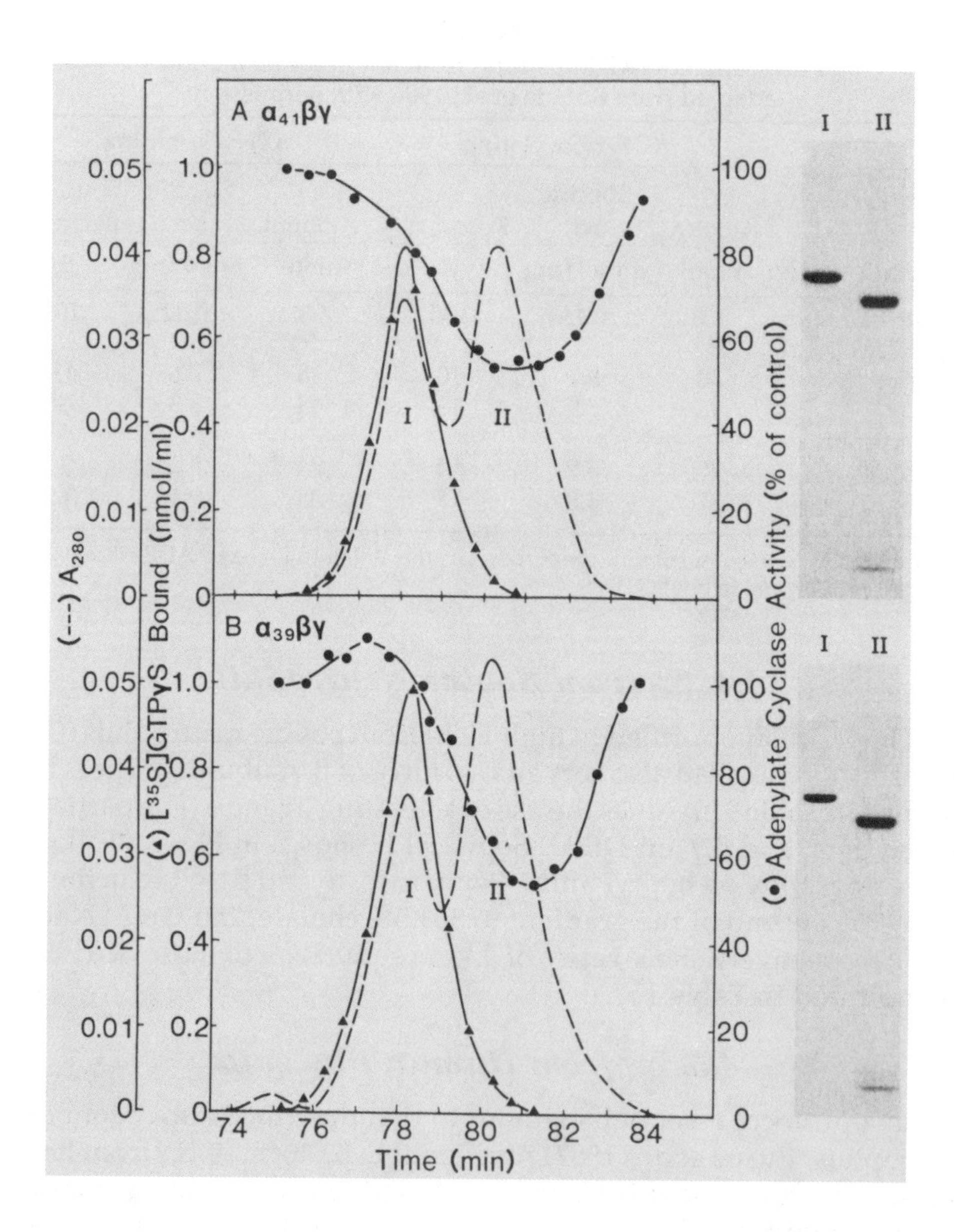

Fig. 4. Resolution of α and β,γ subunits of GTPγ S-treated IAP substrates by high performance gel filtration. In each case two peaks of protein (---) were obtained, I and II. Purified α_{41} ,β,γ (Panel A) or α_{39} ,β,γ (Panel B) was incubated with [^{35}S] GTPγ S + Mg and [^{35}S] GTPγS bound α subunits (s) identified in resolved fractions. Fractions containing β,γ subunit activity (●) inhibited human platelet membrane adenylate cyclase. Analysis of each peak fraction by SDS PAGE is also shown and the resolution of α from β,γ confirmed (reproduced from Katada et al., 1986, with permission).

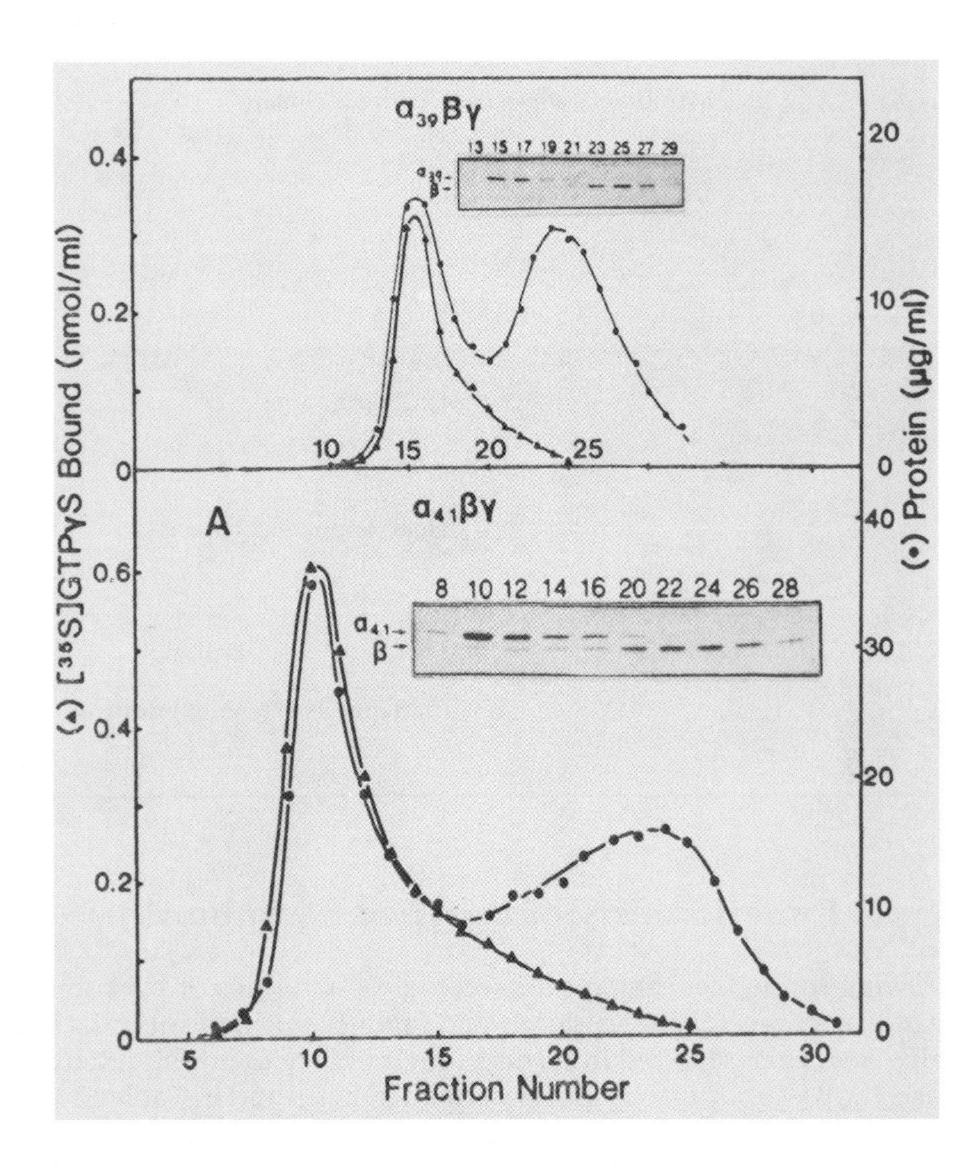

Fig. 5. Resolution of α and β,γ subunits of AMF-treated IAP substrates by heptylamine-Sepharose chromatography. In each case two peaks of protein (●) were obtained, corresponding to α and β subunits (the insets show SDS PAGE analysis of fractions, excluding the γ subunits). Binding of [^{35}S] GTPγ S to α-subunits (▲) is also indicated (reproduced from Katada et al., 1986, with permission).

Table 9
Purification of β,γ Subunits from Bovine Neutrophils (*see* Gierschik et al., 1987)

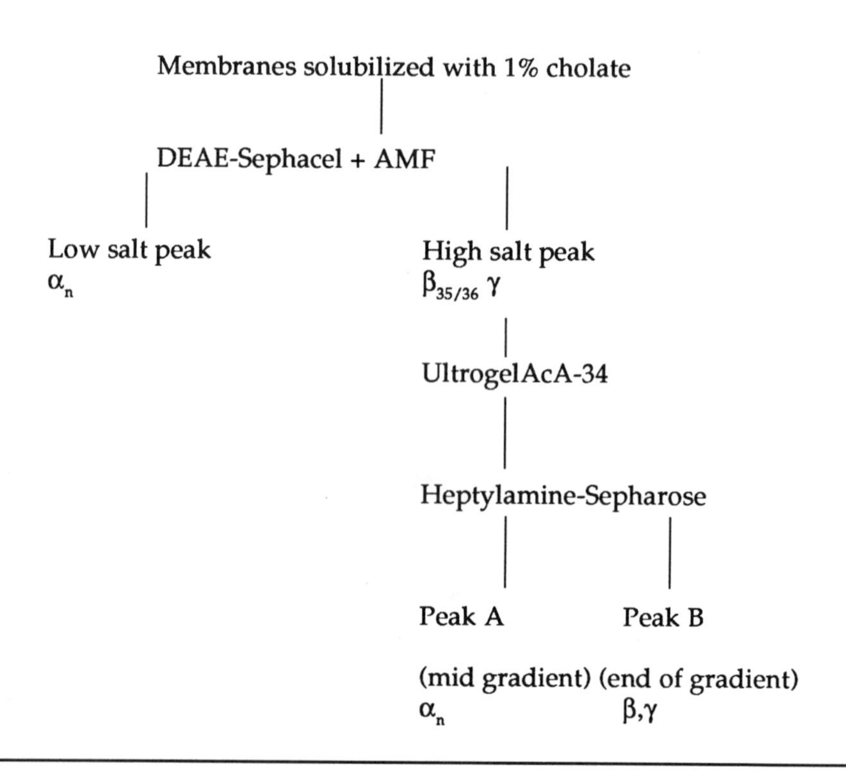

5. Functionality of Purified β,γ Subunits

Owing to the fact that specific biological activities of β,γ subunits have only very recently been identified (Table 2), purified subunits have usually been characterized in terms of their ability to inhibit adenylate cyclase; the β,γ subunit activities described in Fig. 4 and in Table 11 were measured as inhibition of human platelet adenylate cyclase. An additional sensitive assay for β,γ subunits is their ability to deactivate AlF_4-activated G_S α subunits, as shown originally by Northup et al. (1983a). The versatility of this method has been demonstrated by Im et al. (1987), who compared the abilities of β,γ subunits purified from a variety of sources to deactivate rabbit liver G_S preactivated with AlF_4-(*see* Fig. 7).

It is known that β,γ subunits are corequired (with α subunits) for the expression of certain G protein functions, and this has been tested *inter alia* in reconstituted systems, where Sternweis (1986) demonstrated

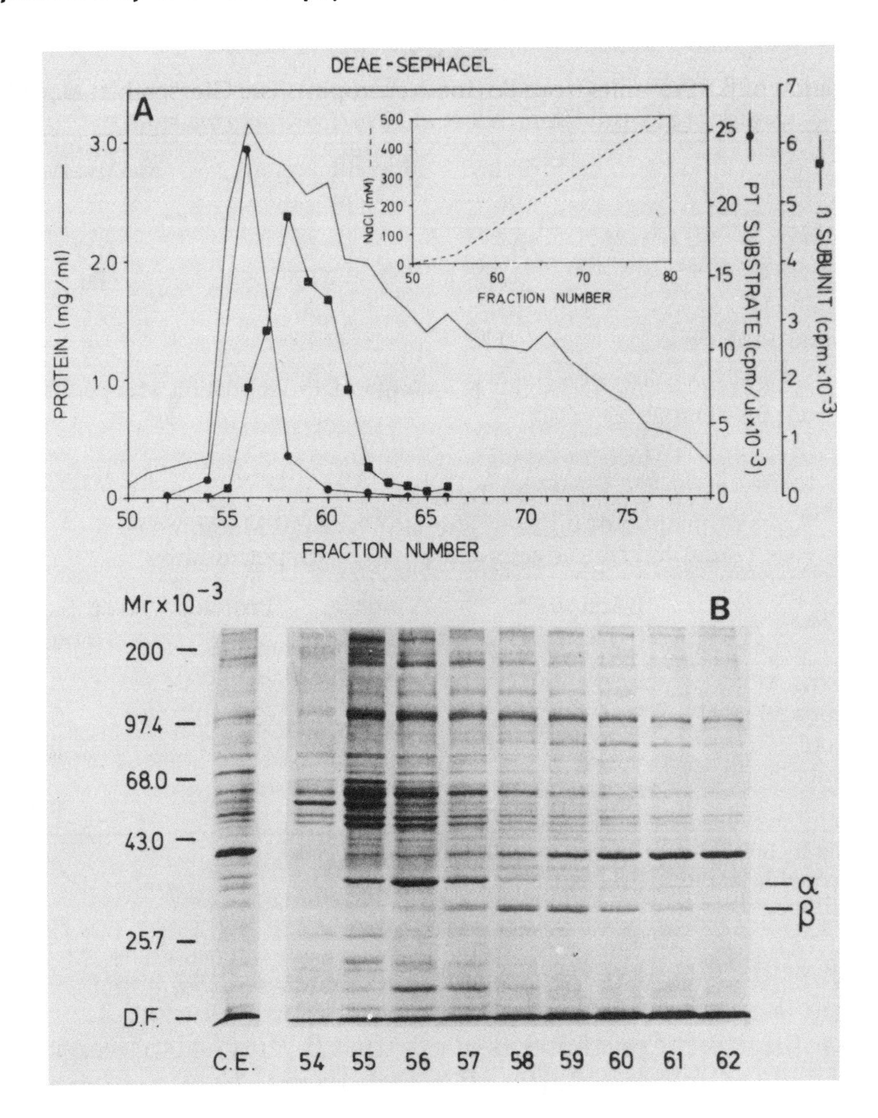

Fig. 6. A. Chromatography of α_n and β,γ through DEAE-Sephacel; α_n was assayed by pertussis-toxin induced ADP-ribosylation in the presence of G_T β,γ subunits, while β,γ was assayed by quantitative immunoblotting of its β_{36} component. B. SDS PAGE analysis of the cholate extract (C.E., 50 μg) and 50 μL aliquots of the indicated fractions obtained form DEAE-Sephacel chromatography. Proteins were stained with Coomassie Blue (adapted from Gierschik et al., 1987, with permission).

that purified $G_o\alpha$ required β,γ for association with phospholipid vesicles. Neer et al. (1984) have suggested that the requirement of β,γ for ADP ribosylation of purified α subunits could serve as an assay of β,γ subunit activity, and this technique has been employed by Im et al. (1987), who used pertussis toxin/[^{32}P] NAD to label bovine neutrophil Gα in the presence of β,γ subunits from different sources as shown in Fig. 8. Puri-

Table 10
Purification of the Major Pertussis Toxin Substrate from Bovine Neutrophils
(adapted from Gierschik et al., 1987, with permission)

Step	Protein, mg	Specific activity,[a] cpm/μg	Recovery, %
Cholate extract	200	149	100
DEAE-Sephacel	61	419	86
Ultrogel AcA-34	10	800	27
Heptylamine-Sepharose	0.28	6013	6

[a]α_n Was quantified in terms of its ability to undergo ADP-ribosylation with pertussis toxin/[^{32}P] NAD + G_T β,γ subunits.

Table 11
Purification of β_{35} from Human Placental Membranes
(adapted from Evans et al., 1987, with permission)

Stage	βActivity,[a] $K_{1/2}$, μg/mL	Volume, mL	Protein, mg	GTPγS binding,[b] nmol/mg
Cholate extract	—	1740	3651[c]	0.41
DEAE-Sephacel pool	6.26	300	300	0.13
AcA 34 pool	0.59	110	36.5	0.50
Heptylamine-Sepharose pool	0.19	80	8.5	0.64

[a]Assayed by inhibition of human platelet adenylate cyclase.
[b]Assayed with 10 μ*M* GTPγS.
[c]Solubilization yield from 10-12g membranes.

fied G_T β,γ subunits have been assayed in terms of their ability to stimulate Gpp(NH)p binding to G_T α subunits, as shown in Fig. 2.

Quantitative measurements of purified β,γ have also been made by the technique of immunoblotting (Gierschik et al., 1985), which also provides a method for studying tissue-specific differences in the amounts of G proteins. More recently, Asano et al. (1988) developed an enzyme immunoassay method for the quantitation of brain β,γ that may also be applicable to characterization of purified subunits.

6. Concluding Remarks—Future Directions

Recent advances in molecular cloning and nucleotide sequencing of the genes encoding β and γ subunits have given us a greater insight into subunit diversity among members of this family of proteins. For example, the existence of two distinct β subunits (β_{36} and β_{35}) has been verified by sequencing the genes encoding the two proteins (Gao et al.,

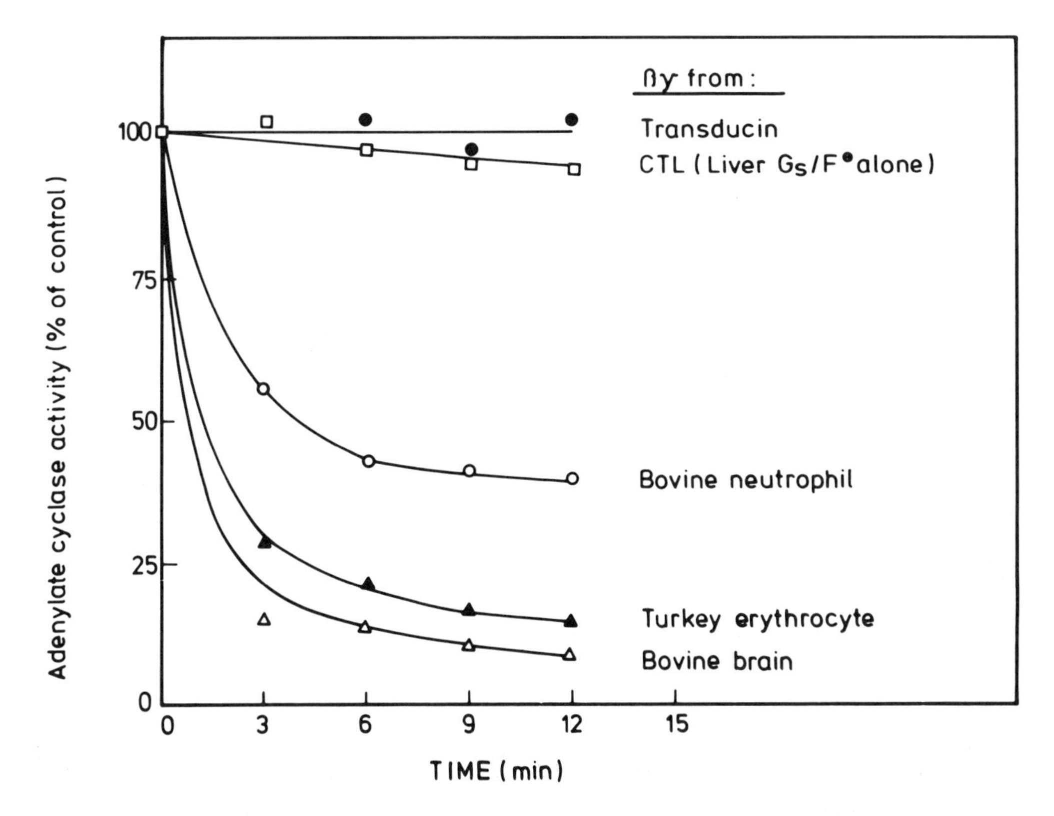

Fig. 7. Deactivation of AlF_4-activated rabbit liver G_S by the indicated different β,γ subunits. Purified activated G_S was incubated with β,γ subunits (1:16 weight ratio) at 4°C and at the indicated times aliquots were removed to measure residual G_S activity using a crude myocardial extract as source of adenylate cyclase (reproduced from Im et al., 1987, with permission).

1987; Amatruda et al., 1988 and references therein). The deduced polypeptides of both are of similar mol wt, although 10% of their amino acid residues differ. Significant differences have also been found in the non-translated regions of the two cDNA sequences and also in the mRNA levels in different tissues. Although both forms of the β subunit are expressed in the majority of tissues, the high degree of conservation of primary sequence between species would suggest that the two proteins may have distinct functional roles. To date, no systematic investigation of the possible functional differences between β_{36} and β_{35} has been reported. It is possible that heterogeneity at the level of β subunits could explain some of the reported differences in the ability of β,γ from brain and retina to inhibit activation of adenylate cyclase by G_S α (Im et al., 1987; Cerione et al., 1987). However, the possibility that heterogeneity at the level of γ subunits contributes to this effect must be seriously consid-

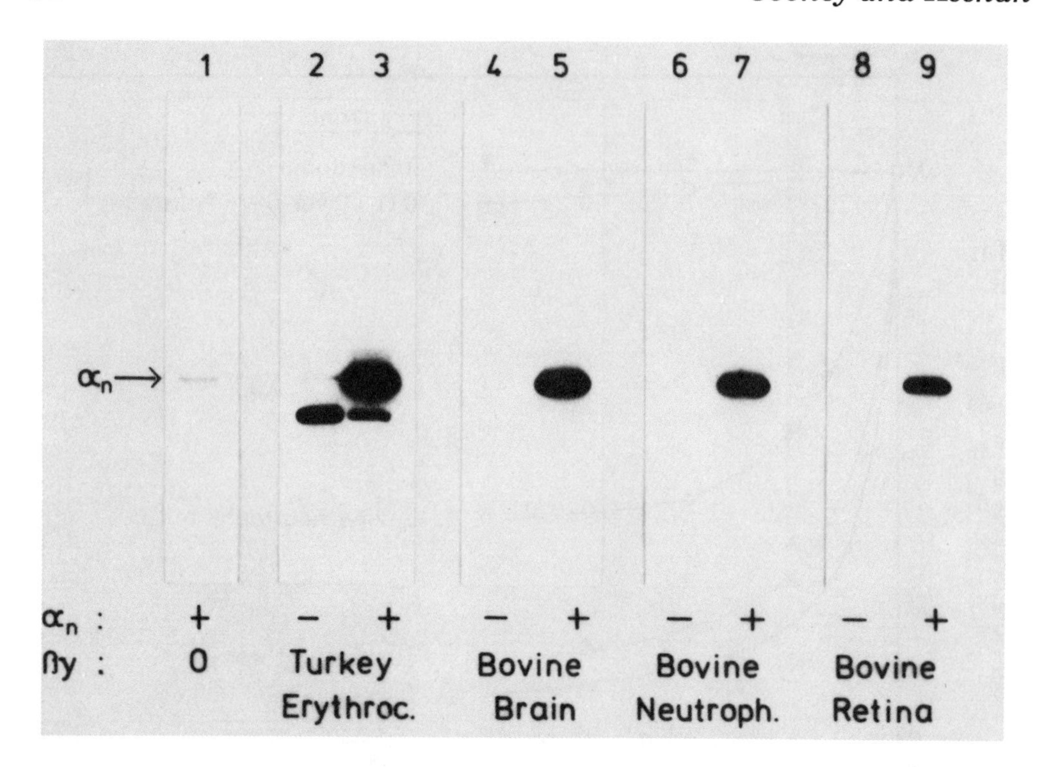

Fig. 8. The β,γ subunit requirement for pertussis toxin-catalyzed ADP-ribosylation of bovine neutrophil α subunit. The incorporation of radioactivity into pure α_n treated with pertussis toxin/[^{32}P] NAD is shown in the presence of β,γ subunits purified from the indicated sources. Incorporated [^{32}p] ADP-ribose was localized by autoradiography of gels after SDS PAGE. In lane 1 purified α_n, visualized by silver staining, is included as a reference. In the case of turkey β,γ (lanes 2 and 3) the additional band labeled even in the absence of α_n may be a peptide derived from turkey erythrocyte G_i α (adapted from Im et al., 1987, with permission).

ered. The extent of diversity of γ subunits among different G proteins is not yet known, since only one γ subunit sequence is available (Hurley et al., 1984; Yatsunami et al., 1985). In summary, therefore, the functional bases for β and γ subunit heterogeneities remain unclear.

Because of the difficulty in resolving β subunits from each other and from γ subunits, the precise role played by each (and by variants of each) in signal transduction awaits the expression of each component in a cell line where the function of one can be examined in the presence or absence of the other. An obvious technical difficulty of such an approach would be the selection of a cell line that does not constitutively express G proteins to an appreciable extent. The rewards, however, in examining such a system would be significant, not only in assessing the differential and variable roles played by each subunit, but also in determining the molecular basis of their biological activities.

7. Acknowledgments

We would like to acknowledge the cooperation of all those who so readily agreed to allow us reproduce their data, and without whom this work would have been incomplete. We would particularly like to thank Marie Doyle and Roddy Monks for their patience and assistance in the preparation of the manuscript.

References

Asano, T., Ogasawara, N., Kitajima, S., and Sano, M. (1986) *FEBS Lett.* **203,** 135–138.

Asano, T., Kamiya, N., Morishita, R., and Kato, K. (1988) *J. Biochem.* **103,** 950–953.

Amatruda, T. T., Gautam, N., Fong, H. K. W., Northup, J. K., and Simon, M. I. (1988) *J. Biol. Chem.* **263,** 5008–5011.

Bokoch, G. M., Katada, T., Northup, J. K., Ui, M., and Gilman, A. G. (1984) *J. Biol. Chem.* **259,** 3560–3567.

Cerione, R. A., Codina, J., Kilpatrick, B. F., Staniszewski, C., Gierschik, P., Somers, R. L., Spiegel, A. M., Birnbaumer, L., Caron, M. G., and Lefkowitz, R. J. (1985) *Biochemistry* **24,** 4499–4503.

Cerione, R. A., Gierschik, P., Staniszewski, C., Benovic, J. L., Codina, J., Somers, R., Birnbaumer, L., Spiegel, A. M., Lefkowitz, R. J., and Caron, M. G. (1987) *Biochemistry* **26,** 1485–1491.

Codina, J., Hildebrandt, J. D., Sekura, R. D., Birnbaumer, M., Bryan, J., Manclark, C. R., Iyengar, R., and Birnbaumer, L. (1984a) *J. Biol. Chem.* **259,** 5871–5886.

Codina, J., Rosenthal, W., Hildebrandt, J. D., Sekura, R. D., and Birnbaumer, L. (1984b) *J. Receptor Res.* **4,** 411–442.

Deterre, P., Bigay, J., Pfister, C., and Chabre, M. (1984) *FEBS Lett.* **178,** 228–232.

Dickey, B. F., Pyun, H. Y., Williamson, K. C., and Navarro, J. (1987) *FEBS Lett.* **219,** 289–292.

Evans, T., Brown, M. L., Fraser, E. D., and Northup, J. K. (1986) *J. Biol. Chem.* **261,** 7052–7059.

Evans, T., Fawzi, A., Fraser, E. D., Brown, M. L., and Northup, J. K. (1987) *J. Biol. Chem.* **262,** 176–181.

Fung, B. K-K., Hurley, J. B., and Stryer, L. (1981) *Proc. Natl. Acad. Sci. USA* **78,** 152–156.

Gao, B., Gilman, A. G., and Robishaw, J. D. (1987) *Proc. Natl. Acad. Sci. USA* **84,** 6122–6125.

Gierschik, P., Codina, J., Simons, C., Birnbaumer, L., and Spiegel, A. (1985) *Proc. Natl. Acad. Sci. USA* **82,** 727–731.

Gierschik, P., Sidiropoulos, D., Spiegel, A., and Jakobs, K. H. (1987) *Eur. J. Biochem.* **165,** 185–194.

Hanski, E., Sternweis, P. C., Northup, J. K., Dromerick, A. W., and Gilman, A. G. (1981) *J. Biol. Chem.* **256,** 12911–12919.

Hescheler, J., Rosenthal, W., Trautwein, W., and Shultz, G. (1987) *Nature* **325,** 445–447.

Hildebrandt, J. D., Codina, J., and Birnbaumer, L. (1984) *J. Biol. Chem.* **259,** 13178–13185.

Hildebrandt, J. D., Codina, J., Rosenthal, W., Birnbaumer, L., Neer, E. J., Yamazaki, A., and Bitensky, M. W. (1985) *J. Biol. Chem.* **260,** 14867–14872.

Hurley, J. B., Fong, H. K. W., Teplow, D. B., Dreyer, W. J., and Simon, M. I. (1984) *Proc. Natl. Acad. Sci. USA* **81,** 6948–6952.
Im, M.-J., Holzhofer, A., Keenan, A. K., Gierschik, P., Hekman, M., Helmreich, E. J. M., and Pfeuffer, T. (1987) *J. Receptor Res.* **7,** 17–42.
Im, M.-J., Holzhofer, A., Bottinger, H., Pfeuffer, T., and Helmreich, E. J. M. (1988) *FEBS Lett.* **277,** 225–229.
Jelsema, C. L. and Axelrod, J. (1987) *Proc. Nat. Acad. Sci. USA* **84,** 3623–3627.
Jones, D. T. and Reed, R. R . (1987) *J. Biol. Chem.* **262,** 14241–14249.
Katada, T., Bokoch, G. M., Northup, J. K., Ui, M., and Gilman, A. G. (1984a) *J. Biol. Chem.* **259,** 3568–3577.
Katada, T., Bokoch, G. M., Smigel, M. D., Ui, M., and Gilman, A. G. (1984b) *J. Biol. Chem.* **259,** 3586–3595.
Katada, T., Kusakabe, K., Oinuma, M., and Ui, M. (1987a) *J. Biol. Chem.* **262,** 11897–11900.
Katada, T., Oinuma, M., and Ui, M. (1986) *J. Biol. Chem.* **261,** 8182–8191.
Katada, T., Oinuma, M., Kusakabe, K., and Ui, M. (1987b) *FEBS Lett.* **213,** 353–358.
Kikuchi, A., Kozawa, O., Kaibuchi, K., Katada, T., Ui, M., and Takai, Y. (1986) *J. Biol. Chem.* **261,** 11558– 11562.
Kuhn, H. (1980) *Nature* **283,** 587–589.
Lerea, C. L., Somers, D. E., Hurley, J. B., Klock, I. B., and Bunt-Milam, A. M. (1986) *Science* **234,** 77–80.
Levitzki, A. (1988) *Trends Biochem. Sci.* **13,** 298–301.
Logothetis, D. D., Kurachi, Y., Galper, J., Neer, E. J., and Clapham, D. E. (1987) *Nature* **325,** 321–326.
May, D. C., Ross, E. M., Gilman, A.G., and Smigel, M. D. (1985) *J. Biol. Chem.* **260,** 15829–15833.
Neer, E. J. and Clapham, D. E. (1988) *Nature* **333,** 129–134.
Neer, E. J., Lok, J. M., and Wolf, L. G. (1984) *J. Biol. Chem.* **259,** 14222–14229.
Neer, E. J., Wolf, L. G., and Gill, D. M. (1987) *Biochem. J.* **241,** 325– 336.
Northup, J. K., Smigel, M. D., Sternweis, P. C., and Gilman, A. G. (1983b) *J. Biol. Chem.* **258,** 11369–11376.
Northup, J. K., Sternweis, P. C., and Gilman, A. G. (1983a) *J. Biol. Chem.* **258,** 11361–11368.
Northup, J. K., Sternweis, P. C., Smigel, M. D., Schleifer, L. S., Ross, E. M., and Gilman, A. G. (1980) *Proc. Natl. Acad. Sci. USA* **77,** 6516–6520.
Robishaw, J. D., Smigel, M. D., and Gilman, A. G. (1986) *J. Biol. Chem.* **261,** 9587–9590.
Roof, D. J., Applebury, M. L., and Sternweis, P. C. (1985) *J. Biol. Chem.* **260,** 16242–16249.
Sternweis, P. C. (1986) *J. Biol. Chem.* **261,** 631–637.
Sternweis, P. C., Northup, J. K., Smigel, M. D., and Gilman, A. G. (1981) *J. Biol. Chem.* **256,** 11517–11526.
Sternweis, P. C. and Robishaw, J. D. (1984) *J. Biol. Chem.* **259,** 13806– 13813.
Strittmatter, S. and Neer, E. J. (1980) *Proc. Nat. Acad. Sci. USA* **77,** 6344–6348.
Yamazaki, A., Tatsumi, M., Torney, D. C., and Bitensky, M. W. (1987) *J. Biol. Chem.* **262,** 9316–9323.
Yamazaki, A., Tatsumi, M., and Bitensky, M. W. (1988) *Methods in Enzymol.* **159,** 702–710.
Yatani, A., Codina, J., Imoto, Y., Reeves, J. P. Birnbaumer, L., and Brown, A. M. (1987) *Science* **238,** 1288–1292.
Yatsunami, K. and Khorana, H. G. (1985) *Proc. Natl. Acad. Sci. USA* **82,** 4316–4320.

Purification of Opioid Binding Proteins

Theresa L. Gioannini and Eric J. Simon

1. Introduction

The existence of cell membrane-bound sites specific for the binding of opiate narcotic analgesics was demonstrated in 1973 by the laboratories of Simon et al. (1973), Terenius (1973), and Pert and Snyder (1973). The binding of radiolabeled opiate agonists and antagonists is saturable, reversible, and of high affinty (10^{-10}–$10^{-7}M$) for drugs that exhibit opiate activity. Furthermore, opiate binding correlates well with pharmacological activity. Considerable evidence has been accumulated suggesting that these binding sites are relevant to the pharmacological activity of opiates and are, therefore, part of the receptors for these drugs. The discovery of endogenous peptides with opiate-like (opioid) activity has resulted in the hypothesis that opiate receptors are the receptor sites for these endogenous peptides. They have, therefore, been renamed opioid receptors to suggest their true function.

As is the case for other neurotransmitter systems, the wide range of pharmacological and physiological effects exerted by different types of opioid ligands is mediated by multiple classes of opioid receptors. Studies in vivo and in vitro (Martin et al., 1976; Gilbert and Martin, 1976; Lord et al., 1977; Smith and Simon, 1980) confirm the existence of at least three major types (mu, delta, kappa), and the possible existence of several other types has been suggested (Vincent et al., 1979; Chang et al., 1981). Opioid receptors are found in the central nervous system and some peripheral tissues of all vertebrates and of some invertebrates. However, these opioid receptors represent a small portion (approx 0.001%) of total cell membrane protein and no highly enriched source of receptors is known. The paucity of available sites and the extreme instability of the binding sites in the presence of various reagents, such as detergents, for a long time hampered progress in the identification at the molecular level of the various components of the opioid receptor system involved in binding and post-binding events. Ultimately, it is the separation of individual protein species followed by the study of their reconstitution into a functioning system that provides the key to confirming the existence of functionally distinct opioid receptors and to an understanding of the mechanism through which the system works.

It has been a primary goal of this laboratory to examine the physical and chemical properties of the opioid receptors on the molecular level. To achieve this, it is necessary to purify the opioid receptor proteins to as near homogeneity as possible. When pursuing the purification of the protein containing the binding site, the most effective method of purification is that of affinity chromatography, which exploits the specificity of ligand–receptor interaction. This chapter will describe the methods used by this laboratory to extract, isolate, and identify an opioid binding protein from bovine striatal membranes. We have combined ligand affinity chromatography with lectin affinity chromatography. The use of the latter became feasible when we discovered that the opioid receptors are glycoproteins (Gioannini et al., 1982b). By this two-column procedure we have successfully purified an opioid binding protein from bovine striatal membranes to apparent homogeneity (Gioannini et al., 1985). Our evidence suggests that the purified receptor is of the mu type. A detailed description of the protocol we have developed is presented in this chapter and will be followed by a brief summary of approaches to the purification of opioid binding sites taken by other laboratories.

2. Solubilization of Active Opioid Binding Sites

The task of detaching opioid receptors from cell membranes and obtaining a species that retains binding activity in solution proved to be a formidable task. Although this laboratory (Simon et al., 1975) in 1975 successfully solubilized a prebound opioid ligand-(etorphine)-macromolecular complex using the nonionic detergent Brij 36T, all attempts to dissociate the complex and obtain opioid binding to soluble protein failed, as did attempts to solubilize an active binding protein by various methods. In recent years, this laboratory (Ruegg et al., 1980,1981; Howells et al., 1982; Gioannini et al., 1982a) and several others (Simonds et al., 1980; Bidlack and Abood, 1980; Cho et al., 1981) have succeeded in developing methods for extracting from membranes receptors that are able to bind ligands reversibly and with high affinity in solution. The procedure initially developed in this laboratory (Ruegg et al., 1980,1981) involved the use of a nonmammalian tissue (toad brain) and the nonionic detergent digitonin. This method resulted in the extraction of active binding sites that represented 50–60% of the membrane-bound site. A modification of this protocol that involved the inclusion of a high concentration of NaCl during solubilization enabled us to extend our success to mammalian tissue. Mammalian tissue solubilized by the modified protocol is routinely used in the purification scheme described here.

2.1. Detergents

An obstacle encountered at the outset and one that has continued to be a source of complications in the purification procedure is the insolubility and inconsistency of commercial digitonin and its interference with a variety of procedures, such as gel electrophoresis and blotting of proteins. The toxicity of this detergent has also plagued our efforts to produce monoclonal antibodies to partially purified opioid binding sites.

Insolubility problems are minimized if digitonin is processed before use by a method described in the literature (Caron et al., 1979). Briefly, a 1.5–2% aqueous solution (w/v) of digitonin is prepared and brought to boiling. The solution is allowed to cool and left to sit overnight. It is then filtered and the clear solution is lyophilized. The resulting fluffy, white powder is water soluble and can be stored indefinitely.

An alternative to digitonin is the zwitterionic detergent CHAPS that has been reported by a number of groups to effectively solubilize opioid receptors (Simonds et al., 1980; Maneckjee et al., 1985). We have obtained inconsistent results using CHAPS alone, but the protocol used for digitonin, i.e., including high concentrations of NaCl during solubilization, has provided yields of active CHAPS-solubilized binding sites comparable to those obtained with digitonin. Indeed, CHAPS may be used in our purification scheme. However, digitonin-solubilized material is routinely used in our laboratory because we have found CHAPS-solublilized material to be unstable upon storage and to give somewhat lower yields of active pure protein.

2.2. Protocol for Solubilization

The methods described are applicable to any tissue containing reasonable levels of opioid receptors. In our laboratory we use bovine striatum as the source of opioid binding sites since fresh bovine brains are readily obtained from the slaughterhouse and the striatum is the brain region that contains by far the highest level of opioid receptors. It is dissected out immediately upon arrival and kept cold. Crude membrane fraction is prepared (Lin and Simon, 1978) by an initial homogenization in 50 m*M* Tris HCl, 1 m*M* K_2EDTA, pH 7.4 at a 1:10 dilution (w/v). After centrifugation for 15 min at 20,000*g* at 4°C, the pellet is resuspended in 0.32*M* sucrose at 1:6 dilution (w/v) and stored at –70°C until needed. For solubilization, membranes are thawed and diluted with an equal volume of 50 m*M* Tris HCl, 1 m*M* K_2EDTA, 0.5*M* NaCl, pH 7.4 (buffer A). This solution is then diluted 1:1 with 1% digitonin in buffer A. The membranes are solubilized by shaking at 4°C for 30 min and then spun at 100,000*g* for 35 min at 4°C. The clear supernatant is removed and either used immediately or frozen at –70°C until needed. When the solubilization is carried out with CHAPS, the final concentration of detergent during solubilization is 5 m*M* CHAPS in buffer A (i.e., the buffer containing 0.5*M* NaCl).

The specific binding activity of detergent-solubilized extracts (as well as for samples obtained from affinity columns) is assayed for opioid receptor activity by incubation with either the antagonist ^{3}H-diprenorphine or the universal ligand ^{3}H-bremazocine, in the presence and absence of unlabeled naloxone (2 μ*M*) for 45 min at 25°C. Since digitonin as well as CHAPS at high concentrations inhibit binding, samples are diluted, when necessary, with buffer A to give a detergent concentration of 0.1% or less of digitonin or 1 m*M* or less of CHAPS. For all samples, drug bound to the solubilized material is separated from free ligand by a modification of the polyethyleneglycol (PEG) precipitation method originally described by Cuatrecasas (1972). PEG (8,000 mol wt) solution and the carrier protein, bovine gamma-globulin, are added to give a final concentration of 11 and 0.2%, respectively. After samples have been vigorously vortexed, they are filtered through GF/B (Whatman) glass fiber filters that are then rinsed with PEG solution (7.5%). Radioactivity collected on the filters is measured by scintillation spectroscopy. Recently, we have found that binding to soluble receptors may be determined by eliminating the PEG precipitation protocol and simply substituting filtration through GF/B filters soaked for at least 1 h in polyethyleneimine (0.3%) followed by a rinse (Bruns et al., 1983). Results similar to those with PEG precipitation are obtained.

3. Ligand Affinity Chromatography

3.1. Synthesis of the Affinity Matrix

3.1.1. The Ligand

As indicated earlier, affinity chromatography offers a number of advantages over conventional methods of chromatography that separate species on the basis of size, charge, hydrophobicity, sugar content, and so on. A matrix is developed based on the specific requirements of the ligand–receptor interaction by incorporating a ligand of moderate affinity into a solid support. For this purpose, we have synthesized a ligand (Gioannini et al., 1984) that is a derivative of the potent antagonist naltrexone, equipped with a free amino terminus, useful for coupling the opioid to the carboxyl end of the spacer arm on the gel CH-Sepharose 4B (Pharmacia) (Fig. 1). The ketone carbonyl in the 6 position of naltrexone is coupled to ethylenediamine via reductive amination in the presence of sodium cyanoborohydride at pH 8.8 (Borch et al., 1971). The compound synthesized, β-6-desoxydiaminoethylnaltrexone (NED, naltrexylethylenediamine), contains a free primary amino group, retains high affinity for membrane-bound receptors, and, like its parent, exhibits a preference for mu receptors. In rat brain membranes the concentration of NED needed to displace 50% of ^{3}H-naltrexone (1 n*M*) binding is 15 n*M*.

3.1.2. The Ligand-Containing Matrix

CH-Sepharose 4B contains a 6-ω-carboxyl group on a methylene chain that allows the free amino group of NED to be easily attached, using the water-soluble carbodiimide, 1-ethyl-3-(3-dimethylaminopropyl) carbodiimide. The final concentration of ligand incorporated by this method is 2–4 μmol/mL of gel. This was determined by incorporating tritiated naltrexone during the synthesis of NED and estimating the concentration of immobilized ligand by counting a portion of the gel in a liquid scintillation spectrometer. We have determined that CH-Sepharose 4B alone does not interact with soluble binding sites. However, a significant proportion of applied opioid binding sites, generally 60–70%, is adsorbed on the matrix into which the ligand has been incorporated. NED CH-Sepharose 4B is stable, reusable, has very little if any nonspecific interactions, and interacts efficiently, and biospecifically with detergent-solubilized opioid binding sites.

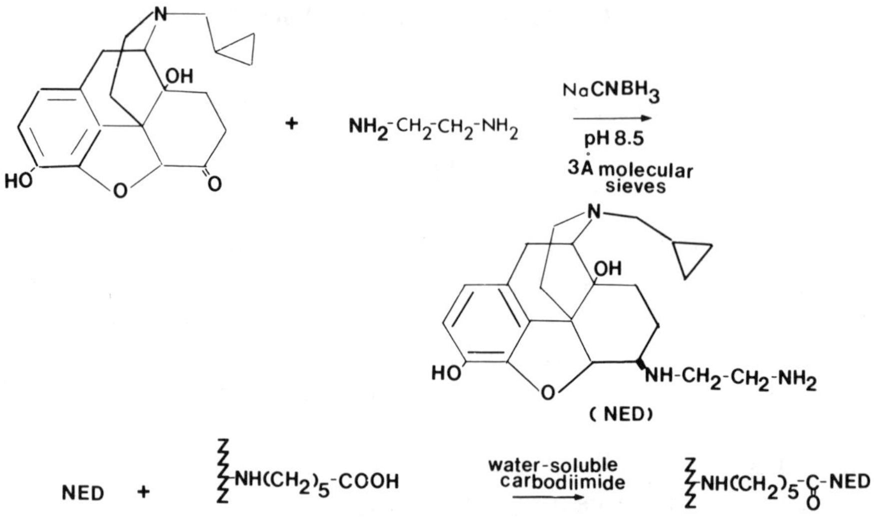

Fig. 1. The synthesis of a ligand affinity matrix for purification of opioid binding sites.

3.2. Adsorption to and Elution from NED CH-Sepharose

In the purification procedure used routinely in this laboratory, both adsorption to and elution of solubilized opioid receptor preparations from the affinity gel are performed in a batchwise manner. Solubilized material is diluted 1:1 with buffer A and then incubated with gel for 1 h at 25°C (5 mL soluble/mL of gel) or overnight at 4°C. The flowthrough is collected by pouring the gel into 2.5×30 cm columns. The gel is then rinsed with buffer A containing 0.05% digitonin (40–50 times the gel volume), followed by one gel volume of buffer A containing 0.2% digitonin. The retained binding sites are eluted by two incubations of the gel with buffer A containing 2.5 μ*M* naloxone and 0.1% digitonin for 60 min at 25°C. The yield of purified binding sites eluted from the gel is usually about 15–25% of applied. It is evaluated on a portion of the eluate by the binding assay described. Before binding, it is necessary to remove the naloxone. This is done by treatment of eluate with BioBeads SM2 (BioRad), which have been previously incubated and washed in buffer containing 0.1% digitonin.

3.3. Properties of Binding Sites Purified on NED CH-Sepharose

A single pass through the affinity column results in a purified protein preparation that binds opioids with an activity that is 4000–7000-fold greater than that of membrane-bound receptors (>900 pmol/mg protein in

Table 1
Purification of an Opioid Binding Protein from Bovine Striatum[a]

Preparation	Protein, mg	Activity,[b] pmol	Step yield, %		Overall yield, %		Specifity activity, B_{max}, pmol • mg^{-1}	Purification factor, x-fold
			Protein	Activity[b]	Protein	Activity[b]		
Membrane homogenate	3270	667	100	100	100	100	0.2	—
Digitonin extract	1308	258	40	38	40	38	0.19	1
Eluate from NED-Sepharose	0.070[c]	64	<0.001	24.8	<0.001	9.6	914	4570
Eluate from WGA-Agarose	0.003[d]	39	4.3	60.9	<0.001	5.8	13,000	65,000

[a]Data are from a typical experiment in which ~50 g of tissue was homogenized and solubilized with digitonin.

[b]Measured by [^{3}H]bremazocine binding at 1.5 n*M* and B_{max} calculated from saturation curves at identical conditions (0.5*M* NaCl, 0.05% digitonin). Values shown represent specific binding at saturation.

[c]Determined by densitometric scanning of Coomassie blue-stained SDS-PAGE gels using bovine serum albumin and carbonic anhydrase as calibration standards.

[d]Determined by amino acid analysis.

the purified vs 0.2 in the membrane) (Table 1). The affinity-purified material binds opioid antagonists stereospecifically in a saturable, reversible manner with affinities slightly lower than those seen with membrane-bound receptors. For example, a saturation curve for bremazocine binding gives a $K_d = 2.5 \pm 0.8$ n*M*, as compared to an affinity of approx. 1 n*M* to membrane-bound receptors. Opioid agonists, such as the selective mu agonist, DAGO, bind with quite low affinity. This loss of affinity of agonist binding is not understood. It may be a result of the loss of an essential lipid component or to uncoupling of the binding site from G protein. There is, in fact, evidence that affinity-purified binding sites are uncoupled from G protein since ^{3}H-diprenorphine, which is significantly inhibited by GTP in both membrane-bound and crude solubilized receptors, is no longer affected by GTP after affinity purification (Y. Sarne and E. J. Simon, unpublished results).

Under nondenaturing conditions on gel exclusion chromatography (Ultragel AcA 34), the opioid binding material elutes with a molecular mass of 300–350kDa and retains its ability to interact with wheat germ agglutinin agarose (WGA). Undoubtedly, a portion of this high molecular mass may be attributed to receptor–digitonin complexes or micelles. The actual amount of digitonin bound per molecule of receptor is not known since conventional methods for calculating detergent binding to protein are inapplicable with digitonin (Steele et al., 1978). On SDS-PAGE, i.e., under denatured conditions, the eluate from the NED-agarose matrix consists of 6–8 protein bands as determined by autoradiography of a previously radioiodinated preparation or by silver-staining of the gel.

4. Lectin Affinity Chromatography

This laboratory has investigated the carbohydrate content of opioid receptors by examining the interaction of digitonin-solubilized binding sites with a number of lectin-agarose gels. We reported (Gioannini et al., 1982) that, of 9 lectins examined, only WGA-agarose consistently retained a significant portion (35–45%) of the applied binding sites, which could be specifically eluted by *N*-acetylglucosamine. These studies demonstrate the glycoprotein nature of the opioid receptors and indicate the presence of either exposed *N*-acetylglucosamine or sialic acid residues. Furthermore, since either free or ligand-bound soluble receptors could be adsorbed and eluted from WGA-agarose, the independence of the ligand and lectin binding sites is established. When crude soluble opioid binding sites are adsorbed on and eluted from WGA-agarose (with *N*-acetylglucosamine), they are enriched 25–35-fold over the crude soluble preparation, as determined by the specific binding of the opioid ligand, ^{3}H-diprenorphine.

Based on these studies, lectin affinity chromatography on WGA-agarose is used as an additional purification step. The affinity-purified binding sites are incubated batchwise at 4°C with WGA-agarose for 2 h (20 mL NED eluate/mL WGA-agarose). The unadsorbed material is collected by placing the beads and supernatant in a column and collecting the flow-through. The gel is washed at 4°C with 20–30 gel volumes of buffer A containing 0.05% digitonin. Finally, active binding sites are eluted twice with one gel volume of 0.5*M* *N*-acetylglucosamine in buffer A containing 0.05% digitonin by incubation at 4°C for 45 min. The eluate is assayed for opioid binding according to the protocol previously described in Section 2. The yield of binding activity from chromatography on WGA-agarose is 61 ± 4% of the binding activity in the affinity-purified material applied to the matrix. This step was shown to provide a further 10–20-fold purification of the binding protein, giving an overall purification of 60,000–75,000-fold.

5. Properties of Purified Protein

When material purified by the above two column procedure is analyzed on SDS-PAGE, only one major protein band is detected, by either silver staining or autoradiography of previously radio-iodinated purified material. Under nonreducing conditions, the protein migrates with an apparent M_r = 53kDa. However, in the presence of high concentrations of DTT, the mobility of the band shifts to 65kDa. This type of shift is indicative of the breakage of internal disulfide bonds in the denatured protein that causes it to be less globular and therefore more restricted in its mobility through the gel. More recent studies with the purified binding protein in the presence of varying concentrations of DTT indicate the breakage of multiple disulfide bonds since increasing concentrations of DTT from 0.6 to 65 m*M* produce a gradual increase in the apparent M_r until 65kDa is reached.

The other bands apparent in the eluate from the NED–agarose gel remain in the flow-through of the lectin column. The purified binding protein binds opioids reversibly and stereospecifically with a specific activity that is 60,000–70,000-fold greater than that of membrane-bound or crude soluble receptors (12,000–15,000 pmol/mg protein vs 0.20 pmol/mg). Theoretical purification to achieve homogeneity, assuming 1 binding site/53kDa glycoprotein, is 63,000-fold.

Affinity crosslinking of bovine striatal membranes with ^{125}I-β-endorphin using the bifunctional crosslinker bis-2-(succiniimidooxy-carbonyloxy) ethyl sulfone (BSCOES) produces four specifically labeled peptides upon autoradiography of SDS-PAGE gels (Howard et al., 1985). The major peptide labeled migrates with an apparent M_r = 65kDa. Evidence suggests that this specifically labeled peptide and the protein purified from the ligand and lectin affinity columns are the same. They migrate with similar apparent molecular masses; the diffuseness of the bands is such that the crosslinked species, despite the additional 3kDa caused by the bound β-endorphin, would be indistinguishable from the unbound protein in this SDS-PAGE system. Both proteins are specifically retained by WGA-agarose and specifically eluted with *N*-acetylglucosamine. Furthermore, the purified binding protein can be bound with ^{125}I-β-endorphin and crosslinked specifically. SDS-PAGE of this material reveals a single band, M_r = 65kDa.

The following information supports the suggestion that the opioid binding site purified is of the mu type. The probability of retaining delta-type binding sites on the matrix is low since they represent only about 10% of the population of opioid receptors in bovine striatal tissue. The ratio of mu:delta:kappa in this tissue was shown to be 50:10:40 (J. M. Hiller and E. J. Simon, unpublished results). In addition, the affinity matrix contains a derivative of naltrexone that is a mu-preferring antagonist. A highly selective ligand for mu receptors, Tyr-D-Ala-Gly-MePhe-Gly-ol (DAGO) displaces 70% of ^{3}H-bremazocine binding in the affinity-purified receptor, though a concentration of 500 n*M* is required, since as mentioned, agonists bind with very low affinity. Finally, ^{125}I-β-endorphin, which binds to mu and delta receptors but not to kappa receptors, can bind to the purified WGA-agarose eluate.

Several reasons may account for the low affinity of purified binding sites for agonists: The high concentrations of NaCl and detergent to which agonist binding is extremely sensitive (Itzhak et al. 1984), possible loss of lipids essential for binding, and uncoupling of the opiate binding protein from G protein.

This laboratory is currently in the process of trying to reconstitute the purified binding protein into liposomes together with purified G_i and G_o proteins. We hope that the restoration of high affinity agonist binding will be achieved. Such a system will also provide a simplified environment for studies of interactions of binding sites with G proteins and second messenger systems.

6. Different Approaches to Purification of Opioid Binding Sites

This section will briefly summarize a number of different approaches that have been developed by several other laboratories to purify opioid receptors and have resulted in varying degrees of purification. The earliest partial purification was reported by Bidlack et al. (1981) and utilized an affinity gel prepared by coupling 14-β-bromoacetamido-morphine to aminohexyl Sepharose. Receptors solubilized from rat brain with Triton X-100 were eluted with 1 μ*M* levorphanol or etorphine. The purified material bound 40 pmol of opioid ligand per mg protein, a 200–400-fold purification over membrane preparations. SDS-PAGE of the affinity-purified preparation revealed three protein bands of M_r = 43, 35, and 23kDa.

Maneckjee et al. (1985) have also reported a partial purification of mu opioid receptors solubilized with CHAPS from rat brain mitochondrial/synaptosomal membranes. The material was purified on a support prepared by coupling Hybromet (Archer et al., 1985), a thebaine analog carrying an HgBr group, to Affigel 401, an affinity matrix containing SH groups. The degree of purification was 506-fold with an overall yield of 0. 6%. In SDS-PAGE, a major band of M_r = 94kDa and minor bands of M_r = 44 and 35 kD were seen. Cho et al. (1986) have also reported purification of an opioid binding protein to apparent homogeneity from rat brain using a modification of an earlier method (1983) that resulted in 3200-fold purification of the protein. The opioid receptors are solubilized with Triton X-100 and then applied to Affigel 102 linked to succinylmorphine, an affinity matrix first developed by Simon et al. (1972). Affinity chromatography is followed by gel filtration on AcA 34, lectin chromatography on WGA-agarose, and isoelectric focusing. This produces material that has a pI of 4.4 and an M_r = 58kDa on SDS-PAGE as determined by autoradiography of iodinated material. To obtain specific opioid binding to the soluble receptor and at each step of the purification, a mixture of acidic lipids had to be added during incubation with the ligand. Why this lipid requirement is observed here but not in the purification methods used by others is not clear.

A 6-succinylmorphine-containing support (Affigel 102 or AF amino-TOYOPEARL 650*M*) was also employed by Ueda et al. (1987,1988) with Triton X-100-solubilized binding sites from rat brain. This step provided a 500–1500-fold purification over membrane-bound receptors that was further improved by the addition of an isoelectic focusing chromatography step. From this a single protein with a pI of 5.6 and an M_r = 58kDa was obtained.

Another partial purification of digitonin-solubilized binding sites was reported by Fujioka et al. (1985), who constructed an affinity matrix using an enkephalin coupled to aminohexyl Sepharose. The adsorbed binding sites were eluted in a batchwise manner with D-Ala-D-Leu-enkephalin and the eluate then fractionated on Sephadex G75. The isolation resulted in a 450-fold purification of material that bound the delta-preferring ligand D-Ala-D-Leu-enkephalin with a K_d = 34 n*M* and a B_{max} = 200 pmol/mg protein. Major bands of M_r = 62 and 39kDa were obtained from SDS-PAGE analysis.

Demoliou-Mason and Barnard (1984) reported solubilization of opioid binding sites from rat brain membranes by a modification of the digitonin procedure. They reported that preincubation of the membranes with Mg^{2+} (10 m*M*) and omission of NaCl permitted the solubilization of binding sites capable of binding mu, delta, and kappa agonists, including peptides. There have been oral reports from this laboratory of the purification to apparent homogeneity of a mu binding site, using two ligand affinity columns. The first step used a novel ligand, D-Ala^2-Leu-enkephalin chloromethylketone (DALECK) (Newman and Barnard, 1984), the second a derivative of codeinone, both immobilized on an agarose support. This work has not yet been published.

A covalently labeled delta-binding opioid protein has been isolated from NG 108-15 cells that contain predominantly delta binding sites by the laboratory of Klee (Klee et al., 1982; Simonds et al., 1985). The protocol for isolation involved

1. Covalent labeling of the receptors with tritiated 3-methylfentanyl- isothiocyanate (SuperFIT), a delta-selective ligand;
2. Solubilization in a mixture of lubrol and CHAPS;
3. Adsorption to and elution from WGA-agarose;
4. Immunoaffinity chromatography on Sepharose 4B containing anti-FIT antibodies; and
5. Preparative SDS-PAGE.

The resulting protein was purified 30,000-fold in a yield of 2–3% and contained 1 mol ligand/mol of protein, i.e., 21,000 pmol/mg protein, and had an apparent M_r = 58kDa on SDS-PAGE.

There has been no purification as yet of a kappa receptor from mammalian sources. However, solubilization with digitonin and partial purification of a receptor from the brain of the frog, *Rana Esculenta*, using ligand affinity and gel permeation chromatography on Sepharose 6B has been reported by Simon et al. (1987). The affinity column consisted of D-Ala^2-Leu^5-enkephalin immobilized on epoxy-activated Sepharose 6B. An enrichment of 4300-fold was obtained. These investigators suggest that this receptor is similar, if not identical, in its properties to the mammalian kappa receptor. Mollereau et al. (1988), on the other hand, who used a dynorphin affinity column for the purification of a digitonin-solubilized opioid binding site from the brain of the frog, *Rana Ridibunda*, suggest that this receptor is different from any of the three major types of opioid receptors present in mammalian tissues.

7. Conclusions

This chapter has described in detail the purification to homogeneity of an opioid binding protein from bovine striatal membranes. Digitonin-solubilized opioid receptors are purified by a combination of ligand and lectin chromatography to produce a single active binding protein that under denatured and reducing conditions migrates with M_r = 65kDa. This laboratory is in the process of producing purified protein on a large scale for the purpose of obtaining amino acid sequence from this protein in order to prepare oligonucleotide probes for the screening of cDNA libraries. Since it has been determined that the *N*-terminal is blocked and fragmentation is necessary before sequencing, a relatively large quantity of receptor is required. A number of technical difficulties have been encountered owing to the notorious characteristics of digitonin that are multiplied upon concentration of the sample for fragmentation as well as to problems routinely encountered in the scaleup of any protocol involving protein purification, e.g., lower yields and greater chances for contamination. Nevertheless, sequences of some peptide fragments have been obtained to which polyclonal antibodies are being generated and, more importantly, oligonucleotide probes are being synthesized for the isolation of cDNA complementary to receptor message from a bovine striatal cDNA library. This will permit sequencing of the cDNA and will yield the complete amino acid sequence of the protein.

In addition, the purified opioid binding site must be reconstituted into liposomes with specific G proteins and into functioning cell systems devoid of opioid receptors, so that the characteristics of the ligand–receptor interaction may be examined in a controlled environment. These types of experiments will provide clues to steps triggered by ligand–receptor interaction as well as further our understanding of the endogenous opioid system and its physiological role.

References

Archer, S., Michael, J., Osei-Gyimah, P., Seyed-Mozaffari, A., Zukin, S., Maneckjee, R., Simon, E. J., and Gioannini, T. L. (1985) *J. Med. Chem.* **28,** 1950–1953.

Bidlack, J. M. and Abood, L. G. (1980) *Life Science* **27,** 331–340.

Bidlack, J. M., Abood, L. G., Osei-Gyimah, P., and Archer, S. (1981) *Proc. Natl. Acad. Sci. USA* **78,** 636–639.

Borch, R. F., Bernstein, M. D., and Durst, H. D. (1971) *J. Amer. Chem. Soc.* **93,** 2897–2904.

Bruns, R. F., Lawson-Wendling, K., and Paigsley, T. A. (1983) *Anal. Biochem.* **132,** 74–81.

Caron, M. G., Srinwasan, Y., Patha, J., Kociolek, K., and Lefkowitz, R. J. (1979) *J. Biol. Chem.* **254,** 2923–2927.

Chang, K.-J., Hazum, E., and Cuatrecasas, P. (1981) *Proc. Natl. Acad. Sci. USA* **78,** 4141–4145.

Cho, T. M., Yamato, C., Cho, J. S., and Loh, H. H. (1981) *Life Sci.* **28,** 2651–2657.

Cho, T. M., Ge, B. L., Yamato, C., Smith, A. P., and Loh, H. H. (1983) *Proc. Natl. Acad. Sci. USA* **80,** 5176–5180.

Cho, T. M., Hasegawa, J.-I., Ge, B. L., and Loh, H. H. (1986) *Proc. Natl. Acad. Sci. USA* **83,** 4138–4142.

Cuatrecasas, P. (1972) *Proc. Natl. Acad. Sci. USA* **69,** 318–322.

Demoliou-Mason, C. D. and Barnard, E. A. (1984) *FEBS Lett.* **170,** 378–382.

Fujioka, T., Inoue, F., and Kurujama, M. (1985) *Biochem. Biophys. Res. Commun.* **131,** 640–646.

Gioannini, T. L., Howells, R. D., Hiller, J. M., and Simon, E. J. (1982a) *Life Sci.* **31,** 1315–1318.

Gioannini, T. L., Foucaud, B., Hiller, J. M., Hatten, M. E., and Simon, E. J. (1982b) *Biochem. Biophys. Res. Commun.* **105,** 1128–1134.

Gioannini, T. L., Howard, A., Hiller, J. M., and Simon, E. J. (1984) *Biochem. Biophys. Res. Commun.* **119,** 624–629.

Gioannini, T. L., Howard, A. D., Hiller, J. M., and Simon, E. J. (1985) *J. Biol. Chem.* **260,** 15117–15121.

Gilbert, P. E. and Martin, W. R. (1976) *J. Pharmacol. Exp. Ther.* **198,** 66–82.

Howard, A. D., de la Baume, S., Gioannini, T. L., Hiller, J. M., and Simon, E. J. (1985) *J. Biol. Chem.* **260,** 10833–10839.

Howells, R. D., Gioannini, T. L., Hiller, J. M., and Simon, E. J. (1982) *J. Pharmacol. Exp. Ther.* **222,** 629–634.

Itzhak, Y., Hiller, J. M., Gioannini, T. L., and Simon, E. J. (1984) *Brain Res.* **291,** 309–315.

Klee, W. A., Simonds, W. F., Sweat, F. W., Burke, Jr., T. R., Jacobson, A. E., and Rice, K. C. (1982) *FEBS Lett.* **150,** 125–128.

Lin, H. K. and Simon, E. J. (1978) *Nature (London)* **271,** 383–384.

Lord, J. A. H., Waterfield, A. A., Hughes, J., and Kosterlitz, H. W. (1977) *Nature* (*London*) **267,** 495–499.

Maneckjee, R., Zukin, R. S., Archer, S., Michael, J., and Osei-Gyimah, P. (1985) *Proc. Natl. Acad. Sci. USA* **82,** 594–598.

Martin, W. R., Eades, C. G., Thompson, J. A., Hoppler, R. E., and Gilbert, P. E. (1976) *J. Pharmacol. Exp. Ther.* **197,** 517–532.

Mollereau, C., Pascaud, A., Baillat, G., Mazarguil, H., Puget, A., and Meunier, J.-C. (1988) *Eur. J. Pharmacol.* **20,** 75–84.

Newman, E. L. and Barnard, E. A. (1984) *Biochemistry* **23,** 5385–5389.

Pert, C. B. and Snyder, S. H. (1973) *Science* **179,** 1011–1014.

Ruegg, U. T., Hiller, J. M., and Simon, E. J. (1980) *Eur. J. Pharmacol.* **64,** 367–368.
Ruegg, U. T., Cuenoud, S., Hiller, J. M., Gioannini, T. L., Howells, R. D., and Simon, E. J. (1981) *Proc. Natl. Sci. USA* **78,** 4635–4638.
Simon, E. J., Dole, W. P., and Hiller, J. M. (1972) *Proc. Natl. Acad. Sci. USA* **69,** 1835–1837.
Simon, E. J., Hiller, J. M., and Edelman, I. (1973) *Proc. Natl. Acad. Sci. USA* **70,** 1947–1949.
Simon, E. J., Hiller, J. M., and Edelman, I. (1975) *Science* **190,** 389–390.
Simon, J., Benyhe, S., Hepp, J., Khan, A., Borsodi, A., Szucs, M., Medzihradszky, K., and Wollemann, M. (1987) *Neuropeptides* **10,** 19–28.
Simonds, W. F., Koski, G., Streaty, R. A., Hjemlmeland, L. M., and Klee, W. A. (1980) *Proc. Natl. Acad. Sci. USA* **77,** 4623–4627.
Simonds, W. F., Burke, T. R., Jr., Rice, K. C., Jacobson, A. E., and Klee, W. A. (1985) *Proc. Natl. Acad. Sci. USA* **82,** 4974–4978.
Smith, J. R. and Simon, E. J. (1980) *Proc. Natl. Acad. Sci. USA* **77,** 281–284.
Steele, J. C., Tanford, C., and Reynolds, J. A. (1978) *Methods Enzymol.* **48,** 11–29.
Terenius, L. (1973) *ACTA Pharmacol. Toxicol.* **32,** 317– 320.
Ueda, H., Harada, H., Misawa, H., Nozaki, M., and Takagi, H. (1987) *Neuroscience Lett.* **75,** 339–344.
Ueda, H., Harada, H., Nozaki, M., Katada, T., Ui, M., Satoh, M., and Takagi, H. (1988) *Proc. Natl. Acad. Sci. USA* **85,** 7013–7018.
Vincent, J. P., Kartolovski, B., Geneste, P., Kamemka, J. M., and Lazdunski, M. (1979) *Proc. Natl. Acad. Sci. USA* **76,** 4578–4582.

μ-Opioid Receptor

Purification and Reconstitution with GTP-Binding Protein

Hiroshi Ueda and Masamichi Satoh

1. Introduction

It was reported in the 1960s that opioids show various actions through specific opioid receptors in in vitro studies (Kosterlitz and Watt, 1968). In 1971 and 1973, opiate receptors were reported by several independent investigators, in a ligand binding study in brain membranes (Goldstein et al., 1971; Pert and Snyder, 1973; Simon et al., 1973; Terenius, 1973). The study of opioid receptors was accelerated by the discovery of endogenous opioids, such as enkephalins (Hughes et al., 1975), β-endorphin, dynorphin, and their related peptides (Goldstein, 1984; Yamashiro and Li, 1984). Biochemical, pharmacological, and behavioral studies provided the evidence that various actions of opiates and opioid peptides in the brain and peripheral tissues are mediated by interactions with different subtypes of opioid receptors, such as μ-, δ-, and κ-opioid receptors (Martin et al., 1976; Lord et al., 1977). Furthermore, the discovery of selective ligands for each subtype of opioid receptor (Paterson et al., 1984) made it easy to study the receptor subtype-specific responses in vivo and in vitro. For instance, in in vivo studies, the μ-opioid receptor is thought to be closely relevant to the production of analgesia in the brain (Satoh et al., 1983; Porreca et al., 1984). On the other hand, there are accumulating findings in vitro that μ- and δ-receptors are functionally coupled to the stimulation of potassium ion channel activity (North et al., 1987), whereas κ-receptor indicates an inhibition of calcium ion channel activity (Gross and Macdonald, 1987). In addition, μ- and δ-receptors are functionally coupled to the inhibition of adenylate cyclase in membranes of the brain and NG108-15 cells (Collier and Roy, 1974; Klee and Nirenberg, 1976).

In the early 1980s, guanine nucleotide binding proteins (G-proteins) were identified, and their physiological roles in the "signal transduction system" were claimed (*see* Gilman, 1987; Neer and Clapham, 1988). According to current models (Gilman, 1987), the stimulation of the receptor by agonist signals is coupled to a stimulation (or inhibition) of various intracellular effector systems through an activation (or inhibition [*see* Ueda et al., 1987,1989]) of G-protein activity. Above all, the discovery of islet-activating protein (IAP), pertussis toxin, is the most important finding in understanding the post receptor mechanisms (Ui, 1984). This toxin is known to uncouple the receptor–G-protein interaction by ADP-ribosylation of α-subunits of G-proteins, such as G_i or G_o (Katada et al., 1986). Such an involvement of G-protein was the case with opioid effects (Burns et al., 1983; Ui, 1984; Kurose and Ui, 1985). The inhibition of adenylate cyclase by opioids in NG 108-15 cells through an activation of G_i is a famous example (Kurose et al., 1983). However, it still remains to be determined whether various actions of opioid effects, including "ionotropic" (ion channel related) actions, are mediated by G-proteins. In addition, whether or not each subtype of opioid receptor shares the same signal transduction system is a most important question. Accordingly, the molecular basis of such a receptor heterogeneity must be studied.

Lowney et al. (1974) and Simon et al. (1975) first reported the solubilization of opioid receptors. Attempts to purify the opioid receptors by use of affinity chromatography were performed by several investigators (Simonds et al., 1985; Maneckjee et al., 1985; Gioannini et al., 1985; Fujioka et al., 1985; Cho et al., 1986; Ueda et al., 1987,1988). However, there seem to exist some differences in properties between the "purified" opioid receptors. Gioannini et al. (1985) reported that purified μ-opioid receptor using an affinity chromatography of β-naltrexylethylenediamine-CH-Sepharose 4B, is 65kDa. Cho et al. (1986) reported that μ-opioid receptor purified using 6-succinylmorphine-Affi-Gel 102 was 58kDa and its pI was 4.4. We also reported the purification of μ-opioid receptor using 6-succinylmorphine-Affi-Gel 102 or 6-succinylmorphine-AF-aminoTOYO-PEARL 650*M* (Ueda et al., 1987a,1988a). Our purified μ-opioid receptor was of 58kDa, but the pI was 5.6. Most recently, we succeeded in the functional reconstitution of purified μ-opioid receptor and purified G-protein. Through these experiments, it appeared that our purified opioid receptor is of μ-subtype and still retains its functional activity. In this chapter, the purification of μ-opioid receptor and reconstitution with G-proteins in our laboratory are described (Ueda et al., 1988a).

2. Solubilization of μ-Opioid Receptor

There are several reports on solubilization of opioid receptors, using Triton X-100 (Cho et al. 1986; Ueda et al., 1987), 3-(3-cholaminopropyl) dimethylammonio-l-propanesulfonate/CHAPS (Simonds et al., 1985), and digitonin (Gioannini et al., 1985; Ueda et al., 1988a). When we use detergents for this purpose, we must be aware of the potential problem that such detergents may inhibit the opioid binding. This problem seems to be more important than the yield of receptors through solubilization, since the binding of solubilized opioid receptor to the opioid ligand-coupled affinity chromatography is a key step in purification. At first, we followed the purification procedures reported by Cho et al. (1986). In such procedures, opioid receptors were solubilized in Triton X-100 after sonication in the presence of sucrose. However, since Triton X-100, even at a concentration of 0.01%, inhibited the [^{3}H]naloxone binding in membrane preparations, and the yield of μ-opioid receptor was only 10% of starting materials (synaptic membranes of rat brains), we replaced Triton X-100 by digitonin for solubilization of opioid receptor.

Digitonin did not inhibit the binding below 0.1% in membrane preparations, and the yield of μ-opioid receptor was 30% in the case of 1% digitonin. This yield was the same as the case of 1% CHAPS, whereas this detergent markedly inhibited the [^{3}H]naloxone binding at concentrations >0.05%. After all, we used highly water-soluble grade of digitonin (Wako Chemicals, Osaka, Japan). In our experiments, sonication at 4°C for 10 min prior to addition of 1% digitonin was carried out in the presence of 0.32*M* sucrose and protease inhibitors, such as 0.002% soybean trypsin inhibitor, 1 μ*M* leupeptin, 0.2 μ*M* phenylmethylsulfonyl fluoride, and 0.01% bacitracin, in order to aid solubilization. We preliminarily found that the pretreatment with sonication could reduce amounts of digitonin required for solubilization of membrane proteins.

3. Purification Procedures

We used 6-succinylmorphine AF-aminoTOYOPEARL 650*M* for affinity column chromatography. Although the displacement of [^{3}H]naloxone binding in rat brain membranes by 6-succinyl-morphine was 5–10 times less potent than that by morphine, it is likely that such a derivative of morphine is still valid for affinity chromatography. In the preparation of this derivative, morphine (free base, 2 g) and succinic acid anhydrate (2 g) were reacted at 120°C for 3 h in benzene with a reflux. The precipitates were dried, dissolved in water, and lyophilized. The products were redissolved in acidic water at pH 1–2. The mixture was added by NaOH to adjust at pH 9.5 in order to remove unreacted morphine and to hydrolyze the phenolic succinylester at 3-position of 3,6-disuccinylmorphine, and then readjusted at pH 5.5 to obtain 6-succinylmorphine in precipitates. 6-Succinylmorphine (2 g) was then reacted at pH 4.7–5.0 overnight at room temperature with 100 mL of AF-aminoTOYOPEARL 650M (TOSOH, Tokyo, Japan), using 1-ethyl-3-(3-dimethylamino-propyl)carbodiimide hydrochloride (1 g). Amounts of morphine derivatives conjugated to the gel were 1–10 μmol/mL gel. Succinylmorphine was extracted in 5*N* HCl from the gel and aliquots (200 μL) were incubated with 1 mL of α-nitroso-β-naphthol (60 mg/100 mL of 80% acetic acid), 0.4 mL of potassium nitrate (10 g/100 mL), and 0.2 mL of sodium nitrite (20 mg/100 mL) at 25°C for 40 min. After the addition of 1 mL of ammonium sulfamate (1 g/100 mL), 2 mL of chloroform was added to remove α-nitroso-β-naphthol. The aqueous solution containing the 6-succinylmorphine derivative (red) was used for colorimetry at 530 nm. Most recently, we improved the preparation procedure for 6-succinylmorphine-conjugated gel. In such a procedure, morphine (2 g) in free base was reacted with succinic anhydrate (0.7 g) in pyridine at room temperature for 2 h. Then, 100 mL of AF-aminoTOYOPEARL 650*M*, that was equilibrated in pyridine, and dicyclohexylcarbodiimide (2 g) were added to the mixture containing succinyl derivatives of morphine, and incubated overnight at room temperature. The gel was then washed with acetone, followed by 20 m*M* Tris-HCl buffer, pH 7.5 (buffer A). The amounts of morphine derivatives conjugated to the gel were 25–50 μmol/mL gel. This procedure seems to have the advantage that all chemical reactions are carried out rapidly under mild condition. In both procedures, to prepare 6-succinylmorphine, the treatment must be quick, since the compound or its solution rapidly becomes yellow or brown, possibly owing to chemical changes in the phenolic ring.

It is well known that the agonist binding is markedly reduced in the presence of GTP or its analogs. According to current models (Gilman, 1987), the binding of GTP (in exchange for GDP) to G-protein accelerates the dissociation of G-protein from the receptor, resulting in reducing the agonist binding but not antagonist binding. In other words, the agonist binding to the receptor in the state of uncoupling with G-proteins is markedly reduced. Therefore, when we use agonist (morphine)-coupled affinity chromatography, the solubilized materials must be applied to the

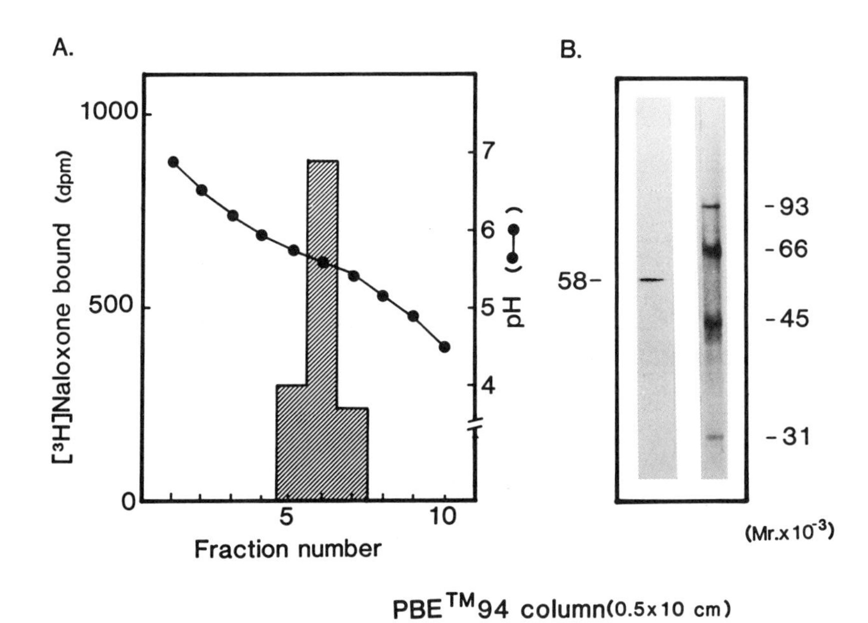

Fig. 1. Isoelectric (PBE) chromatography of affinity-purified μ-opioid receptors (**A**) and the SDS-PAGE (silver-staining) of the active fraction of pI 5.6 (**B**). The numbers in B represent molecular weights of purified μ-opioid receptor and marker proteins. Adopted from Ueda et al. (1988a).

affinity chromatography prior to other chromatographies that may separate G-proteins from the receptor. Accordingly, the solubilized materials were diluted with Tris-HCl buffer and concentrated with an ultrafiltration kit (Labocasette, Millipore) to adjust the expected concentration of digitonin to 0.01% and then applied to the affinity column (2 × 30 cm) at room temperature. The flowrate was 0.3 mL/min. The flow through materials were repeatedly applied to the column. The following procedures of wash and elution of opioid receptors were carried out in buffer A containing 0.01% CHAPS (buffer B). After extensive washes of the column with buffer B (300 mL), 20 mL of 1 m*M* morphine hydrochloride was added to buffer B. The opioid receptor was obtained from 0–100 mL after the addition of morphine hydrochloride.

The affinity purified opioid receptor was concentrated to 0.5 mL and immediately applied to a PBE 94 (Pharmacia) column (0.5 × 10 cm). The isoelectric chromatography was performed with Polybuffer 74. Eluates (2 mL fractions) were collected and used for [³H]naloxone binding assay and for SDS-PAGE. As shown in Fig. 1A, [³H]naloxone binding activity was

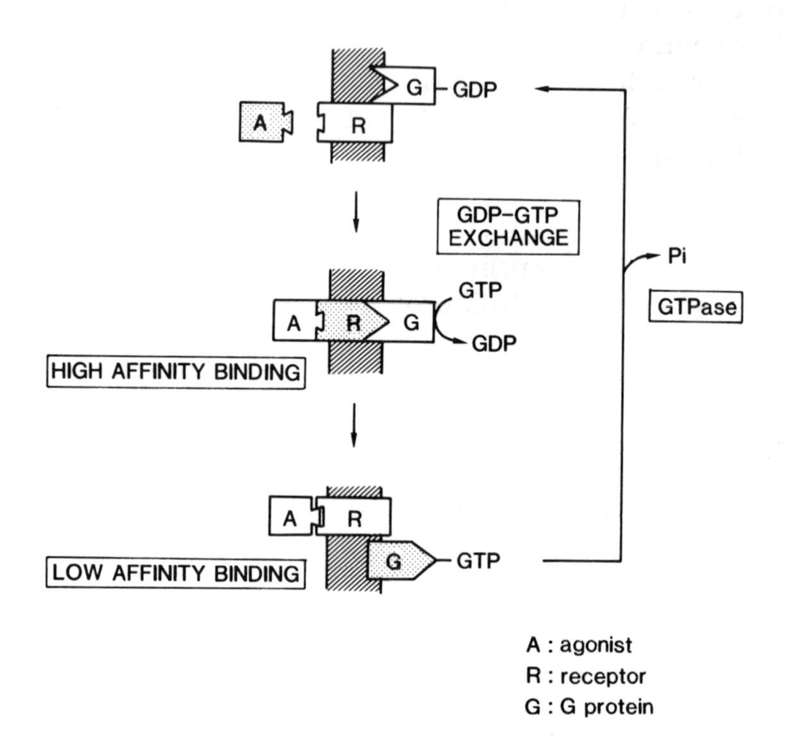

Fig. 2. Proposed model of receptor–GTP-binding protein interaction.

observed at pI (or pH in eluates from the PBE 94 column) 5.6 ± 0.1. When materials in such active fractions were separated in SDS-PAGE under the reduced condition and proteins were silver-stained, a major band was observed at 58kDa (Fig. 1B). This molecular size is in good accord with our previous findings (Ueda et al., 1987) that the μ-opioid receptor purified from rat brains using 6-succinylmorphine-Affi-Gel 102 or membrane-bound μ-opioid receptor was of 58kDa from affinity crosslinking experiments with a selective μ-agonist, [^{3}H][D-Ala^2,MePhe^4,Glyol^5]-enkephalin/DAGO.

4. Reconstitution of Purified μ-Opioid Receptor with GTP-Binding Protein

In order to examine the biological activity of the purified μ-opioid receptor, functional interactions between μ-opioid receptor and G-protein were studied. Two kinds of G-proteins, G_i (G_{i1}) and G_o were purified (>95%) from the cholate extract of rat and porcine brain membranes, respectively, as reported (Katada et al., 1986,1987). The μ-opioid receptor from PBE 94 column (1 pmol) and purified G_i or G_o (10 pmol) were mixed with 50 μg of phosphatidylcholine in 150 μL of buffer A containing 100 m*M* NaCl, 0.1 m*M* EDTA and 5–10 m*M* CHAPS and applied to a Sephadex G-50. Reconstituted vesicles were obtained in the void volume.

According to current models, the signal transduction mechanism between receptor and G-protein is proposed as shown in Fig. 2. The agonist stimulation of receptor primarily leads to a stimulation of GTP binding in exchange for GDP bound to G-protein. When one molecule of the receptor is once stimulated, one molecule of GTP–GDP exchange is believed to be performed. The agonist also stimulates the low-K_m GTPase (or Pi release). This effect is exclusively caused by the receptor- stimulated GTP–GDP exchange, and the resultant accumulation of G-protein in a GTP-bound form, since the k_{cat} of low-K_m GTPase is not changed by receptor stimulation (4 min^{-2}, *see* Gilman, 1987). Thus, the determination of agonist-induced changes in low-K_m GTPase or GTP–GDP exchange activities would be good assays for the functional coupling between re-ceptor and G-protein.

Since our purified μ-opioid receptor had no activity of low-K_m GTPase, it is evident that G-proteins were already removed in this preparation. In reconstituted preparations of 12.5 fmol of μ-opioid receptor and 125 fmol of G_i or G_o, the basal Pi release (GTPase activity) for 20 min incubation was 680 or 969 fmol. When 1 m*M* DAGO, a μ-opioid agonist, was added, the Pi release was approximately doubled (or increased by 700 or 1000 fmol Pi release) in G_i or G_o-reconstituted preparation, respectively. Accordingly, it is evident that 5.6 or 8 times higher numbers of G_i or G_o, respectively, were stimulated by the agonist than that of μ-opioid receptor repeatedly for 20 min. In order to examine how many G-proteins are stimulated through the receptor-stimulation, we measured the [^{3}H]GppNHp (an unhydrolyzable analog of GTP) binding in such reconstituted preparations. As shown in Fig. 3, the basal [^{3}H]GppNHp binding in both G_i and G_o-reconstituted preparations increased as incubation time (**A** in Fig. 3), and DAGO at 100 μ*M* (this concentration of DAGO completely occupied opioid receptor in the receptor binding assay) stimulated the [^{3}H]GppNHp binding in a naloxone-reversible manner (**B** and **C** in Fig. 3). When the DAGO-increase in the binding was plotted (**D** in Fig. 3), it appears that such a DAGO-stimulation was almost completed within 4 min. From double reciprocal plots, the maximal stimulation of [^{3}H]GppNHp binding was calculated to be 62.5 and 34.3 fmol per 25 fmol of μ-opioid receptor for G_i and G_o, respectively. As the [^{3}H]GppNHp bound G-protein is no more stimulated (because GppNHp is not hydrolyzed by GTPase), it is calculated that the

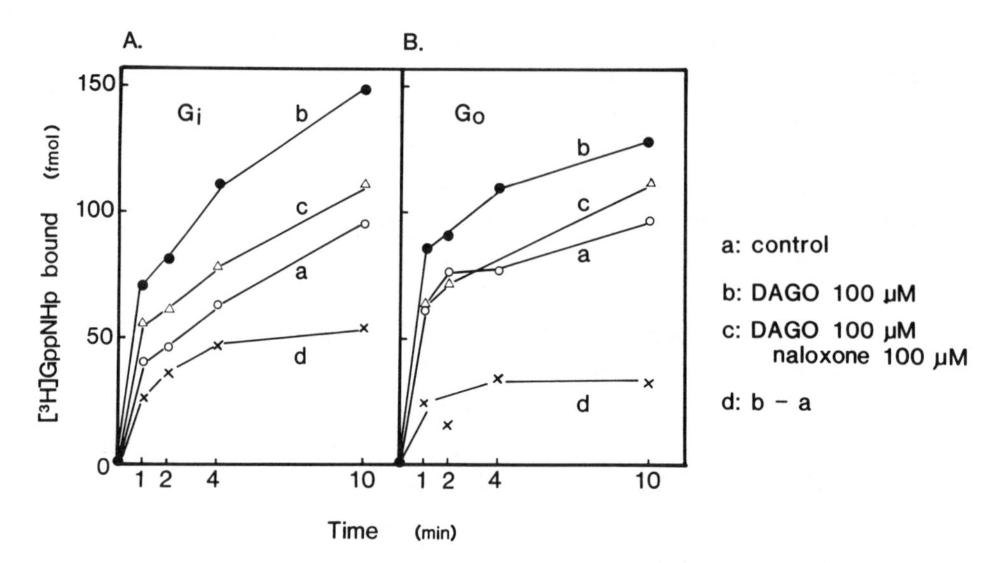

Fig. 3. The μ-agonist stimulation of [^{3}H]GppNHp binding in reconstituted vesicles of purified μ-opioid receptors and GTP-binding proteins. Results represent the [^{3}H] GppNHp binding in exchange for GDP in reconstituted vesicles with G_i (**A**) or G_o (**B**). Adopted from Ueda et al. (1988a).

stimulation of one molecule of μ-receptor leads to a stimulation of 2.5 or 1.37 molecules of G_i or G_o, respectively. Taking into account the stoichiometry in the GTPase assay, it is evident that the G-protein is repeatedly stimulated by agonist-stimulated receptor. Although it is expected that the stoichiometry of coupling between receptor and G-protein depends on their densities in phospholipid vesicles, it is at least likely that the μ-opioid receptor is functionally coupled to several molecules of G_i or G_o and the receptor-mediated stimulation of G-protein is repeated.

The coupling between receptor and G-protein can be also confirmed by the measurement of agonist binding. The purified opioid receptor had a high affinity to [^{3}H]naloxone, a selective μ-opioid antagonist. The K_d was approximately 10 n*M*. However, [^{3}H]DAGO (agonist) showed only a weak binding to the purified receptor. When the displacement of [^{3}H]naloxone binding by DAGO was measured, the IC50 was approximately 100 μ*M* in the absence of G-protein (Fig. 4). The displacement by DAGO was increased 215- or 33-fold by reconstitution with 250 fmol of G_i or G_o per 25 fmol of μ-receptor, respectively. However, the [^{3}H]naloxone binding was not affected by such reconstitutions. When GTPγS was added in the presence of $MgCl_2$ to the G_i- or G_o-reconstituted preparation, such an increase in the agonist binding disappeared. These findings suggest that the agonist binding is increased by coupling to G-protein in a GDP-, but not GTP (GTPγS or GppNHp)-bound form.

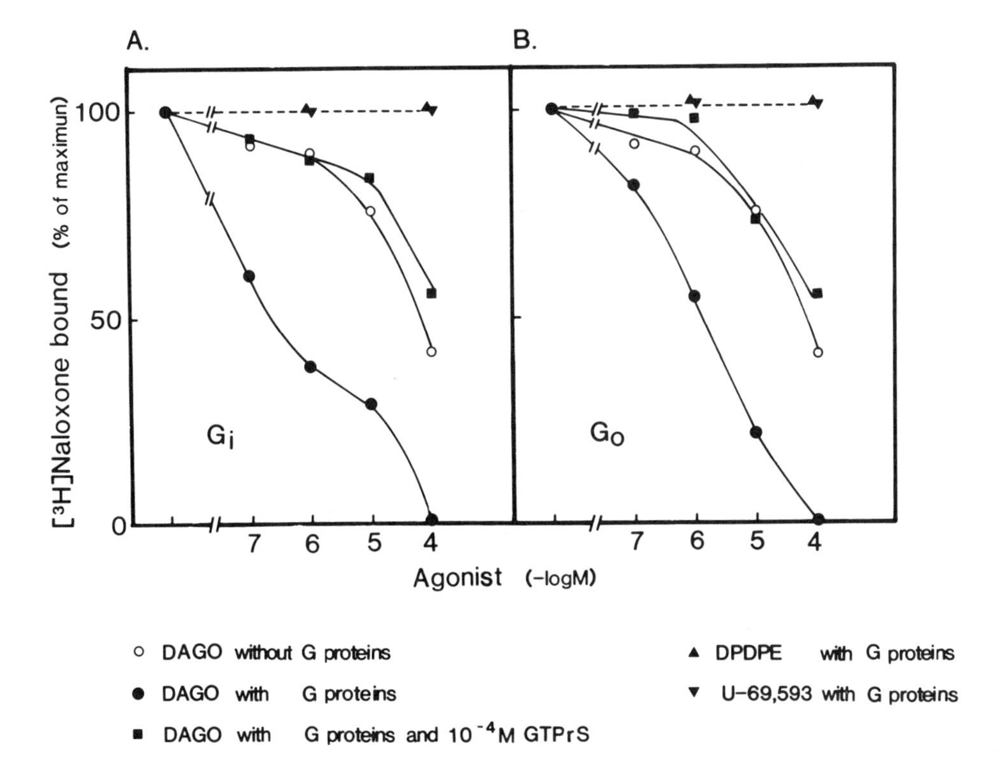

Fig. 4. Displacement of [^{3}H]naloxone binding in reconstituted vesicles of purified μ-opioid receptors and GTP-binding proteins. Results represent the [^{3}H]naloxone binding in the presence of agonists in reconstituted vesicles with or without G_i (**A**) or G_o (**B**). Adopted from Ueda et al. (1988a).

It is evident that purified μ-opioid receptor is functionally coupled to G_i or G_o. The agonist stimulation of receptors led to an enhancement of G-protein activities in reconstituted preparations. Thus, it is likely that the signal transduction in μ-opioid receptor–G-protein interaction is essentially similar to those of muscarinic receptors, α_2-, and β-adrenoceptors (Kurose et al., 1986; Cerione et al., 1986; Pedersen and Ross, 1982). The proposed model of such an interaction is given in Fig. 5. In this model, the agonist binding to its receptor stimulates the GTP–GDP exchange in G-protein through R_A-g_A coupling domains. The formed GTP-bound G-protein is dissociated into α- and βγ-subunits, and thereby the receptor is dissociated from the G-protein. Consequently, the agonist binding becomes lower by lack of g_B-R_B interaction. The GTP bound to G-protein is hydrolyzed into GDP by α-subunits of G-protein, thereafter the G-protein goes back to the resting state. Since the stimulation of one molecule of opioid receptor leads to a stimulation of plural number of G_i or G_o, it may be speculated that another G-protein becomes coupled to the receptor

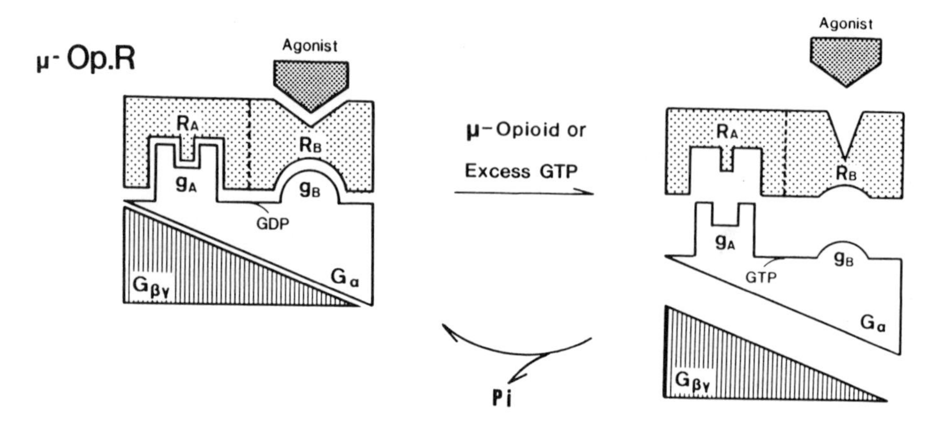

Fig. 5. Proposed mechanism of signal transduction in μ-opioid receptor–GTP-binding protein coupling. Abbreviations: R_A, a receptor domain involved in the functional coupling with GTP-binding protein; R_B, a receptor domain involved in the agonist binding modulation through receptor–GTP-binding protein interaction; g_A and g_B, GTP-binding protein domains to be coupled with R_A and R_B, respectively; G_α and $G_{\beta\gamma}$, subunits of GTP-binding protein.

when the previously coupled G-protein is dissociated from the receptor by agonist-stimulation.

5. Reconstitution of Membrane-Bound μ-Opioid Receptor with Purified G-Protein

As mentioned above, we showed that a purified μ-opioid receptor is functionally coupled to G_i or G_o. Furthermore, we reported that there is no functional coupling between the μ-opioid receptor and v-Ki-*ras* p21 GTP-binding protein expressed from a plasmid encoding 189 amino acids of the Kirsten murine sarcoma virus oncogene and purfied from *E. coli* (Hara et al., 1988). However, there remains an important problem—whether the μ-opioid receptor is coupled to both G_i and G_o, or the functional coupling between them is less selective in respect to the kind of G-protein. We obtained important findings in detailed reconstitution experiments using synaptic membrane preparations and purified G-proteins. Since native G-proteins exist in the membranes, we must remove or inactivate them before use for reconstitution with purified G-proteins. For this purpose, IAP, which is known to uncouple receptors from substrate G-proteins, such as Gi_1,Gi_2, Gi_3, Go, and Gt (transducin), is a useful tool (Ui et al., 1984; Katada et al., 1986; Itoh et al., 1988). Recently, we reported the use of

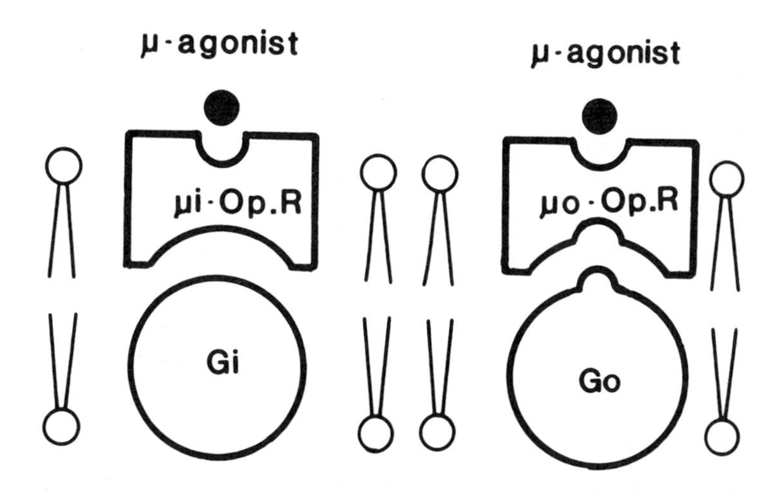

Fig. 6. μ_i and μ_o-Opioid receptor separately coupled to G_i and G_o, respectively.

low concentrations (5–8 μM) of *N*-ethylmaleimide (NEM) to inactivate the basal low-K_m GTPase activity in membrane-bound G-proteins (Ueda et al., 1989, 1990). The NEM-pretreatment of the membrane seems to have the following advantages. The pretreatment (4°C for 15 min) is milder than that with IAP (37°C for 30–60 min). The basal low-K_m GTPase activity was reduced by NEM-treatment, but not by IAP-treatment. Since the basal low-K_m GTPase activity is expected to increase by reconstitution with purified G-proteins, the NEM-treatment is preferable to the IAP-treatment in order to detect the receptor-mediated small changes in the GTPase activity in reconstituted preparations.

The DAGO-stimulation in the low-K_m GTPase activity was abolished by 5 μM NEM-pretreatment of striatal synaptic membranes of guinea pigs. When G_i or G_o was reconstituted into the NEM-treated membranes, the DAGO-stimulation of low-K_m GTPase was recovered. The recovery of the DAGO-stimulation was maximal with 0.5 pmol/20 μg of protein of the membrane/tube, in both cases. When both G_i and G_o (each 0.5 pmol/tube) were simultaneously reconstituted, the DAGO-stimulation was additively recovered. Therefore, the μ-opioid receptor may exist in two different forms, that is, "μ_i" and "μ_o," which are functionally coupled to G_i and G_o, respectively (Ueda et al., 1990; Fig. 6). However, the identification of subtypes of μ-opioid receptor must await the molecular cloning of the receptor.

6. Phosphorylation of μ-Opioid Receptor

Recent studies revealed that phosphorylation of receptors regulates their functions, in relation to the down regulation of receptors (Burgoyne, 1983; Kelleher et al., 1984; Sibley and Lefkowitz, 1985; Matozaki et al., 1986; Ho et al., 1987). In view of these previous works, we studied the effect of phosphorylation of the μ-opioid receptor on the functional coupling between receptor–G-protein by use of reconstitution experiments (Ueda et al., 1988b; Harada et al., 1989). DAGO stimulated the low-K_m GTPase in striatal membranes of the rat, and this stimulation was markedly reduced by phosphorylation of the membrane with A-kinase. However, neither the B_{max} and K_d of [^{3}H]DAGO binding nor the inhibition of the binding by GTPγS was affected by pretreatment with A-kinase. On the other hand, the DAGO-stimulation of GTPase was abolished by IAP-treatment and recovered by reconstitution with G_i. When IAP-treated membranes were further treated with A-kinase, the functional coupling (GTPase stimulation) was no longer recovered. More direct evidence of this mechanism was obtained from the experiment using the μ-opioid receptor purified from rat brains. DAGO showed no stimulation of GTPase activity in reconstituted preparations of G_i and phosphorylated μ-receptor by A-kinase.

These findings suggest a very important mechanism. The guanine-nucleotide sensitivity of agonist binding was not affected by phosphorylation of the μ-receptor, although the agonist-stimulation of G-protein activity owing to the "functional" interaction between the receptor and G-protein was abolished. This would indicate that the μ-receptor possesses two independent domains involved in coupling to the G-protein. As shown in Fig. 7, one of them is related to "functional" coupling or to the transduction of agonist signal to G-protein. This domain R_A is inactivated by phosphorylation. The other is related to the guanine-nucleotide-induced reduction of agonist binding that is attributed to the dissociation of α-subunits of G-proteins from their βγ-subunits by exchange for GDP. In other words, such a domain in the μ-receptor (R_B) may play a role in receiving information (related to the uncoupling of receptor–G-protein interaction) from G-protein.

Conclusion

In reconstitution experiments using μ-opioid receptor and G-protein, we suggested various functional roles of domains within the receptor. Furthermore, we found that the κ-opioid agonist inhibits the low-K_m

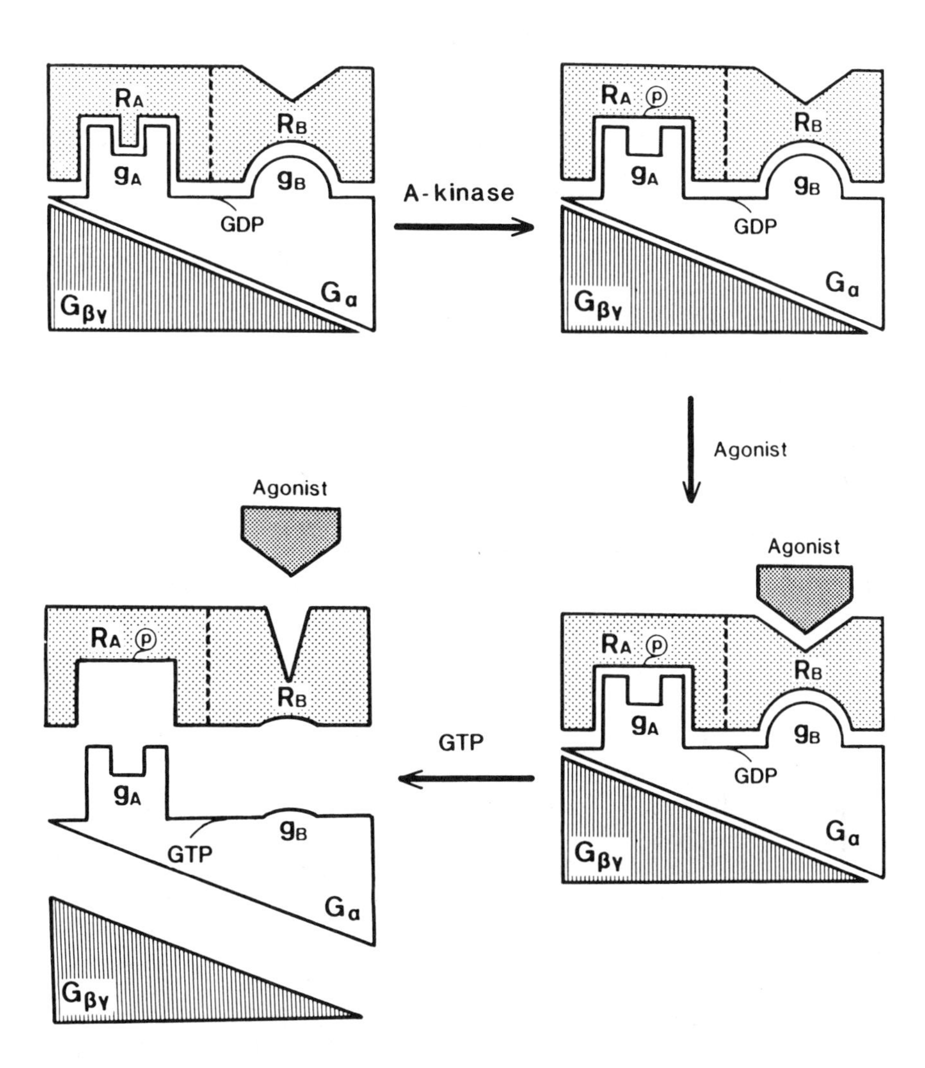

Fig. 7. Proposed model of signal transduction in reconstituted vesicles of phosphorylated μ-opioid receptor–GTP-binding protein. A-kinase represents cyclic AMP-dependent protein kinase. Other abbreviations were as given in the legend of Fig. 5.

GTPase activity through an inhibition of G_i activity in guinea pig cerebellar membranes (Ueda et al., 1987b,1989; Ueda and Satoh, 1988) and that the δ-agonist showed no effect on the G-protein activity in guinea pig striatal membranes, whereas the δ-agonist binding was drastically inhibited by GTPγS (Ueda and Satoh, 1988). Thus, there are diverse modes of coupling of receptor with G-protein. The molecular basis of analysis of receptor domains must await the study of regulation of receptor expression following the molecular cloning of the receptor.

References

Burgoyne, R. D. (1983) *J. Neurochem.* **40,** 324–331.
Burns, D., Hewlett, E. L., Moss, J., and Vaughan, M. (1983) *J. Biol. Chem.* **258,** 1435–1438.
Cerione, R. A., Regan, J. W., Nakata, H., Codina, J., Benovic, J. L., Gierschik, P., Somers, R., Spiegel, A. M., Birnbaumer, L., and Lefkowitz, R. J. (1986) *J. Biol. Chem.* **261,** 3901–3909.
Cho, T. M., Hasegawa, J., Ge, B., and Loh, H. H. (1986) *Proc. Natl. Acad. Sci. USA* **83,** 4138–4142.
Collier, H. O. J. and Roy, A. C. (1974) *Nature (London)* **248,** 24–27.
Fujioka, T., Inoue, F., and Kuriyama, M. (1985) *Biochem. Biophys. Res. Comm.* **131,** 640–646.
Gilman, A. G. (1987) *Ann. Rev. Biochem.* **56,** 615–649.
Gioannini, T. L., Howard, A. D., Hiller, J. M., and Simon, E. J. (1985) *J. Biol. Chem.* **260,** 15117–15121.
Goldstein, A. (1984) in *The Peptides,* vol. 6 (Udenfriend, S. and Meienhofer, J., eds.), Academic, Orlando, FL, pp. 95–145.
Goldstein, A., Lowney, L. I., and Pal, B. K. (1971) *Proc. Natl. Acad. Sci. USA* **68,** 1742–1747.
Gross, R. A. and Macdonald, R. L. (1987) *Proc. Natl. Acad. Sci. USA* **84,** 5469–5473.
Hara, M., Tamaoki, T., and Nakamura, H. (1988) *Oncogene Res.* **2,** 325– 333.
Harada, M., Ueda, M., Wada, Y., Katada, T., Ui, M., and Satoh, M. (1989) *Neurosci. Lett.* **100,** 221–226.
Ho, A. K. S., Ling, Q.-L., Duffield, R., Lam, P. H., and Wang, J. H. (1987) *Biochem. Biophys. Res. Comm.* **142,** 911–918.
Hughes, J., Smith, T. W., Kosterlitz, H. W., Fothergill, L. A., Morgan, B. A., and Morris, H. R. (1975) *Nature (London)* **258,** 577–579.
Itoh, H., Toyama, R., Kozasa, T., Tsukamoto, T., Matsuoka, M., and Kaziro, Y. (1988) *J. Biol. Chem.* **263,** 6656–6664.
Katada, T., Oinuma, M., Kusakabe, K., and Ui, M. (1987) *FEBS Lett.* **213,** 353–358.
Katada, T., Oinuma, M., and Ui, M. (1986) *J. Biol. Chem.* **261,** 8182–8191.
Kelleher, D. J., Pessin, J. E., Ruoho, A. E., and Johnson, G. L. (1984) *Proc. Natl. Acad. Sci. USA* **81,** 4316–4320.
Klee, W. A. and Nirenberg, M. (1976) *Nature (London)* **263,** 609–612.
Kosterlitz, H. W. and Watt, A. J. (1968) *Br. J. Pharmacol.* **33,** 266–276.
Kurose, H., Katada, T., Amano, T., and Ui, M. (1983) *J. Biol. Chem.* **258,** 4870–4875.
Kurose, H., Katada, T., Haga, T., Haga, K., Ichiyama, A., and Ui, M. (1986) *J. Biol. Chem.* **261,** 6423– 6428.
Kurose, H. and Ui, M. (1985) *Arch. Biochem. Biophys.* **238,** 424–434.
Lord, J. A. H., Waterfield, A. A., Hughes, J., and Kosterlitz, H. W. (1977) *Nature (London)* **267,** 495–499.
Lowney, L. I., Schulz, K., Lowery, P., and Goldstein, A. (1974) *Science* **183,** 749–753.
Maneckjee, R., Zukin, R. S., Archer, S., Michael, J., and Osei-Gyimah, P. (1985) *Proc. Natl. Acad. Sci. USA* **82,** 594–598.
Martin, W. R., Eades, C. G., Thompson, J. A., Huppher, R. E., and Gilbert, P. E. (1976) *J. Pharmacol. Exp. Ther.* **197,** 517–532.
Matozaki, T., Sakamoto, C., Nagao, M., and Baba, S. (1986) *J. Biol. Chem.* **261,** 1414–1420.
Neer, E. J. and Clapham, D. E. (1988) *Nature (London)* **333,** 129–134.
North, A., Williams, J. T., Surprenant, A., and Christie, M. J. (1987) *Proc. Natl. Acad. Sci. USA* **84,** 5487–5491.

Paterson, S. J., Robson, L. E., and Kosterlitz, H. W. (1984) in *The Peptides* vol. 6 (Udenfriend, S. and Meienhofer, J., eds.), Academic, Orlando, FL, pp. 147–189.

Pedersen, S. E. and Ross, E. M. (1982) *Proc. Natl. Acad. Sci.* **79,** 7228–7232.

Pert, C. B. and Snyder, S. H. (1973) *Science* **179,** 1011–1014.

Porreca, F., Mosberg, H. I., Hurst, R., Hruby, V. J., and Burks, T. F. (1984) *J. Pharmacol. Exp. Ther.* **230,** 341–348.

Satoh, M., Kubota, A., Iwama, T., Yasui, M., Fujibayashi, K., and Takagi, H. (1983) *Life Sci.* **33 (suppl. I),** 689–692.

Sibley, D. R. and Lefkowitz, R. J. (1985) *Nature (London)* **317,** 124–129.

Simon, E. J., Hiller, J. M., and Edelman, I. (1973) *Proc. Natl. Acad. Sci. USA* **70,** 1947–1949.

Simon, E. J., Hiller, J. M., and Edelman, I. (1975) *Science* **190,** 389–390.

Simonds, W. F., Burke, T. R., Rice, K. C., Jacobson, A. E., and Klee, W. A. (1985) *Proc. Natl. Acad. Sci. USA* **82,** 4974–4978.

Terenius, L. (1973) *Acta Pharmacol. Toxicol.* **33,** 377–384.

Ueda, H., Harada, H., Nozaki, M., Katada, T., Ui, M., Satoh, M., and Takagi, H. (1988a) *Proc. Natl. Acad. Sci. USA* **85,** 7013–7017.

Ueda, H., Harada, H., Misawa, H., Nozaki, M., and Takagi, H. (1987a) *Neurosci. Lett.* **75,** 339–344.

Ueda, H., Harada, H., Wada, Y., Katada, T., Ui, M., Takagi, H., and Satoh, M. (1988b) *Abstracts of 18th Annual Meeting of Society for Neuroscience,* Toronto, p. 1053.

Ueda, H., Misawa, H., Fukushima, N., and Takagi, H. (1987b) *Eur. J. Pharmacol.* **138,** 129–132.

Ueda, H., Misawa, H., Katada, T., Ui, M., Takagi, H., and Satoh, M. (1989) in *Advances in the Biosciences* (Cros, J., Meunier, J.-C., and Hamon, M., eds.), Pergamon Press, Great Britain, pp. 129–132.

Ueda, H., Misawa, H., Katada, T., Ui, M., Takagi, H., and Satoh, M. (1990) *J. Neurochem.,* in press.

Ueda, H. and Satoh, M. (1988) in *Advances in Endocrinology* (Imura, H., ed.), Elsevier, Netherlands, pp. 1137– 1142.

Ui, M. (1984) *Trends Pharmacol. Sci.* **5,** 277–279.

Yamashiro, D. and Li, C. H. (1984) in *The Peptides,* vol. 6 (Udenfriend, S. and Meienhofer, J., eds.), Academic, Orlando, FL, pp. 191–217.

The Neurotensin Receptor from Mammalian Brain

Solubilization and Purification by Affinity Chromatography

Jean-Pierre Vincent, Jean Mazella, Joëlle Chabry, and Nicole Zsurger

1. Introduction

Neurotensin is a peptide of 13 amino acids (pGlu-Leu-Tyr-Glu-Asn-Lys-Pro-Arg-Arg-Pro-Tyr-Ile-Leu) that is mainly localized in the central nervous system (Carraway and Leeman, 1973) and in the gastrointestinal tract (Kitabgi et al., 1976). A number of pharmacological, biochemical, and histochemical data suggest that neurotensin works as a neurotransmitter or neuromodulator in the brain (Nemeroff et al., 1982) and as a hormone in the periphery (Hirsh Fernstrom et al., 1980). Both modes of action imply as a first step the selective interaction of the peptide with specific receptors located on the plasma membrane of target cells. The structural and functional properties of neurotensin binding sites have been characterized in a number of tissue preparations and cell cultures of neural or nonneural origin (Kitabgi et al., 1985). Some of these sites represent functional neurotensin receptors since they are coupled to GTP-binding proteins and regulate intracellular levels of second messengers such as cAMP, cGMP, and inositol phosphates (Bozou et al., 1986; Amar et al., 1985: Gilbert et al., 1986; Goedert et al., 1984; Amar et al., 1986,1987). Although these studies significantly improved our knowledge of the mode of action of neurotensin, it is clear that isolation of a pure and active form of neurotensin receptor would be necessary to fully understand the mechanism of neurotensin action as well as to elucidate the molecular structure of the receptor and to undertake the cloning of its gene. We describe in this chapter the rational and technological aspects of the methodology employed to purify the neurotensin receptor from mouse brain.

2. Properties of Membrane-Bound Neurotensin Receptors

Knowledge of these data is essential in order to make the best choice of the starting material for solubilization and purification of the receptor.

2.1. Binding Characteristics of Central and Peripheral Membrane Preparations

Using ^{125}I-labeled [monoiodo Tyr^3] neurotensin as radioactive ligand (Sadoul et al., 1984), we have found that membranes prepared from brain or gastrointestinal tissues generally contain two different classes of neurotensin binding sites. For example, as illustrated in Fig. 1, the Scatchard plot describing the binding of [^{125}I Tyr^3] neurotensin to adult mouse brain homogenate is curvilinear and can be resolved into two independent linear components (Mazella et al., 1983). Each component represents a single class of noninteracting binding sites. Sites 1 are characterized by a high affinity (0.13 n*M*) and a low capacity (Table 1, *see* p. 134). By comparison, the affinity of sites 2 is lower (2.4 n*M*) and their binding capacity is higher. Sites 1 and 2 cannot be easily differentiated by their structure–function relationships since they exhibit the same order of affinity for a series of neurotensin analogs. Both types of sites recognize the 8–13 C terminal hexapeptide of the neurotensin sequence (Mazella et al., 1983). On the other hand, the affinity of site 1 for neurotensin can be selectively decreased by sodium ions or GTP, whereas sites 2 are much less sensitive to sodium ions and insensitive to GTP. Binding capacities of sites 1 and 2 remain unchanged under these various conditions. However, the best tool to distinguish between sites 1 and 2 is levocabastine. This potent antihistamine-1 drug was introduced by Janssen Pharmaceutica and found to be able to partly inhibit [^{3}H] neurotensin binding to rat brain membrane, although it was devoid of any neurotensin-like activity (Schotte et al., 1986). Our results indicate that a 1 μM concentration of levocabastine completely blocks the binding of [^{125}I Tyr^3] neurotensin to sites 2 without changing the binding properties of sites 1.

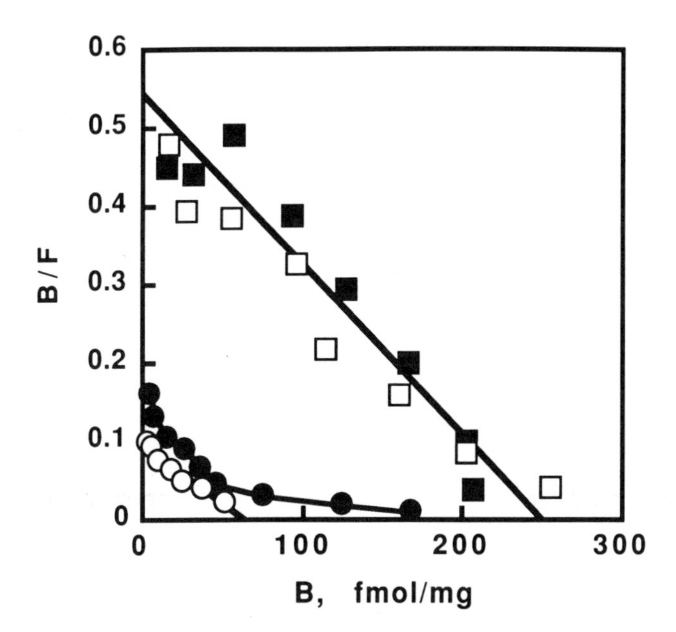

Fig. 1. Binding of ^{125}I-labeled [Tyr^3] neurotensin to mouse brain homogenates. Freshly prepared homogenates from adult(○,●) or newborn (□,■) mouse brain (0.1 mg of protein per assay) were incubated at 25°C, pH 7.5, with increasing concentrations of ^{125}I-labeled [Tyr^3] neurotensin alone or isotopically diluted with unlabeled neurotensin, in the presence (open symbols) or in the absence (closed symbols) of 1 μ*M* levo-cabastine. This incubation medium (25 μL) consisted of a 50 m*M* Tris-HCl buffer containing 0.02% bovine serum albumin, 1 m*M* 1,10-phenanthroline and 0.1 μ*M* *N*-benzyloxycarbonyl-prolyl-prolinal to prevent degradation of the ligand. The specific binding was determined by filtration (Mazella et al., 1983). Data are presented under the form of Scatchard plots. B and F, bound and free concentration of labeled ligand.

This "all or none" effect is obviously the simplest way to differentiate sites 1 from sites 2.

Table 2 shows that almost all membrane preparations of central or peripheral origin contain two types of neurotensin binding sites. The amount and the relative proportions of each site vary from one species to the other. In agreement with results previously obtained from newborn rat brain (Schotte and Laduron, 1987) we have observed that brains of 5–10-d-old mouse (Fig. 1), rabbit, and rat contain 4–6 times more high affinity neurotensin binding sites than brains of adults. Moreover, the low affinity sites are undetectable in brains of these newborn animals.

Table 1
Compared Properties of the Two Different Types of Neurotensin Binding Sites in Homogenates of Adult Mouse Brain

Property	Site 1	Site 2
Dissociation constant (Tris 50 m*M* pH 7.5, 25°C)	Kd1 = 0.13 n*M*	Kd2 = 2.4 n*M*
Binding capacity	Bm1 = 66 fmol/mg	Bm2 = 160 fmol/mg
Effect of Na^+ ions	Kd1 increased	Kd2 increased
Effect of GTP	Kd1 increased	no effect
Effect of levocabastine	no effect	complete inhibition of neurotensin binding
Solubilization in an active form by CHAPS	yes	no
Occurence in cell cultures (NIE 115, neurons)	yes	no
Involvement in transduction mechanisms	yes	no

2.2. Transduction Mechanisms of Neurotensin Receptors

Differentiated mouse neuroblastoma N1E1 15 cells contain neurotensin binding sites of high affinity (Poustis et al., 1984; Table 2). These sites are very similar to sites 1 of mouse brain homogenates since their affinity is negatively modulated by sodium ions and GTP without change of the binding capacity (Bozou et al., 1986). Moreover [^{125}I Tyr^3] neurotensin binding to neuroblastoma cells is totally insensitive to levocabastine. The same type of site is also found in HT29 cells, an adenocarcinoma of human colon (Amar et al., 1986), and in primary cultured neurons of mouse fetal cortex (Checler et al., 1986b).

The association of neurotensin to its receptor in neuroblastoma N1E115 cells triggers three differents effects. First, the intracellular cGMP concentration increases according to a rapid and transient mechanism. This effect is calcium-dependent and leads to a maximal cGMP stimulation of 10-fold over basal level (Amar et al., 1985; Gilbert et al., 1986). A parallel 20–30% decrease of the cAMP basal level is also detected. This inhibitory effect of neurotensin is much easier to study after stimulation of cAMP by prostaglandin E_1. Under these conditions, neurotensin inhibits cAMP production by 50–55% (Bozou et al., 1986). A third consequence of the association of neurotensin to its receptor in N1E115 cells is an increase of intracellular inositol phosphate concentrations. In the presence of lithium ions, which block inositol monophos-

Table 2
Proportions of High and Low Affinity Neurotensin Binding Sites in Various Membrane Preparations and Cell Cultures

Preparation	Maximal Binding capacity, fmol/mg(%)	
	High affinity	Low affinity
Rat brain homogenate		
adult	13 (7%)	171 (93%)
newborn	75 (100%)	0 (0%)
Rat fundus smooth muscle plasma membranes	6.6 (36.7%)	11.4 (63.3)
Rabbit brain homogenate		
adult	50 (100%)	0 (0%)
newborn	200 (100%)	0 (0%)
Guinea pig brain synaptic membranes	11.9 (14%)	73 (86)
Mouse brain homogenate		
adult	67 (38.5%)	107 (61.5%)
newborn	250 (100%)	0 (0%)
Pig brain homogenate	6.4 (12.7%)	44 (87.3%)
Human brain homogenate		
substancia nigra	26.7 (30.3%)	61.3 (69.7%)
frontal cortex	9.9 (11.8%)	74.1 (88.2%)
Dog ileum circular muscle plasma membranes	9.7 (7%)	130 (93%)
Cell cultures		
Adenocarcinoma HT 29, human	120 (100%)	0 (0%)
Neuroblastoma NIE115, mouse	30 (100%)	0 (0%)
Primary cultured neurons, mouse	178 (100%)	0 (0%)

phate hydrolysis, neurotensin produces a rapid and transient stimulation of inositol trisphosphate and inositol bisphosphate levels and a slower increase of the inositol monophosphate concentration (Amar et al., 1987). We have demonstrated that each one of the three effects described above are direct consequences of neurotensin receptor occupancy and are totally insensitive to levocabastine. Neurotensin is also able to stimulate phosphatidylinositol turnover in HT29 cells without changing the intracellular levels of cAMP and cGMP (Amar et al., 1986).

3. Solubilization of the Neurotensin Receptor

3.1. Choice of the Starting Material

Selection of the preparation that will be used as starting material for solubilization and purification of the receptor should take into account several criteria.

1. The receptor concentration should be as large as possible.
2. The tissues used as source of receptor should be easily and continuously available.
3. Detergent solubilization should give high yields of active and stable receptor.
4. In the present work, we wanted to purify functionally relevant neurotensin receptors. All these considerations led us to select brain homogenates of new born mouse as source of receptor. This preparation contains large amounts of sites 1 that have been shown to regulate intracellular levels of second messengers (*see above*). Moreover, mouse breeding is easy to maintain and provides regular amounts of newborn mouse brains.

3.2. Choice of the Solubilization Conditions

3.2.1. Binding Assay of Soluble Receptor

The choice of a convenient assay for binding of neurotensin to its soluble receptor is an extremely important first step. Insofar as the binding assay has not been validated, solubilization experiments cannot be properly interpreted. Thus, failure to detect neurotensin binding activity in a soluble extract can be owing either to the absence of active receptor in the extract or to the inability of the binding assay to detect it. The necessity to find simultaneously suitable conditions for solubilization of active receptor and for measurement of the soluble binding activity is one of the main difficulties in studies dealing with receptor solubilization. The best way to solve this problem is probably to use an assay as simple as possible. In our case, the separation of ^{125}I-labeled [Tyr^3] neurotensin bound to the soluble receptor from free ligand was carried out by gel filtration on Sephadex G-50 (medium). The radioactivity bound to the receptor was eluted in the void volume and counted directly in a γ counter (Mazella et al., 1988).

3.2.2. Choice of Detergent

Brain homogenates from newborn mouse were incubated at 0°C, pH 7.5 with 0.1 to 1% (w/v) concentrations of sodium cholate, 3-[(3 cholamidopropyl) dimethylamonio]-1-propanesulfonic acid (CHAPS), digitonin, Triton X-100 or Lubrol PX. After centrifugation, the binding of [^{125}I Tyr^3] neurotensin to the soluble fraction was measured by gel filtration. These experiments demonstrated that CHAPS was the only detergent that could solubilize neurotensin receptors in an active form. Comparison of the binding properties of membrane-bound and CHAPS-solubilized neurotensin receptors (*see below*) clearly established that the soluble binding sites corresponded exclusively to the high affinity sites found in mouse brain homogenates, i.e. sites 1. These sites were quantitatively solubilized by CHAPS concentrations comprised between 0.6–0.7%.

3.2.3. Stabilization of the Soluble Receptor

In the absence of any stabilizing agent, the half-life (t 1/2,) of the mouse brain receptor solubilized with 0.6% CHAPS was about 3 h at 0°C. The presence of 0.12% cholesterol hemisuccinate (CHS) in the solubilization buffer largely improved the stability of the receptor (t 1/2 = 31 h). Addition of 1 n*M* neurotensin to mouse brain homogenate prior to the solubilization step further increased the stability at 0° C by a factor of 2 (t 1/2 = 65 h). CHAPS-solubilized extracts frozen at –30°C could be stored for weeks without any detectable loss of binding activity provided they contain 10% glycerol. This additive did not increase the half-life of the receptor at 0° C, but protected it against losses of binding activity that occurred upon freezing and thawing.

Taking into account all the data described above, solubilization was carried out as follows. Freshly prepared homogenates of newborn mouse brain were incubated at a concentration of 10 mg of protein per ml in a 20 m*M* Tris-HCl buffer, pH 7.5, containing 10% glycerol, 0.6% CHAPS, 0.12% CHS, and a mixture of protease inhibitors (0.1 m*M* phenylmethylsulfonyl fluoride, 1 μ*M* pepstatin, 1 m*M* iodoacetamide, and 5 m*M* EDTA). After 30 min at 0°C, the incubation medium was centrifuged at 110,000*g* for 15 min. The supernatant that contained the soluble neurotensin receptor was either used immediately or stored at –30°C.

3.3. Properties of the Solubilized Neurotensin Receptor

3.3.1. Binding Properties

Soluble extracts obtained from either newborn or adult mouse brain contain a single type of high affinity neurotensin receptors, as demonstrated by the linearity of the corresponding Scatchard plots (Fig. 2). In both cases, the affinity of the soluble receptor for neurotensin (about 0.3 n*M*) is decreased by sodium ions and is insensitive to levocabastine (Fig. 2). The ability of GTP to negatively modulate the affinity of the receptor for neurotensin has been lost upon solubilization. Apart from this difference, CHAPS seems to have solubilized neurotensin receptors that correspond to the membrane-bound sites 1 (Table 1). Although the levocabastine-sensitive sites 2 are more abundant than sites 1 in adult mouse brain (Fig. 1), sites 2 are no longer detectable after solubilization (Fig. 2).

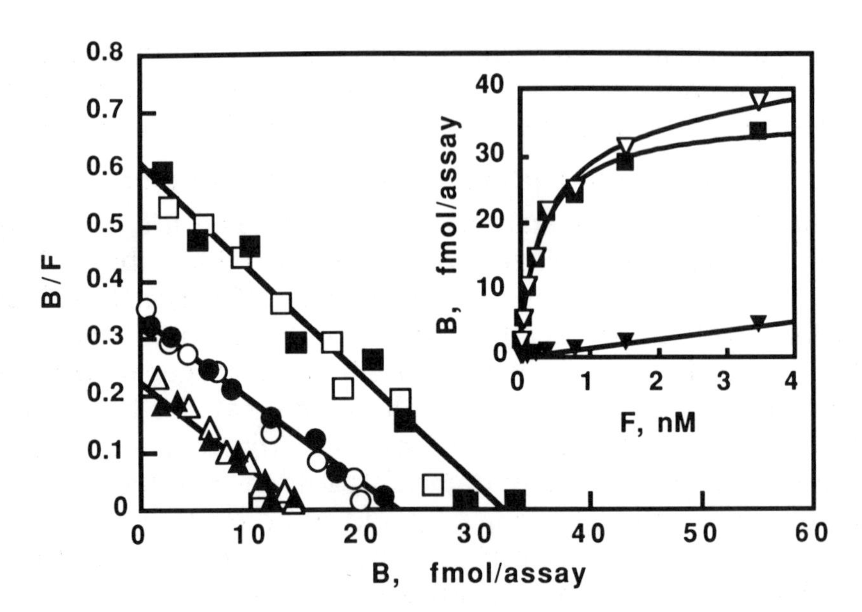

Fig. 2. Binding of ^{125}I-labeled [Tyr^3]neurotensin to soluble neurotensin receptor from mouse brain. Crude soluble extracts from adult (Δ,▲, 0.2 mg of protein) or newborn (○,●, 0.09 mg of protein) mouse brain and the purified neurotesin receptor (□,■, 4 ng of protein) were incubated at 0°C, pH 7.5, with increasing concentration of radiolabeled neurotensin in the presence (open symbols) or in the absence (closed symbols) of 1μ*M* levocabastine. After 1 h of incubation, the bound radioactivity was separated from the free ligand by gel filtration on Sephadex G-50 (Mazella et al., 1988). Specific binding data are presented in the main part of the figure under the form of Scatchard plots. A direct representation of the saturation curve for the purified neurotensin receptor is shown in the inset. ∇, total binding; ▼, nonspecific binding determined in the presence of 1 μ*M* unlabeled neurotensin; ■, specific binding calculated as the difference between total and nonspecific binding. B and F, bound and free concentration of labeled ligand.

3.3.2. Molecular Structure

Gel filtration of the CHAPS-solubilized neurotensin receptor from mouse brain on an Ultrogel AcA34 column calibrated with protein standards gave an approximate mol wt of 110,000 (Mazella et al., 1988). The soluble receptor was also covalently radiolabeled by photoaffinity and by chemical crosslinking. Both techniques resulted in the specific labeling of the same protein band of mol wt about 100,000 (Mazella et al., 1988).

To summarize, neurotensin receptors from mouse brain homogenate can be solubilized in an active and stable form using the zwitterionic detergent CHAPS in the presence of CHS. The soluble receptor is a single polypeptide chain of 100 kilodaltons (kDa) that binds neurotensin with a high affinity. From these data, it is reasonable to undertake the purification of the receptor.

4. Purification of the Neurotensin Receptor

The soluble neurotensin receptor from newborn mouse brain was purified essentially in a single step of ligand affinity chromatography. However, in order to minimize the total amount of protein loaded on the affinity column, the soluble extract was first prepurified by ion exchange chromatography.

4.1. Prepurification Step

The soluble extract was diluted 2.5 times and passed at 4°C through two columns of SP-Sephadex C-25 and hydroxylapatite (100 mLeach) connected in series and equilibrated with the Tris-glycerol buffer containing 0.1% CHAPS, 0.02% CHS, and protease inhibitors. About 50% of proteins, including several brain proteases (Checler et al., 1986a), were retained on the gels, whereas the bulk of the neurotensin binding activity was collected with the flowthrough and used in the next purification step.

4.2. Affinity Chromatography

4.2.1. Preparation of the Affinity Gel

In preliminary experiments, covalent binding of native neurotensin to various chemically activated gels was found to occur in low yields. The reason is probably that the only free amine available for the coupling reaction in the neurotensin sequence is the sidechain of Lys6, the *N*-terminal residue being pGlu. To overcome this difficulty, neurotensin (2–13) was used in place of the native peptide to prepare the affinity column. Interestingly, the affinity of the (2–13) sequence for the neurotensin receptor was found to be better than that of native neurotensin in binding competition experiments. The neurotensin (2–13) solution coupled to the gel was radiolabeled by tracer amounts of [^{3}H] neurotensin (2–13). Glutaraldehyde-activated Ultrogel AcA22 (IBF Biotechnics) was chosen as support because it contained a spacer arm and because ligand leakage was very low after extensive washing of the affinity gel. The yield of the coupling reaction calculated from the radioactivity incorporated into the gel was between 50 and 70% in three different experiments.

4.2.2. Binding Properties of the Affinity Column

One of the most striking properties of the affinity gel is that its binding capacity increased with time. This increase was found to be directly correlated to a decrease of the leakage of the neurotensin (2–13) bound to the gel. The leakage could be easily monitored since the peptide coupled to the gel was radiolabeled. Immediately after its preparation, the amount of neurotensin (2–13) released by the gel equilibrated with the Tris-glycerol buffer containing 0.1% CHAPS and 0.02% CHS was 1 nmol/5 mL of eluate. This amount dropped to less than 10 pmol/5mL of eluate after 6 mo of intense use of the affinity column. The parallel loss of neurotensin (2–13) that remained covalently bound to the gel was proportionally much less important, from 8.3 μmol (initial value) to 4.6 μmol (6 mo later). During this period, the amount of soluble receptor retained by the affinity column increased by a factor of 10. These data show that leakage of the covalently bound ligand is a parameter of primary importance in affinity chromatography. Even low levels of uncoupled ligand can compete very efficiently for binding of the soluble receptor and thus drastically limit the binding capacity of the affinity gel.

4.2.3. Affinity Chromatography of Prepurified Neurotensin Receptor

The binding activity eluted from SP-Sephadex C-25 and hydroxylapatite was loaded on the affinity column (Fig. 3, fractions 1–60) which was then washed with the Tris-glycerol buffer containing 0.1% CHAPS and 0.02% CHS (fractions 61–150). When the affinity column was working at its better level of efficiency, pooled fractions 1–150 contained almost all the loaded proteins but only 10–15% of the total binding activity. Proteins retained nonspecifically on the gel by ionic interactions were eliminated by elution with the washing buffer containing 200 m*M* KCl (fractions 151–170). A small amount (2–5%) of specific neurotensin binding activity was lost during this step. The column was rinsed one more time with 300 mL of washing buffer (fractions 171–240) and the neurotensin receptor was eluted with the washing buffer containing 1*M* NaCl (fractions 241–280). Under the best conditions, after months of repeated use of the affinity column, the NaCl fraction contained 75–80% of the total binding activity loaded on the column. Table 3 summarizes the purification data obtained under these optimal conditions.

After each run, the column was washed successively with 2 vol of 6*M* urea (pH 3), 2 vol of water, 2 vol of 0.5% sodium dodecylsulfate (SDS), and 2 vol of water, then reequilibrated with the washing buffer. The affinity gel was either used immediately for a new purification cycle, or stored at 4°C in the presence of 0.02% sodium azide.

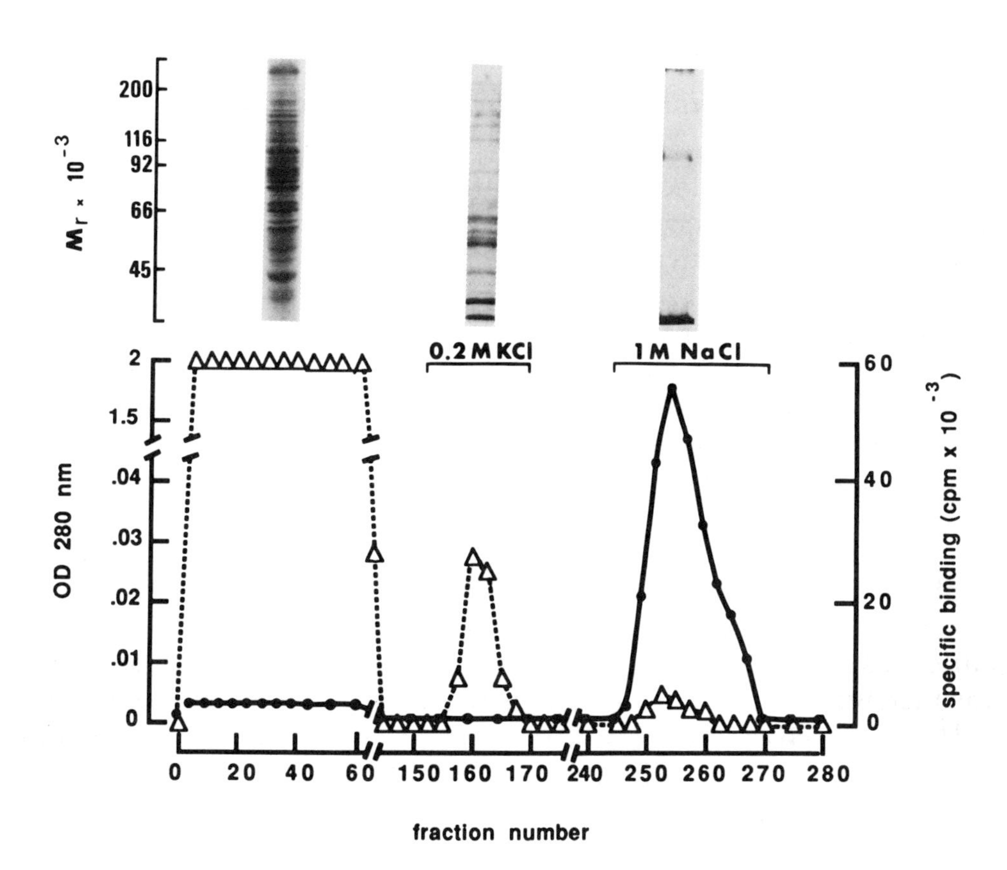

Fig. 3. Affinity chromatography of CHAPS-solubilized neurotensin receptor. Soluble proteins (300 mL, 1.1 mg/mL) eluted without retention from SP-Sephadex C-25 and hydroxylapatite were loaded at 4°C on the Ultrogel AcA22-neurotensin (2–13) column (3 x 15 cm) equilibrated with the Tris-glycerol buffer, pH 7.5, containing 0.1% CHAPS and 0.02% CHS. The column was eluted with (i) the equilibration buffer (fractions 60–150); (ii) the equilibration buffer containing 200 m*M* KCl (fractions 151–170); (iii) the equilibration buffer (fractions 171–240); (iv) the equilibration buffer containing 1*M* NaCl (fractions 241–280). The chromatography was followed by automatic recording of the absorbance at 280 nm (dashed line) and by measuring the specific binding of ^{125}I-labeled [Tyr^3] neurotensin (0.2 n*M*) to aliquots of each fraction previously desalted on Sephadex G-50 (solid line). Fractions corresponding to the flow through (left), the 0.2 *M* KCl eluate (center) and the 1 *M* NaCl eluate (right) were pooled and analyzed by SDS-PAGE. The corresponding Coomassie blue-stained gels are shown in the upper part of the figure. Fraction volume, 5 mL; flowrate, 60 mL/h.

4.3. Structural and Functional Properties of the Purified Receptor

4.3.1. Binding Parameters and Specificity

After elimination of NaCl by gel filtration on Sephadex G-50, the purified material eluted from the affinity column bound ^{125}I-labeled [Tyr3] neurotensin specifically and in a saturable manner (Fig. 2, insert). Binding parameters calculated from the corresponding linear Scatchard plot (Fig. 2) were $K_d = 0.26 \pm 0.09$ n*M* and $B_{max} = 31 \pm 6.5$ fmol/assay. From the amount of protein estimated by silver staining, the binding capacity of the purified receptor was calculated to be 7–8 nmol/mg of protein, corresponding to a purification factor of about 30,000 from the crude soluble receptor. This value is close to the theoretical value of 10 nmol/mg calculated on the basis of 1 neurotensin binding site per receptor of mol wt 100,000.

The specificity of the neurotensin receptor was not affected during solubilization and purification. The binding of ^{125}I-labeled [Tyr^3] neurotensin to the purified receptor was competitively inhibited by acetylneurotensin (8–13), neurotensin (9–13), and neurotensin (1–12) with relative potencies identical to those observed in membrane homogenate or crude CHAPS-solubilized extracts. The common order of decreasing affinity was neurotensin = acetylneurotensin (8–13) >> neurotensin (9–13) >> neurotensin (1–12).

4.3.2. Molecular Properties

Results in Fig. 4 show that affinity chromatography leads to the purification of a major protein band of 100kDa as determined by silver or Coomassie blue staining of the protein fraction eluted from the affinity column by 1*M* NaCl and analyzed by sodium dodecylsulfate-polyacrylamide gel electrophoresis (SDS-PAGE). Iodination of this fraction with ^{125}INa followed by SDS-PAGE and autoradiography gives the same result but allowed to detect some additional minor impurities. Finally, the 100-kDa protein was directly identified as the neurotensin receptor by affinity labeling with [^{125}I Tyr^3] neurotensin in the presence of disuccinimidyl suberate (Fig. 4). A value of 100,000 for the mol wt of the purified neurotensin receptor is in agreement with data obtained by gel filtration and affinity labeling of the crude soluble receptor (Mazella et al., 1988). Using similar experimental approaches, we have found that neurotensin receptors from rat and rabbit brain can also be solubilized and purified by affinity chromatography, and consisted of a single polypeptide chain of apparent mol wt 100,000 (results not illustrated).

Table 3
Purification of the Neurotensin Receptor from Newborn Mouse Brain by Affinity Chromatography

Step	Protein mg	Activity,[a] pmol/mg	Purifaction, -fold	Recovery, %
CHAPs extract	700	0.25	1	100
SP-HA eluate[b]	340	0.50	2	97
Affinity chromatography				
flow through	330	0.06	0.24	12
KCl elute	1	5	20	3
NaCl elute	0.01[c]	8050	32,200	78
Theoretical value[d]		10,000	40,000	

[a]Maximal binding capacity calculated from saturation experiments with [^{125}I Tyr] neurotensin.

[b]Fraction not retained on SP-Sephadex C-25 and hydroxylapatite.

[c]Determined by densitometric scanning of silver-stained SDS-PAGE bands using Bio-Rad protein markers as standards.

[d]Calculated for a 100-kDa protein receptor having a single neurotensin binding site.

5. Conclusion

It is shown in this chapter that an active and stable preparation of soluble neurotensin receptor can be obtained from mouse brain homogenate by using the detergent CHAPS in the presence of CHS. The soluble receptor can be purified essentially in a single step by affinity chromatography on an activated Ultrogel AcA22-neurotensin (2–13) column. Results in Fig. 4 show that the affinity-purified receptor is about 80% pure and that SDS-PAGE can be used as the last purification step. Therefore, the technology described here makes it possible to accumulate sufficient amounts of pure protein to undertake microsequencing of the receptor and cloning of its gene.

Acknowledgments

The authors are grateful to S. St. Pierre (Faculté de Médecine, Université de Sherbrooke, Québec, Canada) for the very generous gift of pure neurotensin and neurotensin (2–13). We wish to thank Valérie Dalmasso for expert secretarial assistance. This work was supported by the Centre National de la Recherche Scientifique, the Institut National de la Santé et de la Recherche Médicale and the Fondation pour la Recherche Médicale.

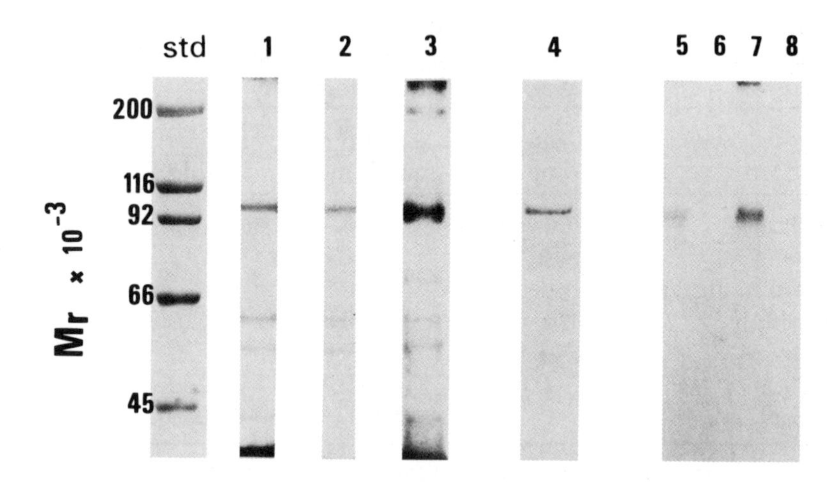

Fig. 4. SDS-PAGE analysis of the purified neurotensin receptor. Fractions eluted from the affinity column with the 1*M* NaCl solution were pooled and submitted to SDS-PAGE analysis in a 8% resolving gel with a 4% stacking gel. Std, protein standards. 1, silver staining; 2, Coomassie blue staining; 3, autoradiogram of an aliquot radioiodinated before electrophoresis; 4, the purified receptor was electrophoresed, the band comprised between 92 and 116kDa was sliced, electroeluted, reanalyzed by SDS-PAGE and silver stained. Five to 8, the purified receptor was desalted on Sephadex G-50 and incubated for 60 min at 0°C with 0.4 n*M* ^{125}I-labeled [Tyr^3] neurotensin in the presence (lanes 6 and 8) or in the absence (lanes 5 and 7) of 1 μ*M* unlabeled neurotensin. The ligand-receptor complex was crosslinked with 10 μ*M* (lanes 5 and 6) or 30 μ*M* (lanes 7 and 8) disuccinimidyl suberate for 15 min at 0°C. Samples were then frozen, lyophilized, and analyzed by SDS-PAGE. The labeled bands were detected by autoradiography of the dried gel.

References

Amar, S., Mazella, J., Checler, F., Kitabgi, P., and Vincent, J. P. (1985) *Biochem. Biophys. Res. Commun.* **129,** 117–125.

Amar, S., Kitabgi, P., and Vincent, J. P. (1986) *FEBS Lett.* **201,** 31–36.

Amar, S., Kitabgi, P., and Vincent, J. P. (1987) *J. Neurochem.* **49,** 999–1006.

Bozou, J. C., Amar, S., Vincent, J. P., and Kitabgi, P. (1986) *Mol. Pharmacol.* **29,** 489–496.

Carraway, R. E. and Leeman, S. E. (1973) *J. Biol. Chem.* **248,** 6854–6861.

Checler. F., Vincent, J. P., and Kitabgi, P. (1986a) *J. Biol. Chem.* **261,** 11274–11281.

Checler, F., Mazella, J., Kitabgi, P., and Vincent, J. P. (1968b) *J. Neurochem.* **47,** 1742–1748.

Gilbert, J. A., Moses, C. J., Pfenning, M. A., and Richelson, E. (1986) *Biochem. Pharmacol.* **35,** 391–397.

Goedert, M., Pinnock, R. D., Downes, C. P., Mantyh, P. W ., and Emson, P. C. (1984) *Brain. Res.* **323,** 193-197.

Hirsch Fernstrom, M., Carraway, R. E., and Leeman, S. E. (1980) in *Frontiers in Neuroendocrinology* , vol 6, (Martini, L. and Ganong, W. F., eds.), Raven, New York, pp. 103–127.

Kitabgi, P., Carraway, R. E., and Leeman, S. E. (1976) *J. Biol. Chem.* **251,** 7053–7058.

Kitabgi, P., Checler, F., Mazella, J., and Vincent, J. P. (1985) *Reviews in Basic and Clinical Pharmacology* **5,** 397–484.
Mazella, J., Poustis, C., Labbé, C., Checler, F., Kitabgi, P., Granier, C., Van Rietschoten, J., and Vincent, J. P. (1983) *J. Biol. Chem.* **258,** 3476–3481.
Mazella, J., Chabry, J., Kitabgi, P., and Vincent, J. P. (1988) *J. Biol. Chem.* **263,** 144–149.
Nemeroff, C. B., Luttinger, D., and Prange, Jr., A. J. (1982) *Handbook of Psychopharmacology* **16,** 363–467.
Poustis, C., Mazella, J., Kitabgi, P., and Vincent, J. P. (1984) *J. Neurochem.* **42,** 1094-1100.
Sadoul, J. L., Mazella, J., Amar, S., Kitabgi, P., and Vincent, J. P. (1984) *Biochem. Biophys. Res. Commun.* **120,** 812–819.
Schotte, A., Leysen, J. E., and Laduron, P. M. (1986) *Naunyn-Schmiedebergs Arch. Pharmacol.* **333,** 400–405.
Schotte, A. and Laduron, P. M. (1987) *Brain Res.* **408,** 326–328.

The Luteinizing Hormone/Human Chorionic Gonadotropin Receptor of Testis and Ovary

Purification and Characterization

Maria L. Dufau

1. Introduction

Biochemical, structural, and molecular characterization of gonadotropin and prolactin receptors and definition of their relationship for interaction with trophic hormone and coupling elements in the membrane facilitates the understanding of biochemical mechanisms that control the reproductive functions and is of general value to gain insight on the mechanisms of action of glycoprotein hormones.

The episodic secretion of LH supports the steroidogenic function through interaction with LH receptors and subsequent stimulation mainly of cAMP dependent events. The trophic actions of LH also include regulation of cell surface receptors for LH and prolactin (Dufau and Catt, 1978; Catt et al., 1980; Baranao and Dufau, 1983). In addition to the positive regulation of membrane receptors caused by physiological increases in endogenous hormone, major elevations in circulating hormone often cause downregulation of LH receptors, and desensitization of steroid responses in target cells (Dufau and Catt, 1978; Catt et al., 1980; Baranao and Dufau, 1983; Dufau, 1988; Conti et al., 1976).

The LH receptor of the rat testis was one of the first protein hormone receptors to be extracted from target-cells and characterized physicochemically (Dufau, Charreau, and Catt, 1973). In studies on both testis and ovary, the LH binding sites were identified as glycoproteins of MW 190,000 and 240,000 for the free receptor and hormone receptor complex, respectively, the latter form of the receptor had identical characteristics in vivo and in vitro (Dufau, Charreau, and Catt, 1973; Dufau et al., 1974, 1975). The mol wt corrected for detergent binding was for the free receptor M_r 160,000 taking in to account that the detergent, Triton X-100, bound to receptor was 0.22g/g protein (Dufau, Charreau, and Catt, 1973).

Testicular and ovarian LH receptors were completely resolved from adenylate cyclase (Dufau et al., 1978). The functional properties of LH receptors were also analyzed after their incorporation into heterologous steroidogenic cells. This was accomplished by the transfer of LH receptor (lipid associated ovarian LH receptor-soluble in the absence of detergent) to an heterologous cell, the adrenal cell, and demonstrated that such receptors were functionally coupled to adenylate cyclase as shown by cAMP and corticosterone responses to gonadotropin (Conti, Dufau, and Catt, 1978; Dufau et al., 1978).

In earlier studies from our laboratory, receptors of rat testis have been purified to homogeneity (Dufau et al., 1975) and appear to be composed of two identical subunits $M_r = 90{,}000$. However, the yield of the purified LH/hCG receptor from rat testis in the early studies were low, mainly owing to receptor degradation through the initial preparative steps and incomplete elution from the affinity support. The availability of the receptor in sufficient quantities and in pure form is necessary to undertake detailed structural and functional studies. The concentration of the LH/hCG receptors in the rat testis is small when compared with the receptor content of pseudopregnant rat ovaries (1 pmol/testis vs 2–4 pmol ovary). Despite the LH/hCG receptor abundance and obvious advantage to pursue characterization, for a number of years the efforts for the purification of the rat ovarian receptor by several laboratories were not fully successful because of marked instability of the receptor owing to proteolytic degradation. These studies on the purification of the receptor from rat and bovine ovary yielded multiple species of reduced biological activity of M_r ranging from 20–240kDa (Metsikko and Rajaniemi, 1982; Dattatreyamurty, Rathnam, and Saxcena, 1983; Bruch, Thobakura, and Bahl, 1986). Since the ovary is the most abundant source of LH receptor and also has the advantage to be a rich source of lactogen receptors, our laboratory elected to design methods to obtain purification of both receptors from ovaries from pseudopregnant rats.

Using relatively simple and rapid techniques, we have purified from the same starting material, the Triton solubilized ovarian LH/hCG and lactogen receptors, to homogeneity by sequential affinity column on lectin-Sepharose and hCG or hGH-Sepharose (Kusuda and Dufau, 1986) and also the Leydig cell LH/hCG receptor by two affinity cycles on hCG Sepharose (Minegishi, Kusuda, and Dufau, 1987). These methods allow the purification of receptors in microgram amounts for microsequence, structural analysis, and reconstitution studies.

In this chapter, we review purification and characterization and structural analysis of LH/hCG receptors.

2. Receptor Purification

2.1. Ovarian LH/hCG Receptors

As the initial step of purification, particulate rich membrane fraction obtained from ovaries of 29-d-old pseudopregnant rats containing (2 pmol/ovary) LH/hCG and (20 pmol/ovary) prolactin receptors are prepared and stored at –60°C until processed. At this point, ovarian membrane pellets from 150 ovaries are processed at one time for purification and all successive steps were followed without interruption until the final purification was achieved. For solubilization, the membrane pellet from 150 ovaries resuspended in 30 mL of PBS containing 20% glycerol, Triton X-100 1%, w/v, and 100 μ*M* phenylmethylsulfonyl fluoride (PMSF) for 60 min at 4°C. This solution was diluted 10-fold with PBS 100 μ*M* PMSF and then centrifuged for 60 min at 240,000*g*. The supernatant was decanted and this crude solubilized extract was used for subsequent purification of LH/hCG and lactogen receptors. The solubilized preparations were subjected to wheat germ lectin-Sepharose chromatography. To scale up the receptor preparation, we used several lectin affinity columns in parallel. A solubilized sample of 50 ovaries is applied to each 80 mL wheat germ column. Because of the lipidic nature of the ovarian triton extracts a considerable quantity of affinity lectin is required in this step of purification. For these reasons, and also because the binding capacity of wheat germ lectin-Sepharose for the lactogen receptor appeared to be relatively lower than for the LH/hCG receptor, it is important to use the optimal concentrations of wheat germ Sepharose to achieve complete purification of the receptor in the subsequent step. The latter probably is a reflection of differences on *N*-acetylglucosamine content between the two receptors (Mitani and Dufau, 1986). Since both receptors bound to the column and were eluted as a single sharp peak with a linear gradient (0.01–0.2*M*) of *N*-acetylglucosamine

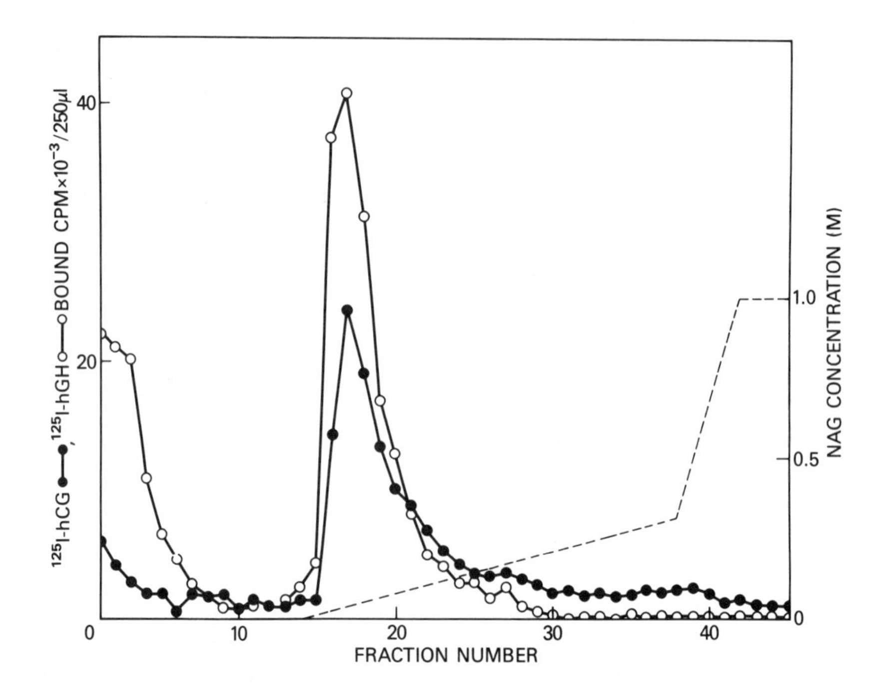

Fig. 1. Initial step for combined purification of ovarian LH/hCG and Prolactin receptors. The ovarian Triton X-100 solubilized preparation was applied to a wheat-germ lectin-Sepharose and eluted with stepwise gradient of *N*-acetylglucosamine (NAG). Aliquots were assayed for receptor activity by ^{125}I-hCG and ^{125}I-hGH binding assay. Protein concentrations (---) were monitored by absorbance at 280 nm.

(Fig. 1) for routine preparation of purified LH/hCG and lactogen receptors a stepwise elution with 0.2*M* *N*-acetylglucosamine can be used (Fig. 2), and aliquots from each fraction eluted by the competing sugar were monitored only for ^{125}I-hCG binding activity by rapid receptor assay. The sharp elution profile obtained for the LH/hCG receptor on wheat germ lectin-Sepharose (Kusuda and Dufau, 1986) contrasted with the several peaks of hCG binding activity observed previously in concanavalin A Sepharose (Wimalasena, Schwab, and Dufau, 1983). The LH/hCG receptor bound to the wheat germ lectin-Sepharose more effectively and gave a considerable higher yield than Con A-Sepharose, since by the latter procedure considerable hCG binding activity was not adsorbed to the column (Wimalasena, Schwab, and Dufau, 1983). Approximately a 20-fold purification for the LH/hCG receptor and 20-fold for the lactogen receptor was obtained with the lectin affinity step (Kusuda and Dufau, 1986). Fractions from wheat germ lectin-Sepharose containing receptor binding activity were pooled

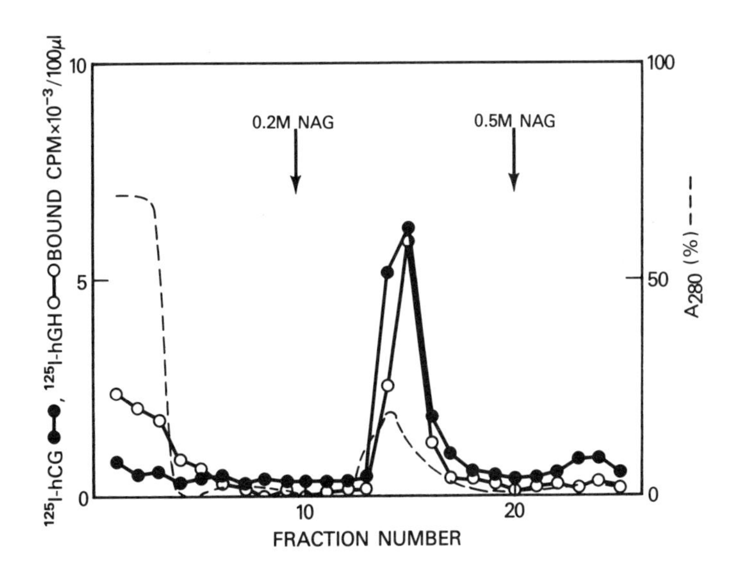

Fig. 2. Initial purification step of solubilized LH/hCG and prolactin receptors by wheat-germ lectin-Sepharose using a stepwise gradient of *N*-acetylglucosamine (NAG). Aliquots were assayed for receptor activity by ^{125}I-hCG and ^{125}I-hGH binding assay. Protein concentrations (— —) were monitored by absorbance at 280 nm.

and subjected to hCG-Sepharose affinity chromatography. Highly purified hCG receptor was eluted as a sharp peak with acetic acid (Fig. 3). Subsequently, the flowthrough fractions containing lactogen receptor activity were subjected to hGH-Sepharose affinity column following steps described by Mitani and Dufau, 1986.

Table 1 summarizes a representative purification cycle of the LH/hCG receptor and the binding constants of purified receptors determined by Scatchard plot analysis of ^{125}I-hCG displacement experiments. Affinity purified LH/hCG receptor bound ^{125}I-hCG specifically with a K_a of $0.94 \times 10^{10} M^{-1}$ and a binding capacity of 5.1–7 nmol/mg protein. The receptor preparations obtained by the above procedure are homogeneous. The differences observed between the theoretical specific activity and that obtained experimentally could be attributable to losses in the binding activity during receptor purification and/or dimerization of the receptor. This contrasted with the considerable stability of lactogen receptors, where the experimental specific activity was very close to the expected theoretical activity (Table 1).

After solubilizing LH/hCG receptors from rat ovarian particulate fractions, the receptors are rapidly degraded to lower M_r forms, presumably by a proteolytic process. This is overcome by including the protease

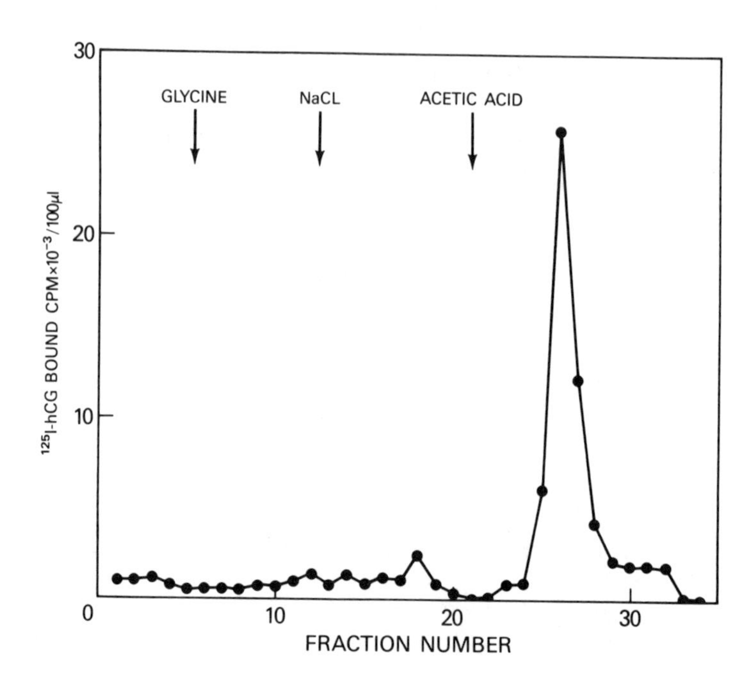

Fig. 3. Purification of ovarian LH/hCG receptor by hCG-Sepharose affinity chromatography. The active fractions eluted from wheat germ lectin-Sepharose with 0.2*M* *N*-acetyl-glucosamine were pooled and subjected to hCG-Sepharose chromatography. The column was washed with 20 mL of PBS containing 0.1% Triton X-100, 5 mL of 50 m*M* glycine/HCl containing Triton X-100 followed by 20 mL PBS containing 2*M* NaCl/0.1% Triton X-100. The LH/hCG receptor was eluted with 25 m*M* acetic acid pH 4 containing 1*M* NaCl and 0.1% Triton X-100, LH/hCG and fraction were neutralized immediately. Pool of samples with lactogen binding activity collected in flowthrough fractions of hCG-Sepharose chromatography (not shown) were applied to the hGH-Sepharose (Kusuda and Dufau, 1986; Mitani and Dufau, 1987). Aliquots from affinity columns were assayed for receptor binding activity by binding assay with ^{125}I-hCG and ^{125}I-hGH.

inhibitor 0.1 m*M* phenylmethyl sulfonyl fluoride to all reagents in the purification sequence and 20% glycerol, in the solubilization buffer, by solubilization of particulate fractions with high specific binding activity and a rapid initial purification step on wheat germ lectin-Sepharose, to be followed by the specific affinity step. Delays between the chromatographic steps in the absence of the hormone (hCG or LH) or freezing and thawing of the preparation resulted in considerable losses of binding activity. Following the cited scheme of purification, the overall recovery is in the range of 20% and 3 µg of receptor can be obtained from 100 rat ovaries.

Table 1
Purification of Ovarian Receptors

Sample	Protein specific activity		K_a		Recovery	
	hCG	PRL	hCG	PRL	hCG	PRL
	(pmol/mg protein)		($\times 10^{-10} M^{-1}$)		%	
Crude homogenate	0.8	—	—	—	100	—
Particulate membranes	1.2	0.7	1.3	0.3	96	—
Triton X-100 solubilized membranes	1.3	7.4	0.7	0.1	48(100)	100
Wheat-germ lectin-Sepharose purified receptors	20.5	95.0	0.6	0.1	33(69)	60
hCG-Sepharose purified receptors	5100–9000		0.9	2.8	10(21)	10
hGH-Sepharose purified receptors	—	18,500	—	3.5	—	15–20

2.2. Testicular LH/hCG Receptor

The Triton X-100-solubilized membrane preparations for use in testicular receptor purification exhibited binding capacity of about 0.5 pmol/testis. The receptors solubilized by Triton X-100 were significantly less stable than the particulate-binding fractions even during exposure to low temperature (2–4°C) or storage at –70°C. Since the soluble hormone-receptor complexes are relatively stable, the detergent-extracted free receptors were subjected to hCG affinity chromatography immediately after solubilization. A large number of nonspecifically bound proteins were removed effectively in the washes, and the binding activity corresponding to the LH/hCG receptor bound to the column was eluted with 25 m*M* acetic acid/1*M* NaCl, pH 4. However, homogeneity of the receptor preparation was observed only after a second purification cycle on hCG-Sepharose (Fig. 4).

About 20,000-fold purification of LH/hCG receptor was obtained by sequential hCG affinity chromatography, and the overall recovery of activity from Triton X-100-solubilized preparations was in the range of 10%. The original soluble receptor preparation and the purified receptor bound a wheat germ lectin-Sepharose column and was effectively eluted with *N*-acetylglucosamine with recoveries of 50%.

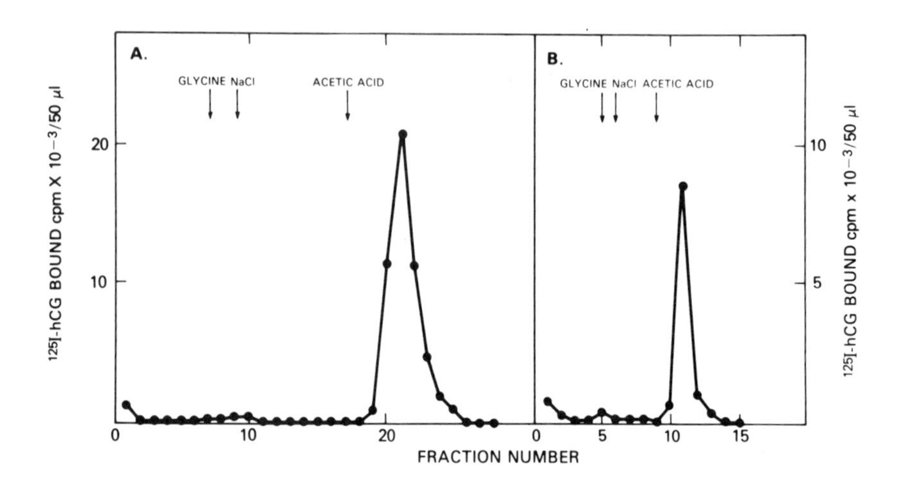

Fig. 4. Purification testicular LH/hCG receptor by sequential hCG-Sepharose affinity chromatography. The active fractions eluted from the first hCG-Sepharose with 25 m*M* acetic acid, 1*M* NaCl were pooled and subjected to a second hCG-Sepharose chromatography.

3. Receptor Characterization

3.1. SDS Analyses of the Free Receptor

Silver or Aurodye staining of polyacrylamide gel of ovarian purified receptor under reducing conditions revealed a single protein band of M_r 80,000 ± 4,000 (Fig. 5) and similar results were observed when the pure receptor was resolved under nonreducing conditions (Kusuda and Dufau, 1986). Generally, no higher M_r bands were observed when the purified receptor was resolved under nonreducing conditions, but sometimes a minor band representing the dimeric are present.

Comparable M_r for the purified ovarian human and porcine LH/hCG receptor was reported by Wimalasena et al., 1985, whereas a slightly higher M_r for the rat ovarian receptor was reported by Keinanen et al., 1987.

Silver and Aurodye staining of the purified testicular LH/hCG receptor following SDS-PAGE analysis showed a single protein species of approx M_{rs} 88,000–92,000 for testicular receptors (Fig. 6). Similar results were observed with nonreducing conditions (Minegishi, Kusuda, and Dufau, 1987).

The receptor always appeared to migrate as a broad band in SDS gel, and this might indicate microheterogeneity of the receptor moiety. Very minor bands of M_r = 66,000 and M_r = 50,000 were not always present and could represent artifactual bands previously observed in SDS-PAGE under reducing condition (Tasheva and Dessev, 1983) or minor quantities

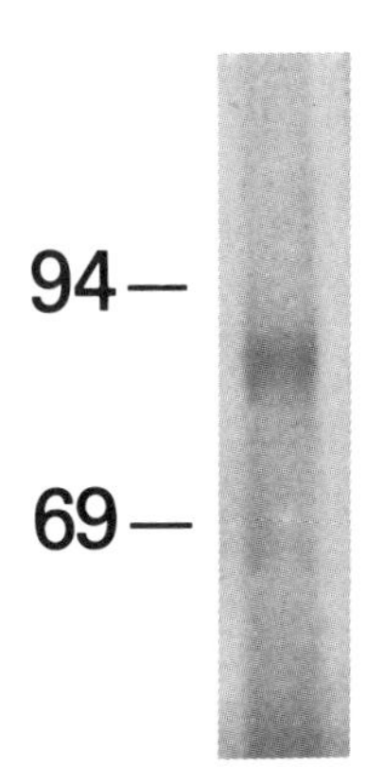

Fig. 5. SDS-PAGE analysis of purified ovarian LH/hCG receptor SDS-PAGE was performed using a Laemmli discontinuous buffer system and 10% acrylamide in the separation gel. Protein was transferred electrophoretically to nitrocellulose and stained with Aurodye forte.

of degraded receptor. Affinity purified LH/hCG receptor bound ^{125}I-hCG specifically with K_a of $0.43 \times 10^{10} M^{-1}$ and a binding capacity of 3.8–5 nmol/mg of protein.

Autoradiographs of ^{125}I-hCG binding to blotted ovarian and testicular Leydig cells LH receptor resolved in SDS-PAGE, showed radioactive bands of M_rs 80,000–85,000 (ovary) and 88,000–92,000 (testis), respectively (Fig. 7), similar to that obtained by staining of the gels of the ovarian and testis and from autoradiographs of iodinated and phosphorylated receptors (Kusuda and Dufau, 1986; Minegishi, Kusuda, and Dufau, 1987; Minegishi, Delgado, and Dufau, 1989).

The M_r value observed for the blotted receptor was consistent with that revealed byAurodye or silver staining. The fact that the blotted receptor was labeled with ^{125}I-hCG provided direct evidence for the biological activity of this molecular species (Kusuda and Dufau, 1988; Minegishi, Kusuda, and Dufau, 1987).

Thus, the M_r of the pure ovarian receptor LH/hCG appears to be somewhat smaller than the M_r of receptor from Leydig cells (M_r = 80,000–85,000) vs M_r 88,000–92,000) (Kusuda and Dufau, 1986, 1988; Minegishi, Kusuda, and Dufau, 1987).

The purified receptor, was characterized by radioiodination with 1,3, 4,6-tetrachloro-3α, 6α-diphenylglycouril (Iodo-Gen). Iodo-Gen is known as a gentle and simple technique for radioiodination of proteins. This reaction is not affected by detergents and can be used over a broad pH and temperature range. Although Triton X-100 has no effect on the chemical

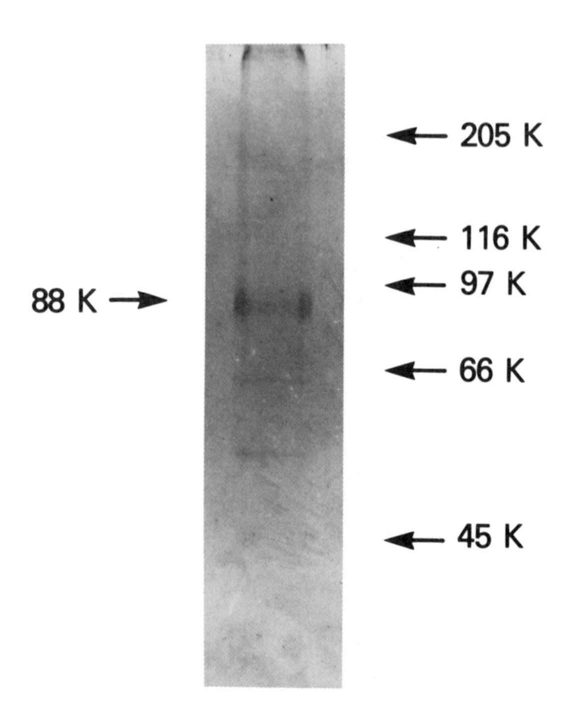

Fig. 6. SDS-PAGE analysis of purified testicular LH/hCG receptor. Purified LH/hCG receptor in sample buffer containing 2% SDS and 2% β-mercaptoethanol was heated at 100°C for 2 min and electrophoresed in a 7.5% SDS-polyacrylamide gel. The protein bands were visualized by silver staining. The positions of the M_r standards, rabbit muscle myosin (205,000), β-galactosidase (116,000), phosphorylase b (97,000), bovine albumin (66,000), and obalbumin (45,000) are indicated on the right of the gel. A band of M_r = 90,000 ± 2,000 is revealed by silver staining (Minegishi, Kusuda, and Dufau, 1987). The very minor bands at M_r 66,000 and 50,000 were not always present and could represent artifactual bands previously observed in SDS PAGE under reducing conditions (Mitani and Dufau, 1986).

reaction, it can be also radioiodinated along with proteins, and, for this reason, the detergent content of the receptor sample is reduced from 0.1 to 0.01% during the final step of purification when subsequent iodination of the receptor is contemplated. Under this condition, the receptors remained are soluble, and radioiodination of Triton X-100 is minimized. Autoradiography of SDS-PAGE analysis under reducing conditions showed a single major band migrating at M_r = 80,000. Under nonreducing conditions, migrated slightly faster, reflecting the folding of the receptor molecules. These studies again demonstrate that the receptors are single protein species (Fig. 8).

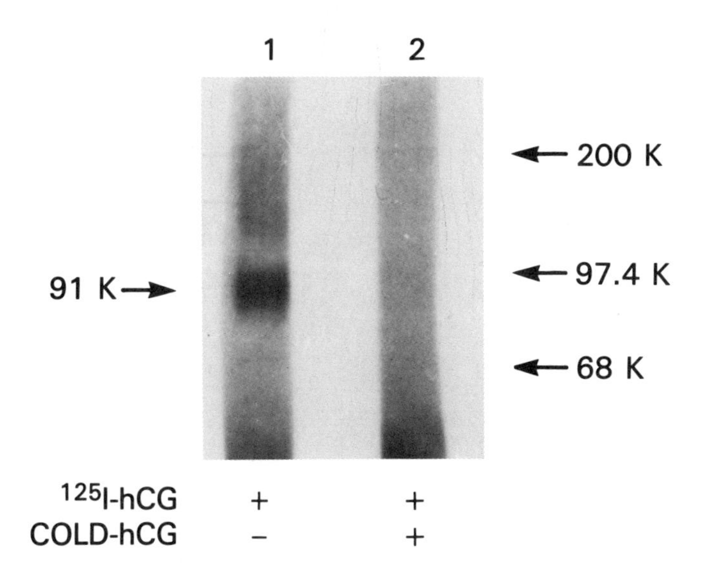

Fig. 7. Blotting of purified testicular LH/hCG receptors. Purified receptors were electrophoresed on a 7.5% SDS-PAGE under nonreducing conditions and transferred electrophoretically onto a nitrocellulose sheet. The blot was incubated with 10^8 dpm of ^{125}I-hCG in the absence (–) or presence (+) of an excess unlabeled hCG for 16 h at 4°C. After washing, the blots were dried for autoradiography. The positions of the molecular weight markers are indicated on the left of the autoradiogram. Incubation of the blot with ^{125}I-hCG labeled revealed a band corresponding to M_r 88,000 (Minegishi, Kusuda, and Dufau, 1987).

3.2. Monomer and Associated Receptor Forms

From our previous findings using sedimentation and gel filtration analysis (Dufau, Charreau, and Catt, 1974) of both free and occupied receptors of M_r 180–190,000 and 240,000, respectively, compared with affinity purified LH/hCG receptor purified from Leydig cells by Affigel 10 displaying on SDS gels an M_r 90,000, we had proposed that the native receptor existed as a dimer of identical subunits (Dufau et al., 1975). In more recent studies resolution of the free testicular and ovarian receptors by fast protein liquid chromatography on Superose 12 column revealed a single peak that eluted in the position of M_r 200,000–240,000 in presence of Triton X-100 (Kusuda and Dufau, 1986; Minegishi, Kusuda, and Dufau, 1988). Since a single protein species is observed with standard SDS-PAGE under

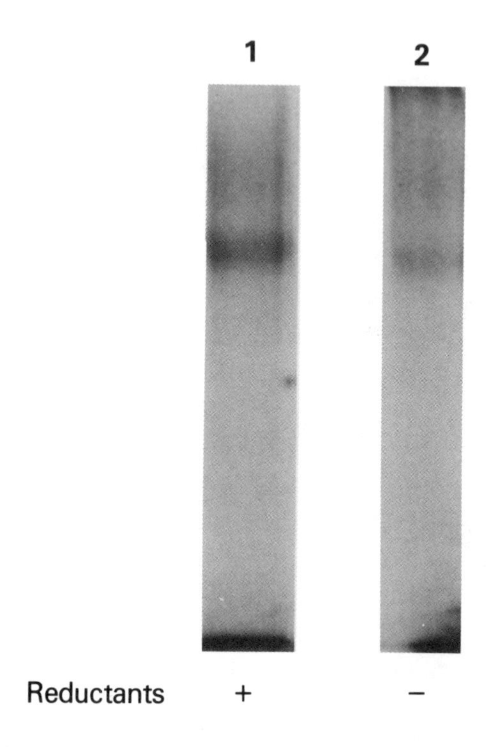

Fig. 8. SDS-PAGE analysis of radioiodinated purified ovarian LH/hCG receptors. Radioiodinated receptors were taken in sample buffer containing 2% SDS in presence (+) or absence (–) of 5% 2-mercaptoethanol, and heated at 100°C for 2 min. Mixtures were then applied on a 10% SDS polyacrylamide gel. The gel was run under constant voltage (150 V) and the dried gel was subjected to autoradiography.

reducing and nonreducing conditions, it is conceivable that the receptor could exist in the membrane as dimeric form composed of noncovalently associated subunits of approx M_r 80,000 (ovary) and 90,000 (testis). Rapid resolution of nondenatured pure LH/hCG receptor in minigels under reduced SDS concentrations (1%) allowed us to observe dimeric forms of the receptors (Kusuda and Dufau, 1986). Autoradiograms of ovarian ^{125}I-hCG bound to electroblotted receptors resolved with SDS-PAGE gave two active receptor species, the monomeric form of the receptor of M_r 80,000, and the dimeric form of M_r about 160,000 (Fig. 9, left) (Kusuda and Dufau, 1986). The monomeric and dimeric forms of the receptor are revealed also by staining of minigels (Fig. 9, right). Although both forms can bind hormone, it is conceivable that the dimeric form could be necessary for signal transduction.

Molecular weights of free and hormone bound LH/hCG receptor derived from gel filtration and sucrose gradient centrifugation indicated binding of one molecule of hormone per molecule of receptor (Dufau, Charreau, and Catt, 1973; Dufau et al., 1974; Dufau, Podesta, and Catt, 1975). Also, analysis of the actual binding of hCG, determined by counting the bound hormone to monomer and dimer on nitrocellulose membrane in relation to receptor mass, has demonstrated that one molecule receptor of biological activity 5.1 nmol/mg protein appears to bind one molecule of hormone per receptor unit. The discrepancy between experimental and theoretical specific activities (5.1 vs 12.5 nmol/mg protein) was previously attributed to losses of binding activity during purification affecting conformational aspects of the binding site of the receptor (*1*). However, since the experimental specific activity of the ovarian receptor is always determined by binding studies in Triton X-100, where the receptor is presumably in the dimeric form, as determined by gel filtration methods, it is conceivable that in the dimeric form only one subunit could be capable to bind hormone. The latter could be attributed to masking of the receptor-binding site of the other subunit as a result of receptor dimerization. Occupancy of receptors is not necessary for dimerization, however, it is possible that occupancy of the receptor subunit (monomer) would favor dimerization.

3.3. Crosslinking Studies

To further characterize the hCG receptor and its interaction with gonadotropin, the purified receptor was crosslinked to labeled hCG (Kusuda and Dufau, 1986; Minegishi, Kusuda, and Dufau, 1987) with the bifunctional crosslinker DSS, and analyzed by SDS-PAGE (only α-subunit is labeled when intact hCG is radioiodinated). Autoradiographs of SDS-PAGE of the crosslinked purified receptor showed two bands at M_r 145,000, and 108,000 for testis and at M_r 130,000, 95,000 (ovary). The two labeled bands correspond to the receptor-hCG αβ complex and the receptor-hCG α-subunit complex, respectively (*see* below, Figs. 10 and 11). The identity of the receptor-hCG αβ complex was demonstrated by DSS crosslinking of the purified receptor incubated with ^{125}I-hCG crosslinked among subunits with dimethyl aminopropyl-carbodimide (EDC) that on SDS-PAGE revealed only the high mol wt band (Kusuda and Dufau, 1986). The high M_r bands 140,000 and 130,000 represents receptor crosslinked to αβ crosslinked hormone, whereas the low M_r bands (M_r 108,000 testis and 95,000 ovary) represent the receptor crosslinked to an individual α-subunit of the hormone. In the latter case, the corresponding unlabeled β-subunit not crosslinked to the α-subunit by DSS is being dissociated during SDS frac-

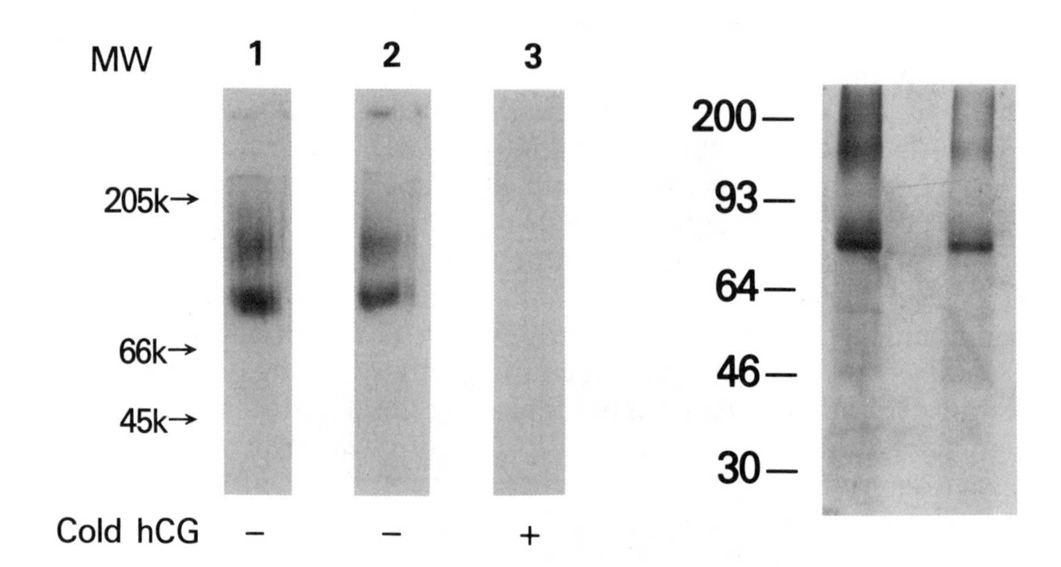

Fig. 9. A, Left, Binding of ^{125}I-hCG to monomeric and dimeric forms of the ovarian receptor. Purified receptors were resolved in 1% SDS (lane 1) or 0.25% SDS (lanes 2 and 3) and analyzed on a 7.5% SDS-polyacrylamide gel using rapid resolution in minigel under nonreducing conditions at 4°C. The receptors were then electroblotted onto a nitrocellulose membrane. The blots were further incubated with ^{125}I-hCG (1.5 × 10^6 dpm, 25 ng) in the absence (lanes 1 and 2) or in the presence (lane 3) of an excess unlabeled hCG and dried for autoradiography. B, Right, Aurodye staining of electroblotted receptor monomeric and dimeric forms. Two different preparations of purified ovarian LH/hCG receptor were resolved with 7.5% SDS polyacrylamide minigel and transferred onto cellulose sheet. Protein bands were stained with Aurodye.

tionation of the hormone. Thus, the relative intensity of the higher molecular band is directly proportional to the efficiency of the crosslink-ing reaction between the hormone subunits (25–40%). Since intact hCG, crosslinked by DSS, and the free or dissociated α-subunit migrated in SDS PAGE at M_r 55,000 and 24,000, respectively (Kusuda and Dufau, 1986), the results are consistent with an M_r value for the testis receptor of 90,000 and the ovarian receptor of 80,000 after accounting for contribution of the intact hormone or free α-subunit. The above findings are consistent with values derived from studies on chemical crosslinking of membrane testicular and ovarian LH/hCG receptors (Rebois et al., 1981; Petaja et al., 1987), 35S-methionine labeled LH receptors in vitro (Aubry et al., 1982) and by immunoblots of membrane and of partially purified ovarian receptors using monoclonal and polyclonal antisera (Podesta et al., 1983; Rosenblit, Ascoli, and Segaloff, 1988).

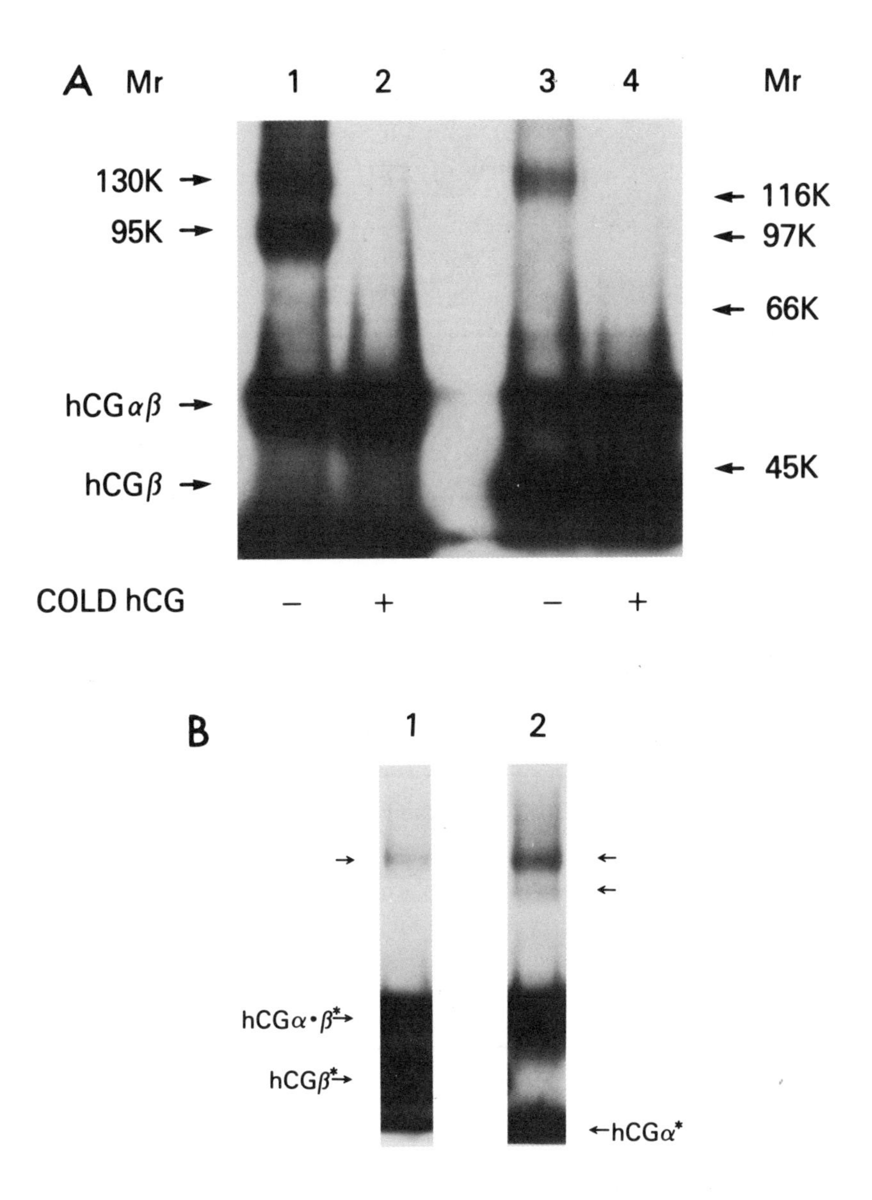

Fig. 10. Crosslinking of ^{125}I-hCG with label in α-subunit or β-subunit to purified LH/hCG receptors. Purified LH/hCG receptors were incubated with ^{125}I-hCG with label in α-subunit (A, lanes 1 and 2; B, lane 2) or in β-subunit (A, lanes 3 and 4; B, lane 1) in the absence (A, lanes 1 and 3; B, lanes 1 and 2) or presence (A, lanes 2 and 4) of 1 μg of unlabeled hCG. The mixtures were crosslinked with DSS. The samples were then heated at 100°C for 2 min at 2% SDS and in the presence (A, lanes 1, 2, 3, and 4) or absence (B, lanes 1 and 2) of 2% β-mercaptoethanol and subjected to SDS-PAGE using a 7.5% polyacrylamide gel. The gel was dried and subjected to autoradiography. The positions of the M_r standards are indicated on the right of the autoradiogram.

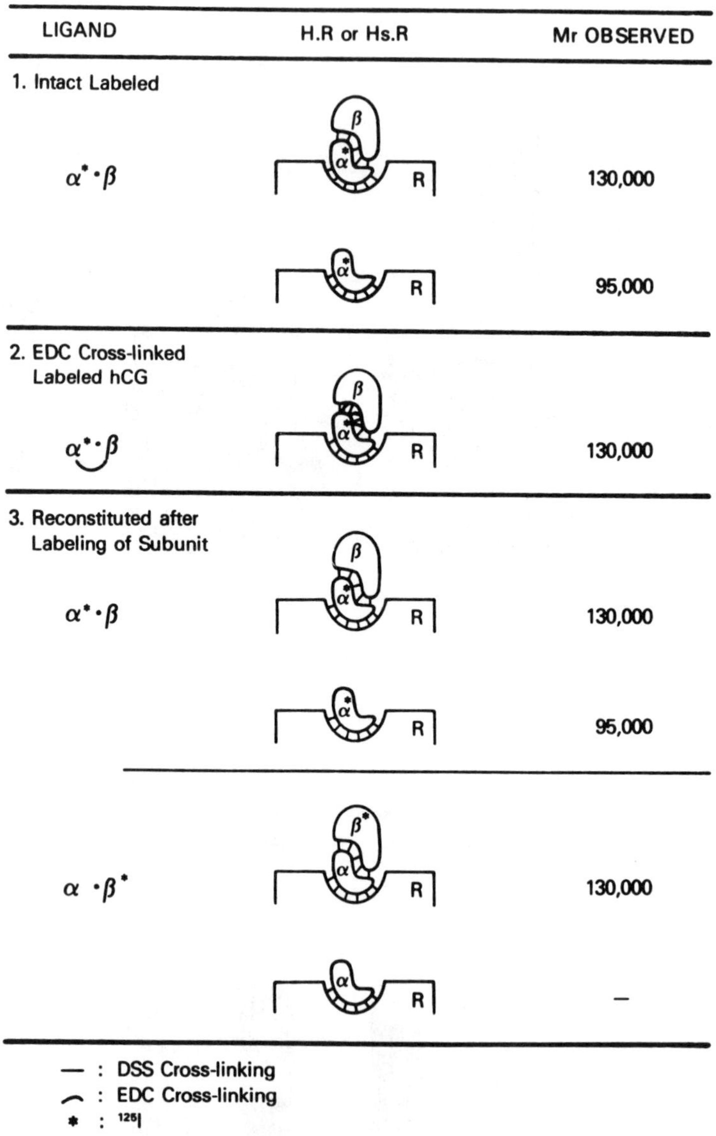

LIGAND	H.R or Hs.R	Mr OBSERVED
1. Intact Labeled		
$\alpha^{*}\cdot\beta$		130,000
		95,000
2. EDC Cross-linked Labeled hCG		
$\alpha^{*}\cdot\beta$		130,000
3. Reconstituted after Labeling of Subunit		
$\alpha^{*}\cdot\beta$		130,000
		95,000
$\alpha\cdot\beta^{*}$		130,000
		–

— : DSS Cross-linking
⌒ : EDC Cross-linking
* : ^{125}I

Fig. 11. Summary of results from experiments crosslinking ^{125}I-intact labeled hCG or reconstituted hormone after labeling of individual subunits to purified LH/hCG re-ceptor using disuccinyl suberate (DSS). The hCG α-subunit is intimately bound to the LH/hCG receptor, whereas the hCG β-subunit appears not to be involved in direct interaction with the receptor. R = receptor; H = hormone; Hs = hormone-subunit; DSS crosslinking; *: ^{125}I-labeled; EDC: hCG hormone subunits crosslinked with amino-propyl-carbodiimide (EDC), the active crosslinked hormone is purified by binding to ovarian membranes and eluted with acid. The EDC crosslinked hormone was used for crosslinking of hormone receptors with DSS.

The ovarian receptor crosslinked to recombined hCG (from individually subunits) with ^{125}I label in the α-subunit revealed two specific bands M_r 130,000 and 95,000, corresponding to receptor-hCG αβ complex and receptor-hCG α-complex, respectively, whereas only the higher molecular weight band was observed for the receptor crosslinked to recombined hCG with label in β-subunit, corresponding to receptor-hCG αβ-complex (Kusuda and Dufau, 1986). No lower molecular weight band of M_r 113,000, corresponding to a receptor-hCG β-subunit complex, that would appear if β-subunit would have been directly interacting with the receptor (Fig. 10). Similar findings were reported using covalent crosslinking of individually labeled β-subunit to rat LH receptor in crude ovarian membranes with glutaraldehyde (Petaja et al., 1987). Thus, these results demonstrated that the major interaction of hCG with its receptor occurs through its α-subunit and that the β-subunit would be involved through its association with and presumably conformational influence on the α-subunit. However, it is possible that the β-subunit of hCG was not crosslinked to the receptor as a result of inaccessibility of relevant amino acids. This is unlikely, since the β-subunit of hCG possesses 4 lysine residues and crosslinks efficiently to the α-subunit. Furthermore, the β-subunit of oLH is known to possess its 2 surface lysine residues available for substitution (Liu et al., 1974). It is also conceivable that an initial association of the β-subunit with the cell surface receptor could occur at an early stage. Such interaction would probably not be detectable in crosslinking studies performed after prolonged equilibration of hormone with the receptor. However, the latter was not revealed in studies where crosslinking of hCG with labeled in the β-subunit was performed at early times (Kusuda and Dufau, unpublished observations).

The direct demonstration of a predominant role of α-subunit is consistent with previous results of preferential masking of immunoreactive sites of α-subunit during receptor binding (Milius, Midgley, and Birken, 1983) and also with the dominant role of sugar residues of the α-subunit in the hormonal signal transduction (Sairam and Bhargavi, 1985).

Taken together, these findings indicate that the α-subunits that are identical within species seem to have an important role in the binding interaction and biological action of glycoprotein hormones. The disimilar β-subunits might serve mainly to induce conformational changes of the α-subunit and to confer the inherent specificity of each hormone for interaction with its receptor.

4. Receptor Phosphorylation

The catalytic subunit of cAMP-dependent protein kinase is able to phosphorylate the purified testicular and ovarian LH/hCG receptor (Minegishi, Kusuda, and Dufau, 1987; Minegishi, Delgado, and Dufau, 1989).

The time course of receptor phosphorylation by the cAMP-dependent protein kinase was relatively slow with minor phosphorylation observed at 5 min and increased gradually at 15 min, reaching maximal at 2 h.

The rate of receptor phosphorylation was 0.08 nmol of P/min/mg protein. The binding affinity of the receptor measured at equilibrium is not influenced by phosphorylation with Ka $0.44 \times 10^{10} M^{-1}$ and $0.43 \times 10^{-1} M^{-1}$ for native and phosphorylated receptors, respectively.

Although DTT is required for hCG receptor phosphorylation by the catalytic subunit of protein kinase, its concentration must be optimally adjusted to prevent the phosphorylation of coincubated hCG. The concentration of DTT used (0.3 m*M*) in phosphorylation studies was optimal to obtain phosphorylation of the receptor and to avoid concomitant phosphorylation of intact hCG. With higher DTT concentrations, phosphorylation of hCG by the catalytic subunit of protein kinase was observed, perhaps reflecting phosphorylation of the β-subunit that most likely occurred because of slight dissociation or unfolding of native hCG during incubation (Keutmann et al., 1983).

hCG caused a significant increase in the ovarian and testicular LH receptor phosphorylation over control levels at 5 min (Fig. 12). The differences between the control and hCG treated receptors were maintained for up to 20 min, whereas at 30 and 60 min, the control and hCG-treated receptor showed similar degrees of phosphorylation.

These results have demonstrated that addition of hCG induce enhancement of phosphorylation of the LH/hCG receptor by increasing the rate rather than the extent of phosphorylation and suggest that early occupancy of the receptor by hCG leads to a conformational change that facilitates its phosphorylation by protein kinase A (23).

In subsequent studies, preincubation of the ovarian or testicular receptor with hormone from 20–60 min prior to phosphorylation, demonstrated that prior occupancy of the pure receptor with hCG reduced significantly the rate of receptor phosphorylation (Minegishi, Delgado, and Dufau, 1989).

Early studies by Katikineni and Catt (Katikineni et al., 1980) have shown two stages of hCG binding to Leydig membranes, initially a "loose state" where the hormone can be dissociated readily from its binding site,

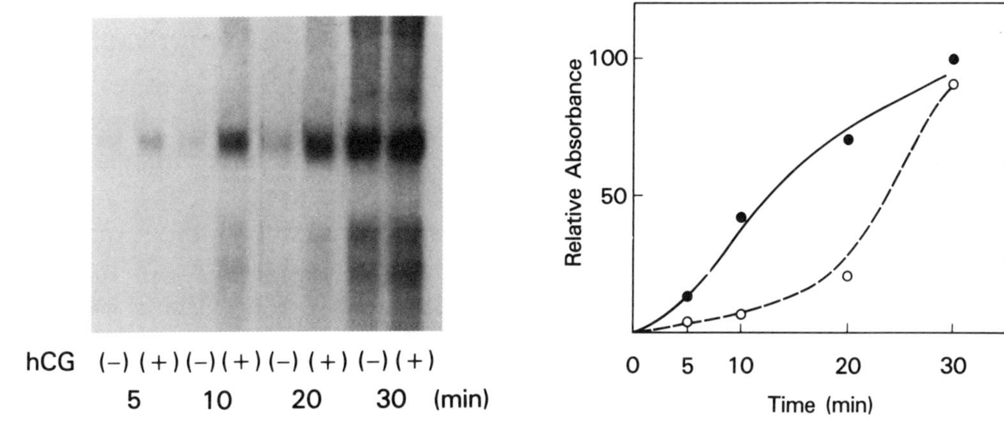

Fig. 12. Time course of phosphorylation of ovarian LH/hCG receptor by protein kinase A in the presence and absence of hCG. Purified receptor were incubated with the catalytic subunit of cAMP-dependent protein kinase (6,23) in the presence or absence of 1 μg of hCG at 25°C for the indicated period of time. $NaDodSO_4$/PAGE analysis (7.5%) of testicular receptors phosphorylation from 5–30 min.

and later becomes "tightly bound" state. It is conceivable that during preincubation the hormone becomes tightly bound to the receptor, and that changes in conformation would render the phosphorylation sites less accessible.

Desensitization of muscarinic responses could be brought about by phosphorylation of the receptor, leading to reduced affinity for binding to agonist ligands (Kwatra et al., 1987). On the other hand, phosphorylation by cAMP dependent protein kinase and protein kinase C did not affect the binding characteristics of the β_2-adrenergic receptor, but the former reduced the ability of the receptor to stimulate GTPase activity of Ns (Benovic et al., 1985).

Similarly, it is likely that in the case of the LH/hCG receptor, phosphorylation may reduce the ability to interact with biochemical effectors, although it is conceivable that reduced transduction could be preceded by stimulatory actions at early times.

The phosphorylated testicular and ovarian receptors bound effectively to hCG Sepharose and wheat germ lectin. This is in good agreement with our findings demonstrating that phosphorylation of the receptor did not affect the binding activity of the receptor for hCG. The phosphorylated receptor resolved as a single band at 90,000 and 85,000, for testis and ovary, respectively (Minegishi, Kusuda, and Dufau, 1987; Minegishi, Delgado, and Dufau, 1989), Fig. 13. Such differences in the mol wt of LH/hCG receptor among both tissues have been observed in our previous studies using nonphosphorylated receptor (*see above*).

Phosphopeptide maps obtained by reverse phase FPLC following trypsinization of both Leydig cell and ovarian receptor from both testis and ovary phosphorylated by protein kinase A, were identical. Six peaks contained phosphoserine, a major peak was found to be phosphorylated in threonine as well (Minegishi, Delgado, and Dufau, 1989). The similarity between the two maps could reflect the homologous amino acid sequences and phosphorylation sites for cAMP dependent protein kinase. Thus, the observed differences in size of the testicular (90,000–92,000) and ovarian (80,000–85,000) LH/hCG receptor may be the result of posttranslational modification rather than difference in primary sequence.

5. The Receptor Glycosyl Residues

For investigation of the contribution of glycosylation to the receptor-protein moeities, the ovarian and testicular receptors (native, iodinated, or phosphorylated) were subjected to enzyme deglycosylation. Endoglycosidase D and H are known to hydrolyze the glycosidic band between the two *N*-acetyl glucosamine residues than link the carbohydrate to the asparagine of the protein backbone. No demonstrable effect were observed on receptor molecules following treatment with Endoglycosidase H (Kusuda and Dufau, 1988). This result suggested that the receptor had no high mannose type glycosylation and are consistent with earlier findings that show weak binding to concanavalin A-Sepharose column. Also, no changes were observed by treatment of receptors with Endoglycosylase D, which cleaves only complex type carbohydrate moieties from *N*-linked glycoprotein (Kusuda and Dufau, 1988). The lack of effect of Endoglycosidase D treatment can be attributed to the narrow specificity of this enzyme.

Treatment with neuraminidase of both receptors increased the mobility of M_r 82,000 (testis) and 77,000 (ovary). The presence of sialic acid in the receptor moieties was also suggested by [^{3}H]NaBH$_4$ labeling of ovarian LH/hCG receptors (Metsikko, 1984). Subsequent treatment with o-glycanase had no effect on the receptor molecule. Since some type of oligosaccharides, such as GlcNAc-β-(1,3)gal NAc, cannot be hydrolyzed by o-glycanase (Umemoto, Bhavanandan and Davidson, 1977), further studies for o-linked sugars of the receptors are necessary when enzymes that can hydrolize other types of o-linked sugars become available (Fig. 14).

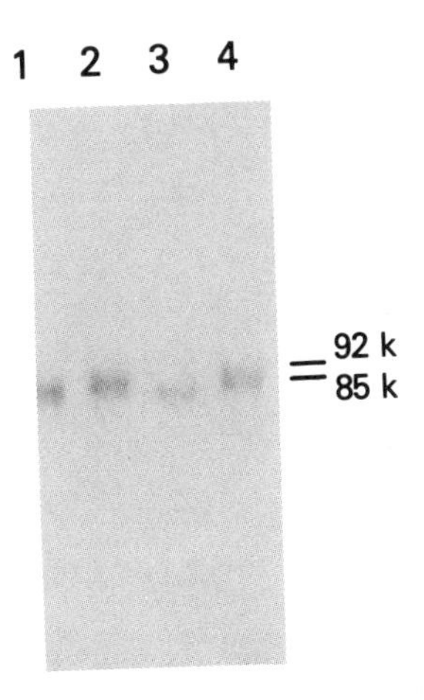

Fig. 13. Phosphorylation of purified LH/hCG receptor. Purified receptor was incubated with the catalytic subunit of protein kinase A at 25°C for 30 min (Minegishi, Kusuda, and Dufau, 1987). The phosphorylated ovarian and testicular receptors were further incubated with hCG-Sepharose 90 min (1,2) and eluted with 25 m*M* acetic acid, pH 4, containing 1*M* NaCl and 0.1% Triton X-100 (3,4). The phosphorylated receptors were incubated with wheat-germ lectin-Sepharose and eluted with 0.3*M* *N*-acetylglucosamine containing 0.1% Triton X-100. Ovarian (1,3) and testicular (2,4) samples were electrophoresed on a 5–10% gradient $NaDodSO_4$/PAGE. k = KDa.

Endoglycosidase F hydrolyzes the glycosidic bond structure adjacent to asparagine, whereas *N*-glycanase hydrolyzes Asn-*N*-linked bonds and yielding carbohydrate free receptor molecule. Deglycosylation of receptors with Endo F or *N*-glycanase, resulted in a single polypeptide band M_r 58,000 for both testicular and ovarian receptors (Fig. 14). The effectiveness of endoglyconase F suggest that the LH/hCG receptor is devoided of Tri- and Tetraantennary complexes.

Since the LH/hCG receptor of testis and ovary are only available in microgram quantities, the use of phosphorylated receptors has facilitated structural studies (Fig. 14). Their use has considerable advantages over that of iodinated receptor preparations that have been shown to contain some degradative products, presumably generated during the iodination procedure. In addition, the iodinated receptor showed a propensity to form aggregates (Kusuda and Dufau, 1988).

These results have shown that LH/hCG receptors contain sialyalated *N*-linked glycosidic chains in both the testis and ovary, and that *N*-linked glycosylation is responsible for the size differences in the LH/hCG receptor from testis and ovary (23).

In studies directed to determine whether the receptor glycosyl residues are involved in the interaction with the hormone milder conditions of enzyme treatment that were effective in deglycosylating the purified

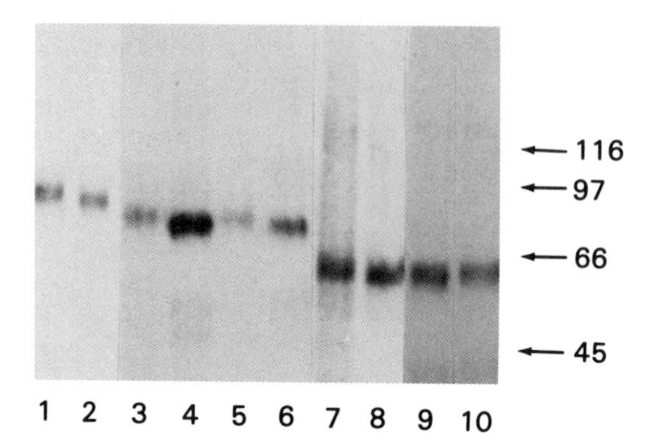

Fig. 14. Enzymatic deglycosylation of purified LH/hCG receptor. Effects of neuraminidase, O-glycanase endo F and *N*-glycanase on phosphorylated LH/hCG receptor. Phosphorylated purified receptors, denatured by heating for 3 min in 0.5% $NaDodSO_4$ plus 0.1*M* 2-mercaptoethanol, were incubated with 100 µg of neuraminidase alone at 37°C for 1, 2, and 4 h or after a 2-h incubation were subjected to additional treatment with 4.5 milliunits of O-Glycanase at 37°C 2 h in 0.2*M* sodium phosphate (pH 7.0) containing 1 m*M* $CaCl_2$, 10 m*M* D-galactono-γ-1actone, and 1.25% Nonidet P-40 for 24 h at 37°C. For other enzyme treatments, receptor was incubated with 0.2 U of *N*-Glycanase in 0.2*M* sodium phosphate (pH 8.6) containing 10 m*M* 1,10-phenanthroline hydrate and 1.25% Nonidet P-40 for 24 h at 37°C or 1 milli international unit of endo F in a solution of 0.15*M* sodium phosphate (pH 7.4), 50 m*M* EDTA, and 1.5% Nonidet P-40 at 37°C for 5 h. Enzyme reactions were terminated by adding sample buffer containing 2% SDS and 5% 2-mercaptoethanol and subsequently heated at 100°C for 2 min. The mixtures were electrophoresed in 5–10% gradient SDS-PAGE. The gels were dried and subjected to autoradiography. Left to right: Lanes 1 and 2 are controls testis and ovary; lanes 3 and 4, neuraminidase treatment; lanes 5 and 6, sequential treatment with neuraminidase and O-glycanase; lanes 7 and 8, *N*-glycanase treatment; lanes 9 and 10, endoglycosidase F.

LH/hCG receptor were used. These included enzyme digestion of phosphorylated nondenatured receptors, and reduced incubation time and temperature, maneuvers known to be essential to preserve optimal bindability of transblotted receptors. Complete reduction on the mol wt were observed with neuraminidase or *N*-glycanase treatment of the nondenatured phosphorylated receptor for only 1 h at 22°C (Fig. 15, left and right, to be compared with Fig. 14).

Neuraminidase treatment showed no apparent differences in ^{125}I-hCG binding to blotted purified receptors when compared with control incubation in the absence of enzyme. In contrast, *N*-glycanase treatment markedly reduced hormone binding, Fig 16. These results indicate that an *N*-linked glycosyl residue may participate in the receptor-hormone binding (23).

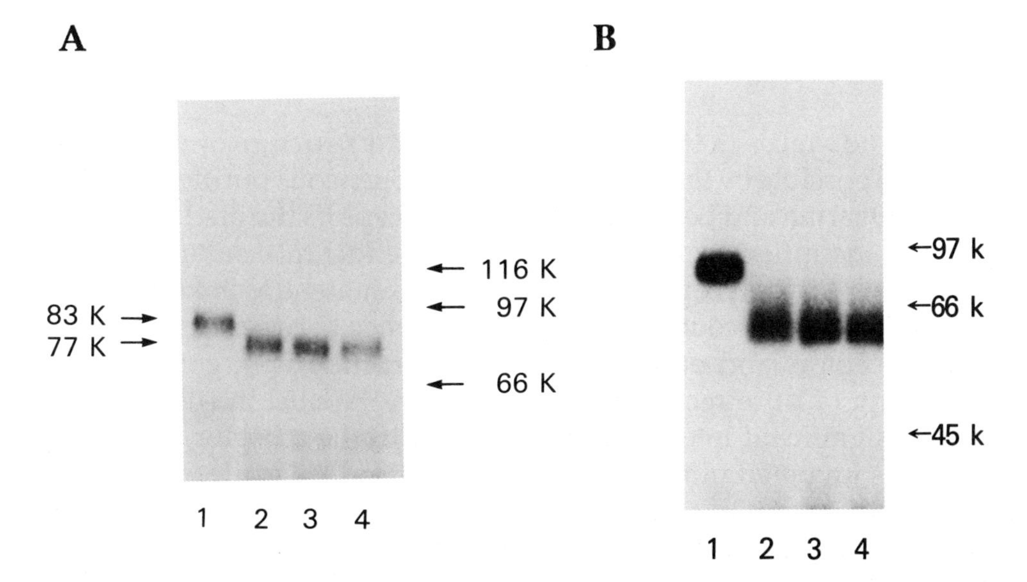

Fig. 15. Time course of neuraminidase (A, left) and *N*-glycanase (B, right) treatment of phosphorylated ovarian LH/hCG receptors. Phosphorylated receptors were incubated under nondenaturing conditions with 100 mg neuraminidase or *N*-glycanase IU and no other additions at 22°C. Enzyme reactions were terminated by adding sample buffer containing 2% SDS and 5% 2-mercaptoethanol and delectrophoresed on SDS-PAGE followed by autoradiography. Left, neuraminidase and right *N*-glycanase treatments (lane 1, control; and lanes 2, 3, and 4 correspond to incubation times of 1, 2, and 4 h, respectively).

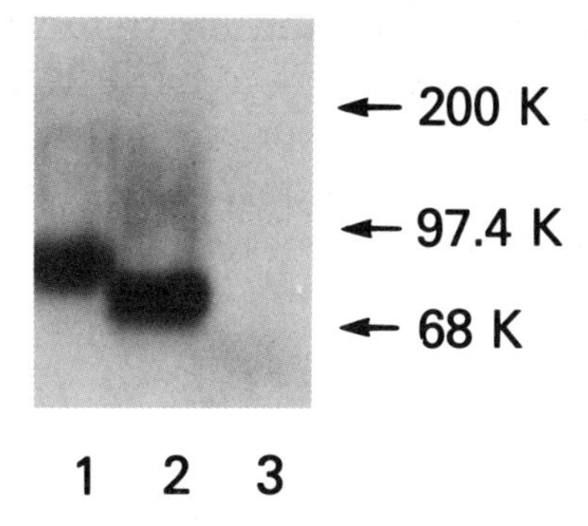

Fig. 16. Effect of neuramidase (left) and *N*-glycanase (right) treatment on ^{125}I-hCG binding. Nonphosphorylated purified ovarian hCG receptors were treated with neuraminidase or *N*-glycanase for 1 h, as described in Fig. 7. Samples were electrophoresed on a 7.5% SDS-PAGE under nondenatured and transferred electrophoretically onto nitrocellulose sheet. The blot was incubated with 10^8 dpm of ^{125}I-hCG for 16 h at 4°C. After washing, the blots were dried for autoradiograph. Control, lane 1; neuraminidase lane 2, and *N*-glycanase treatment, lane 3. Parallel blotting of control and enzyme treated receptor incubated with ^{125}I-hCG in presence of excess unlabeled hCG showed no appreciable binding.

6. Conclusions

Major advances in the characterization and structure of the receptor were made possible by the availability of homogeneous purified LH/hCG receptor. Ovarian and Leydig cell LH/hCG receptors purified to homogeneity were identified as noncovalent dimers of identical subunits. These receptors are sialoglycoproteins with predominately *N*-linked glycosyl residues that may account for the size difference between testicular and ovarian receptors and participate in the interaction with gonadotropin. Crosslinking of pure receptor with hCG with ^{125}I label in either subunit indicated significant interaction of α-hCG with the receptor, whereas β-hCG seems involved mostly through association and conformational influence on the α-subunit. Comparison of M_r derived from SDS with those from FPLC suggested that the native LH receptor are dimers of identical subunits. Autoradiographs of blotted receptors demonstrated that both monomeric and dimeric forms can bind hCG. Receptors from both tissues can be phosphorylated by the catalytic subunit of cAMP-dependent PK, and phosphopeptide maps were identical. Occupancy of the receptor by hCG significantly increase the rate but not the extent of phosphorylation. However, prolonged preincubation of receptors with hCG reduced the subsequent rate of receptor phosphorylation, indicating that receptor occupancy by agonist leads to a conformational change that facilitates its phosphorylation during initial binding and reduces the rate of phosphorylation after more prolonged exposure to hCG. Aggregation or dimerization of lactogen and hCG/LH receptors could promote clustering and/or crosslinking of receptors in the membrane, favoring the initial transduction steps in the action of these hormones.

References

Aubry, M., Collu, R., Ducharme, J. R., and Crine, P. (1982) *Endocrinology* **111,** 2129–2131.

Baranao, J. L. S. and Dufau, M. L. (1983) *J. Biol. Chem.* **258,** 7322–7330.

Benovic, J. L., Pike, L. J., Cerione, R. A., Staniszewski, C., Yoshimasa, T., Codina, J., Caron, M. G., and Lefkowitz, R. J. (1985) *J. Biol. Chem.* **260,** 7094–7101.

Bruch, R. C., Thobakura, N. R., and Bahl, O. P. (1986) *J. Biol. Chem.* **261,** 9450–9460.

Catt, K. J., Harwood, J. P., Clayton, R. N., Davies, T. F., Chan, V., and Dufau, M. L. (1980) *Recent Prog. Horm. Res.* **36,** 557–622.

Conti, M., Harwood, J. P., Hsueh, A. J., Dufau, M. L., and Catt, K. J. (1976) *J. Biol. Chem.* **251,** 7729–7731.

Conti, M, Dufau, M. L., and Catt, K. J. (1978) *Biochem. Biophys. Acta* **541,** 35–44.

Dattatreyamurty, H., Rathnam, P., and Saxena, B. B. (1983) *J. Biol. Chem.* **258,** 3140–3158.

Dufau, M. L., Charreau, E. H., and Catt, K. J. (1973) *J. Biol. Chem.* **248,** 6973–6978.

Dufau, M. L., Charreau, E. H., Ryan, D., and Catt, K. J. (1974) *FEBS Letters* **39,** 149–153.
Dufau, M. L., Podesta, E. J., and Catt, K. J. (1975) *Proc. Nat. Acad. Sci. USA* **72,** 1272–1275.
Dufau, M. L., Ryan, D. W., Baukal, A., and Catt, K. J. (1975) *J. Biol. Chem.* **250,** 4822–4825.
Dufau, M. L. and Catt, K. J. (1978) *Vitam. Horm. (NY)* **36,** 461–600.
Dufau, M. L., Hayashi, K., Sala, G., Baukal, A., and Catt, K. J. (1978) *Proc. Natl. Acad. Sci. USA* **75,** 4769–4779.
Dufau, M. L. (1988) *Ann. Rev. Physiol.* **50,** 483–508.
Katikineni, M., Davies, T. F., Huntaniemi, I. T., and Catt, K. J. (1980) *Endocrinology* **107,** 1980–1988.
Keinanen, K. P., Kellokumpu, S., Metsikko, T. K., and Rajaniemi, H. J. (1987) *J. Biol. Chem.* **262,** 7920–7926.
Keutmann, H. T., Ratanabanang-Koon, K., Pierce, M. W., Kilzmann, K., and Ryan, R. J. (1983) *J. Biol. Chem.* **258,** 14521–14531.
Kusuda, S. and Dufau, M. L. (1986) *J. Biol. Chem.* **261,** 16161–16166.
Kusuda, S. and Dufau, M. L. (1988) *J. Biol. Chem.* **263,** 3046–3049.
Kwatra, M. M., Leung, E., Maan, A. C., McMahon, K. K., Ptasienski, J., Green, R. D., and Hosey, M. M. (1987) *J. Biol. Chem.* **262,** 16314–16321.
Liu, W.-K, Yang, K.-P., Nakagawa, Y., and Ward, D. N. (1974) *J. Biol. Chem.* **249,** 5544–5550.
Metsikko, M. K. and Rajaniemi, H. J. (1982) *Biochem. J.* **208,** 300–316.
Metsikko, M. K. (1984) *J. Biol. Chem.* **219,** 583–591.
Milius, R. P., Midgley, A. R., and Birken, S. (1983) *Proc. Natl. Acad. Sci.* **80,** 7375–7389.
Minegishi, T., Delgado, C., and Dufau, M. L. (1989) *Proc. Natl. Acad. Sci.* **86,** 1470–1474.
Minegishi, T., Kusuda, S., and Dufau, M. L. (1987) *J. Biol. Chem.* **262,** 17138–17143.
Mitani, M. and Dufau, M. L. (1986) *J. Biol. Chem.* **261,** 1309–1315.
Petaja, Kellokumpu, S., Keinanen, K., Metsikko, K., and Rajaniemi (1987) **7,** 809–827.
Podesta, E. J., Solano, A. R., Attar, R., Sanchez, M. L., and Vedia, L. M. Y. (1983) *Proc. Natl. Acad. Sci. USA* **50,** 3986.
Rebois, R. V., Omedeo-Sale, F., Brady, R. O., and Fishman (1981) *Proc. Natl. Acad. Sci. USA* **78,** 2066–2069.
Rosenblit, N., Ascoli, M., and Segaloff, D. H. (1988) *Endocrinology* **123,** 2284–2290.
Sairam, M. R., and Bhargavi, G. N. (1985) *Science* **229,** 65–67.
Tasheva, B. and Dessev, G. (1983) *Anal. Biochem.* **129,** 98–102.
Umemoto, J., Bhavanandan, V. P., and Davidson, E. A. (1977) *J. Biol. Chem.* **252,** 8609–8614.
Wimalasena, J., Schwab, S., and Dufau, M. L. (1983) *Endocrinology* **113,** 618–624.
Wimalasena, J., Moore, P., Wiebe, J. P., Abel, J. J., and Chan, J. P. (1985) *J. Biol. Chem.* **280,** 10689–10697.

Purification of LH/hCG Receptor

Om P. Bahl and Hakimuddin T. Sojar

1. Introduction

Lutropin (LH) and human choriogonadotropin (hCG) bind to a common specific receptor on the ovarian and testicular plasma membranes (Catt et. al, 1976). The availability of the purified receptor and a knowledge of its size and the number of subunits is essential before any structural characterization can be undertaken. Several investigations have been undertaken to isolate the LH/hCG receptor from a variety of tissues and animals, including rat ovaries (Pandian and Bahl, 1977; Bruch et al., 1986; Kusuda and Dufau, 1986; Keinänen et al., 1987) and testes (Dufau et al., 1975; Bellisario and Bahl, 1975) and bovine (Dattatreyamurty et al., 1983; Saxena et al., 1986) and porcine ovaries (Wimalasena et al., 1986). The various purification schemes employed in these investigations have involved both the conventional and lectin affinity and immunoaffinity chromatographic methods. The determination of the size and the number of subunits has been carried out by electrophoresis of the purified receptor and/or of the chemically or photoaffinity crosslinked membrane receptor with ^{125}I-hCG in polyacrylamide gel in SDS under nonreducing and reducing conditions. The receptor has been visualized by silver staining in the case of the purified receptor or by autoradiography of the ^{125}I-hCG crosslinked receptor or ^{125}I-hCG bound receptor after Western blotting (ligand blotting). These studies have yielded conflicting results

Receptor Purification, vol. 1

indicating the receptor to be made up of 1–4 noncovalently or covalently linked homo- or heteropolypeptide chains. The molecular size of the receptor as reported has varied between 12 and 300 k. The present studies indicate that the receptor occurs primarily as a monomer of M_r of 70 k. The lower molecular components were probably owing to proteolytic degradation or other contaminants and the high mol wt forms were the consequence of self-association of the receptor. It is clear that, whereas the commonly used experimental techniques for studying membrane receptors have yielded meaningful results in the case of other well characterized receptors, these have given ambiguous data when applied to LH/hCG receptor.

One of the problems that has hampered progress in this area and seems to be unique to LH/hCG receptor, is the tendency of the receptor to undergo association during its purification, on storage or on ^{125}I-labeling chemically. The phenomenon of homoaggregation has also been observed during covalent crosslinking and photoaffinity labeling studies of the receptor.

Consequently, we attempted to understand the molecular basis of these conflicting results, particularly to understand the phenomenon of association and dissociation of LH/hCG receptor. Such an understanding not only would help in improving the recovery of the purified receptor, but also facilitate in resolving the question of its subunit structure and size. We believe that the cause of association is the presence of free thiol groups in the receptor as cysteine residues which undergo rapid oxidation forming intermolecular disulfide bonds and thereby resulting in oligomers (Scheme 1). The evidence in support of this finding was derived from several lines of experimentation. Appropriate precautions, therefore, to prevent oxidation of the thiol groups during the purification procedure can prevent the association of the receptor into homooligomers. This can be achieved by the protection of thiol groups by carrying out the purification of the receptor in the presence of S-alkylating agents or low levels of mercaptoethanol, DTT, cysteine, and sodium sulfite.

Another problem encountered in the isolation of the LH/hCG receptor has been its extreme lability, particularly that of the rat lutropin receptor. This, however, has been overcome by simply using 20% glycerol in the buffers employed in the solubilization and purification of the receptor. Thus, by controlling the instability and aggregation of the LH/hCG receptor, it has been possible to purify reproducibly rat ovarian LH/hCG receptor in good yields.

The present communication describes a method for the purification of LH/hCG receptor based on these findings. The evidence is presented to

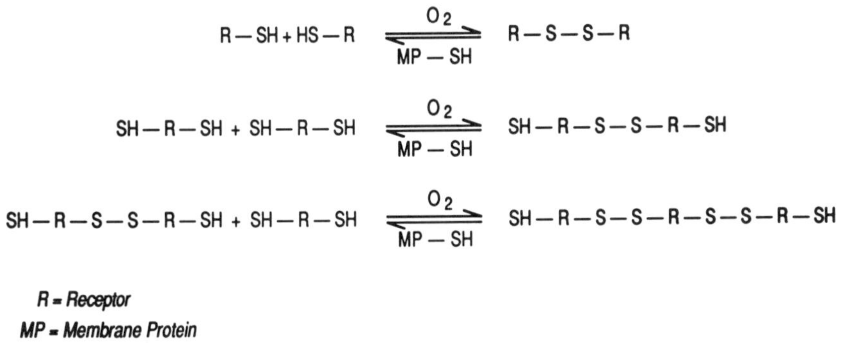

$$R{-}SH + HS{-}R \underset{MP-SH}{\overset{O_2}{\rightleftharpoons}} R{-}S{-}S{-}R$$

$$SH{-}R{-}SH + SH{-}R{-}SH \underset{MP-SH}{\overset{O_2}{\rightleftharpoons}} SH{-}R{-}S{-}S{-}R{-}SH$$

$$SH{-}R{-}S{-}S{-}R{-}SH + SH{-}R{-}SH \underset{MP-SH}{\overset{O_2}{\rightleftharpoons}} SH{-}R{-}S{-}S{-}R{-}S{-}S{-}R{-}SH$$

R = Receptor
MP = Membrane Protein

Scheme 1. Proposed mechanism of association of rat ovarian LH/hCG receptor. a. Dimer formation only if the receptor has a single sulfhydryl group. b, c. Dimer, trimer, and oligomer formation with two sulfhydryl groups in the receptor.

show the existence of free thiol groups and the effect of their protection on the ligand binding property of the receptor. Further characterization of the receptor is also discussed.

2. Occurrence of Free Thiol Groups in LH/hCG Receptor

The presence of the sulfhydryl groups in the receptor has been established in several different ways, including:

1. By specific adsorption of the soluble receptor on pCMB-agarose and its elution with 25 m*M* cysteine;
2. By quantitative analysis of cysteine as S-carboxymethylcysteine;
3. By oxidation with H_2O_2 to form oligomers owing to intermolecular oxidation of SH-groups and subsequent conversion of the oligomers into the monomer by reduction;
4. By prevention of aggregation of the receptor caused by molecular O_2 or H_2O_2 by prior *N*-ethylmaleylation or S-carboxymethylation; and
5. By incorporation of ^3H-label on treatment with tritiated *N*-ethylmaleimide or iodoacetic acid.

When 1% Triton X-100 solubilized receptor preparation from rat ovarian tissue was passed through a column of pCMB-agarose, 56% of the receptor activity was held on the column that could be eluted specifically with 25 m*M* cysteine containing buffer (Fig. 1). However, when the receptor was *N*-ethylmaleylated prior to treatment with pCMB-agarose, only about 10% of the receptor was held on the column (Table 1), indicating that prior protection of the SH-group prevented the interaction of the receptor with pCMB-agarose. The slight reaction with the affinity ligand after NEM

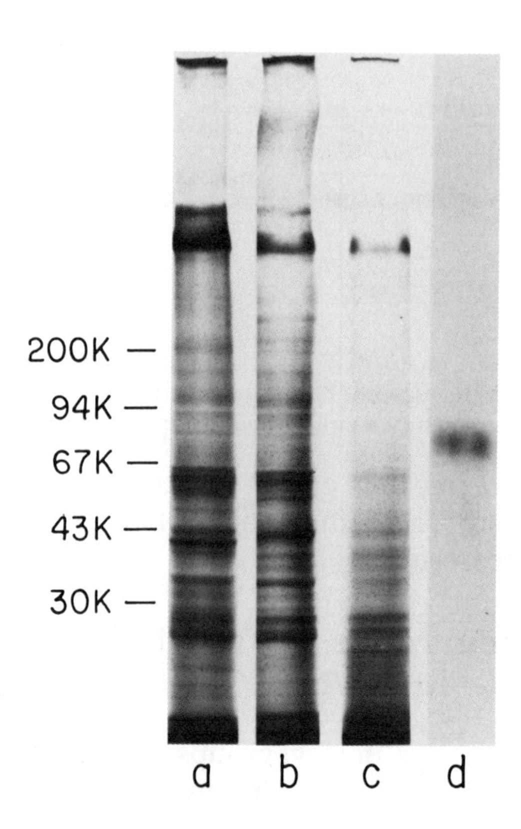

Fig. 1. SDS-PAGE and ligand blotting of the fractions from the pCMB-agarose column chromatography of the soluble receptor. Chromatography of crude rat ovarian LH/hCG receptor on *p*-chloromercuribenzoate-agarose column. The crude soluble receptor prepared from the ovarian membranes equivalent to one ovary, in the absence or the presence of 5% NEM, was applied to 1 mL of pCMB-agarose gel. Fractions of 10 µg of protein each were subjected to 7% nonreduced SDS-PAGE. The gel was subjected to silver staining (lanes a–c) or ligand blotting (lane d). Lanes: a: crude soluble receptor; b: flow through; c: cysteine eluate; d, autoradiogram after ligand blotting of the cysteine eluate. ^{14}C-labeled mol wt markers: lysozyme, 14.4 k; carbonic anhydrase, 30 k; ovalbumin, 43 k; bovine serum albumin, 67 k; phosphorylase b, 94 k; and myosin, 200 k.

treatment was probably a result of the inaccessibility of certain SH-groups of the receptor to NEM. The specifically eluted fraction from pCMB-agarose on SDS-PAGE and ligand blotting using ^{125}I-hCG under nonreducing conditions showed on autoradiography a single receptor band with an apparent mol wt of 70 k (Fig. 1, lane d). It is also obvious from the figure that in this fraction several other membrane proteins having mol wt less than 90 k were also thiol proteins (Fig. 1, lane c).

Table 1
Chromatography of Crude Rat Ovarian LH/hCG Receptor on *p*-Chloromorcuribenzoate-Agarose Column[a]

Fraction	In the absence of NEM		In the presence of NEM	
	% Receptor[b]	% Total protein[c]	% Receptor[b]	% Total protein[c]
Flow through	19	34	25	53
Buffer B wash	3	3	2	18
Buffer B + 1*M* NaCl wash	20	24	55	20
Buffer B + 25m*M* cysteine HCl, pH 6.5	56[d]	39	10[d]	9

[a]The crude soluble receptor prepared from the ovarian membranes equivalent to one ovary, in the absence or the presence of 5% NEM was applied to 1 mL of pCMB-agarose gel.
[b]Determined by radioreceptor assay.
[c]Determined by Bradford method.
[d]Determined from the standard RRA carried out in the presence of cysteine HCl, pH 6.5.

The treatment of the crude soluble receptor with H_2O_2 ranging in concentration from 1–5 m*M* converted the monomeric form of the receptor into oligomers. Reduction of the oligomeric forms with β-mercaptoethanol yielded a monomer indicating the oligomers were formed because of the intermolecular oxidation of the SH-groups into -S-S-bonds (Fig. 2). The oligomerization with H_2O_2 could be avoided by pretreatment of the crude receptor preparation with NEM or IAA. It may be noted that the reduced form of the receptor was detected by immunoblotting after SDS-PAGE. The monomeric form of the native receptor, but not the reduced form and the various oligomers interacted with ^{125}I–hCG as observed on ligand blotting (Fig. 2).

The cysteine and cystine content of the receptor was determined by amino acid analysis of the S-carboxymethyl receptor prepared by its S-carboxymethylation before and after reduction, respectively (Table 2). The values for free cysteine and that for cysteine as cystine were found to be 0.4 and 1.9%, respectively. These values amounted to 2–3 residues of cysteine and 6 residues of cystine (-S-S-bonds) per 60 k polypeptide chain, an approximate estimated mol wt of the deglycosylated receptor. When the native receptor was treated with ^{3}H-NEM or ^{3}H-IAA the label was incorporated in the receptor (Fig. 3), again indicating the presence of cysteine. Finally, when the receptor was prepared in the presence of var-ious S-alkylating agents or thiol protecting agents, the receptor was obtained in the monomeric form (Figs. 4 and 5). The S-alkylating agents used were NEM or IAA and the various thiol protecting agents that proved to be effective, included 5–10 m*M* cysteine or DTT and 1–5 m*M* Na_2SO_3.

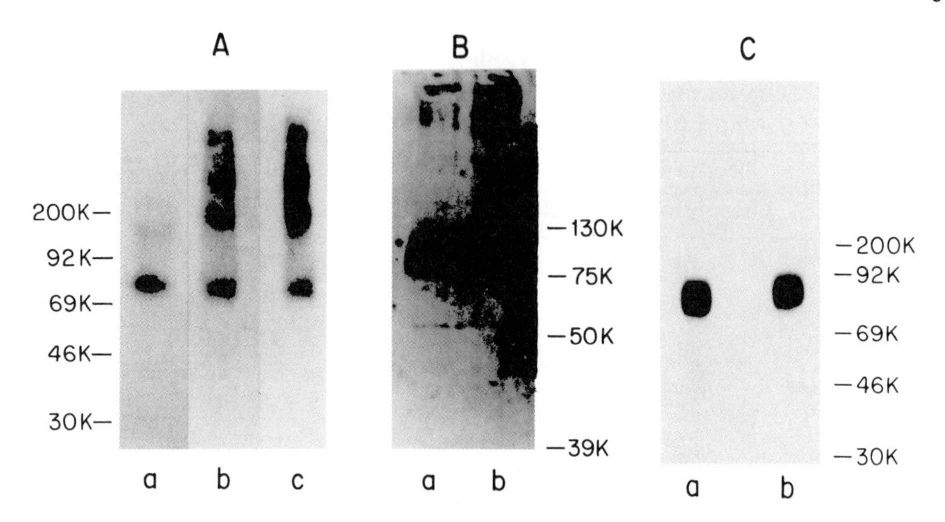

Fig. 2. Ligand blotting of H_2O_2 treated crude rat ovarian LH/hCG receptor. A. Crude soluble receptor prepared without NEM and incubated with various concentrations of H_2O_2 using about 1/10th of the amount of the total membrane protein from a single ovary in each tube. Lanes: a, none; b, 1 m*M* H_2O_2; c, 2 m*M* H_2O_2; B. Reduction with DTT of the crude receptor prepared in presence of 5 m*M* H_2O_2 and immunoblotting under reducing conditions. Lanes: a, without any H_2O_2; b, with 5 m*M* H_2O_2 and reduction. Molecular lit. markers: soybean trypsin inhibitor, 27k; carbonic anhydrase, 39 k; ovalbumin, 50 k; bovine serum albumin, 75 k and phosphorylase b, 130k. C. Crude receptor prepared in the presence of 5 m*M* NEM and treated with H_2O_2. Lanes: a, none; b, 5 m*M* H_2O_2. For mol wt markers, *see* legend to Fig. 1.

3. Effect of Zn(II) on the Receptor Association

Zn(II) has been found to stabilize SH-groups by forming a tetrahedral coordinate comple (Giedroc et al., 1987). In the absence of any other sulfhydryl reagent, zinc acetate alone at the levels of 0.1 μ*M* and 0.1 m*M* was able to prevent association of the LH/hCG receptor. The crude soluble receptor prepared in the presence of zinc acetate showed a single sharp band of M_r 70 k (Fig. 6, lanes a and b) under nonreducing conditions, indicating that Zn(II) did indeed have a protective effect against oxidation of SH-groups and thus against receptor association. The absence of aggregated forms would indicate that Zn(II) did not form a bridge between intermolecular thiol groups in the receptor. It may be noted that the aggregated forms retain their ligand binding property (Sojar and Bahl, 1989). The effect of Zn(II) on the stability or the prevention of aggregation is highly significant. Zn(II) has been found to form coordinate complexes involving four SH-groups or SH-group and imidazole *N* in proteins. For instance, in the case of regulatory subunits of aspartate carbamoyl transferase (Monaco et al., 1978) the presence of Zn(II) was found to be necessary for their association

Table 2
Amino Acid Composition of Rat Ovarian LH/hCG Receptor (% mol/mol)

Amino acid	Dattatreyamurty et al.	Kusuda and Dufau	This paper
Asp	9.0	8.1	6.2
Ala	7.5	7.8	5.7
Arg	5.6	3.3	2.1
Ile	4.4	3.3	4.1
Gly	7.4	14.5	18.9
Glu	10.9	8.3	10.1
1/2 Cys	—	—	0.4[a]
1/2 Cys	2.5[b]	2.5	2.3[c]
Ser	7.3	18.5	7.2
Tyr	3.6	2.1	3.3
Thr	5.8	6.2	5.3
Val	9.5	4.8	3.6
His	2.4	4.0	5.0
Phe	4.0	2.8	3.7
Pro	6.0	5.0	4.7
Met	2.0	2.2	3.1
Lys	6.4	4.3	7.3
Leu	8.8	2.3	7.5

[a]Determined as S-carboxymethylcysteine after S-alkylation of the receptor without reduction.
[b]Determined as cysteic acid.
[c]Determined as S-carboxymethylcysteine after reduction and S-alkylation of the receptor.

with the catalytic subunits. Similarly, Zn(II) probably forms an intramolecular complex involving SH-groups in the rat LH/hCG receptor. The physiological significance of Zn(II) in the receptor function is not known at present.

4. Purification of Rat Lutropin LH/hCG Receptor

Although conventional, lectin, immunoaffinity, and affinity chromatographic methods individually or in combination have been employed for the purification LH/hCG receptor, affinity chromatographic procedure is a method of choice. Since the receptor is present in minute quantities, it is important that the number of purification steps is kept to a minimum to avoid excessive losses. The earlier purification procedures lacked in the use of 20% glycerol in the buffers and, therefore, the recovery of the receptor was generally poor owing to the labile nature of the receptor. Dattatreyamurty et al., 1983 reported the purification of LH/hCG receptor from bovine *corpora lutea* by a procedure involving the solubiliza-

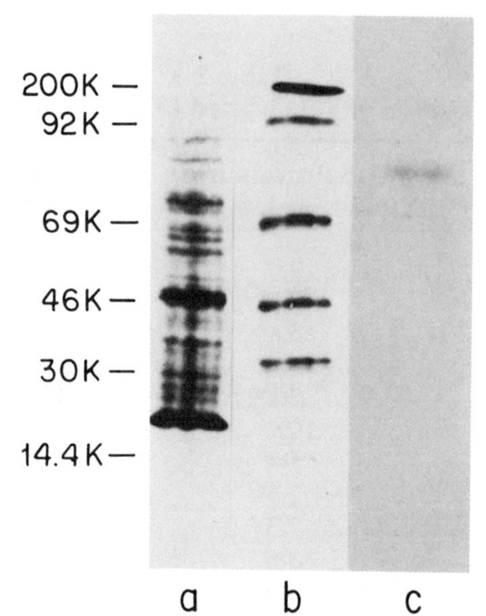

Fig. 3. Autoradiogram of the rat ovarian LH/hCG receptor prepared in the presence of ^{3}H-NEM. The crude receptor prepared by the solubilization of the rat ovarian membrane with 50 m*M* phosphate buffer, pH 7.4 containing 1% Lubrol PX, 0.1 m*M* ^{3}H-NEM, and 20% glycerol was subjected to affinity chromatography on hCG-Sepharose. The purified receptor was eluted with pH 4.0 buffer containing 0.01% Lubrol PX. Lanes: a, ^{3}H-labeled crude soluble receptor; b, mol wt markers; c, purified ^{3}H-labeled receptor. For mol wt markers, *see* legend to Fig. 1.

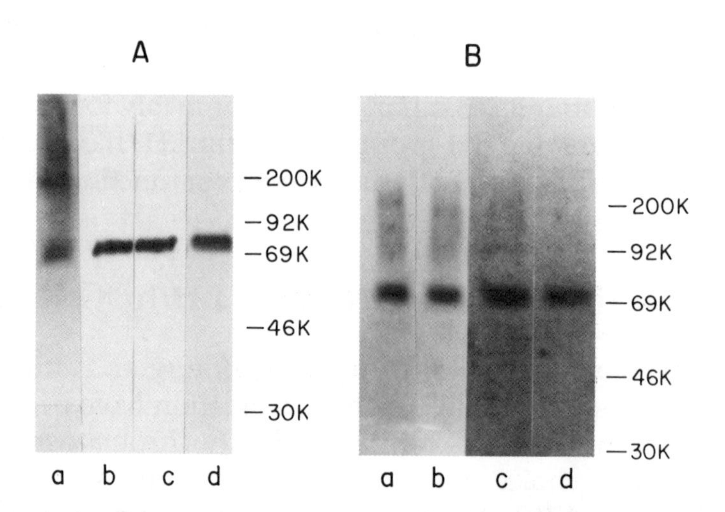

Fig. 4. Ligand blotting of crude rat ovarian LH/hCG receptor prepared in the presence of various sulfhydryl protecting reagents. A. Lanes: a, none; b, 5 m*M* NEM; c, 5 m*M* sodiun sulfite; d, 5 m*M* nodoacetic acid. For mol wt markers, *see* legend to Fig. 1. B. ^{125}I-labeled receptor purified in the presence of 5 m*M* NEM. B. Lanes: a, without cysteine; b, with 5 m*M* cysteine; c, 10 m*M* cysteine; d, 25 m*M* cysteine. For mol wt markers, *see* legend to Fig. 1.

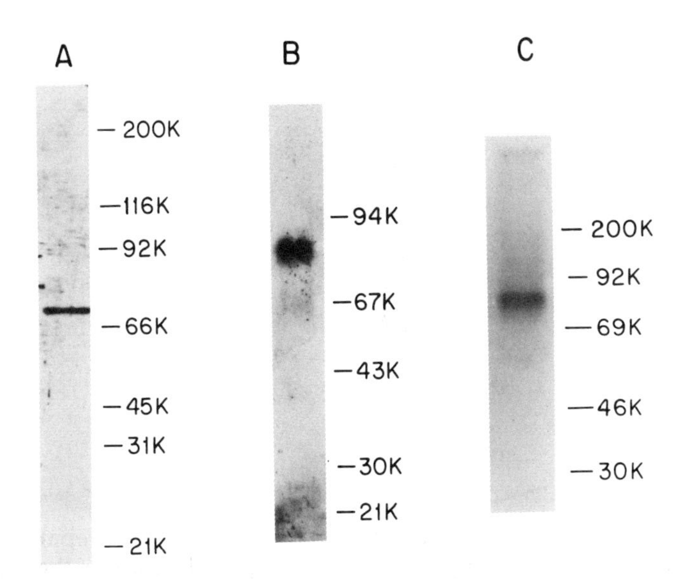

Fig. 5. SDS-PAGE of the purified rat ovarian LH/hCG receptor. Receptor was purified and subjected to 7.5% SDS-PAGE. A. Nonreduced receptor stained with silver stain. Molecular wt. markers: Soybean trypsin inhibitor, 21 k; bovine carbonic anhydrase, 31 a; hen egg white ovalbumin, 45 k; bovine serum albumin, 66 k; rabbit muscle phosphorylase b, 92 k; *E. coli* b-galactosidase, 116 k; myosin, 200 k. B. Reduced receptor stained with silver stain. Molecular wt. markers: Soybean trypsin inhibitor, 21kDa; carbonic anhydrase, 30kDa; ovalbumin, 43 k; bovine serum albumin, 67k; phosphorylase b, 94 k. C. ^{125}I-labeled receptor purified in the presence of 5 m*M* NEM. For mol wt markers, *see* legend to Fig. 1.

tion of the receptor with 0.5% Triton X-100, and extraction of the crude soluble receptor with petroleum ether to remove lipids, followed by gel filtration on Sepharose B and zone electrophoresis on cellulose columns. The receptor thus obtained was in mg quantities. It had an M_r of 240–280 k in SDS-PAGE under reducing conditions. The receptor was found to be composed of two identical disulfide-linked units of 120 k each. Each of the units was further made up of two subunits of 85 k and 38 k. Metsikko and Rajaniemi (1979, 1980) isolated a protein of M_r 100 k from rat ovaries by indirect immunoaffinity chromatography that was biologically inactive. In contrast, Wimalasena et al. (1983) and Wimalasena and Dufau (1982) described the isolation by lectin affinity chromatography of rat ovarian LH/hCG receptor in 5 molecular species ranging from 12–165 k that retained hormone binding activity.

In earlier studies from our laboratory, Bruch et al. (1986) reported a preparation of rat ovarian LH/hCG receptor by using reverse immunoaffinity and affinity chromatography having M_r of 240 k made of four spe-

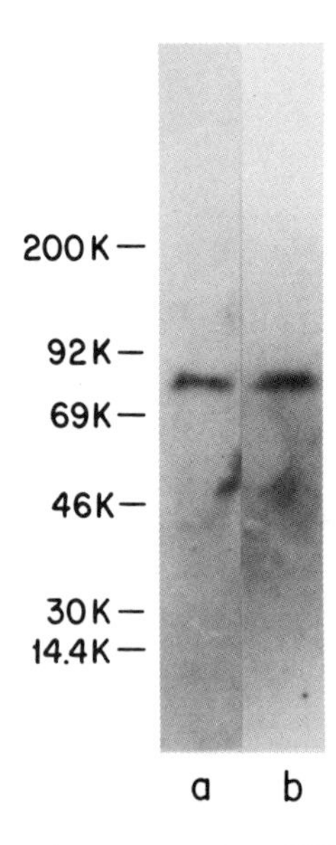

Fig. 6. Ligand blotting of crude rat ovarian LH/hCG receptor prepared in the presence of Zn(II). The crude receptor was prepared as described in legend to Fig. 3. Lanes: a, 0.1 n*M* zinc acetate; b, 0.1 m*M* zinc acetate. For mol wt markers, *see* legend to Fig. 1.

cies, two major 79 and 64 k, and two minor 55 and 47 k, as indicated by SDS-PAGE under reducing conditions. More recently, Keinänen et al. (1987) reported the purification of rat ovarian LH/hCG receptor by affinity chromatography with a mol wt of 90 k on SDS-PAGE under nonreducing conditions. However, under reducing conditions, the SDS-PAGE of the receptor showed faint bands of mol wts 58, 55, 43, 42, 35, and 33 k. Kasuda and Dufau (1986) in their recent publication have reported a rat ovarian LH/hCG receptor purification by affinity chromatography of mol wt 73 k, although it occurred primarily as a noncovalently bonded dimer according to the authors.

4.1. Purification Method

The present method developed in our laboratory takes into consideration our recent findings on the presence of free sulfhydryl groups in the rat ovarian LH/hCG receptor and their rapid oxidation to form aggregates. Briefly, the method involves the solubilization of the crude membranes from pseudopregnant rat ovaries with a detergent containing buffers, treatment of the crude soluble receptor with hCG-Sepharose affinity adsorbent, and elution of the receptor at low pH. The pseudopregnancy in the rats was achieved by successive injections of 50 U of eCG/rat followed 55–65 h later by 25 U of hCG/rat (Fig. 7). About a week later, the rats were sacrificed, and ovaries were removed. If the homogenization and

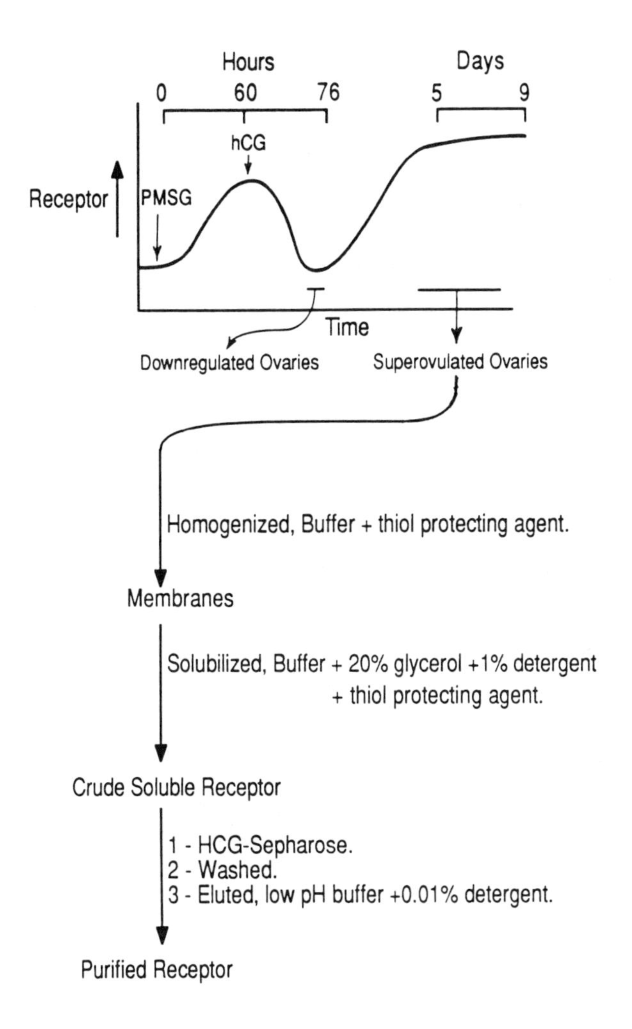

Fig. 7. Superovulation of rat ovaries and purification of LH/hCG receptor. Immature female Sprague-Dawley rats (21 d old, 40–50 g) were injected subcutaneously with 50 IU of equine choriogonadotropin (eCG) in 0.9% NaCl containing 0.1% bovine serum albumin followed 56–65 h later by injection of 25 IU of hCG. Animals were sacrificed by cervical dislocation after 7 d of hCG injection to obtain superovulated ovaries. Ovaries were quickly removed, freed of fat and connective tissue, and rinsed with ice-cold 50 m*M* Tris-HCl buffer, pH. 7.4. Receptor was purified according to the scheme outlined.

solubilization buffers did not include a thiol protecting agent NEM, IAA, DTT, cysteine, or sodium sulfite at 5–10 m*M*, the receptor preparation obtained invariably showed on SDS-PAGE multiple bands of oligomers. In addition to a sulfhydryl protecting agent the buffers also included 20% glycerol and a nonionic detergent concentration of 0.01–1%. The highest concentration of the detergent (0.5–1%) was used for the solubilization, whereas the least concentration of 0.01% in the pH, 4-buffer was employed for the elution of the receptor from the affinity ligand, i.e. hCG-Sepharose. Since the receptor is not stable at low pH, the pH of the eluate was immediately adjusted to 7.4. The detergents that have been effectively used for the solubilization of the receptor are Triton X-100, Lubrol PX, or emphulphogene, the last two being devoid of phenolic groups and, therefore, are not susceptible to labeling during ^{125}I-labeling of the receptor. For this reason, Lubrol PX is preferable to Triton X-100. The detergent was removed by passing through a column of Extracti gel D (Pierce Chemical Co.) and the concentration of the receptor solution was carried out by using Centricon microconcentrator. The last traces of the detergent were eliminated by precipitation of the receptor with acetone at –20°C before labeling. It is worth noting that the receptor prepared in the presence of a thiol protecting agent yields on ^{125}I-labeling a preparation that on SDS-PAGE shows a single band (Fig. 5C). As discussed earlier, in the absence of this precaution, multiple high mol wt labeled bands are observed on SDS-PAGE. Figure 5A,B shows the homogeneity of the receptor preparation obtained according to the procedure outlined above.

The affinity-ligand used in this procedure was obtained by coupling hCG to activated tresyl-agarose. There are several advantages of using tresyl-agarose. For instance, the linkage between hCG and agarose is quite stable and the activity of the coupled hCG is 100-fold greater than hCG coupled to cyanogen bromide activated Sepharose (Bruch et al., 1986; Sojar and Bahl, 1989).

5. Immunological Properties of LH/hCG Receptor

A polyclonal antireceptor antibody was produced by immunizing a rabbit with a total of 70 µg of the purified receptor at two time intervals, an initial multiple site subcutaneous injections of 40 µg followed 6 wk later a booster of 30 µg of the receptor. The antibody thus obtained had a titer of 1:100 for 30% binding of the ^{125}I-labeled receptor. It showed a single band on immunoblotting when the crude soluble or purified receptor was subjected to SDS-PAGE, indicating that the antibody was quite specific. The specificity of the antibody was further established when the rat liver extract failed to react with the antibody. On immunoblotting it did not show any band. The antibody reacted both with the purified native as well as with the reduced rat LH/hCG receptor and gave a single band on immunoblotting (Fig. 8), again indicating the homogeneity of the receptor preparation. It may be noted that the antibody was also capable of reacting with

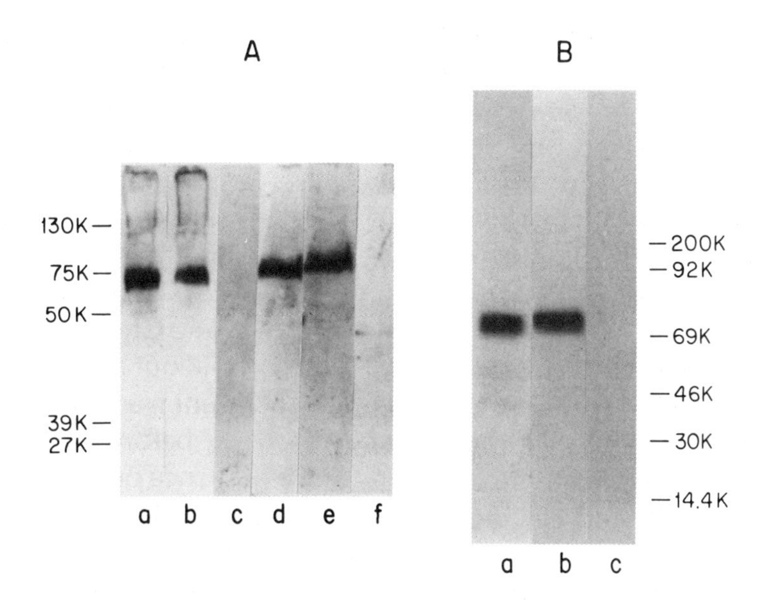

Fig. 8. Alkaline phosphatase conjugated immunoblot of the purified soluble rat ovarian LH/hCG receptor and ligand blotting of purified LH/hCG receptor. For the method of preparation of the crude receptor, *see* legend to Fig. 3. **A.** Lane a, purified receptor under nonreducing conditions; b, crude soluble receptor under nonreducing conditions; c, liver extract under nonreducing conditions; d, purified receptor under reducing conditions; e, soluble receptor under reducing conditions; f, liver extract under reducing conditions. Molecular weight markers: lysozyme, 17 k; soybean trypsin inhibitor, 27 k; carbonic anhydrase, 39 k; ovalbumin, 50 k; bovine serum albumin, 75 k; and phosphorylase b, 130 k. **B.** Effect of FSH and hCG/LH on ligand blotting of the purified receptor. Lanes: a, none; b, in presence of unlabeled FSH; c, in presence of unlabeled hCG or LH. For mol wt markers, *see* legend to Fig. 1.

the reduced receptor. Therefore, the latter could be visualized on SDS-PAGE by immunoblotting. The ligand blotting did not work in this case, since, on reduction, the receptor lost its ability to bind to ^{125}I-hCG.

6. Characterization of the Receptor

Rat ovarian LH/hCG receptor is a glycoprotein with an apparent mol wt of 70–80 k on SDS-PAGE under nonreducing and reducing conditions, respectively. The receptor occurs predominantly as a monomer, as shown by ligand blotting when it is solubilized in the presence of a thiol protecting agent. In the absence of a thiol reagent, oligomers are formed because of rapid oxidation of SH-groups, as shown by multiple bands observed on SDS-PAGE and ligand blotting. The oligomers retain their property of binding to ^{125}I-hCG. In addition to 2–3 thiol groups, the rat ovarian LH/

hCG receptor contains 6 disulfide bridges per 60 k of the polypeptide chain. Also, it may be noted that ^{3}H-label is incorporated in the receptor when it is prepared in the presence of ^{3}H-NEM owing to *N*-ethylmaleylation. Similarly, when the receptor is prepared in the presence of 0.1 μM zinc acetate, the sulfhydryl groups are probably protected by Zn, which forms an intramolecular tetrahedral complex. The binding of the *N*-ethylmaleylated receptor or Zn-receptor complex to hCG remains unaltered, as shown by the data in Fig. 9A,B. However, it is not known at present whether the *N*-ethylmaleylation or S-carboxymethylation or complexing of the receptor with Zn in any way affects its biological activity.

The rat ovarian LH/hCG receptor has been found to contain 0.4 mol% of cysteine and 1.9 mol% cystine. This would be equivalent to 2–3 residues of cysteine and 6 residues of cystine per 60 k mol wt of the polypeptide chain. The cysteine and cystine contents were determined as S-carboxymethylcysteine. The S-carboxymethylation of the receptor was carried out under conditions that were first established on a micro scale using hCG-β, a protein of known cystine content. The amino acid composition of the receptor is given in Table 2 and agrees fairly with that reported by Kusuda and Dufau (1986), except in two residues, Lys and Leu. It differs considerably from that reported by Dattatreyamurthy et al. (1983). The glycine value seems to be high and is most likely an artifact as a contaminant in the buffers used in the preparation of the receptor. The receptor does not have unusually high content of hydrophobic amino acids as one would expect from a membrane protein with multiple transmembrane fragments.

7. Proteolytic and Glycosidic Cleavage of LH/hCG Receptor

The polypeptide chain of the receptor was cleaved at a specific glutamyl residue by endoproteinase Glu-C into two components, 46 and 36 k (Fig. 10B, lane b). The 36 k component contained the carbohydrate since on deglycosylation with endo-F it yielded two components of mol wts 27 and 25 k (Fig. 10, lane c). Thus, the entire or part of 36 k component must be extracellularly located. The native LH/hCG receptor under the reduced state (mol wt 80 k) on deglycosylation with endo-F yielded two components with mol wts of 73 and 64 k (Fig. 10A, lane b), indicating the presence of approximately 20% carbohydrate distributed in two or more *N*-linked carbohydrate chains.

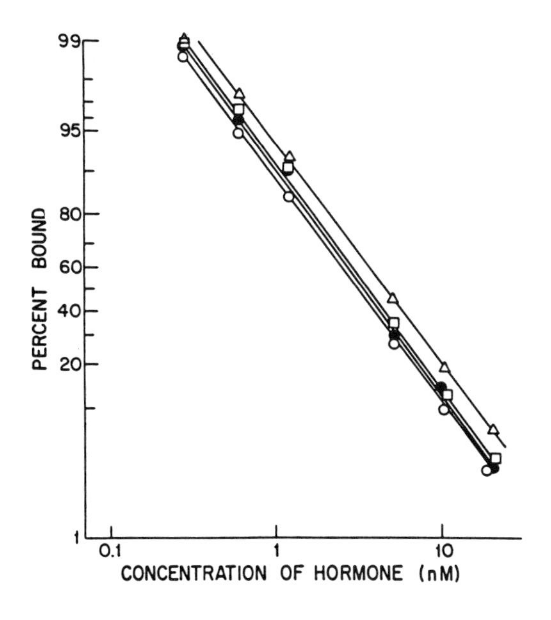

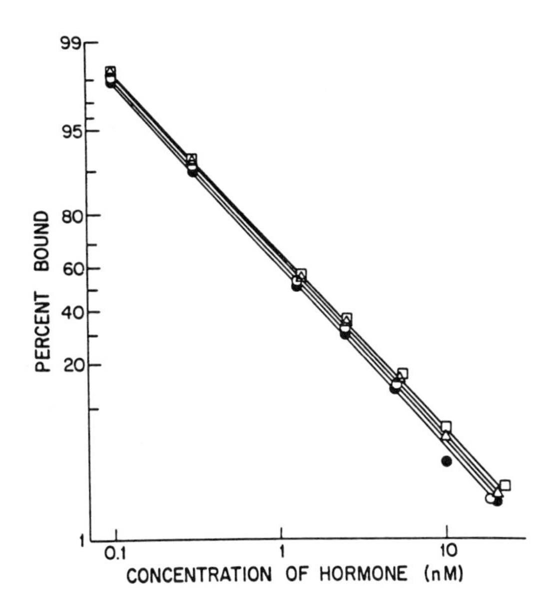

Fig. 9. Effect of various reagents on the hormone-receptor binding. Dose response curves of the receptor: A, the rat ovarian membrane receptor assay. None, ○; 5 m*M* NEM, ●; 5 m*M* sodium sulfite, Δ; 5 m*M* cysteine-HCl, □; B, the soluble radioreceptor assay in the presence of Zn(II). In the absence of NEM and Zn(II), ○; in the absence of NEM and in the presence of 100 n*M* Zn(II), ●; in the presence of 5 m*M* and in the absence of Zn(II), Δ; in the presence of 5 and 100 n*M* Zn(II), □.

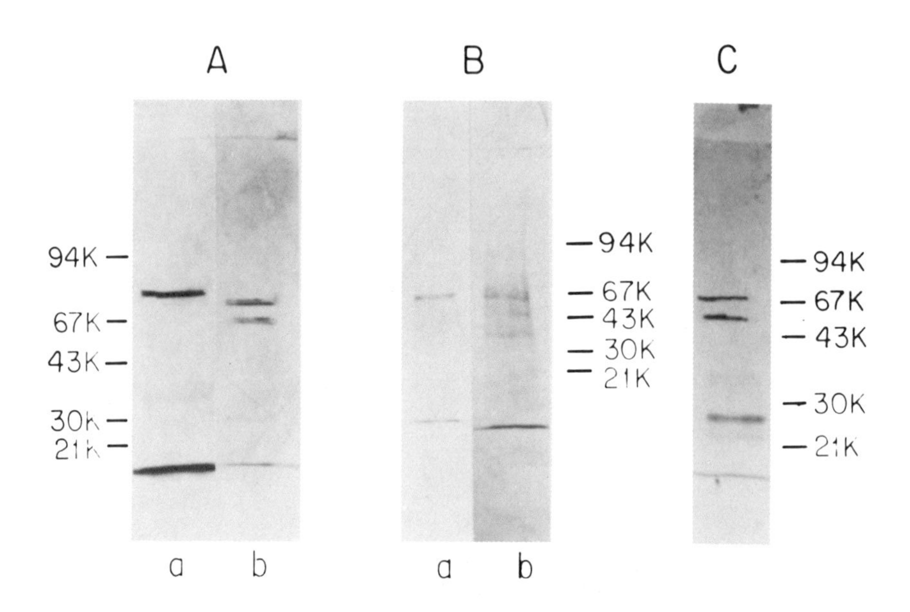

Fig. 10. SDS-PAGE of the purified rat ovarian LH/hCG receptor treated with endoproteinase Glu-C and Endo-F. A. Lanes: a, purified ovarian LH/hCG receptor; b, purified ovarian receptor treated with endo-proteinase Glu-C. B. Lanes a, purified ovarian LH/hCG receptor treated with Endo-F. C. Purified ovarian LH/hCG receptor treated sequentially with Endo-F and endoproteinase Glu-C. For mol wt markers, *see* legend to Fig. 8A.

8. Conclusions

The present findings on the existence of free cysteine residues in rat ovarian LHL/hCG receptor and their role in the receptor association will hopefully help eliminate the current confusion with regard to the size and the subunit structure of the receptor. Our data indicate that the receptor occurs predominantly as a monomer of mol wt 70 or 80 k on SDS-PAGE carried out under nonreducing and reducing conditions, respectively. The association of the receptor into oligomers is caused by the rapid oxidation of intermolecular thiol groups present in the receptor. The association can be controlled by using a thiol protecting agent, such as NEM, IAA, cysteine, DTT, and sodium sulfate. Also, the previous reports on the presence of high or low mol wt forms of the receptor can be rationalized by the present findings. The low mol wt components in the receptor are probably owing to either proteolytic degradation or intermolecular disulfide bond formation between the thiol groups of the receptor and other low mol wt thiol proteins that are released under reducing conditions. This may explain the results of Keinänen et al. (1987), who observed a number of low molecular components on SDS-PAGE under reducing conditions. The receptor is quite labile and can be stabilized by using 20% glycerol in the buffers employed for its solubilization and purification. A single step purification procedure has been developed in which the receptor association and lability have been controlled. Consequently, the procedure gives a homogeneous preparation of the receptor in the monomeric form in good yields. The homogeneity of the receptor is established by SDS-PAGE visualized by silver staining, immuno-, and ligand blotting. It may be noted that the receptor prepared in the presence of S-alkylating agents has its sulfhydryl groups covalently blocked. Although this modification does not alter the ligand binding property of the receptor, its effect on the biological function of the receptor is not known. The effect of Zn(II) on the stability of the SH groups of this receptor is quite interesting. The zinc–receptor complex retains fully its ligand binding potency. The physiological significance of the Zn(II) in the function of the receptor is not known at present. Endo-proteinase Glu-C cleaves the polypeptide chain at a specific Glu residue giving two fragments of 45 and 36 k. The latter fragment contains the *N*-linked carbohydrate as shown by its deglycosylation into 27 and 25 k components, and is, therefore, extracellularly located. The deglycosylation of the receptor with endo-F occurs in two steps, converting the 80 k reduced receptor into 73 and 64 k components, indicating the presence of two or more *N*-linked carbohydrates. These findings should pave the way for further understanding of the LH/hCG receptor.

Abbreviations

The abbreviations used are: hCG, human chorionic gonadotropin, choriogonadotropin; LH, lutropin, luteinizing hormone; PMSF, *p*-phenylmethylsulfonyl fluoride; NEM, *N*-ethyl-maleimide; IAA, iodoacetic acid; PCMB-agarose, *p*-chloromercuribenzoate-agarose; DTT, dithiothreitol; BSA, bovine serum albumin; SDS-PAGE, sodium dodecyl sulfate-polyacrylamide gel electrophoresis.

Acknowledgments

The authors wish to express their appreciation to James Stamos for preparing the figures and Ursula Brunn for typing the manuscript.

This work was supported by a grant from the National Institutes of Health, US Dep. of Health and Human Services #HD 08766.

References

Bellisario, R. and Bahl, O. P. (1975) *J. Biol. Chem.* **250,** 3857–3844.
Bruch, B. C., Thotakura, N. R., and Bahl, O. P. (1986) *J. Biol. Chem.* **261,** 9450–9460.
Catt, K. J. Ketelslegers, J.-M., and Dufau, M. L. (1976) in *Methods in Receptor Research,* vol. 1 (Blecher, M. ed.), Marcel Dekker, New York, pp. 175–250.
Dattatreyamurty, B., Bathnam, P., and Saxena, B. B. (1983) *J. Biol. Chem.* **258,** 3140–3158.
Dufau, M. L., Ryan, D. W., Baukat, A., and Catt, K. J. (1975) *J. Biol. Chem.* **250,** 4822– 4825.
Giedroc, D. P., Keating, K. M., Williams, K. R., and Coleman, J. E. (1987) *Biochemistry* **26,** 5251–5259.
Keinänen, K. P., Kellokumpu, S., Metsikko, M. K., and Rajaniemi, J. (1987) *J. Biol. Chem.* **262,** 7920–7926.
Kusuda, S. and Dufau, M. (1986) *J. Biol. Chem.* **261,** 16161–16168.
Metsikko, M. K. and Rajaniemi, J. (1979) *FEBS Lett.* **106,** 193–196.
Metsikko, M. K. and Rajaniemi, J. (1980) *Biochem. Biophys. Res. Commun.* **95,** 1730–1736.
Monaco, H. L., Crawford, J. L., and Lipscomb, W. N. (1978) *Biochemistry* **15,** 5276–5280.
Pandian, M. R. and Bahl, O. P. (1977) *Arch. Biochem. Biophys.* **182,** 420–436.
Saxena, B. B., Dattatreyamurty, B., Ota, H., Milkov, V., and Rathnam, P. (1986) *Biochemistry* **25,** 7943–7950.
Sojar, H . T. and Bahl, O. P. (1989) *J. Biol. Chem.* **264,** 2552–2559.
Wimalasena, J. and Dufau, M. L. (1982) *Endocrinology* **110,** 1004–1012.
Wimalasena, J., Schwab, S., and Dufau, M. L. (1983) *Endocrinology* **113,** 618–624.
Wimalasena, J., Abel, J. A. Jr., Wiebe, J. P., and Chen, T. T. (1986) *J. Biol. Chem.* **261,** 9416–9420.
Zang, Q. Y. and Menon, K. M. J. (1988) *J. Biol. Chem.* **263,** 1002–1007.

The Follicle Stimulating Hormone Receptor

Bosukonda Dattatreyamurty and Leo E. Reichert, Jr.

1. Introduction

Regulation and maintenance of Sertoli cell functions essential for spermatogenesis are primarily dependent on follitropin (FSH), a glycoprotein hormone that is elaborated by the anterior pituitary gland. The initial event in follitropin action is the specific binding of the hormone by high affinity follitropin receptors present on the Sertoli cell plasma membrane. We have studied this process extensively using membrane preparations of variable degrees of purity from rat and bovine calf testis (Bhalla and Reichert, 1974; Abou-Issa and Reichert, 1976,1977). Such studies have proven useful in understanding the functional requirements for specific interactions of follitropin to hormone-specific membrane receptors, and these have been reviewed previously (Reichert et al., 1983,1984; Reichert and Dattatreyamurty, 1989). A soluble follitropin binding component derived from bovine calf testis in the absence of detergent has also been identified (Dias and Reichert, 1982) and the modulation of follitropin binding to receptors by monovalent salts and divalent cations has been reported (Andersen and Reichert, 1982). We also established that the hydrophobic effect is the most important force participating in the interaction of follitropin with its receptor in the membrane and after solubilization (Andersen et al., 1983; Sanborn et al., 1987). As with many simple proteins and small peptides, follitropin is cellularly processed after binding to its receptors on Sertoli cells through mechanisms involving internalization by receptor-mediated endocytosis and degradation via a lysosomal pathway (Fletcher and Reichert, 1984). It has been well-established that follitropin stimulates Sertoli cell adenylate cyclase consequent to its interaction with receptors (Means, 1973; Heindel et al., 1975). The involvement of GTP in signal transduction was first shown by Rodbell and his associates (Rodbell et al., 1971). Our studies have demonstrated that GTP enhances signal transduction from follitropin receptor to the catalytic unit of adenylate cyclase (Abou-Issa and Reichert, 1979). When Gpp(NH)p, a nonhydrolyzable GTP analog, was used instead of GTP adenylate cyclase became active and the rate of enzyme activation was increased by follitropin (Abou-Issa and Reichert, 1978). These findings support the notion that under physiological conditions GTP is hydrolyzed to GDP at the regulatory site thereby terminating the hormonal signal. The role of follitropin, therefore, is to facilitate exchange of GTP for GDP, thereby activating adenylate cyclase (Fletcher and Reichert, 1986). Recent studies have shown that, in addition to cAMP, second messengers such as diacylglycerol and inositol triphosphate may be involved in the actions of a variety of hormones that exert their effects through membrane receptors, including lutropin (Dimino et al., 1987; Davis et al., 1987). We have verified the presence of an active phosphoinositide pathway in Sertoli cells by demonstrating positive effects of ALF_4 on inositol phosphate accumulation (Quirk and Reichert, 1988). Follitropin, however, had no effect on accumulation of inositol phosphate in a Sertoli cell population that responded to this hormone with increased conversion of androstenedione to estradiol (Quirk and Reichert, 1988). In other receptor systems it has been shown that activation of the phosphatidylinositol pathway has a synergistic effect on adenylate cyclase (Sugden et al., 1985; Yoshimasa et al., 1987). Synergism between two second messenger systems in relation to the mechanism of follitropin action, therefore, would not seem precluded by lack of a direct effect of follitropin on phosphatidylinositol hydrolysis.

Two approaches have been generally used to study the biochemical characterisitcs of the FSH receptor. The *first approach* requires chemical crosslinking or photoaffinity labeling of hormone-receptor complexes, followed by SDS-polyacrylamide gel electrophoresis in one or two dimensions and autoradiography. Although this approach made a significant contribution toward elucidating the subunit structure of insulin receptors (Jacobs et al., 1980; Massague et al., 1981), similar experiments to characterize receptors for follitropin (Branca et al., 1985; Smith et al., 1985,1986; Shin and Ji, 1985), lutropin (Ji and Ji, 1981; Hwang and Menon, 1984; Ascoli and

Segaloff, 1986) and thyrotropin (Kohn et al., 1983; Buckland et al., 1985; McQuade et al., 1986) have yielded variable results. Perhaps this may result from inherent limitations of the techniques used in this approach, particularly when they are applied to characterize receptors for noncovalently linked heteroglycodimers. The *second approach* requires a carefully designed purification scheme for the isolation of follitropin receptor from a conveniently available receptor source.

Several practical points need to be considered in the latter approach. Since follitropin receptor is a membrane protein and can only be solubilized by detergents or compounds with detergent-like properties, its purification involves several problems not usually encountered when studying secreted or cytoplasmic proteins. Receptors are present in very low concentrations in gonadal tissues of most species, and, therefore, a suitable receptor source and its large scale availability is essential. Optimum conditions must be established to extract the receptor from its natural environment and to maintain its stability and biological and chemical properties for a period of time that allows for its further study. One must ensure that what has been solubilized and purified represents the physiologically relevant receptor. Earlier attempts to purify follitropin receptors, although not fully successful, provided some insight into the problems associated with receptor purification. A widely used nonionic detergent, Triton X-100, was utilized in our earlier studies to solubilize follitropin receptors from bovine calf testis membranes. A significant purification of detergent-solubilized receptor was achieved by affinity chromatography on oFSH coupled to Affigel-10 (Abou-Issa and Reichert, 1980). The receptor preparation so obtained, however, had only a 1000-fold increase in binding capacity for FSH than that of the initial detergent extract and required an additional 100-fold purification to approach maximal theoretical hormone-binding. Bluestein and Vaitukaitis (1981) have used a different approach for affinity chromatography to purify follitropin receptor. Based on a previous finding that human chorionic gonadotropin (hCG) binds weakly to follitropin receptors of testis, these investigators devised an hCG-coupled CH-Sepharose 4B matrix and utilized it to adsorb detergent-solubilized follitropin receptor that was then eluted with radiolabled human FSH or 1*M* NaCl (Bluestein and Vaitukaitis, 1981). The degree of purity of the receptor preparation, however, could not be assessed because of its instability. Despite a significant increase in specific activity and percent recovery of solubilized follitropin receptor through application of affinity chromatography, a continuous problem in purification has been the rapid destabilization of follitropin receptor, once solubilized by detergents. Importantly, the addition of solutes, such as sucrose or glycerol, to aqueous or detergent solutions of labile proteins has often proved effective in stabilizing macromolecules, with retention of biological activity. Previous studies from this laboratory have shown that in the presence of glycerol, the ^{125}I-hFSH binding activity of Triton X-100 solubilized calf testis receptors was significantly preserved (Dias et al., 1981). Thus, use of glycerol permitted studies on the detergent-soluble FSH receptor not previously possible (Dias and Reichert, 1982; O'Neill and Reichert, 1984). Glycerol, however, is required in relatively high concentrations (30%) for its stabilizing effect, and this complicates subsequent purification maneuvers such as gel filtration chromatography. Recently, we have used a new approach for the solubilization of follitropin receptors eliminating the need for glycerol during the critical early stages of purification. The approach is dependent on large-scale isolation of light membrane fraction from bovine calf testis homogenate and optimizing conditions for solubilizing follitropin receptors by utilizing low ratios of Triton X-100 to membrane protein and removal of interfering lipids by petroleum ether extraction (Dattatreyamurty et al., 1986). In this chapter, we will describe the experimental details of follitropin receptor solubilization by detergent and its purification from bovine calf testis. We will also focus on some aspects of receptor, such as its molecular assembly and oligomeric structure.

2. Description of Purification Procedure and Results

A flowchart of the procedure developed for the purification of follitropin receptor from bovine calf testis is given in Fig. 1.

2.1. Choice of Receptor Source

Although rat testis has been a convenient model for the study of follitropin receptor physiology (Ketelslegers et al., 1978; O'Shaugnessy and Broun, 1978; Francis et al., 1981), it is not a practical source for receptor purification. In several species, including farm animals, follitropin receptor concentration is greater in immature than mature testis (Abou-Issa and Reichert, 1977; Maghuin-Rogister et al., 1978; Dias and Reeves, 1982). Bovine calf testis had the highest concentration of follitropin receptors among the species examined (Abou-Issa and Reichert, 1977), and was a practical, relatively inexpensive, and readily available source of receptor for solubilization and purification.

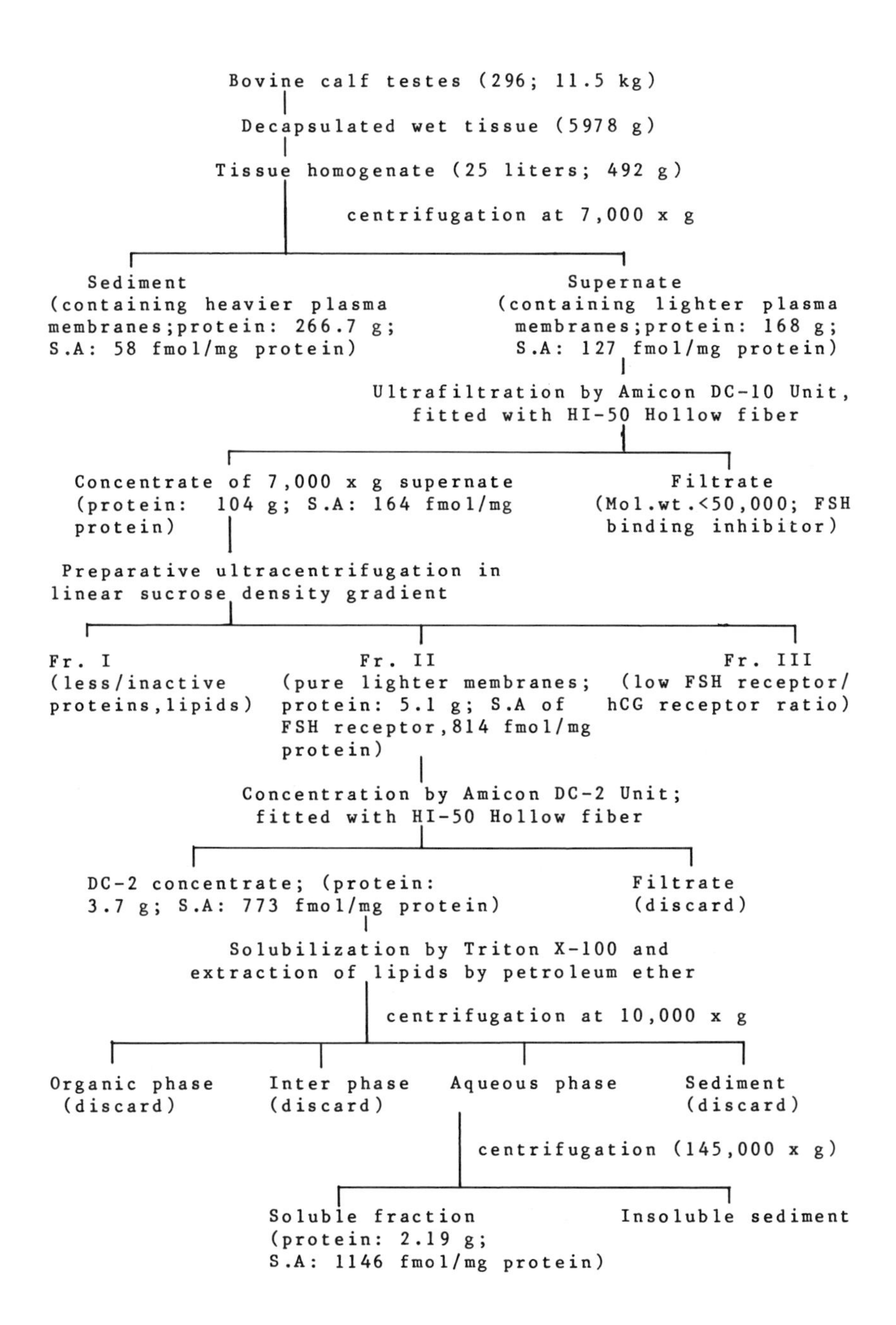

Fig. 1A. A flowchart of a typical procedure developed for the large scale isolation of follitropin receptor-rich lighter membranes from bovine calf testes, and for the solubilization of follitropin receptors from these membranes (Dattatreyamurty et al., 1986).

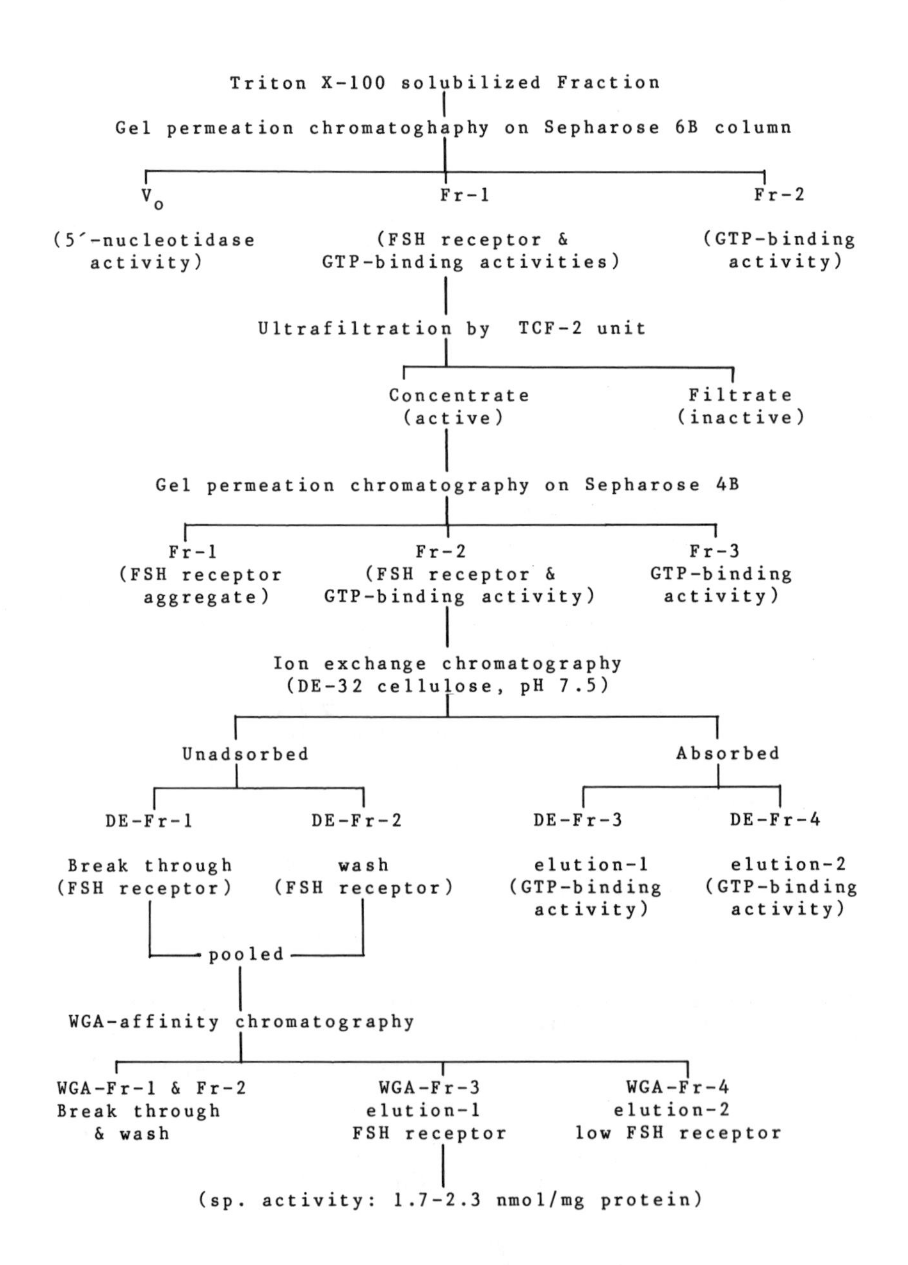

Fig. 1B. A flowchart of a typical procedure developed for the purification of follitropin receptor from detergent-soluble fraction of calf testis lighter membranes.

2.2. Isolation of Light Membranes from Bovine Calf Testis

Bovine calf testis were obtained from a local abattoir and kept at–20°C until used. Approximately 11 kg of testis were thawed and processed in each batch. All processing was conducted between 0–4°C. Testis were decapsulated, weighed, and separated into 200 g batches. To each 200 g batch was added 0.8 L Tris-HCl *buffer-A*, pH 7.4 (10 m*M* Tris-HCl, containing 1 m*M* $MgCl_2$, 10 μ*M* *p*-hydroxymercuribenzoate Na, 0.001% mercaptoethanol, and 15% sucrose) and homogenized for 10 s using a Polytron model homogenizer. Then, taking 300 mL of this homogenate at a time, further homogenization was performed for 15 s at maximum speed, using a Polytron homogenizer adapted with a small generator. After 25 L of total homogenate had been prepared, it was centrifuged for 45 min in liter capacity swing-out buckets at 7000 × g (Sorvall RC-3 refrigerated centrifuge, 4°C). The 7000 × g supernatant, containing light plasma membranes, was separated by decantation from the 7000 × g sediment containing heavier membranes. Approximately 22 L of 7000 × g supernatant was recovered, and this was concentrated to 1.1 L by using an Amicon DC-10 ultrafiltration unit fitted with an HI-50 hollow-fiber cartridge that has a molecular exclusion limit of less than 50,000. The passing fraction containing components of molecular weight less than 50,000 included low molecular weight lutropin- and follitropin-receptor binding inhibitors, but no follitropin-receptor activity. To the concentrated 7000 × g supernatant, was added an equal volume of Tris-HCl *buffer B*, pH 7.2 (10 m*M* Tris-HCl containing 1 m*M* $MgCl_2$ and 0.001% NaN_3). This was concentrated again to its original volume, thereby reducing the sucrose concentration to approximately 7.5% and stored in 70 mL aliquots at –80°C.

The average yields of 7000 × g supernatant containing light plasma membranes and 7000 × g sediment containing heavier membranes from 11 kg bovine calf testes (about 300 testes) were 168 and 267 g protein, respectively. Specific binding of ^{125}I-hFSH to membranes in these testis fractions was determined by Scatchard analysis. The Scatchard plot for follitropin receptors in the 7000 × g sediment indicated the presence of two classes of binding sites having high affinity ($4.6 \times 10^{10}\ M^{-1}$) and low capacity ($1.5 \times 10^{-10}\ M$), and low affinity ($1.4 \times 10^{8}\ M^{-1}$, and high capacity ($6.4 \times 10^{-9}\ M$). In contrast, the 7000 × g supernatant contained only a single class of high affinity ($1.2 \times 10^{10}\ M^{-1}$) and low capacity ($2.02 \times 10^{-10} M$) receptors. The concentration of hCG receptors in 7000 × g sediment and supernatant fractions (19 and 18 fmol/mg protein, respectively) were much less than that of the follitropin receptors (127 and 58 fmol/mg protein).

A noteworthy feature of follitropin receptor activity in the crude homogenates of testicular tissue is its rapid decline owing to proteolytic degradation and the presence of follitropin receptor-binding inhibitors (Dias and Reichert, 1984; Reichert and Abou-Issa, 1977). These inhibitors, if not removed, cosolubilize and copurify with the receptor and are responsible for decreasing the follitropin-binding activity of the detergent-solubilized receptor. As described above, an initial concentration of the 7000 × g supernatant by Amicon DC-10 filtration was carried out in the presence of a protease inhibitor sodium *p*-hydroxymercuribenzoate. An average protein recovery of 62% was obtained, with no loss of FSH- or hCG-binding activities. This rendered FSH receptor activity in the resulting concentrate more stable and suitable for solubilization and also removed significant amounts of inactive proteins as well as FSH-binding inhibitor.

Two criteria have been used to examine the concentrated 7000 × g supernatant and the 7000 × g sediment testis fractions for their suitability for follitropin receptor isolation. *First,* we examined several preparations of the concentrated 7000 × g supernatant for the binding of ^{125}I-hFSH and ^{125}I-hCG. All preparations showed the presence of only high affinity receptors for FSH and hCG with average specific activities of 164 and 26 fmol, respectively. This is in contrast to the presence of both high and low affinity receptors for FSH and hCG present in 7000 × g sediment preparations. *Secondly,* we solubilized follitropin receptors from the concentrated 7000 × g supernatant as well as 7000 × g sediment preparations, and compared the recoveries of receptor activity from these preparations. Aliquots of 7000 × g sediment fraction, each containing 8.8 mg protein/mL, were treated with increasing final concentrations of Triton X-100 (0.01–2%). Detergent-soluble and insoluble fractions were separated (*see below* for experimental details), analyzed for protein content and ^{125}I-hFSH binding activity. In contrast to the concentrated 7000 × g supernatant, the 7000 × g sediment fraction containing heavier plasma membranes required Triton X-100 concentrations higher than 0.5% for the solubilization of follitropin receptors, a concentration that has a deleterious affect on follitropin receptor activity. When Triton X-100 concentrations less than 0.5% were used, the recovery of follitropin receptor activity extracted from the 7000 × g sediment fraction was only 30% of that recovered from the lighter membrane fraction (concentrated 7000 × g supernatant). Based on these observations, we preferred to further process the concentrated 7000 × g supernatant for the isolation of follitropin receptors.

2.3. Preparative Ultracentrifugation in Sucrose Density Gradient

The concentrated 7000 × g supernatant was further fractionated by preparative ultracentrifugation in a linear sucrose density gradient. This method offers the most direct means to isolate pure light plasma membrane fraction. The procedure, in brief, is as follows (for details *see* Dattatreyamurty et al., 1986; Cline and Ryel, 1971). A linear gradient of sucrose concentration from 10–30% or densities of 1.0381–1.127 g/mL was prepared in a Ti-14 rotor with the aid of a Beckman gradient pump (Model 141). The concentrated 7000 × g supernatant was homogeneously suspended in Tris-HCl *buffer-B* and layered on top of the sucrose gradient. The rotor was accelerated to 40,000 rpm and the centrifugation was performed for 2.5 h at 4°C. The rotor was then decelerated to maintain a speed of 3000 rpm and the contents of the rotor were eluted in 10 mL fractions by displacement with a 40% sucrose solution. Each fraction was analyzed for sucrose (Siegal and Monty, 1966) and protein concentrations (Lowry et al., 1951), for the presence of 5'-nucleotidase activity, for hormone- and NaF-induced adenylate cyclase activities (Abou-Issa and Reichert, 1979), and for the specific binding of ^{125}I-hFSH, ^{125}I-hCG, and [^{3}H]Gpp(NH)p.

In a typical purification sequence, 16 sucrose density gradient (SDG) centrifugations were carried out to fractionate a total vol of 1.1 L of the concentrated 7000 × g supernatant, obtained from an 11 kg batch of bovine calf testes (Fig. 2). Significant amounts of lipid and inactive proteins (85%) were separated in *SDG Fraction I* between sucrose concentrations of 6–23% or densities of 1.0218–1.0945 g/mL. *SDG Fraction II*, eluted between sucrose concentrations of 23–27.5% or densities of 1.0945–1.1152 g/mL, contained high specific ^{125}I-hFSH-binding activity and also some ^{125}I-hCG-binding activity. A heavier membrane fraction was eluted in SDG Fraction III, between sucrose concentrations of 27.5–32% or densities of 1.1152–1.1366 g/mL, and contained high specific ^{125}I-hCG-binding activity and a low degree of ^{125}I-hFSH-binding activity. A noteworthy advantage of sucrose density gradient centrifugation was the apparent partial separation of membrane fractions having FSH- and hCG-binding activity. When the fractions were analyzed at equal protein concentrations, a maximum ^{125}I-hFSH binding activity was observed in SDG Fraction II. On the other hand, ^{125}I-hCG binding activity was maximum in SDG Fraction III.

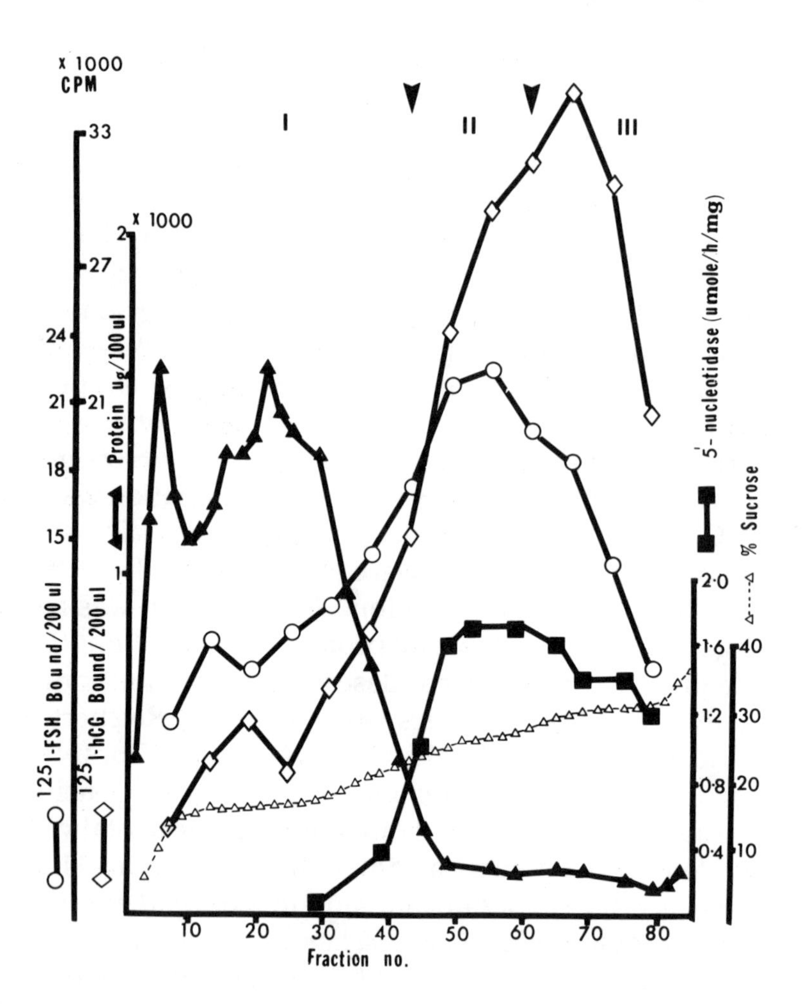

Fig. 2. Preparative centrifugation in linear sucrose density gradient of concentrated 7000 × g supernatant. Approximately 6.5 g protein was layered over a 10–30% linear sucrose gradient (Δ---Δ) in a zonal rotor (Ti-14)of 665 mL capacity with the aid of a Beckman Model 141 gradient pump. Experimental details regarding centrifugation and elution of rotor contents are as described previously (Dattatreyamurty et al., 1986). Fractions were pooled into three major pools (SDG Fractions I, II, and III) as indicated by arrows in the figure based on protein content and receptor activity profiles of individual fractions.

2.4. Functional Properties of Plasma Membranes in SDG Fractions II and III

The 5'-nucleotidase activity was maximum in fractions that were enriched with plasma membranes. The activities of the membrane-bound testicular adenylate cyclases and their responses to hCG and FSH were determined in the fractions separated by sucrose density gradient centrifugation. Adenylate cyclase activity stimulation by hCG and FSH was readily demonstrable in SDG Fractions II and III that were enriched with plasma membranes. Comparable responses of adenylate cyclase were produced by 10 m*M* fluoride and 4.5 μg of pure hCG. A direct correlation appeared between hCG-induced adenylate cyclase and hCG-binding activity present in the SDG Fractions II and III. *SDG Fraction II*, which contained maximum ^{125}I-hFSH binding, showed a significant induction of adenylate cyclase by FSH. A higher adenylate cyclase response to FSH, however, was observed in *SDG Fraction III*, and this could be attributed to LH/hCG receptor-mediated cyclase response to the LH contamination present in the FSH preparation used.

It has been shown that the actions of most hormones and fluoride ion are mediated through G-protein that binds GTP or a nonhydrolyzable analog, Gpp(NH)*p*. The presence of G-protein in the membrane fractions was examined by studying the specific binding of [^{3}H]Gpp(NH)*p* and the activation of adenylate cyclase by fluoride. A maximum [^{3}H]Gpp(NH)*p* binding was observed in SDG Fractions II and III, which contained higher hormone-binding activities. There appeared a close correlation between [^{3}H] Gpp (NH)*p*-binding and fluoride-induced adenylate cyclase activity, suggesting that a significant portion of testicular G-protein is functionally related to adenylate cyclase present in SDG Fractions II and III.

Specific binding of ^{125}I-hFSH and ^{125}I-hCG to membranes present in SDG Fractions I and II was calculated on the basis of Scatchard analyses. The results indicated that the bulk of inactive protein (85%) was removed into SDG Fraction I. On the other hand, seven replicate preparations of SDG Fraction II consistently showed only high affinity receptors for FSH ($K_a = 1.02 \times 10^{10} M^{-1}$), and for hCG ($K_a = 4.31 \times 10^{10} M^{-1}$), with average specific activities of 814 and 88 fmol, respectively.

2.5. Concentration of SDG Fraction II

Approximately 3.2 L of SDG Fraction II, obtained from 16 sucrose gradient centrifugations, was concentrated about 20-fold on an Amicon DC-2 unit using an HI-50 hollow fiber cartridge that has an exclusion limit of less than 50,000 Daltons. The DC-2 concentrate was diluted tenfold with Tris-HCl buffer-B to reduce excess sucrose, the presence of which poses problems in subsequent extraction of receptor by the detergent. A minimum concentration of 2.5% sucrose was maintained in the sample. The diluted sample was again concentrated 20-fold. The resulting DC-2 concentrate contained high affinity follitropin receptors, with no detectable loss of hormone-binding capacity during the process of concentration.

2.6. Solubilization of Follitropin Receptors by Triton X-100

Perhaps the greatest obstacle to the isolation of follitropin receptors is the identification of procedures that disrupt the membrane lipid matrix and extract receptors without causing denaturation. A widely used non-ionic detergent, Triton X-100 was employed for the solubilization of receptors from DC-2 concentrate. Two different approaches were used to establish optimum conditions for the extraction of follitropin receptors. In the first approach, we examined the stability of FSH and hCG receptors of testicular membranes in the presence of high and low concentrations of Triton X-100. DC-2 concentrate was solubilized in Tris-HCl *buffer-B*, at a protein concentration of 8.6 mg/mL, in the presence of excess Triton X-100 (2%), sonicated and centrifuged at 100,000 × g for 1 h. The supernatant was then treated with graded amounts of moist Bio-beads (SM_2) for up to 150 min. Sample aliquots were adjusted to equal protein and Triton concentration before they were analyzed for ^{125}I-hFSH and ^{125}I-hCG binding activities. A progressive decrease in Triton concentration as a result of detergent removal by Bio-beads was dependent on the mass of the Bio-beads added and period of incubation. The stability of FSH receptors appears to be different from that of hCG receptors in the presence of low concentrations of Triton X-100. A minimum of 0.44% Triton X-100 was adequate to keep the hCG receptors soluble without affecting its hormone-binding property. In fact, in the presence of 0.5% Triton X-100, soluble hCG receptor showed a significant increase in its hormone-binding activity, a situation previously reported with LH-hCG receptors from bovine *corpora lutea* membranes (Dattatreyamurty et al., 1983). This concentration of Triton X-100 almost totally destroyed the hormone-binding activity of the soluble FSH receptors.

In another set of experiments, we used increasing concentrations of Triton X-100, up to 0.5%, for the solubilization of follitropin receptors from DC-2 concentrate. The purpose of this experiment was to determine the minimum concentration of Triton X-100 that effectively solubilizes FSH receptors without affecting its hormone-binding activity. Aliquots of DC-2 concentrate, each containing 9.1 mg protein were made in 1 mL of Tris-HCl buffer-B that contained final detergent concentrations of 0, 0.02, 0.05, 0.1 and 0.5%, respectively. Samples were either centrifuged directly or after sonication at 100,000 × g for 1 h. There was no change in the specific FSH-binding activity of the sample aliquots following sonication of the samples. The resulting supernatants were adjusted to equal protein and Triton concentrations before being analyzed for ^{125}I-hFSH binding. Extraction of follitropin receptor activity from DC-2 concentrate was maximum at a protein concentration of 9.1 mg in 1 mL of 0.1% Triton X-100.

We also determined the solubilization of FSH receptor activity as a function of Triton X-100. The DC-2 concentrate was solubilized in Tris-HCl *buffer B,* at protein concentrations of 19.3, 9.65, and 3.86 mg/mL, in the presence of increasing concentrations (final) of Triton X-100 from 0.01–2.0%. The samples were sonicated and centrifuged at 100,000 × g for 1 h. The resulting supernatants were adjusted to equal protein and Triton concentrations before they were analyzed for ^{125}I-hFSH binding. A progressive increase in the amount of protein solubilized as well as total recovery of FSH-binding activity in the soluble fraction was observed in the presence of Triton concentrations from 0.01 to 0.1%. In the presence of 0.25% and higher of Triton X-100, the amount of total protein solubilized was greater, but the specific FSH-binding activity of the soluble receptor decreased significantly. We also examined receptor activity extracted from the DC-2 concentrate as a function of the ratio of detergent to protein (v/w) (Fig. 3). At the relatively high protein concentrations of 19.3 and 9.65 mg/mL, an optimum detergent to protein ratio was observed at about 0.01. At the lower protein concentration of 3.86 mg/mL, however, the optimum shifted to a higher value of detergent to protein ratio (0.025). Thus, the data show values of optima with decreasing detergent to protein ratio until a constant value (0.01) was reached at high protein concentrations. This suggests that the initial protein concentration should be high when attempts are made to solubilize active receptors by detergent. The solubilized sample must retain full receptor activity despite the certain dilution that normally occurs during molecular sieving chromatography.

Thus, a series of painstaking, systematic studies indicated that an optimum ratio of protein to Triton X-100 is of critical importance for the extraction of labile receptors from testicular membranes with maintenance of hormone-binding activity. An optimum ratio of protein to Triton X-100 was found to be 19 mg of protein to 1 mL of 0.2% Triton X-100. An interesting feature of bovine testis FSH receptor solubilized by Triton X-100 under these conditions, was its slight but reproducibly higher FSH-binding activity (1.34-fold) and affinity ($K_a = 2.27 \times 10^{10} M^{-1}$) than that of membrane-bound receptor. Increased receptor affinity with increased hormone-binding capacity following membrane solubilization has also been reported for prolactin, and thyrotropin receptors (Shiu and Friesen, 1974; de Bruin

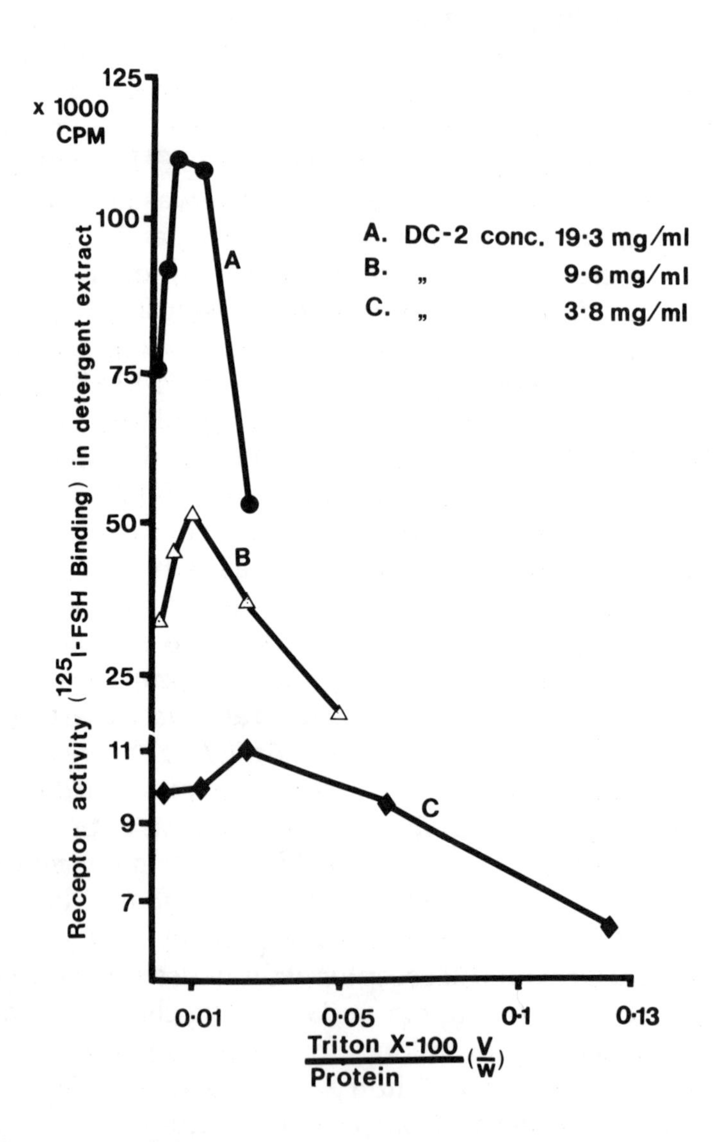

Fig. 3. Plots of follitropin receptor activity extracted from DC-2 concentrate as a function of detergent to protein ratio. Lighter membrane fraction (DC-2 concentrate) was solubilized at indicated sample concentrations, in the presence of increasing concentrations (final) of Triton X-100, 0.01–2.0%. Soluble fractions were adjusted to equal Triton and protein contents before analyzing for ^{125}I binding.

and van der Heide, 1984); hCG receptors solubilized from rat gonads, however, had lower affinity for hCG than did membrane-bound receptors (Dufau et al., 1973, 1977; Lee and Ryan, 1974).

2.7. Treatment of DC-2 Concentrate with Organic Solvents During Extraction of Follitropin Receptors by Detergent

Organic solvents have been shown to facilitate solubilization of membrane proteins (Morton, 1950; Boyn and Clement-Cormier, 1984). Certain groups of proteins are so hydrophobic that they become tightly embedded within the organic phase of the lipid bilayer. The exposure of membranes to organic solvents under conditions that extract lipids or perturb the internal arrangements of lipids can facilitate the extraction of membrane proteins. We examined two different organic solvents, petroleum ether and *n*-butyl alcohol, for their potential usefulness for extraction of lipids from DC-2 concentrate. Aliquots of DC-2 concentrate were treated with Triton X-100 under optimum conditions, as described above. The samples were sonicated and further treated with different volumes [ratio (v/v), 0.1 – 3.2] of chilled petroleum ether/*n*-butyl alcohol. The organic phase, the interphase, the aqueous phase and insoluble sediment were separated by centrifugation at 10,000 × g (in Sorvall RC-3B centrifuge, SM-24 rotor) for 1 h at 4°C. Equal aliquots of aqueous sample from each group, were further centrifuged (30,000 × g to 145,000 × g) for 1 h. The supernatants were separated and analyzed for protein content and for ^{125}I-hFSH binding. The soluble fraction extracted from DC-2 concentrate by Triton alone showed an increase of 25% in specific ^{125}I-hFSH binding activity. Treatment of the detergent-solublized fraction from DC-2 concentrate with 0.2 volume (v/v) of petroleum ether removed free lipids in the organic phase and resulted in a further increase of 15% in the specific FSH-binding activity of detergent-soluble protein recovered in the aqueous phase. The specific FSH-binding activity and protein recovery, however, declined significantly when detergent-extracted protein was treated with higher volumes (0.4–3.2 v/v) of petroleum ether. Also, treatment of sample with *n*-butyl alcohol at sample to solvent ratios of 1:0.1–1:2 (v/v) resulted in a total loss of follitropin receptor activity. This may be a result of loss of essential phospholipids. Earlier, we showed that phospholipid hydrolysis by phospholipase C markedly reduced FSH binding to the membrane-bound receptors from bovine calf testes, indicating an essential role of phospholipids in follitropin interaction with its receptor (O'Neill and Reichert, 1984). A noteworthy result from the above procedure was a substantial degree of separation between FSH-binding and hCG-binding activities of the extracted material. This was a result of the incomplete solubility of hCG receptors under optimum conditions for solubilization of FSH receptors. Thus, the detergent-soluble fraction from DC-2 concentrate after centrifugation at 145,000

× g, retained full specific FSH-binding activity, whereas hCG-binding activity was only 47% of that originally present.

The soluble nature of follitropin receptors so obtained was judged by experimental criteria such as failure to sediment at 145,000 × g for 90 min, passage through 0.22 μm Millipore filters and retardation on molecular seiving through Sepharose 6B (*see below*). An average recovery of detergent-soluble fraction from 11 kg of bovine calf testes was 2.19 g, and its specific FSH-binding activity was 1146 fmol/mg protein, which represents a 24-fold increase over that of starting material (testicular homogenate). Approximately 86% of the follitropin receptors originally present in the light membranes were extracted by detergent and this is certainly a considerable improvement over the 20–40% recovery obtained in previous reports (Dufau et al., 1977; Closset et al., 1977). Importantly, the soluble follitropin receptor was stable for 4 d when stored at 1°C or for >6 mo at –80°C, in the absence of glycerol. Another important feature of detergent-soluble follitropin receptor of bovine testis is its functionality as reflected by FSH-simulation of adenylate cyclase. This observation suggests that a high proportion of follitropin receptors present in detergent-soluble testicular fraction are functionally related, and that the fraction could be an ideal precursor isolated for follitropin receptors.

2.8. Gel Filtration on a Sepharose 6B Column

The DC-2 concentrate, at a final protein concentration of 20 mg/mL, was solubilized by detergent (Triton X-100) under above described optimum conditions. The detergent-soluble fraction was fractionated in aliquots of 480 mg protein (about 1/4 of total material obtained from 11 kg bovine calf testes) on a gel bed of 5 cm × 46 cm Sepharose 6B. The column was equilibrated with 10 m*M* Tris-HCl buffer, pH 7.2, containing 1 m*M* $MgCl_2$, 0.001% NaN_3, 2.5% sucrose, 8 μ*M* *p*-hydroxymercuribenzoate sodium, 0.0016% 2-mercaptoethanol, and 0.1% Triton X-100 at 4°C. The column was eluted with the same buffer at a flowrate of 15 mL/h. The fractionation of Triton X-100-soluble light plasma membrane (DC-2 concentrate) protein resulted in three major fractions (Fig. 4). Most 5'-nucleotidase activity was eluted in the void volume. The first retarded Sepharose 6B-Fraction-1(*6B-Fr-1*) contained FSH-binding activity. Some Gpp(NH)*p*-binding activity was also coeluted in this fraction. Most Gpp(NH)*p*-binding activity was, however, eluted in the more retarded Sepharose 6B-Fraction-2 (*6B-Fr-2*) and was clearly separated from FSH-binding activity. Several lines of experimental evidence suggest that 6B-Fr-1 was a functional complex consisting of FSH receptor and a guanine nucleotide regulatory protein, probably G_s (Dattatreyamurty et al., 1987).

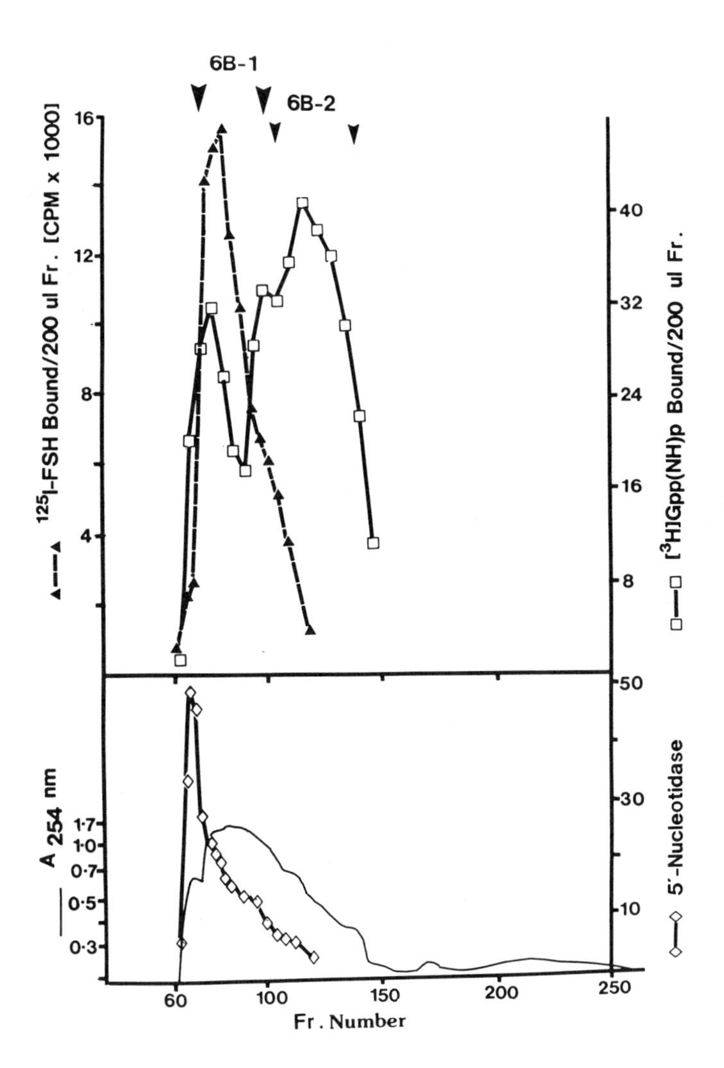

Fig. 4. Elution profiles of follitropin receptor and [^{3}H]Gpp(NH) *p*-binding activities after chromatography on Sepharose-6B. Detergent (Triton X-100)-solubilized protein, approximately 185 mg was applied on a gel bed of 2.8 × 71 cm (or 480 mg on a gel bed of 5 × 46 cm). Column was equilibrated and eluted with 10 m*M* Tris-HCl buffer, pH 7.2, containing 1 m*M* $MgCl_2$, 0.001% NaN_3, 2.5% sucrose, 8 µ*M* *p*-hydroxymercuribenzoate sodium, 0.0016% 2-mercaptoethanol, and 0.1% Triton X-100. Selected fractions were pooled into two major fractions (6B-Fr-1 and 6B-Fr-2) as indicated by arrows in the figure.

2.9. Concentration of Active Fraction (6B-Fr-1) by Amicon TCF-2 Ultrafiltration Unit

Detergents at sufficiently elevated concentrations are known to uncouple the functional units of adenylate cyclase (Haga et al., 1977; Keenan et al., 1982). We therefore exposed 6B-Fr-1 to a higher concentration of

Triton X-100, while maintaining the optimum ratio between sample protein and Triton previously shown to preserve FSH receptor activity (Dattatreyamurty et al., 1986). This was achieved through concentration of active material (6B-Fr-1) by ultrafiltration using an Amicon TCF-2 unit fitted with Diaflo YM-30 membrane (mol wt cut off <30,000 Daltons). A total vol of 75 mL active fraction (6B-Fr-1) obtained after each gel filtration on Sepharose 6B was concentrated 2.5-fold and centrifuged at 2500 rpm for 15 min at 4°C. The supernatant was immediately filtered through a Sepharose 4B column.

2.10. Gel Filtration on a Sepharose 4B Column

The concentrated 6B-Fr-1was fractionated in aliquots of 110 mg of protein on a gel bed of 5 cm × 36 cm Sepharose 4B. The column was equilibrated with 10 m*M* Tris-HCl buffer, pH 7.2, containing 1 m*M* $MgCl_2$, 0.001% NaN_3, 5% glycerol and 0.0975% Triton X-100 at 4°C. The same buffer was used to elute the column at a flowrate of 15 mL/h. Fractionation of 6B-Fr-1concentrate on Sepharose 4B resolved in three major fractions. An apparent FSH receptor aggregate, having low specific activity was eluted in the unretarded fraction, Sepharose 4B-Fraction-1(4B-Fr-1). Most receptor activity with a high specific FSH-binding was eluted in the retarded fraction, Sepharose 4B-Fraction-2 (4B-Fr-2). Some Gpp(NH)*p*-binding activity was also present in this fraction. The latter, however, was apparently not associated with receptor, since the receptor in this fraction was insensitive to GTP effect (Dattatreyamurty et al., 1987). Significant Gpp(NH)*p*-binding activity was separated from FSH-binding activity and eluted in a more retarded fraction, Sepharose 4 B-Fraction-3 (4B-Fr-3). Most 5'-nucleotidase activity of 6B-Fr-1 concentrate was eluted in 4B-Fr-1. Fractions 4B-Fr-2 and 4B-Fr-3 contained only trace and no detectable 5'-nucleotidase activity, respectively.

2.11. DEAE-Cellulose Chromatography

Previously, anion exchange resins (DEAE-cellulose/DEAE-Sephacel) were shown to be effective in adsorbing insulin receptors (Cuatrecasas, 1972; Siegel et al., 1981), LH/hCG receptors (Dattatreyamurty, 1983) or Gpp(NH)*p*- binding protein (G_s-protein) (Northrup et al., 1980; Codina et al., 1984), and thus have become widely utilized in the purification of receptors and G-protein. Clearly, the fraction (4B-Fr-2) isolated after gel filtration on Sepharose 4B contained high specific FSH-binding activity, and, in addition, some Gpp(NH)*p*-binding activity as well as low (but significant) concentrations of LH/hCG receptors as contaminants.

DEAE-cellulose-32 was extensively washed and equilibrated in 50 m*M* Tris-HCl buffer pH 7.5, containing 1 m*M* $MgCl_2$, 0.001% NaN_3, 30% glycerol, and 0.0488% Triton X-100. An appropriate volume of glycerol was added to fraction 4B-Fr-2 to adjust the final concentrations of glycerol and Triton X-100 in the sample to 30% and 0.0609%, respectively. Equal volumes of packed resin and fraction 4B-Fr-2 were incubated together, while the mixer was rotating, end over end, for 1 h at 4°C. After the incubation period, the sample-treated resin was centrifuged at 1500 rpm for 15 min at 4°C. The supernatant containing unadsorbed sample material was carefully separated, and the resin was washed with two bed volumes of equilibrating buffer. The adsorbed proteins were eluted from the resin, first by 0.2*M* ammonium acetate in equilibrating buffer, pH 7.5, and then by 0.2*M* ammonium acetate in equilibrating buffer, pH 6.3. Each treatment was repeated once. During each treatment, the resin was suspended in one-half bed volume of elution buffer, rotated end over end for 10 min at 4°C, and centrifuged at 1500 rpm to separate the supernatant containing the eluted proteins. Fractions containing unadsorbed proteins and adsorbed and eluted proteins were dialyzed against equilibrating buffer before they were analyzed for protein content, 5'-nucleotidase activity, ^{125}I-hFSH-binding, ^{125}I-hCG-binding, and [^{3}H]Gpp(NH)*p*-binding activities. The results indicated that most hCG-binding, Gpp(NH)*p*-binding activities, and 5'-nucleotidase enzyme activity were adsorbed by DEAE- cellulose and could be eluted (DE-Fr-3 and DE-Fr-4) by ammonium acetate. FSH-binding activity of 4B-Fr-2, however, was unadsorbed by DEAE-cellulose (DE-Fr-1 and DE-Fr-2). The Gpp(NH)*p*-binding activity in these fractions was extremely low and an excellent separation between the two activities was achieved by DEAE-cellulose chromatography.

2.12. Lectin Affinity Chromatography

Affinity chromatography on immobilized lectin is based on specific and reversible molecular interactions between lectin and specific sugars. Like most cell surface proteins, membrane receptors generally contain covalently bound carbohydrate chains. Immobilized lectins have become useful for the purification of insulin receptors (Fujita-Yamaguchi et al., 1983), and to some degree, for the isolation of TSH receptor (Kress and Spiro, 1986), and LH/hCG receptor (Keinanen et al., 1987). Therefore, we were encouraged to explore the potential application of this method to the purification of follitropin receptor. We used agarose-linked wheat germ agglutinin (WGA), a lectin that specifically binds to *N*-acetyl glucosamine and sialic acid residues. WGA-agarose was extensively washed with 50 m*M* HEPES buffer, pH 7.5, containing 10 m*M* $MgCl_2$, 1 m*M* $CaCl_2$, 0.001% NaN_3, 0.02% Triton X-100, 30% glycerol (buffer-I). Fractions DE-Fr-1 and DE-Fr-2 obtained from DEAE-cellulose chromatography were pooled. To 20 mL of the pooled fraction was added 23.34 mL of 90 m*M* HEPES buffer, pH 7.5, containing 9 m*M* $MgCl_2$, 2 m*M* $CaCl_2$, 0.001% NaN_3, 30% glycerol. The total sample was divided into two equal aliquots and each sample aliquot was added to a 50 mL capacity culture tube containing 10 mL (packed volume) of washed WGA-agarose. The tube containing sample and affinity matrix was rotated end over end in the cold (4°C) for 16 h. The sample and affinity matrix slurry was then either poured into a column or directly processed in batchwise fashion. The affinity matrix was first washed with 2 bed volumes of 100 m*M* NaCl in *buffer-I*, and then with 1 bed volume of 50 m*M* HEPES buffer, pH 7.5, containing 0.02% Triton X-100 and 30% glycerol (*buffer-II*). The column was sequentially eluted with 0.2*M* and 0.4*M* *N*-acetyl glucosamine in buffer-II. Alternatively, in batch-wise processing, the specific elution by appropriate concentrations of *N*-acetyl glucosamine was accomplished by resuspending the affinity matrix in 1 bed volume of buffer-II containing *N*-acetyl glucosamine. The suspension was then rotated end over end for 10 min, as before, and centrifuged at 2,000 rpm for 10 min to separate the supernatant containing the eluted active receptor. The entire elution process with 0.2*M* or 0.4*M* *N*-acetyl glucosamine was repeated twice at 4°C. The flow-through, buffer-wash, and eluted fractions were extensively dialyzed against 10 m*M* Tris-HCl buffer, pH 7.2 containing 10 m*M* $MgCl_2$, 0.001% NaN_3, 30% glycerol,and 0.02% Triton X-100, before they were analyzed for protein content, 5'-nucleotidase activity, ^{125}I-hFSH-binding, ^{125}I-hCG-binding, and [^{3}H]Gpp(NH)*p*-binding activities. More than 80% of the total receptor activity of the sample applied adsorbed to immobilized lectin, of which 75% could be eluted with *N*-acetyl glucosamine. Trace amounts of Gpp(NH)*p*-binding activity present in the sample applied were unadsorbed by the lectin. Fraction WGA-Fr-3 eluted by 0.2*M* *N*-acetyl glucosamine, contained the highest specific FSH-binding activity and represented highly purified receptor.

3. Characteristics of the Purified Follitropin Receptor

The final purified receptor (*WGA-Fr-3*) bound 1.7–2.3 nmol ^{125}I-hFSH/mg protein and was purified about 11,000-fold over that present in preparations of calf testis membranes (7000 × g supernatants). An average yield of about 500 µg receptor preparation could be obtained in each batch of purification from 11 kg bovine calf testes. No Gpp(NH)*p*-binding or 5'-nucleotidase activity could be detected in the preparation. The fact that wheat germ lectin could bind follitropin receptor and could be specifically eluted by *N*-acetyl glucosamine provides evidence of the glycoprotein nature of the receptor.

3.1. Criteria of Purity

Several lines of evidence established the purity of the final receptor preparation.

1. Upon SDS-polyacrylamide gel electrophoresis (SDS-PAGE), the preparation gave a single prominent band in the region of mol wt about 240,000 (Fig. 5);
2. Ligand (Western) blotting with ^{125}I-hFSH followed by autoradiography identified specific hormone-binding to the 240,000 mol wt band;
3. The affinity crosslinking of the purified receptor to ^{125}I-hFSH produced a complex that, upon SDS-PAGE, revealed a band in the region of 300,000 under nonreducing conditions. The band, however, was abolished after incubation of the receptors with ^{125}I-hFSH in the presence of excess unlabeled FSH;
4. Purification of partially purified receptor preparation (DE-Fr-1) by hormone-affinity chromatography on immobilized oFSH-Affigel-10 (rather than WGA agarose) yielded a receptor preparation that, upon analysis by SDS-PAGE under reducing conditions, gave an electrophoretic profile (mol wt about 60,000) consistent with that of receptor preparation (WGA-Fr-3) purified by WGA affinity chromatography.

3.2. Molecular Assembly and Apparent Oligomeric Nature of the Follitropin Receptor

Purified follitropin receptor obtained after lectin affinity chromatography (WGA-Fr-3) was iodinated with ^{125}I, using Chloramine-T (Greenwood et al., 1963). Following treatment of radioiodinated follitropin receptor with 2% SDS for 90 s at 100°C, and analysis by SDS-PAGE and auto-radiography, a prominent band with an apparent mol wt of about 240,000 was observed (Fig. 5). This band was not affected by 8*M* urea treatment (4 h at 37°C) prior to analysis by SDS-PAGE, suggesting that the receptor does not contain noncovalently associated subunits. However, treatment with the reducing agent dithiothreitol (DTT, 3 to 70 m*M*), and SDS (2%) for 90 s at 100°C, induced a loss of the 240,000 band, with the appearance of a 60,000 mol wt band. Exposure of receptor preparation to higher concen-trations (200 m*M*) of DTT, under identical conditions did

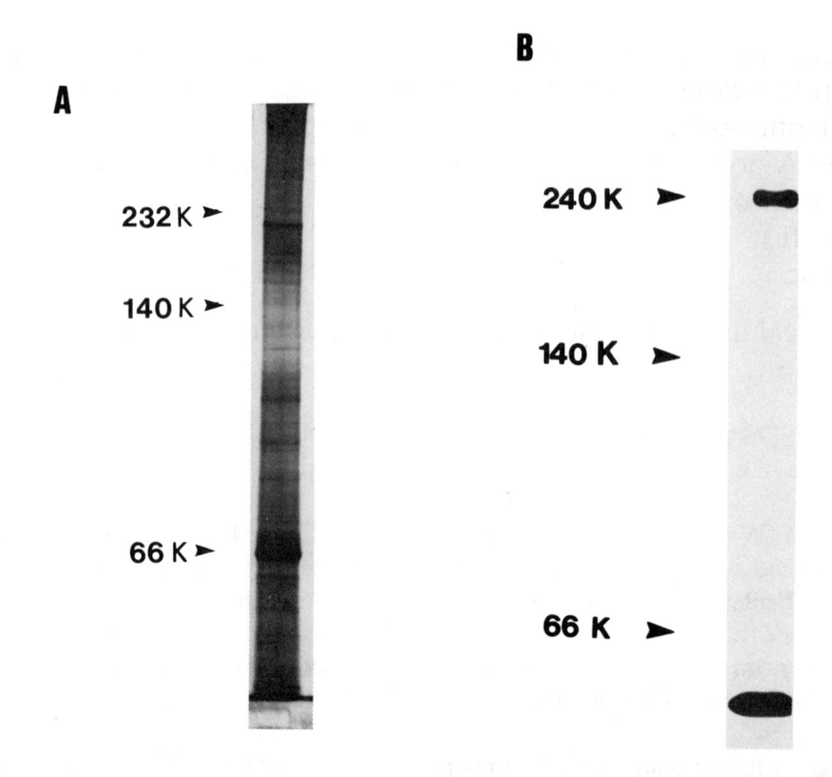

Fig. 5. A. SDS-PAGE analysis of detegent-soluble fraction from DC-2 concentrate. B. SDS-PAGE analysis of purified ^{125}I-labeled follitropin receptor preparation. Follitropin receptor preparation after wheat germ lectin affinity chromatography, was radioiodinated with ^{125}I. Detergent-soluble fraction from DC-2 concentrate (A) or radiolabeled receptor preparation (B) in sample buffer containing 2% SDS, was heated at 100°C for 90 s and subjected to SDS-PAGE in either 7.5% (A) or 6% (B) acrylamide gels. Gels were either silver stained (A) or subjected to autoradiography (B). Positions of the mol wt standards are indicated by arrows in the figure.

not further dissociate mol wt band of about 60,000. Moreover, the band of about 60,000 mol wt, gave a single spot on autoradiography after SDS-PAGE in 7.5% acrylamide gel, followed by electrophoresis in the second dimension in 15% acrylamide gel. These results indicated that the follitropin receptor may be an oligomeric complex of four subunits of similar size, about 60,000, linked by interchain disulfide bonds.

4. Conclusions

Immature bovine testis have proven to be a convenient and practical source of FSH receptors for purification. Problems of destabilization of the receptor upon solubilization from sucrose-density gradient purified testis membranes have been overcome by utilizing an optimal ratio of nonionic detergent Triton X-100 to protein (v/w) of 0.01 and inclusion of petroleum ether in the extraction medium. Purification of the solubilized receptor, utilizing standard methods of protein chemistry (gel filtration, ion-exchange chromatography, affinity chromatography) allows recovery of approximately 500 mcg of pure receptor per 11 Kg bovine calf testis (11,000-fold purification) having a binding capacity (single class of high affinity, low capacity receptors) of about 2.3 nmol ^{125}I-hFSH/mg protein. The holoreceptor, which gives a single band on SDS-PAGE electrophoresis, has an apparent mol wt of 240,000 and is dissociated by dithiothreitol (but not urea) into subunits of 60,000 mol wt. The purified receptor is suitable for chemical studies relating structure to function, as well as for sequence studies leading to construction of cDNA probes.

References

Abou-Issa, H. and Reichert, Jr., L. E. (1976) *J. Biol. Chem.* **251,** 3326–3337.

Abou-Issa, H. and Reichert, Jr., L. E. (1977) *J. Biol. Chem.* **252,** 4166–4174.

Abou-Issa, H. and Reichert, Jr., L. E. (1978) *Biochim. Biophys. Acta* **526,** 613–625.

Abou-Issa, H. and Reichert, Jr., L. E. (1979) *Endocrinology* **104,** 189–193.

Abou-Issa, H. and Reichert, Jr., L. E. (1980) *Biochim. Biophys. Acta* **631,** 97–103.

Andersen, T. T. and Reichert, Jr., L. E. (1982) *J. Biol. Chem.* **257,** 11551–11557.

Andersen, T. T., Curatolo, L. M., and Reichert, Jr., L. E. (1983) *Mol. Cell. Endocrinol.* **33,** 37–52.

Ascoli, M. and Segaloff, D. L. (1986) *J. Biol. Chem.* **261,** 3807–3815.

Bhalla, V. K. and Reichert, Jr., L. E. (1974) *J. Biol. Chem.* **249,** 43–51.

Bluestein, B. I. and Vaitukaitis, J. L. (1981) *Biol. Reprod.* **24,** 661–669.

Boyn, B. D. and Clement-Cormier, Y. (1984) in *Membranes, Detergents, and Receptor Solubilization* (Venter, J. C. and Harrison, L. C., eds.), Liss, New York, pp. 47–63.

Branca, A. A., Sluss, P. M., Smith, R. A., and Reichert, Jr., L. E. (1985) *J. Biol. Chem.* **260,** 9988–9993.

Buckland, P. R., Strickland, T. W., Pierce, J. G., and Smith, B. R. (1985) *Endocrinology* **116,** 2122–2124.

Cline, G. B. and Ryel, R. B. (1971) *Meth. Enzymol.* **22,** 168–204.

Closset, J., Maghuin-Rogister, G., Ketelslegers, J. M., and Hennen, G. (1977) *Biochem. Biophys. Res. Commun.* **79,** 372–379.

Codina, J., Hildebrandt, J. D., Sekura, R. D., Birnbaumer, M., Bryan, J., Manclark, C. R., Iyengar, R., and Birnbaumer, L. (1984) *J. Biol. Chem.* **259,** 5871–5886.

Cuatrecasas, P. (1972) *Proc. Natl. Acad. Sci. USA* **69,** 1277–1281.

Dattatreyamurty, B., Rathnam, P., and Saxena, B. B. (1983) *J. Biol. Chem.* **258,** 3140–3158.

Dattatreyamurty, B., Schneyer, A., and Reichert, Jr., L. E. (1986) *J. Biol. Chem.* **261,** 13104–13113.

Dattatreyamurty, B., Figgs, L. W., and Reichert, Jr., L. E. (1987) *J. Biol. Chem.* **262,** 11737–11745.

Davis, J. S., Weakland, L. L., Farese, R. V., West, L. A. (1987) *J. Biol. Chem.* **262,** 8515–8521.

de Bruin, T.W.A. and van der Heide, D. (1984) *Mol. Cell. Endocrinology* **37,** 337–348.

Dias, J. A., Huston, J. S., and Reichert, Jr., L. E. (1981) *Endocrinology* **109,** 736–742.
Dias, J. A. and Reeves, J. J. (1982) *J. Reprod. Fertil.* **66,** 39–45.
Dias, J. A. and Reichert, Jr., L. E. (1982) *J. Biol. Chem.* **257,** 613–620.
Dias, J. A. and Reichert, Jr., L. E. (1984) *Biol. Reprod.* **31,** 975–983.
Dimino, M. J., Snitzer, J. and Noland, T. A., Jr. (1987) *Biol. Reprod.* **36,** 97–102.
Dufau, M. L., Charreau, E. H., and Catt, K. J. (1973) *J. Biol. Chem.* **248,** 6973– 6982.
Dufau, M. L., Ryan, D. W., and Catt, K. J. (1977) *FEBS Lett.* **81,** 359–362.
Fletcher, P. W. and Reichert, Jr., L. E. (1984) *Mol. Cell. Endocrinol.* **34,** 39–49.
Fletcher, P. W., and Reichert, Jr., L. E. (1986) *Endocrinology* **119,** 2221–2226.
Francis, G. L., Broun, T. J., and Bercu, B. B. (1981) *Biol. Reprod.* **24,** 955–961.
Fujita-Yamaguchi, Y., Choi, S., Sakamoto, Y., and Itakura, K. (1983) *J. Biol. Chem.* **258,** 5045–5049.
Greenwood, F. C., Hunter, W. M., and Glover, J. S. (1963) *Biochem. J.* **89,** 114–123.
Haga, T., Haga, K., and Gilman, A. G. (1977) *J. Biol. Chem.* **252,** 5776–5782.
Heindel, J. J., Rothenberg, R., Robinson, G. A., and Steinberger, A. (1975) *J. Cyclic Nucleotide Res.* **1,** 69–79.
Hwang, J. and Menon, K. M. J. (1984) *J. Biol. Chem.* **259,** 1978–1985.
Jacobs, S., Hazum, E., and Cuatrecasas, P. (1980) *J. Biol. Chem.* **255,** 6937–6940.
Ji, I. and Ji, T. H. (1981) *Proc. Natl. Acad. Sci. USA* **78,** 5465–5469.
Keenan, A. K., Gal, A., and Levitszki, A. (1982) *Biochem. Biophys. Res. Commun.* **105,** 615–623.
Keinanen, K. P., Kellokumpu, S., Metsikko, M. K., and Rajaniemi, H. J. (1987) *J. Biol. Chem.* **262,** 7920–7926.
Ketelslegers, J. M., Hetzel, W. D., Sherins, R. J., and Catt, K. J. (1978) *Endocrinology* **103,** 212–222.
Kohn, L. D., Valente, W. A., Laccetti, P., Cohen, J. L., Aloj, S. M., and Grollman, E. F. (1983) *Life Sci.* **32,** 15–30.
Kress, B. C. and Spiro, R. G. (1986) *Endocrinology* **118,** 974–979.
Lee, C. Y. and Ryan, R. J. (1974) in *Gonadotropins and Gonadal Function* (Moudgal, N. R., ed.), Academic, New York, pp. 444–459.
Lowry, O. H., Rosebrough, N. J., Farr, A. L., and Randall, R. J. (1951) *J. Biol. Chem.* **193,** 265–275.
Maghuin-Rogister, G., Closset, J., Combarnous, Y., Hennen, G., Dechenne, C., and Ketelslegers, J. M. (1978) *Eur. J. Biochem.* **86,** 121–131.
Massague, J., Pilch, P. F., and Czech, M. P. (1981) *J. Biol. Chem.* **256,** 3182–3190.
McQuade, R., Thomas, Jr., C. G., and Nayfeh, S. N. (1986) *Archiv. Biochem. Biophys.* **246,** 52–62.
Means, A. R. (1973) *Adv. Exptl. Med. Biol.* **36,** 431–448.
Morton, R. K. (1950) *Nature* **166,** 1092–1095.
Northup, J. K., Sternweis, P. C., Smigel, M. D., Schleifer, L. S., Ross, E. M., and Gilman, A. G. (1980) *Proc. Natl. Acad. Sci. USA* **77,** 6516–6520.
O'Neill, W. C. and Reichert, Jr., L. E. (1984) *Endocrinology* **114,** 1135–1149.
O'Shaugnessy, P. J. and Broun, P. S. (1978) *Mol. Cell. Endocrinol.* **12,** 9–15.
Quirk, S. M. and Reichert, Jr., L. E. (1988) *Endocrinology* **123,** 230–237.
Reichert, Jr., L. E. and Abou-Issa, H. (1977) *Biol. Reprod.* **17,** 614–621.
Reichert, Jr., L. E., Dias, J. A., O'Neill, W. C., and Andersen, T. T. (1983) in *The Anterior Pituitary Gland* (Bhatnager, A. J., ed.), Raven, New York, pp. 139–148.
Reichert, Jr., L. E., Andersen, T. T., Dias, J. A., Fletcher, P. W., Sluss, P. M., O'Neill, W. C., and Smith, R. A. (1984) in *Hormone Receptors in Growth and Reproduction* (Saxena, B. B., Catt, K. J., Birnbaumer, L., and Martini, L., eds.), Raven, New York, pp. 87–101.

Reichert, Jr., L. E. and Dattatreyamurty, B. (1989) *Biol. Reprod.* **40,** 13–26.
Rodbell, M., Birnbaumer, L., Pohl, S. L., and Krana, M. J. (1971) *J. Biol. Chem.* **246,** 1877–1882.
Sanborn, B. B., Andersen, T. T., and Reichert, Jr., L. E. (1987) *Biochemistry* **26,** 8196–8200.
Shin, J. and Ji, T. (1985) *J. Biol. Chem.* **260,** 12822–12827.
Shiu, R. P. C. and Friesen, H. G. (1974) *J. Biol. Chem.* **249,** 7902–7911.
Siegal, L. M. and Monty, K. J. (1966) *Biochim. Biophys. Acta* **112,** 346–362.
Siegel, T. W., Ganguly, S., Jacobs, S., Rosen, O. M., and Rubin, C. S. (1981) *J. Biol. Chem.* **256,** 9266–9273.
Smith, R. A., Branca, A. A., and Reichert, Jr., L. E. (1985) *J. Biol. Chem.* **260,** 14297–14303.
Smith, R. A., Branca, A. A., and Reichert, Jr., L. E. (1986) *J. Biol. Chem.* **261,** 9850–9853.
Sugden, D., Vanecek, J., Klein, D. C., Thomas, T. P., and Andersen, W. B. (1985) *Nature* **314,** 359–361.
Yoshimasa, T., Sibley, D. R., Bouvier, M., Lefkowitz, R. J., and Caron, M. G. (1987) *Nature* **327,** 67–70.

Thyrotropin Receptor

Characterization and Purification

Takashi Akamizu, Michele De Luca, and Leonard D. Kohn

1. Introduction

There is general agreement today that the TSH receptor has a high affinity TSH-binding component (Fig. 1) that is a membrane glycoprotein (Kohn et al., 1985; Kress and Spiro, 1986). The existence of a glycolipid (Fig. 1) that can modify the binding properties of this glycoprotein, as well as act as a coupler of the high affinity binding site to G proteins related to adenylate cyclase complex, has also been argued (Kohn, 1978; Kohn et al., 1985); its role is more controversial. No consensus currently exists, however, as to the detailed structure of the thyrotropin receptor, despite the efforts of many laboratories. This results from the current unsettled state of receptor purification and characterization; hence, the present summary can only be viewed as a state of the art and procedures as they currently exist. To place them in perspective, some discussion of the various procedures used to characterize the receptor will be included. The prime focus will concentrate on the glycoprotein component and on one set of techniques and their problems as a representative picture. In addition, cloning studies will be presented that are necessary and appear to be clarifying the problem.

1.1. The TSH Receptor: What Is It?

The receptor is that site on the cell surface that recognizes the hormone, binds it, and, in binding, initiates a signal to the cell. The signal usually associated with the TSH receptor is the cAMP signal (Robison et al., 1971; Dumont, 1971). However, evidence exists that the TSH-receptor

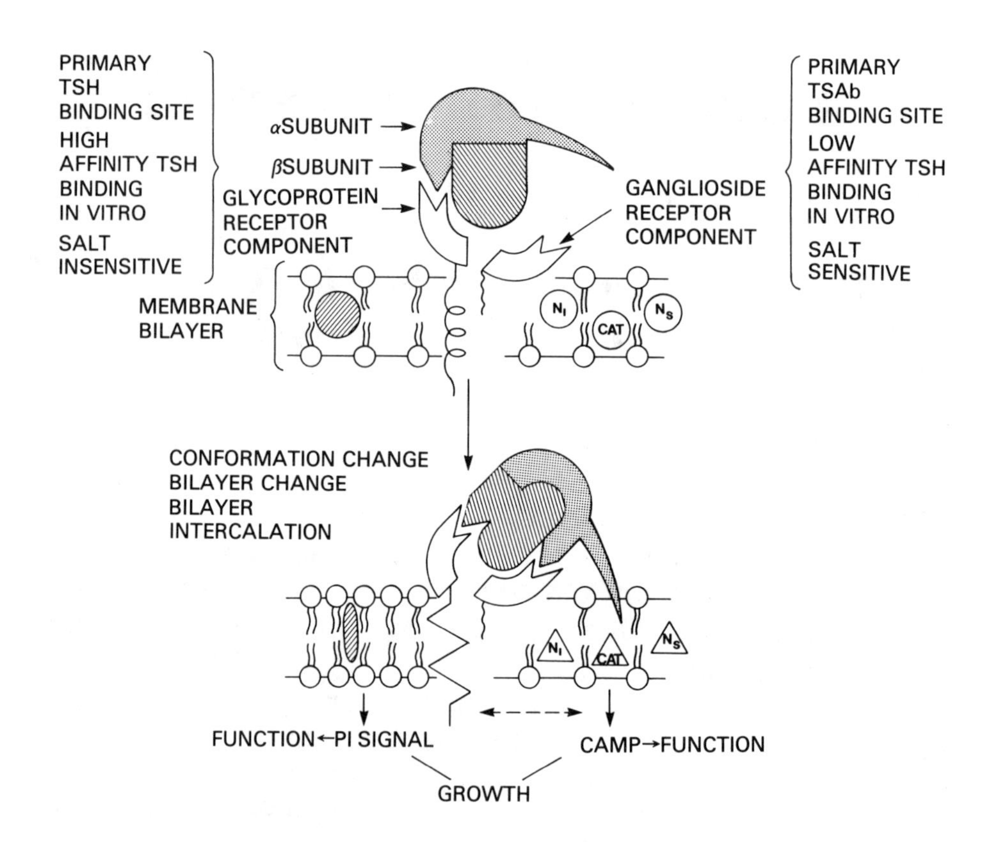

Fig. 1. Model of TSH receptor based on binding studies with ^{125}I-TSH and on both purification experiments and studies of monoclonal antibodies to the TSH receptor. A membrane glycoprotein is the high affinity, cell surface TSH binding site. Trypsinization of cells results in the loss of that portion of this receptor component that is external to the bilayer and a coincident loss in TSH binding and TSH stimulated adenylate cyclase activity. A membrane glycolipid, a ganglioside, can be identified in purification studies; its role is argued to both modulate receptor affinity and specificity and, more importantly, couple the membrane glycoprotein to the G protein complex. The N_i protein is selected particularly since both TSH and ganglioside related ADPribosylation activities are linked to a pertussis toxin sensitive N_i protein. Monoclonal TBIAb antibodies interact with the glycoprotein component, TSAbs with gangliosides, mixed antibodies with both receptor domains. Two signals are invoked by TSH: cAMP and phosphoinositide/Ca^{2+}/arachidonic acid (PI); the former is involved in iodide uptake and thyroid hormone secretion, the latter in iodination of thyroglobulin and iodide efflux from the cell into the follicular lumen. The receptor interaction induces a conformational change in the TSH, involves a change in the bilayer, and causes a portion of the TSH molecule to intercalate within the bilayer.

interaction can result in a phosphoinositide-Ca^{2+}-arachidonic acid signal process (Weiss et al., 1984; Corda et al., 1985; Philp and Grollman, 1986; Bone et al., 1986; Burch et al., 1986; Santisteban et al., 1986; Marcocci et al., 1987; Tahara et al., 1989), in ionic shifts leading to pH and membrane potential changes in the cell (Grollman et al., 1977), and in ADP ribosylation of membrane components (De Wolff et al., 1981; Vitti et al., 1982; Corda et al., 1987; Riberio-Neto et al., 1987). Further, it is evident that there are a multiplicity of G-proteins interposed between at least some signals and the recognition or binding event (Axelrod, 1986; Casey and Gilman, 1988). Understanding this relationship is critical to the purification and characterization process. Thus, in the light of our current increasing sophistication of receptor structure, we must consider that more than one TSH receptor will exist, each coupled to different signal systems. We must, in addition, explain the mechanism by which the molecule with recognition properties couples to the G-proteins; a structure analogous to the adrenergic receptor 7 transmembrane, G protein-binding domain must be anticipated (Lefkowitz and Caron, 1988).

Early ideas of characterization and purification rested on the presumption that adenylate cyclase activity was a valid measure of the receptor. Today, given our knowledge of the complexity of the adenylate cyclase complex, the multiplicity of G-proteins that may interact with the TSH receptor, and evidence these are separate molecular entities from the binding or recognition moiety (Axelrod, 1986; Casey and Gilman, 1988), that presumption must be discarded. In the absence of evidence the re-ceptor has a phosphorylation site or phosphorylation activity, as in the case of the insulin, IGF-I, or EGF receptors, for example, only one function remains integral to the receptor and is readily measured—the binding or recognition event. Purification or characterization at this time depends, therefore, on the TSH binding assay.

2. The TSH Binding Assay: A Problem at the Start

Unfortunately for TSH receptor purification studies, this assay has been controversial. Early studies of TSH binding noted that it was best measured in a low salt environment and at lower pH values than "physiologic" (Amir et al., 1973; Tate et al., 1975). Binding under these "less physiologic" conditions exposed a higher number of binding sites, resulted in curvilinear Scatchard plots, suggesting a multiplicity of different sites with different affinities, and exposed specificity curves that were less dramatic with respect to other glycoprotein hormones than were believed to be "physiologically correct." Nevertheless, it has become evident in studies of thyroid cells in culture that these low salt and low pH conditions do expose physiologic receptor binding sites that are coupled to ion fluxes (Grollman et al., 1977) and adenylate cyclase activity (Kohn et al., 1986c); such conditions are now the conditions of choice in whole cell bioassays in, for example, measuring the interaction of autoantibodies in Graves' patients with the TSH receptor (Kohn et al., 1986C). In addition, evidence has accumulated that hCG can readily bind to solubilized preparations of the TSH receptor (Kohn et al., 1985; Jeevanram et al., 1989) and, given to rats, can induce thyroid hormone release from the thyroid as well as TSH, despite markedly different effects on cAMP levels in the thyroid (Jeevanram et al., 1989). Resolving the controversy of specificity and conditions is, however, no cause for joy in purification studies. The fact that binding can be significantly inhibited in buffers with higher salt concentrations and exhibits lower apparent specificity with respect to its glycoprotein hormone analogs under these conditions makes purification studies more difficult by limiting conditions and by limiting the ease with which a "specific" receptor can be identified.

2.1. The Soluble Receptor Assay

Purification of the TSH receptor requires solubilization of the membrane. In turn, this requires an assay to measure soluble receptors. Two assays have been used, one involving a precipitation procedure involving polyethylene glycol (PEG) and the second a solid phase assay. Representative assays of this type are described below.

2.1.1. PEG Assay

2.1.1.1. Filter Procedure. The binding assay is performed in a final volume of 150 microliters containing: 20 mM (2-N-morpholino) ethanesulfonic acid or HEPES, pH 6.0 or 7.0, respectively; 0.05% crystalline bovine serum albumin; ^{125}I-TSH, 25,000–250,000 cpm plus or minus, as appropriate, unlabeled TSH; and appropriate aliquots of the solubilized receptor preparation. The binding incubation is continued for 1 h at 0–4°C, after which the following is added: 5 µL of bovine gamma-globulin, 1 mg/mL; 75 µL 16–20% PEG 6000. The incubation is continued for 15 min and filtered over 25 mm Millipore filters (Tate et al., 1975) that have been prewashed with three aliquots of incubation buffer containing 2.5% bovine serum albumin. After filtration of the assay sample, the filters are washed three times with aliquots of the in-cubation buffer containing 4% PEG 6000. All wash solutions are at 0–4°C. Filters are counted in a gamma counter.

2.1.1.2. Precipitation Procedure. This binding assay is performed by mixing and incubating, for a 1 h period at 37°C, 100 µL aliquots, each, of the solubilized receptor preparation, a ^{125}I-TSH solution with or without unlabeled TSH, and a buffer sufficient to yield a final concentration of 10 m*M* TrisCl, pH 7.4, 50 m*M* NaCl, 0.1% crystalline bovine serum albumin, and 0.1% detergent. Ice cold in-cubation buffer, 650 µL, containing 1 mg crystalline bovine serum albumin and 50 µL of normal serum or an IgG solution, 18 mg/mL, is added followed by 1 mL of ice cold 30% PEG 6000 containing 1*M* NaCl. After mixing and centrifuging for 30 min at 4°C, the supernate is aspirated and the pellet counted in a gamma counter.

The concentration of PEG is critical since higher concentrations can precipitate the TSH itself. A final PEG concentration should be chosen, which maximizes the difference between specific (plus membranes) and nonspecific (no membranes) control samples.

2.1.2. Solid Phase Assay

Dynatech microtiter plates are precoated by adding 0.1 mL of a 20 µg/mL solution of poly-L-lysine (70,000 mol wt, Sigma) solution in water to each well. After incubation at room temperature for 1 h, the solution is removed by suction and 0.1 mL of solubilized thyroid membranes are added to each well (1–100 µg protein). Control wells are incubated with 0.5% bovine serum albumin or gelatin in the same buffer. After incubation for 4 h or longer (overnight), the nonattached membrane solution is removed by suction following a 5 min centrifugation at 2000 × *g*. Each well is then incubated for 30 min at room temperature with 0.1 mL binding buffer containing 0.5% bovine serum albumin; the buffer can be either 20 m*M* Tris acetate at pH 6.7 or 20 m*M* TrisCl at pH 7.4 containing 50 m*M* NaCl. The buffer is aspirated and replaced with 50 µL of the same buffer containing ^{125}I-labeled TSH, 40,000 cpm. After 2 h at 4°C, if the Tris-acetate buffer is used, or 1 h at 37°C, if the TrisCl/NaCl buffer is used, the microtiter plates are centrifuged (2000 × *g*) for 5 min and the unbound radioactivity removed by placing the plates upside down over utility wipes. The wells are washed three times with incubation buffer without radiolabeled TSH, heat dried under an IR lamp, and individually counted in a gamma counter.

2.2. Membrane Solubilization

To purify the receptor, the usual starting place is a solubilized thyroid membrane preparation. The most popular method of solubilization uses a detergent mixture, for example, one including triton, deoxycholate, or NP-40. An alternative procedure uses a chaotropic agent, lithium diiodosalicylate. Representative procedures are as follows.

2.2.1. Lithium Diiodosalicylate Procedure

Stock membrane preparations stored at –70°C are centrifuged at 25,000 × *g* for 10 min and the pellets mixed with 0.1 *M* recrystallized lithium diiodosalicylate containing 100 m*M* TrisCl, pH 7.4, and 200 U/mL of the protease inhibitor, aprotinin, to a final concentration of 2.5–5 mg protein/mL. The mixture is homogenized with 10–15 strokes of a loose-fitting, motor-driven, Teflon™ coated homogenizer. The mixture is incubated, with stirring, for 1.5 h at room temperature, centrifuged at 38,000 × *g* for 30 min, dialyzed 12 h against 20 m*M* sodium bicarbonate, pH 9.4, and stored at –70°C until further use. Triton X-100, to achieve a final concentration of 0.1%, can be added after the lithium diiodosalicylate solubilization and centrifugation procedures; in this case the bicarbonate dialysis buffer should also contain 0.1% triton.

2.2.2. Detergent Solubilization Procedure

Stock membrane preparations are pelleted as above, mixed with ice cold 10 m*M* TrisCl, pH 7.4, containing 50 m*M* NaCl, 1% NP-40, 0.2 m*M* phenylmethylsulfonylfluoride (PMSF) and 200 U/mL aprotinin, then homogenized as above. Suspensions are kept on ice for 30 min and then centrifuged at 100,000 × *g* for 1 h at 4°C. Membrane preparations for the above solubilization procedures should include PMSF and aprotinin in the buffers during their preparation and storage since protease fragmentation of receptors is a major and universal complicating problem of identification and purification procedures for receptors.

The problem of detergents is their potential interference with the binding assay. This problem is illustrated in Tate et al., 1975. Thus, different detergents yielded different yields of TSH-binding activity despite otherwise identical conditions. It is not clear whether the differences represent recovery, loss of activity resultant from the procedure, or interference in the assay itself by the detergent. The initial value, whatever it is, becomes the reference starting point.

3. Purification

The most frequently used first step of purification is the TSH affinity column. Other procedures include immunoprecipitation with monoclonal antibodies to the TSH receptor or affinity chromatography on Sepharose-linked anti-TSH receptor preparations, and purification with lectins coupled to Sepharose. These procedures will be discussed below. Independent of the procedures, however, it is important to recognize that the starting source of membranes can complicate the issue.

Most sources in the literature use tissues from whole animals. Autolysis from the time of excision of the gland to initial purification in the presence of protease inhibitors is an important but difficult to control problem.

More recently, studies with cells in culture uncovered a second problem, receptor turnover and degradation, which is rapid in the presence of TSH. This is illustrated in Fig. 2. Thus if FRTL-5 rat thyroid cells are maintained with no TSH for 7 d and are then given radiolabeled methionine in the presence or absence of TSH, it is apparent that receptor protein, which can be immunoprecipitated by monoclonal antibodies to the TSH receptor, increases more rapidly as a function of time in the presence rather than in the absence of TSH (Fig. 2A). Further, if the same cells are equilibrium labeled with [^{35}S]methionine for 5 d in the absence of TSH, then placed in media free of labeled methionine and in the presence or absence of TSH, TSH causes a major loss in radiolabeled protein that can be immunoprecipitated by monoclonal antibodies to the TSH receptor (Fig. 2B). More important from a purification viewpoint, if one examines the immunoprecipitation pattern by SDS gels before the TSH is added and 24 h after the TSH is added, different proteins are immunoprecipitated (Fig. 2C). As a first approximation, therefore, we must consider that processing is occurring with different forms of the receptor apparent in the different conditions or even that different receptors exist under the two conditions.

3.1. TSH Affinity Chromatography

3.1.1. Preparation of TSH-Sepharose

Cyanogen bromide activated Sepharose 4B (Pharmacia), 1 g, is washed with 200 mL 1 m*M* HCl, then incubated 2 h at room temperature with 10 mL of 100 m*M* bicarbonate buffer, pH 8.3, containing 0.5*M* NaCl and approx 5mg TSH. The buffer is then replaced with one containing 1*M* glycine rather than TSH and the gel suspension is incubated a further 2 h at room temperature. The gel suspension is then washed, alternatively, with 100 m*M* bicarbonate buffer, pH 8.4, and 100 m*M* acetate buffer, pH 4.0, both containing 1*M* NaCl. The gel can be stored at 4°C in the pH 8.4 buffer containing several drops of toluene or 0.05% sodium azide. Inclusion of a ^{125}I-TSH tracer with the cold TSH allows a calculation of the TSH bound.

3.1.2. TSH Affinity Chromatography

Aliquots of the TSH-Sepharose are poured into Biorad disposable columns (0.5 × 10 cm) and washed thoroughly with a binding buffer.

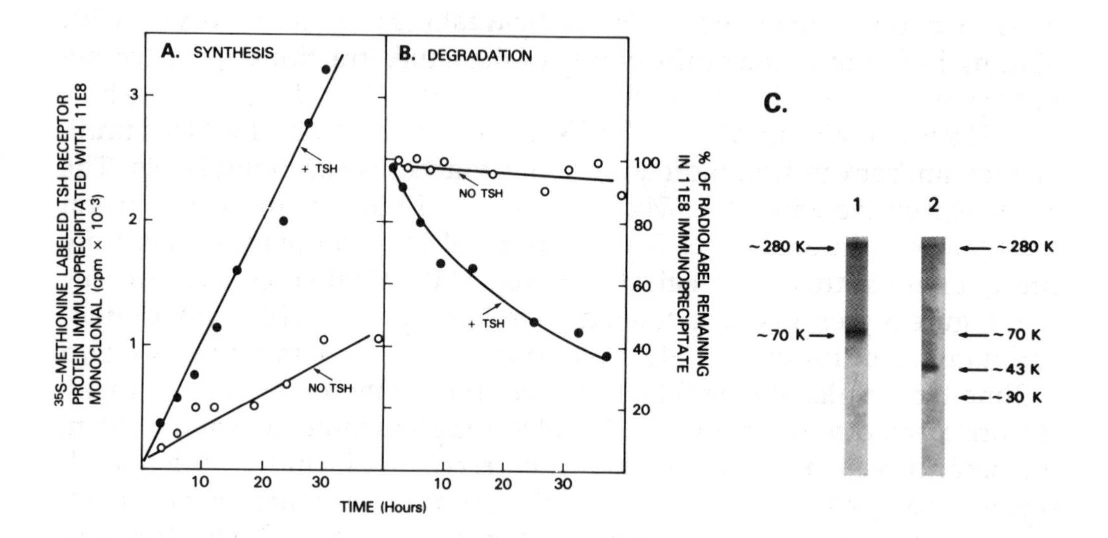

Fig. 2. TSH regulation of the synthesis (A) and degradation (B) of the TSH receptor membrane glycoprotein in FRTL-5 thyroid cells as measured by [^{35}S]methionine incorporation into protein immunoprecipitable with the 11E8 monoclonal antibody to the TSH receptor. In (A) and (B) cells were grown to confluency and then changed to a medium with no TSH for a 7-d period. In (A) different groups of cells were challenged with fresh medium containing *TSH* or containing *no TSH* along with the radiolabeled methionine. In (B) the cells were equilibrium labeled with [^{35}S]methionine for 4 d before the medium was removed, the cells washed, and fresh medium containing *TSH* or no *TSH* was added. At the times indicated the cells were washed, solubilized with a Triton/deoxycholate mixture, and immunoprecipitated with Sepharose-coupled 11E8 antibody. The Sepharose was washed and counted. In (C) the radiolabeled material bound to the 11E8-Sepharose at the start of the experiment in Fig. 2B (0 time), and after 24 h of TSH was recovered by incubation with sodium dodecyl sulfate and subjected to SDS gel electrophoresis. In the absence of TSH, the major proteins detected are at 70 kDa and about 280 kDa, whereas, 24 h after TSH, the major immunoprecipitable band is slightly lower than 45 kDa.

Representative buffers are 50 m*M* Tris acetate, pH 7.4 or 6.4, containing 0.1% triton X-100 or 0.1% NP-40, 200 U/mL aprotinin, and 0.2 m*M* phenylmethylsulfonyl fluoride (PMSF). Aliquots of radiolabeled membrane preparations (surface iodination with lactoperoxidase or [^{35}S]methionine if membranes are from thyroid cells), ranging from 1–20 mg protein, are passed through the column overnight using a peristaltic pump set such that the membrane fraction is recycled through the column at least 10 times. Unbound material is collected by washing the gel extensively with binding buffer until the radioactivity of the eluted fractions approaches a background level. The gel is then washed with 100 m*M* bicarbonate buffer, pH 8.5, containing 0.1% Triton or NP-40, aprotinin and 1*M* NaCl until, once again, background radioactivity approaches background levels. The gel is then eluted with 100 m*M* glycine HCl, pH 2.5, containing 0.1% triton or NP-40 and the aprotinin. Several drops of 1*M* sodium bicarbonate are present in each tube into which the glycine HCl eluate is collected.

Figure 3 presents a representative elution pattern of lithium diiodosalicylate solubilized ^{125}I-labeled bovine thyroid membranes and [^{35}S]methionine-labeled FRTL-5 rat thyroid cell membranes. In all cases, the first peak contains about 0.5–1% of the applied radiolabel and protein, the second peak about 0.1–0.5%. Total recovery, including the nonadsorbed eluate, is 100±3%. The autoradiographs of gels containing the starting material, the unbound material, and peak fractions from the high salt eluate and the low pH eluate are depicted in Fig. 3. The enrichment of a high molecular weight protein is evident, however, a multiplicity of proteins is still present, particularly in the high salt eluate. This is particularly true if a triton or NP-40 solubilized membrane fraction is used rather than the lithium diiodosalicylate solubilized preparation.

The use of high salts as the elution agent is based on the observation that TSH binding to thyroid membrane receptors is salt sensitive. This procedure can be modified by using TSH as the eluting material; the problem here is cost and material source. Another trick used is to elute with a IgG from a patient with active Graves' disease since this is presumed to have antibodies to the TSH receptor. The purification by salt elution is not specific since a multiplicity of bands is still evident even after immunopurification (*see below*) as illustrated in Fig. 4A; the pH 2.5 elution is better in terms of fewer bands (Fig. 4B). In work by Remy et al. (1987), a 48 kd protein was eluted by TSH. Elution with TSH, in the hands of others, resulted in a multiplicity of protein bands. In both cases, however, TSH or salt elutions resulted in the calculation of a 100–500-fold purification; this is less than might be expected for a receptor estimated by some to have fewer than 1000 binding sites per cell. Elution with IgG must be treated cautiously since we now recognize that these preparations contain a wide array of antibodies to membrane determinants (*see below*) that have nothing to do with the TSH receptor. As one might expect, there are, once again, a multiplicity of eluted bands by Graves' IgG; further, different IgG preparations give different results.

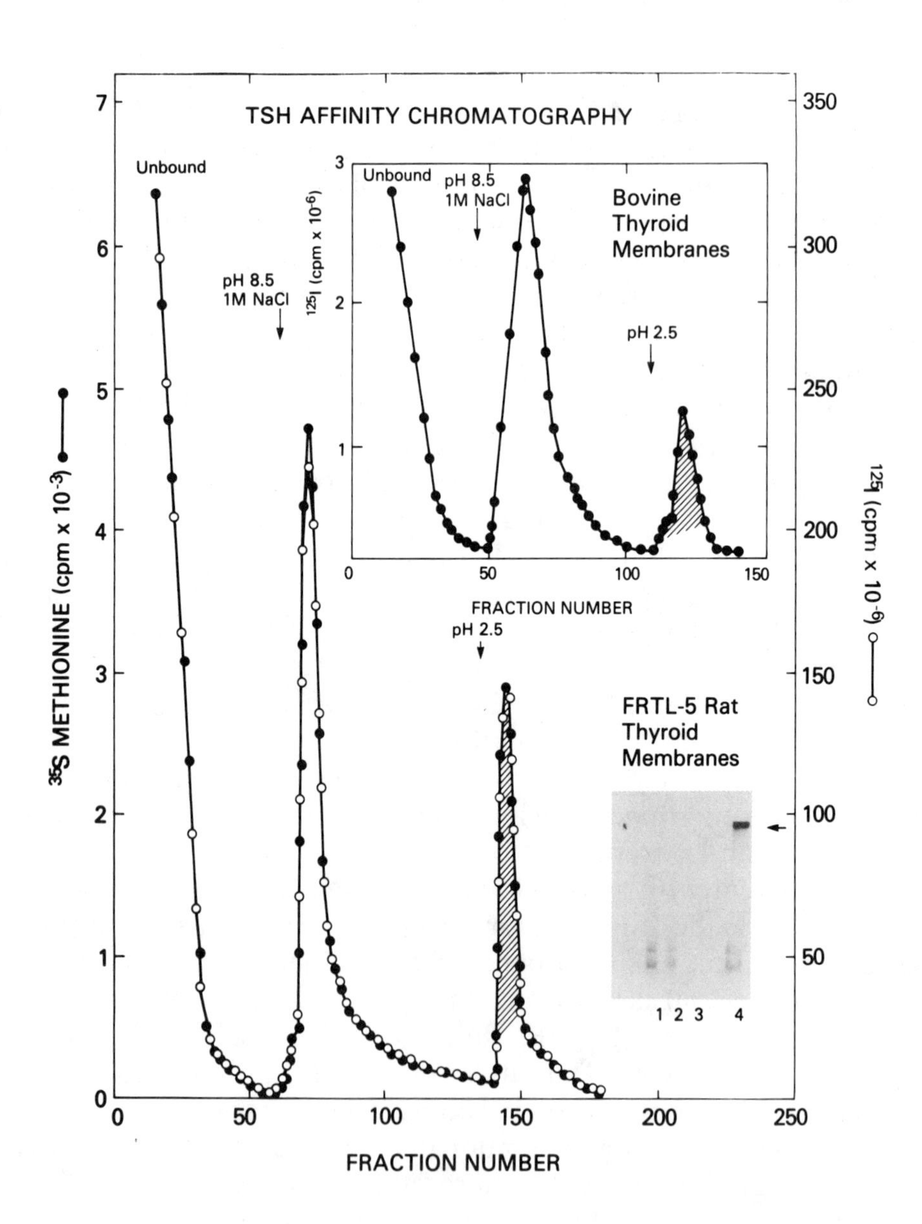

Fig. 3. TSH affinity chromatography of [^{35}S]methionine labeled FRTL-5 cell and ^{125}I-labeled bovine thyroid membranes solubilized by the lithium diiodosalicylate procedure and chromatographed as described in the text. Elution is with high salt and then low pH. The gels represent the unbound material (1), the salt eluate (2), the baseline tube before the low pH eluate (3), and the peak tube of the low pH eluate (4).

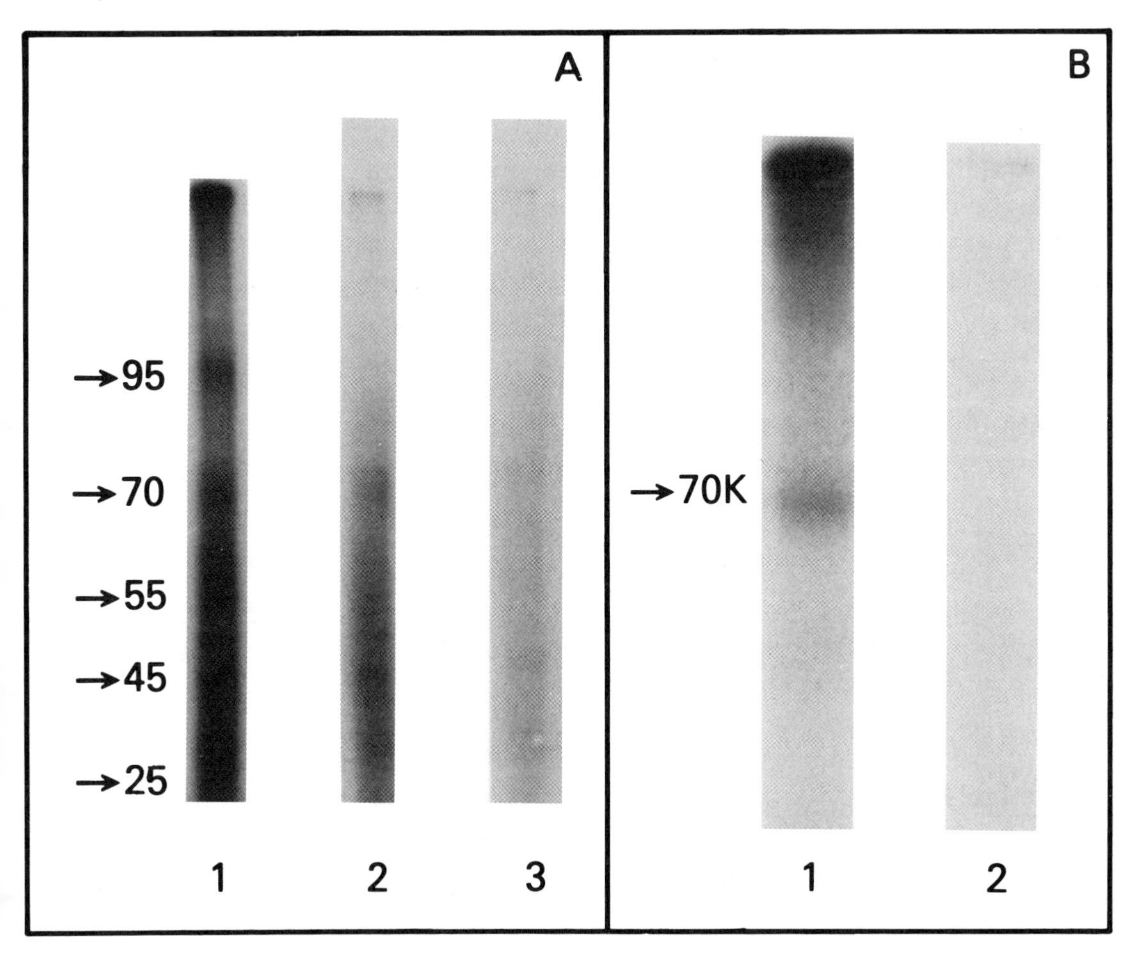

Fig. 4. Immunopurification by a monoclonal antibody to the TSH receptor (11E8) of TSH receptor components eluted from a TSH affinity column by high salt (A) and low pH (B). The immunopurification on the 11E8-Sepharose immunoadsorbant was performed in the absence of TSH (gel 1 in A and B) or in the presence of $1 \times 10^{-8}M$ (gel 2 in A and B) or $1 \times 10^{-7}M$ TSH (gels in A).

3.2. Immunopurification with Monoclonal Antibodies to the TSH Receptor

3.2.1. Preparation of the Immunoadsorbant

Monoclonal antibodies to the TSH-receptor have now been described by several laboratories. It is generally agreed that there are several groups of antibodies as originally described (Kohn et al., 1983a,1984,1986a): stimulating antibodies (TSAbs) that elevate CAMP levels but are poor inhibitors of TSH binding; inhibitory antibodies (TBIAbs) that are competitive antagonists of TSH binding and function; and mixed antibodies with the ability to both elevate cAMP levels and block TSH binding. For purposes of immunopurification, the inhibiting or mixed antibodies are best, or a mixture of monoclonals can be used. In our own laboratory, antibody 11E8, a TBIAb, and antibody 52A8, a mixed antibody, have been used most frequently. Both are mouse IgGs; the former was made against solubilized bovine thyroid membranes; the latter against solubilized human thyroid membranes. In interpreting results, however, it is important to remember that even monoclonal antibodies have different epitopes and crossreactivities with other membrane components (Kohn et al., 1983a,b, 1984,1985, 1986a; Rotella et al., 1986).

The antibodies are purified from ascites by diluting them sevenfold in buffer and subjecting them to a polyethylene glycol precipitation procedure. Thus, 0.473 mL 50% PEG (mol wt 3350) is added per mL diluted ascites and the suspension incubated overnight in the cold. The mixture is centrifuged at 20,000 × *g* for 20 min at 4°C and the recovered pellet solubilized with 20 m*M* TrisCl, pH 8.0. After filtration through a 0.22 micron filter, this preparation is purified by high performance liquid chromatography using a standard procedure for isolating mouse monoclonal IgG. The purified IgG is aliquoted and stored at –70°C until use.

The Sepharose-coupled antibody preparation is made following the same procedure as for the Sepharose-TSH with the following exceptions: the IgG is 2 mg/mL of gel and the coupling buffer was at pH 8.7.

3.2.2. Immunopurification

Immunopurification can be conveniently performed as a batch rather than column procedure. The gel, 100 µL to 1 mL, is placed in a 1.5 mL or 15 mL screwcapped tube with 50 µL or 1–4 mL amounts of the solubilized membrane preparations eluted from the TSH-affinity column. The final volume of 1–5 mL is made by adding buffer to yield a final concentration of 50 m*M* tris acetate, pH 7.4, containing 100 m*M* NaCl, 0.1% triton X-100, and 200 U/mL aprotinin. The tubes are rotated gently in the cold overnight. The gel suspension can be centrifuged, the supernatant solution removed, and the gel washed three times with 50 m*M* Tris acetate, pH 7.4, containing 100 m*M* NaCl, 0.1% Triton X-100, and 200 U/mL aprotinin before the bound fraction is eluted by one of several solutions: Laemmli sample buffer (0.125*M* TrisCl, pH 6.8, containing 2% SDS, 10% glycerol, and 5% 2-mercaptoethanol); 100 m*M* glycine HCl, pH 2.5, containing 0.1% triton X-100 and 200 U/mL aprotinin; or 6*M* guanidine HCl containing 0.2*M* 2-mercaptoethanol.

Analysis of the pH 2.5 eluate from the TSH-Sepharose column, to which was applied lithium diiodosalicylate solubilized membranes from FRTL-5 thyroid cells maintained in the absence of TSH for 7 d, reveals, after immunopurification with monoclonal antibody 11E8, an enrichment of a protein that barely penetrates the gel but that behaves like a single major component of approx 280 kDa on guanidine HCl columns (Fig. 5). This approx 280 kDa protein is prevented by TSH from binding to the 11E8 monoclonal antibody; 50% inhibition is evident at 5×10^{-10} *M* TSH, 100% at 1×10^{-8} *M* TSH (Fig. 6A). No inhibition is evident with albumin, mouse IgG, insulin, glucagon, or prolactin. It does not react with antithyroglobulin preparations, nor does it inhibit in thyroglobulin-antithyroglobulin radioimmunoassays. This protein has properties, therefore, of the high mol wt TSH-binding protein evident in early studies of solubilization and Sepharose chromatography. The protease problem is illustrated in Fig. 6B; incubating the approx 290 kDa protein from Fig. 6A for 24 h without aprotinin results in the formation of a 70 kDa band.

If the same procedure is used for detergent solubilized membranes from FRTL-5 thyroid cells maintained in TSH for 5–10 d before use, a major protein eluted in the pH 2.5 eluate has a mol wt of approx 45 kDa; it is enriched further by immunoprecipitation with the 52A8 monoclonal antibody to the TSH receptor and even identified as reactive on Western blots with 52A8. This protein, after V8 protease digestion, blotting, and microsequencing of two peptides, turns out to be gamma-actin (Fig. 7). This illustrates the caution that must be applied to analyzing proteins identified by TSH binding/purification studies as TSH receptor—let alone the multiplicity of bands identified in crosslinking studies.

Analysis of the salt eluate from the TSH affinity columns (Fig. 3) by immunopurification with monoclonal antibodies to the TSH receptor reveals enrichment of several bands, particularly a 70 kDa protein in FRTL-5 thyroid cells maintained with no TSH (Fig. 2C), but still a multiplicity of bands. Current studies using microsequencing techniques are in progress to define these as TSH receptor or TSH receptor related proteins.

3.3. Lectin Purification

The identification of TSH receptor as a glycoprotein led to several attempts to use lectins as purification agents. A major glycoprotein of the thyroid is, however, thyroglobulin. In most studies, those using Con A-Sepharose, for example, the TSH receptor was swamped by the thyroglobulin.

Nevertheless, a recent innovative approach was taken by Kress and Spiro (1986). They used Bandeiraea (Griffonia) simplicifolia I affinity chromatography in the presence or absence of TSH to identify several radioactive glycoproteins, mol wt 316,000, 115,000, and 54,000, which might be

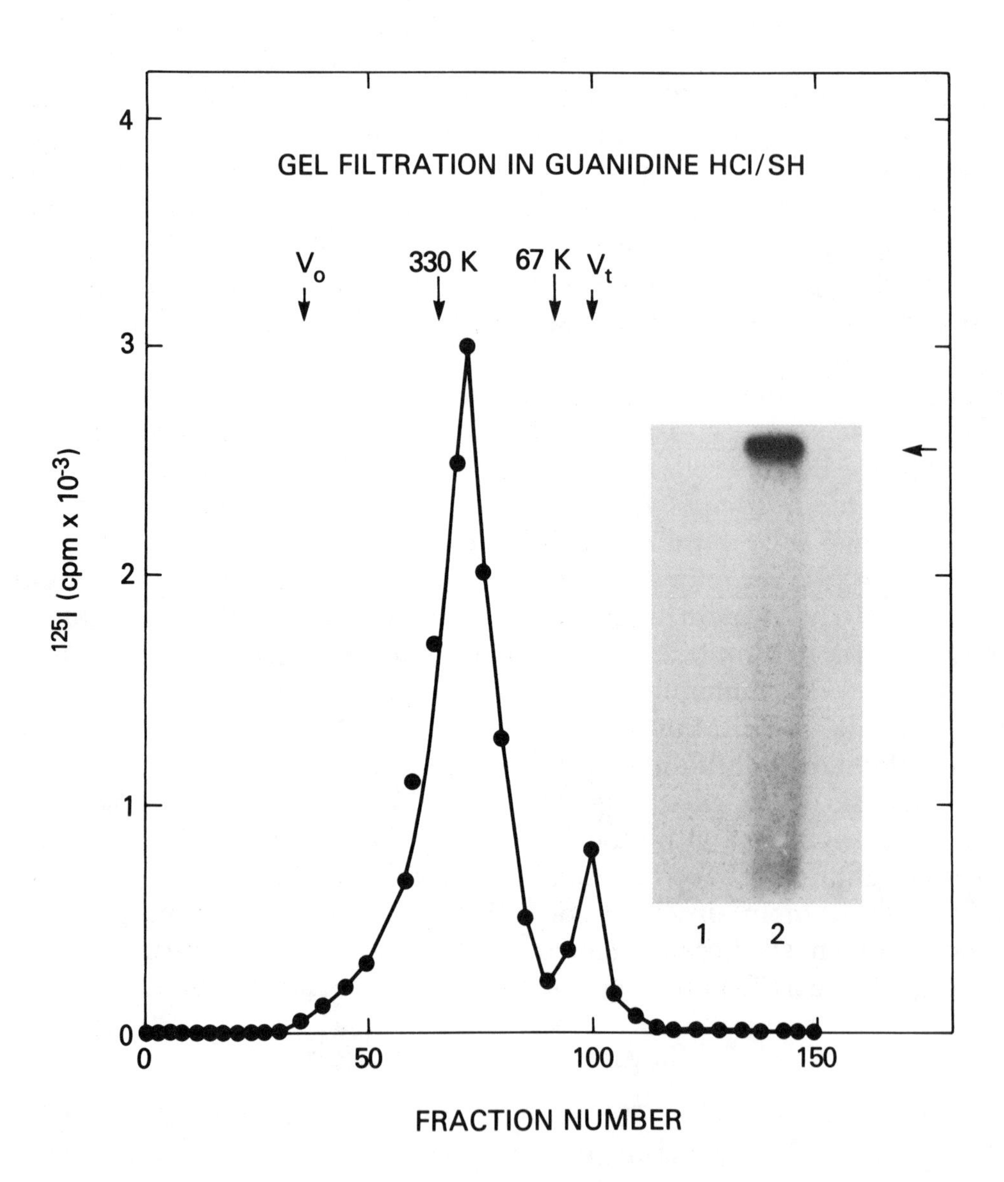

Fig. 5. Sepharose gel filtration in 6 *M* guanidine HCl of the ^{125}-I labeled bovine thyroid membrane fraction from Fig. 3 after being eluted from TSH-Sepharose by low pH and then purified by 11E8-Sepharose. The insert is an SDS gel of the peak fraction. Standards to determine the mol wt reference points were thyroglobulin and albumin.

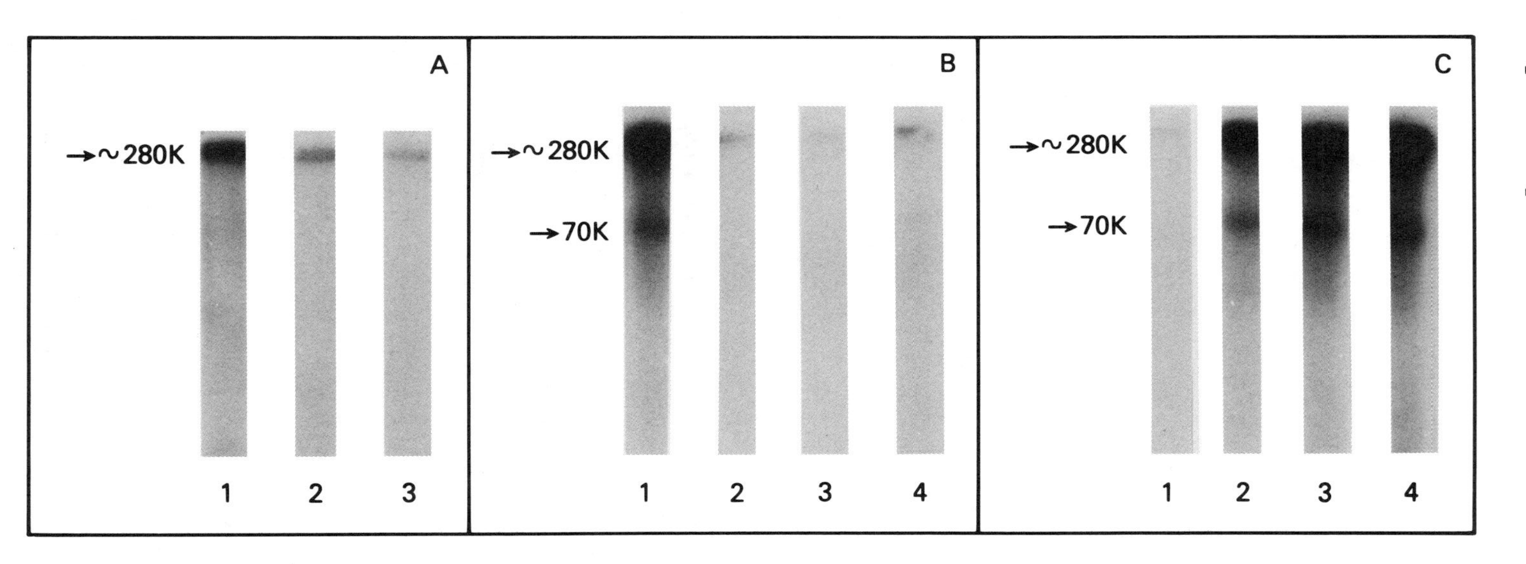

Fig. 6. (A) Effect of TSH on the interaction of the 280 kd protein (purified by gel filtration on Sepharose and in guanidine HCl [Fig. 4]) with the 11E8 mono-clonal antibody. The protein was adsorbed to 11E8-Sepharose in the absence of TSH (1) or in the presence of 1×10^{-8} *M* (2) or 1×10^{-7} *M* (3) TSH. (B) Effect of incubation in the absence of aprotinin on the 280 kd protein immunopurified in Figs. 4 and 6A. The material from Fig. 6A was incubated with no aprotinin in the buffer for 24 h at room temperature, then again subjected to 11E8-Sepharose immunopurification in the absence of TSH (1) or in the presence of 1×10^{-8} (2) or 1×10^{-7}(3) *M* TSH. The interaction of the material with a control Sepharose preparation coupled to normal mouse IgG is presented (4) to show the immunopurification step is specific. (C) Immunopurification of the material in Fig. 6B by three other monoclonal antibodies to the TSH receptor coupled to Sepharose, 129H8 (2), 122G3 (3), or 307H6 (4), by comparison to normal mouse IgG coupled to Sepharose.

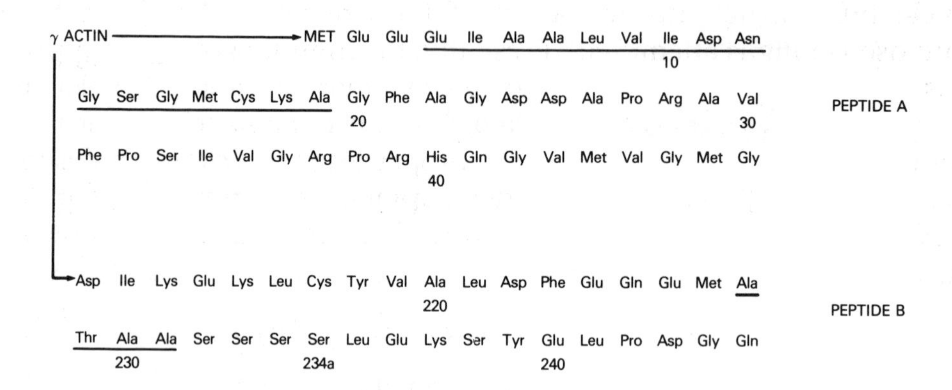

Comparison of the Amino Acid Sequence of Peptide A and Different Actins

Residue No.	2	3	4	5	6	7	8	9	10	11	12	13	14	15	16	17	18	19
peptide A			Glu	Ile	Ala	Ala	Leu	X	Ile	Asp	Asn	Gly	X	Gly	Met	X	Lys	Ala
human γ actin	Glu	Glu	Glu	Ile	Ala	Ala	Leu	Val	Ile	Asp	Asn	Gly	Ser	Gly	Met	Cys	Lys	Ala
human β actin			Asp	Ile	Ala	Ala	Leu	Val	Val	Asp	Asn	Gly	Ser	Gly	Met	Cys	Lys	Ala
γ at β actin			Asp	Ile	Ala	Ala	Leu	Val	Val	Asp	Asn	Gly	Ser	Ala	Met	Cys	Lys	Ala
TSH-BINDING GLYOPROTEIN	Glu 391	Lys 392	Glu 393	Val 394	Ala 395	Ala 396	Leu 397											

Fig. 7. (A) The sequence of two peptides from a 43 kDa protein isolated from the pH 2.5 eluate of a TSH affinity column to which was applied detergent solubilized FRTL-5 thyroid cell membranes from cells treated with TSH for 24 h (*see* Fig. 2C). The protein was isolated from slab gels, digested with V8 protease, since it had a blocked amino-terminus, and the peptides isolated on blots and microsequenced. The sequences match those from gamma actin. (B) The sequences of the peptide A are compared with gamma and beta-actins showing this particular sequence is gamma-actin specific. The sequence does have a homology, with a sequence from a TSH-binding glycoprotein identified in cloning studies (*see* Fig. 9 below).

TSH receptor. Again, however, no further information exists as to the relation of these proteins to TSH the "physiologic" TSH receptor.

4. Cloning

The above studies emphasize a current problem. Once identified and purified, a protein must be both structurally characterized and proven to be TSH receptor by reconstitution studies. Identification of a protein as reactive with TSH and with antibodies, even monoclonal antibodies to the TSH receptor, cannot be accepted as unequivocal evidence the protein is TSH receptor. Today, this is best done by cloning technology. Most frequently, sequences from the purified protein are used to construct oligonucleotides, which, in turn, are used for screening. Alternatively, immunoscreening is attempted. Presently, the immunoscreening approach has identified a TSH binding glycoprotein that appears to be relevant to exophthalmos (Kohn et al., 1989a,b,c), but it remains to be functionally coupled to signal transduction. The properties of this clone are described below. Also cited is an exciting development concerning cloning of the hCG receptor that may, in turn, lead to characterization of TSH receptor structure.

4.1. The Immunoscreening Approach

A human thyroid carcinoma lambda gt11 expression library has been immunoscreened with an IgG preparation from a patient with active Graves' disease (Chan et al., 1989). The IgG preparation was highly positive in assays measuring thyroid stimulating antibodies (TSAbs), thyrotropin binding inhibiting antibodies (TBIAbs), and thyroid growth promoting antibodies (TGPAbs). Although the IgG preparation contained no detectable microsomal or thyroglobulin antibodies, it was presumed that the IgG preparation would react to a multiplicity of autoantigens other than the TSH receptor. It is not surprising, therefore, that this approach identified 25 different human cDNA clones. The problem, then, was to define these cDNAs and relate their gene products to Graves' disease.

In a separate approach (Kohn et al., 1989a,b,c), a FRTL-5 rat thyroid cell lambda gt11 expression library was immunoscreened with a polyclonal rabbit antibody developed against FRTL-5 thyroid cell TSH receptor preparations purified by affinity chromatography on TSH-Sepharose and reactive with monoclonal TSH receptor antibody-Sepharose as above (Kohn et al., 1986a; Chan et al., 1987b,1988). This rabbit antibody immunoprecipitated the same multiplicity of TSH-binding proteins as did the monoclonal antibodies to the TSH receptor and could inhibit TSH-induced increases in CAMP levels and thymidine incorporation into DNA (growth) in FRTL-5 cells. Immunoscreening with the rabbit antibody to the TSH receptor preparation resulted in a single rat cDNA clone identical in size, 1.25 Kb, and restriction mapping to one of the 25 cDNA clones identified with the Graves' IgG preparation and exhibiting cross-hybridization with it (Fig. 8). These results suggested the cDNA coded for an autoantigen related to the TSH receptor.

The 1.25 kb cDNA insert from both clones identified a 2 Kb mRNA in human, rat, ovine, and rat thyroid tissues or cells, in guinea pig adipocytes and human IM9 lymphocytes but not in human or rat tissues such as brain, liver, kidney, or muscle (Chan et al., 1989; Kohn et al., 1989a,b,c).

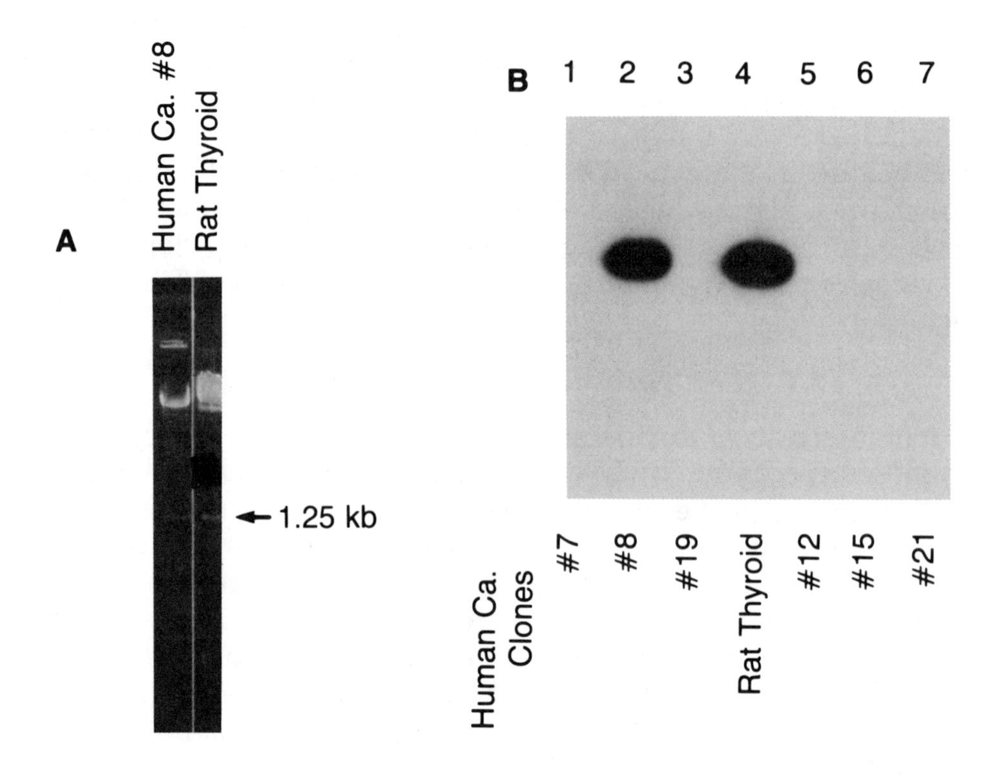

Fig. 8. (A) The 1.25 kb clones identified by immunoscreening a thyroid carcinoma library and a rat thyroid library by, respectively, IgG from a patient with active Graves' disease and a rabbit polyclonal antibody prepared against rat FRTL-5 thyroid cell TSH receptor preparations purified by TSH-Sepharose affinity chromatography and 11E8 reactivity. (B) Hybridization of the ^{32}P-labeled rat thyroid 1.25 kb insert with inserts purified from clones 7, 8, 19, 12, 15, and 21 isolated by immunoscreening the thyroid carcinoma library with patient IgG by comparison to "self-hybridization" with its own unlabeled insert. Inserts were electrophoresed on 1.2% agarose gels.

Although the presence or absence of demonstrable mRNA in poly A+ RNA preparations thus appeared to correlate with the presence or absence of TSH receptor expression in tissues, it was evident that the level of expression in different tissues did not relate to functional expression of TSH receptor. Thus, the IM9 lymphocyte transcript was detectable in total RNA preparations, whereas poly A+ RNA was required for Northern analysis in thyroid cells (Chan et al., 1989). Further, conditions that enhanced the growth of tissues or cells increased expression of the transcript (*see below*). This raised the possibility that the clone had its importance in the TSH-related growth process and autoimmunity.

The full length sequence of the human thyroid carcinoma cDNA clone, together with its predicted amino acid sequence, is presented in Fig. 9. Several important features can be discerned. First, there are two potential *N*-linked glycosylation sites (Asn-X-Ser/Thr, underlined). Second although a hydrophobicity plot of the sequence reveals multiple hydrophobic areas, only one is compatible with a membrane spanning domain, residues 374–390. The clone does, therefore, appear to identify a transmembrane glycoprotein. Third, the clone has an open reading frame of 1827 nucleotides, which encodes a protein of 69,812 daltons.

The radiolabeled protein product of the full length clone, obtained after in vitro transcription/translation experiments, was an autoantigen in that it reacted with the sera of a multiplicity of Graves' patients, not simply the screening antisera, but not with normal sera similarly diluted (Fig. 10A) (Kohn et al., 1989a,b,c). The radiolabeled protein was also able to bind to Sepharose-TSH; this binding was inhibited by free TSH but not by free insulin, ACTH, prolactin, cholera toxin, or even hCG (Fig. 10B) (Kohn et al., 1989a,b,c). The radiolabeled protein could also be immunoprecipitated by a monoclonal antibody to the human TSH receptor, 52A8; immunoprecipitation by 52A8 was specifically inhibited by free TSH (Fig. 10C) (Kohn et al., 1989a,b,c).

As noted earlier, monoclonal antibody 52A8, made against the human TSH receptor (Kohn et al., 1983a,1984,1985,1986a; Chan et al., 1987a,b), is a "mixed" antibody since it can be both a competitive antagonist, inhibiting TSH binding, and also a competitive agonist of TSH, stimulating adenylate cyclase and growth activities in FRTL-5 rat thyroid cells. These are a strain of continuously cultured cells that maintain many of the functional properties of the thyroid in vitro, including TSH regulation of growth and function (Ambesi-Impiombato, 1986; Kohn et al., 1986a). The 52A8 antibody binds to rat FRTL-5 thyroid cells (Fig. 11A); this binding can be inhibited by TSH but not by the same concentrations of insulin or hCG (Kohn et al., 1989a,b,c). Twelve peptides representative of different portions of the amino acid sequence predicted were synthesized (Fig. 9) and tested for reactivity with monoclonal 52A8 (Kohn et al., 1989a,b,c). Peptide 3, residues 212 through 228, was able to bind to 52A8 (Fig. 11B), to inhibit the ability of TSH to increase cAMP levels in FRTL-S thyroid cells (Fig. 11C) and the ability of TSH to increase tritiated thymidine incorporation into FRTL-5 cell DNA (Fig. 11D) (Kohn et al., 1989a,b,c). In both cases, the peptide activity was concentration dependent.

Using 3T3 cells and L cells, measurements of transient expression of TSH binding were detected with radiolabeled TSH 48 h later using "sense" but not "antisense" constructs of the full length clone with an SV 40 promo-

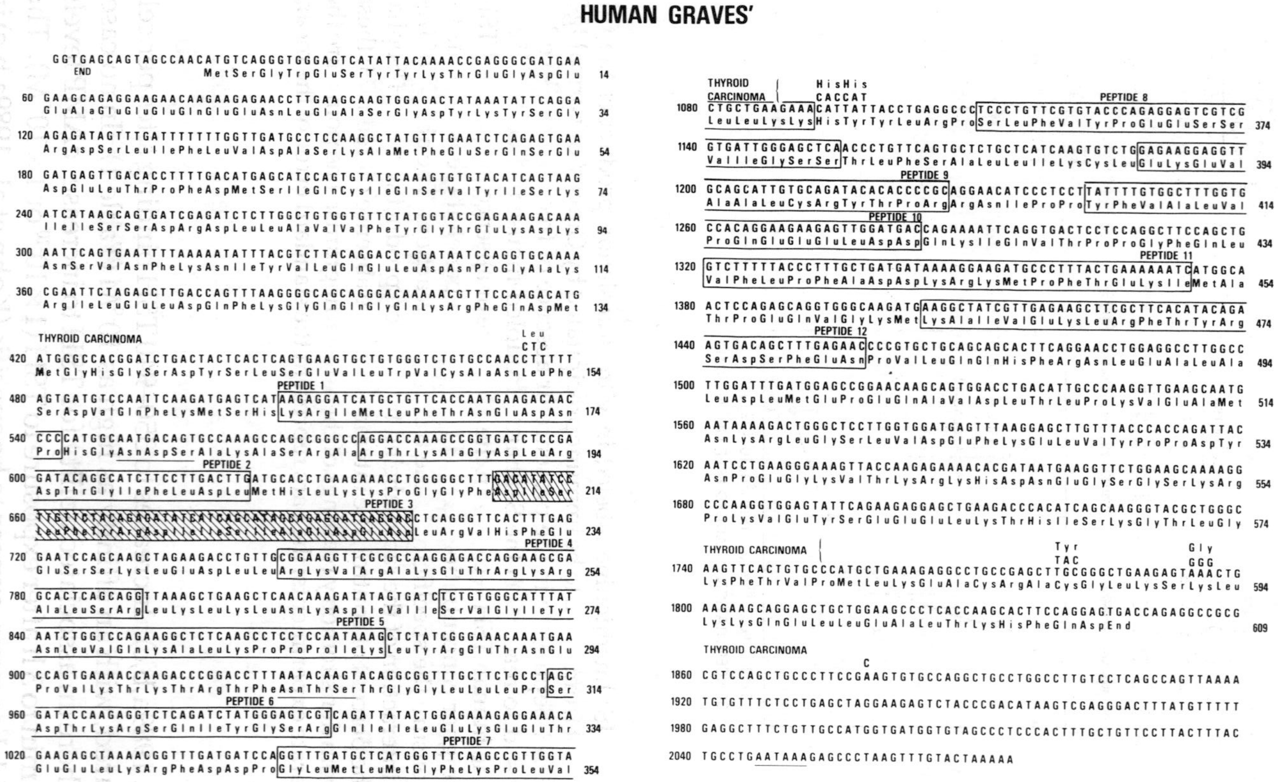

HUMAN GRAVES'

```
      GGTGAGCAGTAGCCAACATGTCAGGGTGGGAGTCATATTACAAAACCGAGGGCGATGAA
       END              MetSerGlyTrpGluSerTyrTyrLysThrGluGlyAspGlu        14
  60  GAAGCAGAGGAAGAACAAGAAGAGAACCTTGAAGCAAGTGGAGACTATAAATATTCAGGA
      GluAlaGluGluGluGlnGluGluAsnLeuGluAlaSerGlyAspTyrLysTyrSerGly        34
 120  AGAGATAGTTTGATTTTTTTGGTTGATGCCTCCAAGGCTATGTTTGAATCTCAGAGTGAA
      ArgAspSerLeuIlePheLeuValAspAlaSerLysAlaMetPheGluSerGlnSerGlu        54
 180  GATGAGTTGACACCTTTTGACATGAGCATCCAGTGTATCCAAAGTGTGTACATCAGTAAG
      AspGluLeuThrProPheAspMetSerIleGlnCysIleGlnSerValTyrIleSerLys        74
 240  ATCATAAGCAGTGATCGAGATCTCTTGGCTGTGGTGTTCTATGGTACCGAGAAAGACAAA
      IleIleSerSerAspArgAspLeuLeuAlaValValPheTyrGlyThrGluLysAspLys        94
 300  AATTCAGTGAATTTTAAAAATATTTACGTCTTACAGGACCTGGATAATCCAGGTGCAAAA
      AsnSerValAsnPheLysAsnIleTyrValLeuGlnGluLeuAspAsnProGlyAlaLys       114
 360  CGAATTCTAGAGCTTGACCAGTTTAAGGGGCAGCAGGGACAAAAACGTTTCCAAGACATG
      ArgIleLeuGluLeuAspGlnPheLysGlyGlnGlnGlyGlnLysArgPheGlnAspMet       134

      THYROID CARCINOMA                                           Leu
                                                                  CTC
 420  ATGGGCCACGGATCTGACTACTCACTCAGTGAAGTGCTGTGGGTCTGTGCCAACCTTTTT
      MetGlyHisGlySerAspTyrSerLeuSerGluValLeuTrpValCysAlaAsnLeuPhe       154
                                       PEPTIDE 1
 480  AGTGATGTCCAATTCAAGATGAGTCATAAGAGGATCATGCTGTTCACCAATGAAGACAAC
      SerAspValGlnPheLysMetSerHisLysArgIleMetLeuPheThrAsnGluAspAsn       174
 540  CCCCATGGCAATGACAGTGCCAAAGCCAGCCGGGCCAGGACCAAAGCCGGTGATCTCCGA
      ProHisGlyAsnAspSerAlaLysAlaSerArgAlaArgThrLysAlaGlyAspLeuArg       194
                       PEPTIDE 2
 600  GATACAGGCATCTTCCTTGACTTGATGCACCTGAAGAAACCTGGGGGCTTTGACATATCC
      AspThrGlyIlePheLeuAspLeuMetHisLeuLysLysProGlyGlyPheAspIleSer       214
                                             PEPTIDE 3
 660  TTGTTCTACAGAGATATCATCAGCATAGCAGAGGATGAGGACCTCAGGGTTCACTTTGAG
      LeuPheTyrArgAspIleIleSerIleAlaGluAspGluAspLeuArgValHisPheGlu       234
                                                          PEPTIDE 4
 720  GAATCCAGCAAGCTAGAAGACCTGTTGCGGAAGGTTCGCGCCAAGGAGACCAGGAAGCGA
      GluSerSerLysLeuGluAspLeuLeuArgLysValArgAlaLysGluThrArgLysArg       254
 780  GCACTCAGCAGGTTAAAGCTGAAGCTCAACAAAGATATAGTGATCTCTGTGGGCATTTAT
      AlaLeuSerArgLeuLysLeuLysLeuAsnLysAspIleValIleSerValGlyIleTyr       274
                                          PEPTIDE 5
 840  AATCTGGTCCAGAAAGCTCTCAAGCCTCCTCCAATAAAGCTCTATCGGGAAACAAATGAA
      AsnLeuValGlnLysAlaLeuLysProProProIleLysLeuTyrArgGluThrAsnGlu       294
 900  CCAGTGAAAACCAAGACCCGGACCTTTAATACAAGTACAGGCGGTTTGCTTCTGCCTAGC
      ProValLysThrLysThrArgThrPheAsnThrSerThrGlyGlyLeuLeuLeuProSer       314
                          PEPTIDE 6
 960  GATACCAAGAGGTCTCAGATCTATGGGAGTCGTCAGATTATACTGGAGAAAGAGGAAACA
      AspThrLysArgSerGlnIleTyrGlySerArgGlnIleIleLeuGluLysGluGluThr       334
                                                      PEPTIDE 7
1020  GAAGAGCTAAAACGGTTTGATGATCCAGGTTTGATGCTCATGGGTTTCAAGCCGTTGGTA
      GluGluLeuLysArgPheAspAspProGlyLeuMetLeuMetGlyPheLysProLeuVal       354

      THYROID  |      HisHis
      CARCINOMA|      CACCAT                                 PEPTIDE 8
1080  CTGCTGAAGAAACATTATTACCTGAGGCCCTCCCTGTTCGTGTACCCAGAGGAGTCGTCG
      LeuLeuLysLysHisTyrTyrLeuArgProSerLeuPheValTyrProGluGluSerSer       374
1140  GTGATTGGGAGCTCAACCCTGTTCAGTGCTCTGCTCATCAAGTGTCTGGAGAAGGAGGTT
      ValIleGlySerSerThrLeuPheSerAlaLeuLeuIleLysCysLeuGluLysGluVal       394
                                        PEPTIDE 9
1200  GCAGCATTGTGCAGATACACACCCCGCAGGAACATCCCTCCTTATTTTGTGGCTTTGGTG
      AlaAlaLeuCysArgTyrThrProArgArgAsnIleProProTyrPheValAlaLeuVal       414
                             PEPTIDE 10
1260  CCACAGGAAGAAGAGTTGGATGACCAGAAAATTCAGGTGACTCCTCCAGGCTTCCAGCTG
      ProGlnGluGluGluLeuAspAspGlnLysIleGlnValThrProProGlyPheGlnLeu       434
                                                       PEPTIDE 11
1320  GTCTTTTTACCCTTTGCTGATGATAAAAGGAAGATGCCCTTTACTGAAAAAATCATGGCA
      ValPheLeuProPheAlaAspAspLysArgLysMetProPheThrGluLysIleMetAla       454
1380  ACTCCAGAGCAGGTGGGCAAGATGAAGGCTATCGTTGAGAAGCTTCGCTTCACATACAGA
      ThrProGluGlnValGlyLysMetLysAlaIleValGluLysLeuArgPheThrTyrArg       474
                          PEPTIDE 12
1440  AGTGACAGCTTTGAGAACCCCGTGCTGCAGCAGCACTTCAGGAACCTGGAGGCCTTGGCC
      SerAspSerPheGluAsnProValLeuGlnGlnHisPheArgAsnLeuGluAlaLeuAla       494
1500  TTGGATTTGATGGAGCCGGAACAAGCAGTGGACCTGACATTGCCCAAGGTTGAAGCAATG
      LeuAspLeuMetGluProGluGlnAlaValAspLeuThrLeuProLysValGluAlaMet       514
1560  AATAAAAGACTGGGCTCCTTGGTGGATGAGTTTAAGGAGCTTGTTTACCCACCAGATTAC
      AsnLysArgLeuGlySerLeuValAspGluPheLysGluLeuValTyrProProAspTyr       534
1620  AATCCTGAAGGGAAAGTTACCAAGAGAAAACACGATAATGAAGGTTCTGGAAGCAAAAGG
      AsnProGluGlyLysValThrLysArgLysHisAspAsnGluGlySerGlySerLysArg       554
1680  CCCAAGGTGGAGTATTCAGAAGAGGAGCTGAAGACCCACATCAGCAAGGGTACGCTGGGC
      ProLysValGluTyrSerGluGluGluLeuLysThrHisIleSerLysGlyThrLeuGly       574

      THYROID CARCINOMA |               Tyr                  Gly
                                        TAC                  GGG
1740  AAGTTCACTGTGCCCATGCTGAAAGAGGCCTGCCGAGCTTGCGGGCTGAAGAGTAAACTG
      LysPheThrValProMetLeuLysGluAlaCysArgAlaCysGlyLeuLysSerLysLeu       594
1800  AAGAAGCAGGAGCTGCTGGAAGCCCTCACCAAGCACTTCCAGGAGTGACCAGAGGCCGCG
      LysLysGlnGluLeuLeuGluAlaLeuThrLysHisPheGlnAspEnd                   609

      THYROID CARCINOMA
                        C
1860  CGTCCAGCTGCCCTTCCGAAGTGTGCCAGGCTGCCTGGCCTTGTCCTCAGCCAGTTAAAA
1920  TGTGTTTCTCCTGAGCTAGGAAGAGTCTACCCGACATAAGTCGAGGGACTTTATGTTTTT
1980  GAGGCTTTCTGTTGCCATGGTGATGGTGTAGCCCTCCCACTTTGCTGTTCCTTACTTTAC
2040  TGCCTGAATAAAGAGCCCTAAGTTTGTACTAAAAA
```

Fig. 9. The complete nucleotide and deduced amino acid sequence of a cDNA from the human thyroid library and identified by immunoscreening as described in the text (Chan et al., 1989). The polyadenylation signal AATAAA is underlined. Two potential glycosylation sites are also underlined. A substitution of T for G (arrow) in the sequence of the comparable cDNA from a human Graves' thyroid library defines an in-frame stop codon and the full length sequence. The blocks denote sequences that were used to synthesize peptides for use in the studies of Fig. 11. The block denoting the sequence of the bioactive peptides, as determined in the experiments of Fig. 11, is cross-hatched.

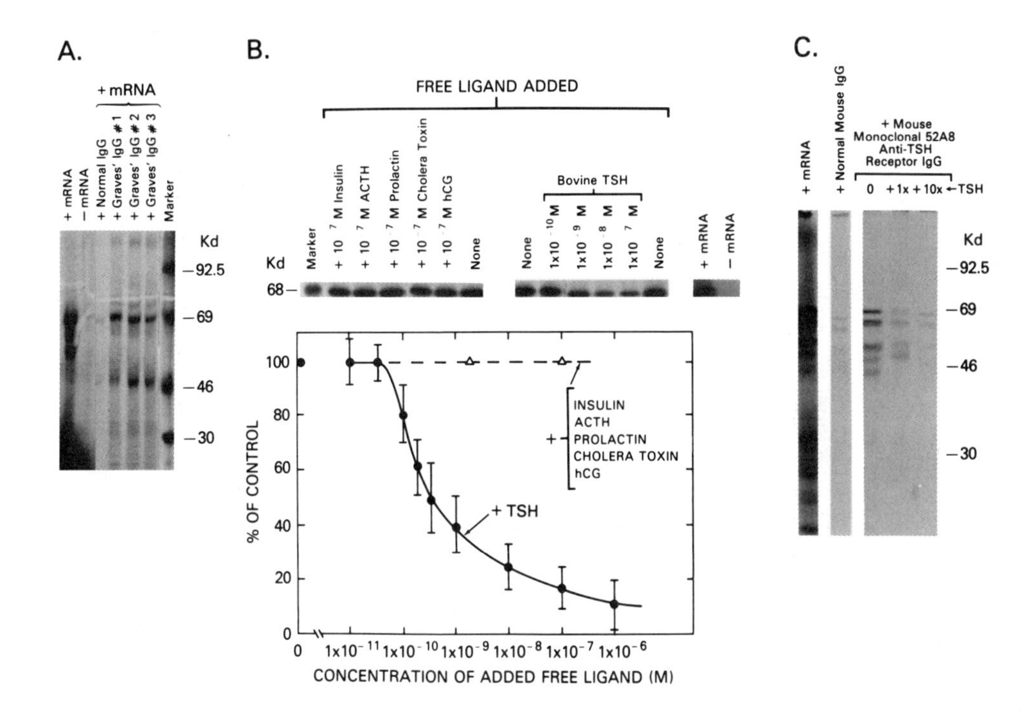

Fig. 10. Ability of [S^{35}]methionine-labeled in vitro translation product of the full length sense transcript from Fig. 2 to be immunoprecipitated by Graves' or normal IgG (A), to react with TSH-Sepharose in the presence or absence of free TSH or the other noted ligands (B), and to react with the Sepharose coupled mouse monoclonal 52A8 antibody (6–8) to the human TSH receptor in the presence or absence of $1 \times 10^{-7}M$ TSH (1×) or a 10-times higher concentration of TSH (C). In each panel, there is a minus (–) mRNA control to ensure the specificity of the translation reaction. The ability of free TSH to inhibit the interaction of the radiolabeled translation product with TSH-Sepharose is plotted as the percent of control, the interaction in the absence of free TSH; these data are derived from densitometry readings of the autoradiograms. The points are the mean of three separate experiments; the bars represent the standard error of the mean.

tor (Kohn et al., 1989a,b,c). Increases to 350–940 TSH binding sites per cell could be measured, with an affinity as high as $10^{-9}M$ (Table 1). In no case, however, was there an associated ability for TSH to increase cAMP levels when tested 15 min to 3 h after exposure to TSH (1×10^{-10} or $1 \times 10^{-9}M$). The TSH binding did, however, appear to be specific (Kohn et al., 1989a,b,c) in that TSH binding was not inhibited by prolactin, albumin, thyroglobulin, insulin, or glucagon (Fig. 12).

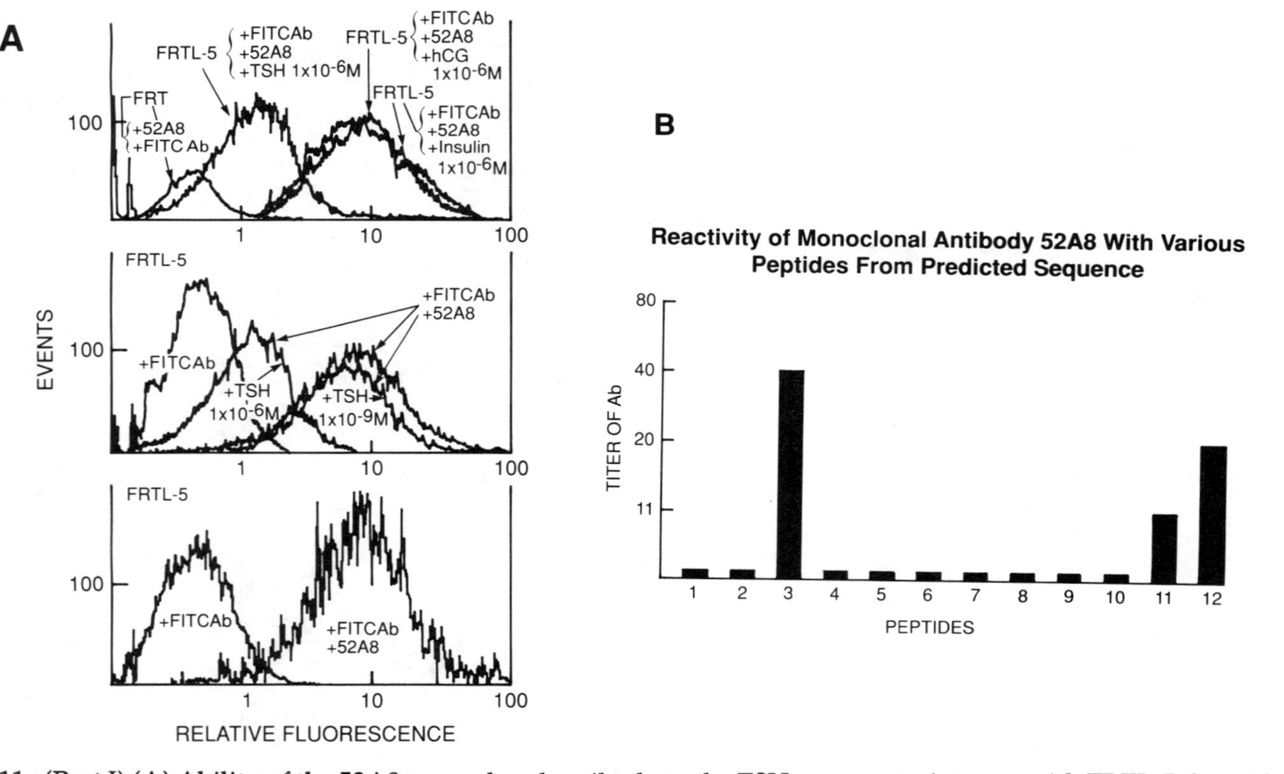

Fig. 11. (Part I) (A) Ability of the 52A8 monoclonal antibody to the TSH receptor to interact with FRTL-5 thyroid cells in the presence or absence of TSH, insulin, or chorionic gonadotropin (hCG) as measured by fluorescence activated cell sorting (FACS). The bottom panel demonstrates the specificity of the reaction in that the reactivity requires the presence of 52A8 and does not occur with the fluorescein isothiocyanate coupled (FITC) second antibody. The middle panel shows that the TSH inhibition is concentration dependent. The top panel shows that insulin or hCG do not duplicate the TSH inhibition at a high concentration of ligand. Also evident is the absence of the 52A8 reaction with FRT cells, which do not have a surface expressed glycoprotein component of the TSH receptor. (B) The ability of synthetic peptides matching the different portions of the predicted sequence in Fig. 9 (*see* blocks) to react with monoclonal antibody 52A8 as measured in a solid phase radioimmunoassay.

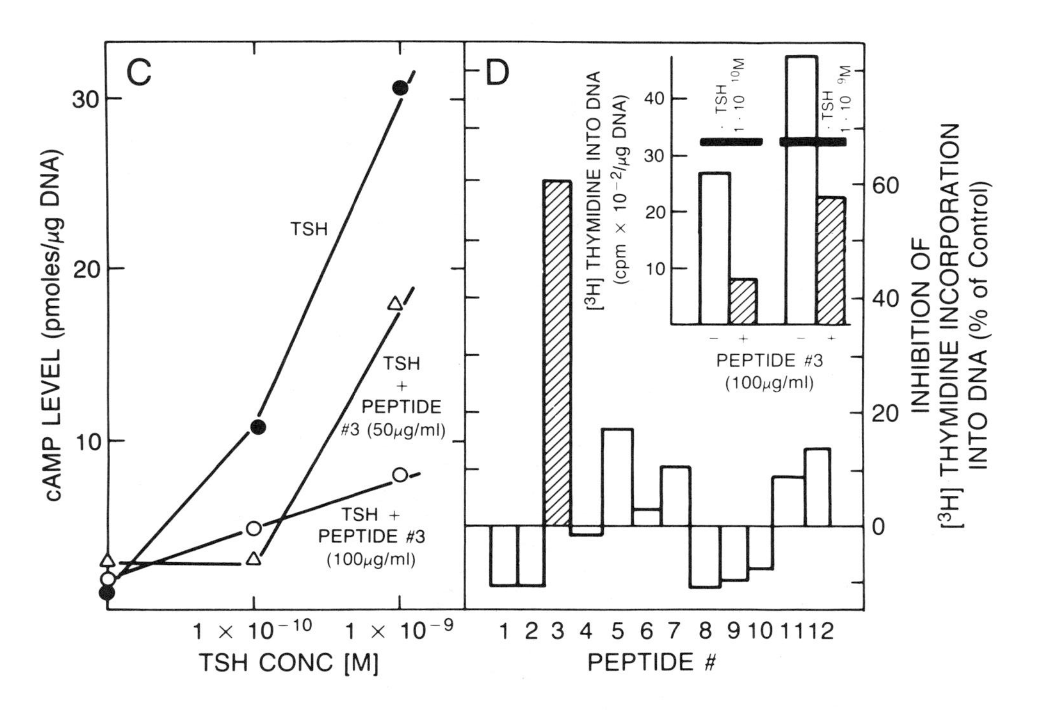

Fig. 11. (Part II) (C) Ability of peptide 3 to inhibit TSH-stimulated cAMP elevations in FRTL-5 thyroid cells. Experiments were performed as described in Kohn et al., 1986c. Other peptides tested at the same concentrations had no affect (data not shown). (D) Ability of the different peptides, 100 μg/mL, to inhibit 1×10^{-10} *M* TSH stimulated tritiated thymidine incorporation into the DNA of FRTL-5 thyroid cells, an assay that has been shown to measure growth under these conditions (Kohn et al., 1986c). The insert tests peptide 3 at different TSH concentrations.

Table 1
Radioiodinated TSH Binding to Cells Transfected with Sense (pSVL-FS2) and Antisense (pSVL-FA) Constructs of the Full Length Clone Whose Sequence Is Presented in Fig. 2[a]

Cells	Construct	Kd	Sites/cell
L-cells	pSVL-FS2	$5 \times 10^{-9} M$	920
L-cells	pSVL-FA	—	<50
3T3 L-cells	pSVL-FS2	$3 \times 10^{-9} M$	330
3T3 L-cells	pSVL-FA	—	<50

[a]pSVL (Pharmacia) was digested with SmaI and dephosphorylated with alkaline phosphatase. Full length insert was purified from agarose gel and filled in with dNTPs to create blunt ends. Vector and insert were ligated and transformed into HB101. Positive clones were selected by ampicillin resistance. Transfection used a dextran-DEAE procedure. After addition of heparin and washing, cells were cultured for 48 h. TSH binding was measured as described in Fig. 12.

This clone thus defines a TSH-binding glycoprotein that has a single transmembrane domain and a binding site for TSH and a monoclonal antibody to the TSH receptor that lies between the two glycosylation sites (Fig. 13). The binding site is negatively charged, in line with studies that suggest that the TSH-binding domain in the receptor is negatively charged and salt sensitive (Kohn, 1978) in its reactivity with the positive charge on TSH. Its size, 70K, is compatible with a major TSH-binding protein seen in FRTL-5 thyroid cells (Kohn et al., 1986a; Chan et al., 1987a,b; Chan et al., 1988) and in crosslinking studies (Kohn et al., 1983b). The absence of a G protein binding domain evident in studies of adrenergic receptors (Lefkowitz and Caron, 1988) is of concern in defining this as the cAMP-coupled TSH receptor, as is the absence of coupling to the adenylate cyclase signal. This is particularly true given work defining the cloning of the hCG receptor (*see below*) where such a domain is evident (Segaloff et al., 1989). Nevertheless, coupling to G proteins in this case could reflect a requirement for a coupling component such as the thyroid specific ganglioside or a missing subunit. The possibility that there is more than one TSH-binding component in cells, that not all are coupled to the adenylate cyclase signal system, or that these other TSH-binding proteins are important both as bioactivators and autoantigens must be considered.

Recombinant *Autograhpha californica* nuclear polyhedrosis virus containing the thyroid CDNA was produced by standard methods (Allaway et al., 1989). Insect cells transfected with the recombinant virus produced a 70K protein in large amounts; surprisingly, however, immunofluorescence and fractionation data indicate that the 70K protein is predominantly present in the nuclei of the infected insect cells and has undergone little or

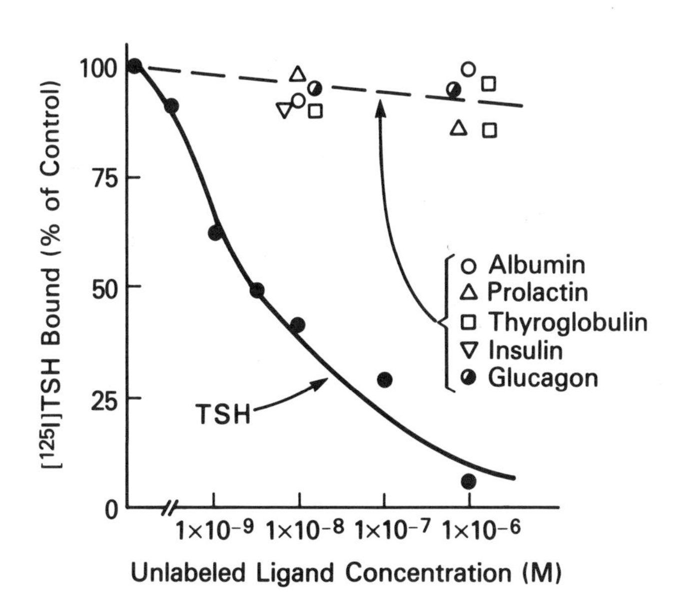

Fig. 12. Specificity of radioiodinated TSH binding to 3T3 fibroblasts transfected as in Table 1 with the full length sense construct (pSVL-FS2) and measured 48 h later. Data are expressed as percent of maximal TSH binding; Table 1 details the results of Scatchard analyses of TSH binding. The transfection used a dextran-DEAE procedure to obtain transient expression. Binding was measured using radioiodinated TSH. A positive control measured TSH binding to FRTL-5 thyroid cells, a negative control to FRT cells and liver cells. TSH binding was measured by incubating cells with 1×10^{-6} cpm radioiodinated TSH together with various concentrations of unlabeled TSH or the noted ligands for 60 min at 37°C. Bound/free separations were performed by centrifugation.

no glycosylation. The possibility that the clone is a cDNA binding protein that is important in autoimmunity and the growth process is supported by two startling and independent observations.

First, the sequence of the clone is identical to the sequence of the cDNA encoding the p70 (Ku) autoantigen (Reeves and Sthoeger, 1979). This antigen is present in large amounts in some patients having systemic lupus erythematosus and related disorders. It is associated with chromosomes of interphase cells but is dissociated from condensing chromosomes in early prophase. Analysis of this protein as a potential growth related autoantigen (Reeves and Sthoeger, 1979) indicates the predicted amino acid sequence does contain two regions with periodic repeats of either

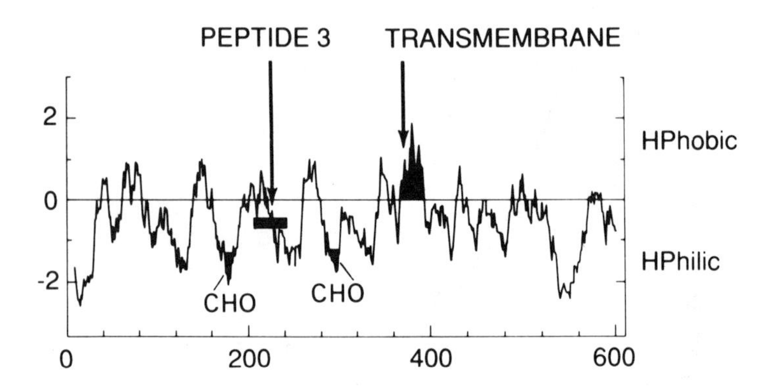

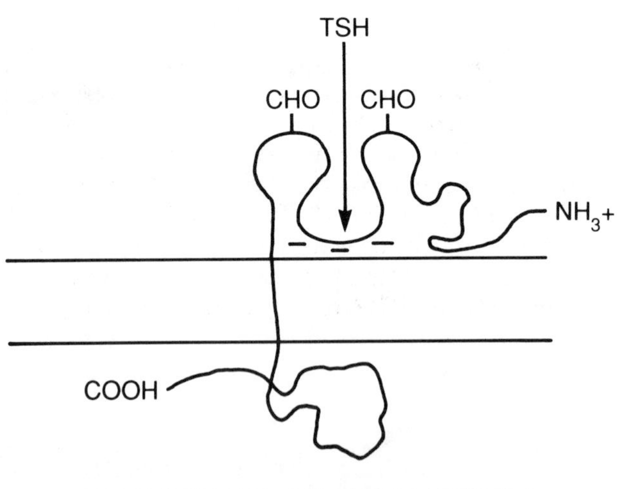

Fig. 13. Model of TSH-binding transmembrane glycoprotein predicted from the data in Figs. 2 and 4. The hydrophobicity plot at the top is derived from the data of Fig. 9. The two glycosylation sites (CHO) are noted as is the transmembrane domain. The region of peptide 3 that interacts with TSH and monoclonal antibodies to the TSH receptor is noted. The model at the bottom places this in the context of the lipid bilayer. A TSH-binding protein with a single transmembrane domain is predicted; the TSH binding site, which is negatively charged, is between the glycosylation sites.

leucine alone, or leucine alternating with serine every seventh position. The latter repeat displays sequence and secondary structural similarities with the leucine zipper regions of the c-*myc* and v-*myc* oncogene products.

Second, a recent report used a monoclonal antibody to Ki-67, a nuclear antigen associated with cell growth, to measure cell proliferation in rat FRTL-5 thyroid cells (Grammerstorf and Wenzel, 1989). In that report the authors concluded the nuclear fluorescence measured by the antibody was a useful alternative to the metaphase index assay of FRTL-5 thyroid cell growth. Ki-67 antigen is p70 (Ku) and the clone defined in Fig. 9. There is, therefore, a direct link between thyroid cell growth and this autoantigen and an indirect link to thyroid cell and other tumors.

Using somatic cell human–hamster hybrids (Chan et al., 1988; Kohn et al., 1989a,b,c), 6 loci were identified on 5 different human chromosomes, namely 22q11-13, Xq, lq, 8, and 10. The presumptive functional chromosome, 22q11-13, is associated with the Sis proto-oncogene. In mice (Chan et al., 1988), the gene is mapped to a single locus on chromosome 15, also appears to be associated with the Sis proto-oncogene, and is on the same chromosome as the thyroglobulin and c-*myc* genes.

These results are intriguing with respect to the possibility that the clone is an important TSH-binding autoantigen in Graves' disease. Thus, recent work (Kendall-Taylor et al., 1988) in patients with Graves' disease and exophthalmos suggests there is a significant increased frequency of P blood group by comparison to patients with Graves' disease who do not have exophthalmos. P blood group antigen is associated with the P_1 antigen and, in particular, to the alpha-4-Gal transferase converting paragloboside to the P_1 antigen. Current studies (McBride, personal communication) indicate that it maps to the same locus 22q11-13 and is probably near and closely linked, but not identical to, the gene coding for the cDNA described in these studies. Paragloboside and the P_1 antigen are glycolipids that can be viewed as structural analogs of gangliosides. This association and these data are very exciting in that they offer the first clue, at a molecular level, as to the relation of Graves', its ophthalmopathy, and the TSH-binding membrane glycoprotein identified by the cDNA in these studies. A TSH binding component, rather than the TSH receptor domain important in adenylate cyclase stimulation, has been linked to exophthalmos in vitro (Rotella et al., 1986).

Second, the identification of genomic material on the X chromosome is obviously provocative with respect to the dramatically increased frequency of Graves' in females. Whether this is more than fortuitous will, however, require much more understanding of the genomic significance of these multiple gene copies.

The cDNA coding for the autoantigen defined above was cloned from a Graves' thyroid lambda gt11 expression library and sequenced. The sequence is nearly identical to the human thyroid carcinoma clone. These results would support the preliminary conclusion that a grossly abnormal protein structure is not the causative factor in forming Graves' autoantibodies.

Graves' disease has been argued to reflect an immune system irregularity at the lymphocyte level. In this respect, the identification of the 2 Kb mRNA in IM9 lymphocytes (Chan et al., 1988; Kohn et al., 1989a,b,c) was not only compatible with the identification of TSH receptor activity in lymphocytes but also raised the possibility that this might be an autoantigen common to thyroid and lymphocytes. This appears to be true (Kohn et al., 1989). Thus, although low levels of transcript were evident in normal tonsilar T cells, phytohaemagglutinin (PHA) increased the expression nearly tenfold. Maximal increases were evident 16 h after PHA and were blocked by cyclosporin. Transformed T cell lines had a high level of transcript without PHA stimulation and expression of the transcript in transformed lines was not changed by PHA. These results are compatible with data accumulated on the p70 (Ku) or Ki-67 autoantigen, which has been linked to proliferation.

Normal tonsillar B cells express even higher levels of the 2 Kb transcript than normal T cells; and mitogen stimulation by *Staphlococcus aureus* Cowan Strain 1 and phorbol myristate acetate (PMA) further elevates transcript levels, approx threefold. Once again, transformed B cell lines have higher levels of the transcript and mitogen stimulation does not increase transcript levels in the transformed lines.

Sequencing of the cDNA coding for the autoantigen after its isolation from an IM9 lymphocyte lambda gt11 expression library indicates that the lymphocyte and thyroid autoantigens are again effectively identical. This result plus the existence of a single gene in the mouse makes it likely that the same gene codes for the transcript in thyroid and immune cells.

Lymphocytes have TSH receptors as defined in binding studies but the receptor is not coupled to the cyclase signal system. This would be compatible with the thesis that TSH binding in these cells involves a glycoprotein membrane component, but coupling requires a second factor (Kohn, 1978; Kohn et al., 1986; Chan et al., 1987a,b) or that this TSH-binding glycoprotein is coupled to a different signal transduction system. One must question why this protein exists in lymphocytes; is its presence key to the development of autoimmune Graves'? The transfection studies in the insect virus system that define a nuclear form of the transcript, and a link to lupus may be relevant. The observation that transcript expression in thyroid and lymphocytes increases with mitogen stimulation or in transformed cells and that nuclear fluorescence measured by antibody to the antigen correlates with thyroid cell growth argues the autoantigen has an important role in growth as evidenced in receptor-oncogene relationships exhibited by the clone.

4.2. Cross-Hybridization Approach

In a preliminary report at the Serono Symposium on Glycoprotein hormones, evidence for cloning of the hCG receptor was presented (Segaloff et al., 1989); this has now been reported by two separate groups (McFarland et al., 1989; Loosfelt et al., 1989). This protein has a hydrophilic tail of about 40 kDa and a intramembrane region containing seven transmembrane, hydrophobic areas with sequence relationships to the adrenergic receptor G protein binding domain. Preliminary results of transfection experiments indicate that this protein does, in fact, increase cAMP levels when given hCG.

This clone has now been used in a cross-hybridizing approach (Frazier-Seabrook et al., 1989a,b) in an attempt to clone the cAMP-coupled TSH receptor. In the initial report (Frazier-Seabrook et al., 1989a) using this approach and a 600 bp portion of the hydrophobic tail of the LH/hCG receptor, a thyroid cDNA clone with greater than 80% homology to the LH/hCG receptor was isolated; this clone, which detects a gene approximately equally expressed in thyroid and testis, appears to be the thyroid LH/hCG receptor (Frazier-Seabrook et al., 1989b). Subsequent studies using this approach and a portion of the G-related, 7-transmembrane domain as the screening probe have, however, identified a receptor with a lesser degree of homology to the LH/hCG receptor but which does appear to detect a gene expressed in thyroid tissue only. This clone is under active investigation as the cAMP-coupled TSH receptor.

The converse cross-hybridizing approach has been used with the clone identified in Fig. 9. A testis specific and related clone has been identified. The significance of this remains unclear.

5. Summary

The present report summarizes the current state of TSH receptor purification. The net suggests that a multiplicity of TSH-binding proteins exist, many of which can react with monoclonal antibodies to the TSH receptor. Nevertheless, cloning studies pursuing these proteins suggest that within the year a detailed structure of the glycoprotein component of the TSH receptor will be resolved, as will be the structure of several TSH binding proteins important to Graves' disease, and apparently related to the TSH receptor structure. The apparent multiplicity of TSH-binding proteins raises important questions as to the basic nature of Graves' disease. Perhaps we should reexamine previous presumptions concerning the structure of the TSH receptor, its mechanism of action in effecting cell processes, and its role as an autoantigen causing Graves' disease.

References

Allaway G. P., Prabhakar, B. S., Vivino A. A., Chan, J. Y. C., Kohn L. D., and Notkins, A. L. (1989) in *Proceedings 23rd Annual Meeting European Society for Clinical Investigation.* Springer, Berlin, Heidelberg, New York.

Ambesi-Impiombato, F. S. (1986) *US Patent 4,608,341.*

Amir S. M., Carraway, T. F., Kohn, L. D., and Winand, R. J. (1973) *J. Biol. Chem.* **248,** 4092–4100.

Axelrod, J. (1986) in *Transduction of Neuronal Signals* (Magistretti, P. J., Morrison, J. H., and Reisine, T. D., eds.), Foundation for the study of the nervous system, Geneva.

Bone, E. A., Alling, D. W., and Grollman, E. F. (1986) *Endocrinology* **119,** 2193–2200.

Burch, R. M., Luini, A., Mais, D. E., Corda, D., Vanderhoek, J. Y., Kohn, L. D., and Axelrod, J. (1986) *J. Biol. Chem.* **261,** 11236–11241.

Casey, P. J. and Gilman, A. G. (1988) *J. Biol. Chem.* **263,** 2577–2581.

Chan, J., DeLuca, M., Santisteban, P., Isozaki, O., Shifrin, S., Aloj, S., Grollman, E. F., and Kohn, L. D. (1987a) in *Thyroid Autoimmunity* (Pinchera, A., Ingbar, S. H., and McKenzie, J. M., eds.), Plenum Press, New York, pp. 11–26.

Chan, J., Lerman, M. I., Prabhakar, B., Isozaki, O., Santisteban, P., Kuppers, R., Oates, E. L., Notkins, A. L., and Kohn,L. D. (1989) *J. Biol. Chem.* **264,** 3651– 3654.

Chan, J., Santisteban, P., DeLuca, M., Isozaki, O., Grollman, E. F., and Kohn, L. D. (1987b) *Acta Endocrinologia (Copenh) Suppl* **281:115,** 166–172.

Chan, J. Y. C., Santisteban, P., Kozak, C. A., Kuppers, R., and Kohn, L. D. (1988) *Endocrinology* **121** (Suppl.), T52.

Corda, D., Marcocci, C., Kohn, L. D., Axelrod, J., Luini, A. (1985) *J Biol. Chem.* **260,** 9230–9236.

Corda, D., Sekura, R. D., and Kohn, L. D. (1987) *Eur. J. Biochem.* **166,** 475–481.

De Wolf, M. J. S., Vitti, P., Ambesi-Impiombato, F. S., and Kohn, L. D. (1981) *J. Biol. Chem.* **256,** 12287– 12296.

Dumont, J. E. (1971) *Vitam. Horm. (NY)* **29,** 287–412.

Frazier-Seabrook, L., Robbins, L. S., Segaloff, D. L., Seeburg, P. H., and Cone, R. D. (1989b) in American Thyroid Association 64th Annual Meeting, *Endocrinology* **122** (Suppl), Abs. 100, T51.

Grammerstorf, S. and Wenzel, B. F. (1989) in *FRTL-5 Today* (Ambesi-Impiombato, F. S. and Perrild, H., eds.), Excerpta Medica International Congress Series **818,** Excerpta Medica, Amsterdam, pp. 181–182.

Grollman, E. F., Lee, G., Ambesi-Impiombato, F. S., Meldolesi, M. F., Aloj, S. M., Coon, H. G., Kaback, H. R., and Kohn, L. D. (1977) *Proc. Natl. Acad. Sci. USA* **74,** 2352–2356.

Jeevanram, R., Blithe, D., Liu, L., Wehmann, R. and Nisula, B. (1989) in *International Sym-*

posium on Glycoprotein Hormones (Chin, W. and Boime, I., eds.), Plenum Press, New York, in press.

Kendall-Taylor, P., Stephenson, A., Stratton, A., Papiha, S. S., Perros, P., and Roberts, D. F. (1988) *Clinical Endocrinology* **28,** 601–610.

Kohn, L. D. (1978) in *Receptors and Recoqnition Series A. Vol. 5* (Cuatrecassas, P. and Greaves, M. F., eds.), Chapman and Hall, London, pp. 133–212.

Kohn, L. D., Aloj, S. M., Tombaccini, D., Rotella, C. M., Toccafondi R., Marcocci, C., Corda, D., and Grollman, E. F. (1985) in *Biochemical Actions of Hormones* (Litwack, G., ed.), Marcel Dekker, NY, pp. 457–512.

Kohn, L. D., Alvarez, F., Marcocci, C., Kohn, A. D., Chen, A., Hoffman, W. E., Tombaccini, D., Valente, W. A., DeLuca, M., Santisteban, P., and Grollman, E. F. (1986a) *Ann. N.Y. Acad. Sci.* **475,** 157–173.

Kohn, L. D., Bone, E. A., Chan, J. Y. C., Corda, D., Isozaki, O., Luini, A., Marcocci, C., Santisteban, P., and Grollman, E. F. (1986b) in *Transduction of Neuronal Signals* (Magistretti,P. J., Morrison, J. H., and Reisine, T. D., eds.), Foundation for the study of the nervous system, Geneva.

Kohn, L. D., Isozaki, O., Chan, J., Akamizu, T., Bellur, S., DeLuca, M., Santisteban, P., Varutti, A. M., Grimaz, A., Ikuyama, S., Saji, K., and Owens, G. (1989a) in *FRTL-5 Today* (Perrild, H. and Ambesi-Impiombato, F. S., eds.), Elsevier, Amsterdam, in press.

Kohn, L. D., Isozaki, O., Chan, J. Y., Akamizu, T., Bellur, S., Santisteban, P., Ikuyama, S., Saji, M., Doi, S., Tahara, K., Kosugi, S., Sabe, H., Honjo, T., and Mori, T. (1989b) in *International Symposium on Glycoprotein Hormones* (Chin, W. and Boime, I., eds.), Plenum Press, New York, in press.

Kohn, L. D., Saji, M., Akamizu, T., Ikuyama, S., Isozaki, O., Kohn A. D., Santisteban, P., Chan, J. Y. C., Bellur, S., Rotella, C. M., Alvarez, F. V., and Aloj, S. M. (1989C) in *Control of the Thyroid: Regulation of Its Normal Growth and Function* (Ekholm, R., Kohn, L. D., and Wollman, S., eds.), Plenum Press, New York, in press.

Kohn, L. D., Tombaccini, D., DeLuca, M., Bifulco, M., Grollman, E. F., and Valente, W. A. (1984) in *Monoclonal Antibodies to Receptors: Probes for Receptor Structure and Function, Receptors and Recognition, Series B* (Greaves, M. F., ed.), **17,** 201–234.

Kohn, L. D., Valente, W. A., Grollman, E. F., Aloj, S. M., and Vitti, P. (1986c) *U.S. Patent 4,609,622.*

Kohn, L. D., Valente, W. A., Laccetti, P., Cohen, J., Aloj, S. M., and Grollman, E. F. (1983b) *Life Sciences* **32,** 15–30.

Kohn, L. D., Yavin, E, Yavin, Z., Laccetti, P., Vitti, P., Grollman, E. F., and Valente, W. A. (1983a) in *Monoclonal Antibodies: Probes for the Study of Autoimmunity and Immunodeficiency* (Haynes, B. F. and Eisenbarth G. S., eds.), Academic Press, New York, pp. 221–258.

Kress, B. C. and Spiro R. G. (1986) *Endocrinology* **118,** 974–985.

Lefkowitz, R. J. and Caron M. G. (1988) *J. Biol. Chem.* **263,** 4993–4996.

Loosfelt, H., Misrahi, M., Atger, M., Salesse, R., Thi, M. T. V. H.-L., Jolivet, A., Guiochon-Mantel, A., Sar, S., Jallal, B., Garnier, J., and Milgrom, E. (1989) *Science* **245,** 525–528.

Marcocci, C., Luini, A., Santisteban, P., and Grollman, E. F. (1987) *Endocrinology* **120,** 1127–1136.

McFarland, K. C., Sprengel, R., Phillips, H. S., Kohler, M., Rosemblit, N., Nikolics, K., Segaloff, D. L., and Seeburg, P. H. (1989) *Science* **245,** 494–499.

Philp, N. J. and Grollman E. F. (1986) *FEBS Letters* **202,** 193–199.

Reeves, W. H. and Sthoeger, Z. M. (1989) *J. Biol. Chem.* **264,** 5047–5052.

Remy, J. J., Salamero, J., and Charreire (1987) *Endocrinology* **121,** 1733– 1741.
Riberio-Neto, F., Birnbaumer, L., and Field, J. B. (1987) *Mol. Endo.* **1,** 482–490.
Robison, G. A., Butcher, R. W., and Sutherland, E. W. (1971) Cyclic AMP, Academic, New York.
Rotella, C. M., Zonefrate, R., Toccafondi, R., Valente, W. A., and Kohn, L. D. (1986) *J. Clin. Endocrinol. Metab.* **62,** 357–367.
Santisteban, P., DeLuca, M., Corda, D., Grollman, E. F. and Kohn, L. D. (1986) in *Frontiers in Thyroidology. Vol.* 2 (Medeiros-Neto, G. and Gaitan, E., eds.), Plenum Press, New York, pp. 837–840.
Segaloff, D. L., Sprengel, R., McFarland, K. C., Rosemblit, N., Kohler, M., Nikolics, K., and Seeburg, P. H. (1989) in *International Symposium on Glycoprotein Hormones* (Chin, W. and Boime, I., eds.), Plenum Press, New York, in press.
Tahara, K., Saji, M., Aloj, S. M., and Kohn, L. D. (1989) in *Control of the Thyroid: Regulation of Its Normal Growth and Function* (Ekholm, R., Kohn, L. D., and Wollman, S., eds.), Plenum Press, New York, in press.
Tate, R. L., Schwartz, H. I., Holmes, J. M., Kohn, L. D., and Winand, R. J. (1975) *J. Biol. Chem.* **250,** 6509–6515.
Vitti, P., De Wolff, M. J. S., Acquaviva, A. M., Epstein, M., and Kohn L. D. (1982) *Proc. Natl. Acad. Sci. USA* **79,** 1525–1529.
Weiss, S. J., Philp, N. J., and Grollman, E. F. (1984) *Endocrinology* **114,** 1108–1117.

Thyrotropin Receptor Purification

Petu J. Leedman and Leonard C. Harrison

1. Introduction

Thyroid stimulating hormone (thyrotropin, TSH) derived from the anterior pituitary gland exerts its action by binding to specific receptors on thyroid follicular cells (Moore and Wolff, 1974; Kotani et al., 1975; Amir et al., 1976; Rees Smith et al., 1977; Pekonen and Weintraub, 1979). Binding initiates a sequence of processes (including stimulation of adenylate cyclase, production of cyclic AMP and activation of protein kinases, stimulation of phospholipid turnover and mobilization of intracellular calcium) that are presumed to mediate the bioeffects of TSH, i.e., iodothyronine hormone synthesis and release and thyroid growth (for review, *see* Field, 1986).

Apart from its crucial role in transducing signals that are vital for survival, the TSH receptor, in humans, is the target for an autoimmune response that accounts for the major form of thyroid pathology: Graves' disease is considered to be attributable to polyclonal autoantibodies to the TSH receptor that inhibit binding of TSH, mimic the bioeffects of TSH, and induce hyperthyroidism and thyroid enlargement (goiter) (for review, *see* Rees Smith et al., 1988).

Graves' immunoglobulins (Igs) have been shown to precipitate specific proteins (Heyma and Harrison, 1984; Parkes et al., 1985) but the relationship of these to the subunits of the receptor remains in question. Purification of the receptor and definition of its subunit structure is necessary if we are to begin to understand the structure–function relationships within the receptor as they apply to the physiological actions of TSH and the pathological actions of TSH receptor antibodies.

Attempts to define the subunit structure of the TSH receptor have produced inconsistent results. Receptor structure has been characterized by various techniques including crosslinking (Buckland et al., 1986; McQuade et al., 1986a,b; Gennick et al., 1987; McQuade et al., 1987; Remy et al., 1987) and photoaffinity labeling (Buckland and Rees Smith, 1984; Kajita et al., 1985a), immunoprecipitation (Heyma and Harrison, 1984; Parkes et al., 1985; Chan et al., 1987), target size analysis (Nielsen et al., 1984), gel filtration (Czarnocka et al., 1981; Iida et al., 1981; Koizumi et al., 1982; Iida et al., 1983; Kajita et al., 1985a), affinity purification with either lectin- (Drummond et al., 1982; Iida et al., 1983; Kress and Spiro, 1986; Leedman et al., 1989) and/or TSH-agarose conjugates (Tate et al., 1975; Rickards et al., 1981; Koizumi et al., 1982; Iida et al., 1983; Heyma and Harrison, 1984; Parkes et al., 1985; Iida et al., 1987; Remy et al., 1987; Leedman et al., 1989) or monoclonal antibodies (Chan et al., 1987). Results from these studies suggest the receptor may comprise one to three protein subunits with molecular weight (M_r) estimates ranging from 25,000–300,000. In addition, the receptor may contain a ganglioside component (Mullin et al., 1976; Gardas and Nauman, 1981). Although its significance remains controversial, studies with monoclonal antibodies directed against the receptor suggest that the ganglioside is involved in signal transduction and is a target for thyroid stimulating autoantibodies (Chan et al., 1987).

Receptors have been characterized from several species including human (Koizumi et al., 1982; Iida et al., 1983; Leedman et al., 1989), porcine (Rickards et al., 1981; Parkes et al., 1985), and bovine (Tate et al., 1975; Koizumi et al., 1982; Kress and Spiro, 1986) thyroid glands, guinea pig fat cells (Iida et al., 1987), guinea pig thyroid (Manley et al., 1974; Buckland and Rees Smith, 1984), cultured Fischer rat thyroid cells (FRTL-5 cells) (Bartalena et al., 1987; Chan et al., 1987; Gennick et al., 1987), a human lymphocyte cell line (IM9) (Pekonen and Weintraub, 1978), and human thyroid hybridoma cells (GEJ cells) (Remy et al., 1987).

Two major obstacles confront attempts to purify the receptor to homogeneity. First, receptors are present in small numbers on thyroid cells (~10^3–10^4 per cell) (Teng et al., 1975; Pekonen and Weintraub, 1978). Second, the receptor is labile and undergoes temperature- and time-dependent degradation (Tate et al., 1975; Pekonen and Weintraub, 1979; Koizumi et al., 1982; Iida et al., 1983), which significantly reduces the final recovery. In this review, we will outline the approaches that have been used to purify the TSH receptor and review the most important measures that may improve the recovery of functional receptors.

2. Chromatographic Techniques Employed for Purification of the TSH Receptor

The nonionic detergent Triton X-100 has been the most widely used agent for solubilizing thyroid membranes prior to chromatography. Greater than 40% of the original TSH binding activity can be recovered after extraction of thyroid membranes in Triton X-100 (Dawes et al., 1978; Czarnocka et al., 1979). Concentrations of Triton X-100 of 0.1% (Czarnocka et al., 1979), 0.5% (Dawes et al., 1978; Iida et al., 1987) and 1.0% (Drummond et al., 1982; Iida et al., 1983; Heyma and Harrison, 1984; Rapoport et al., 1984; Kress and Spiro, 1986; Leedman et al., 1989) have been shown to be effective. Other detergents, less commonly used, include Triton N-101 (Koizumi et al., 1982), Triton X-114 (Remy et al., 1987), Lubrol 12A9 (Davies Jones et al., 1985a; Kajita et al., 1985a,b), lithium diiodosalicylate (Tate et al., 1975) and sodium deoxycholate (Furmaniak et al., 1986). In an attempt to avoid the interaction between detergents and solubilized proteins Czarnocka et al. (1981) used a 1% butanol/water mixture to extract bovine thyroid membrane glycoproteins, but recovered only ~14% of the original TSH binding activity.

We routinely solubilize thyroid membranes (prepared as described by Heyma and Harrison, 1984) at 4°C in a 10 m*M* Tris-HCl, pH 7.4, 50 m*M* NaCl (Tris-NaCl) buffer containing 1% Triton X-100 (v/v) and a cocktail of protease inhibitors consisting of 1000 kallikrien inhibitory U/mL aprotinin, 2 m*M* phenylmethylsulphonyl fluoride (PMSF), 100 U/mL bacitracin, and 20 µg/mL leupeptin. TSH receptors solubilized in this manner retain ^{125}I-TSH binding activity for several months when stored at –70°C.

2.1. Lectin

Several studies have demonstrated that the TSH receptor contains a glycoprotein component (Czarnocka et al., 1981; Drummond et al., 1982; Iida et al., 1983; Davies Jones and Rees Smith, 1984; Kress and Spiro, 1986; Leedman et al., 1989). The binding of TSH receptors to immobilized lectins is reversible and can be inhibited by specific sugars. Elution from lectins is simple and can be performed at neutral pH without alteration to the ionic concentration of the sample. The use of lectins for the purification of membrane receptors has been extensively reviewed by Hedo (1984).

Lectins have been used to partially purify solubilized thyroid membranes for use in immunoprecipitation studies (Rapoport et al., 1984), to adsorb TSH binding activity from solubilized thyroid mem-branes (Tate et al., 1975), to separate high and low affinity TSH binding components in solubilized thyroid membranes (Drummond et al., 1982), and to demonstrate that the TSH receptor contains a glycoprotein component by adsorption to Concanavalin A-Sepharose (Con A-Sepharose) (Davies Jones and Rees Smith, 1984).

Three groups have utilized lectins to specifically purify the receptor. Iida et al. (1983) used Con A-Sepharose 4B to purify solubilized human thyroid membranes and obtained a 4–12-fold purification with ~20% recovery of TSH binding activity. More recently, Kress and Spiro (1986) performed extensive studies on bovine and human thyroid membranes comparing the efficacy of different lectin-agarose conjugates. They observed that the binding of TSH receptors to different lectins displayed significant species specificity; for example, although *Bandeiraea simplicifolia I*-agarose bound 84% of TSH binding sites present in bovine thyroid membranes less than 10% in human membranes was retained, suggesting that the human TSH receptor is relatively deficient in terminal α-D-galactosyl groups. In contrast, ~65% of TSH binding sites in bovine and human thyroid membranes were retained on wheat germ agglutinin-agarose and could be eluted with *N*-acetyl-D-glucosamine. With sequential TSH- and lectin-agarose chromatography they purified bovine TSH binding sites ~800-fold. In addition, they confirmed the presence of terminal α-D-galactosyl, β-D-galactosyl, sialyl, and α-D-mannosyl sugar residues in the bovine TSH receptor by demonstrating its binding, respectively, to *Bandeiraea simplicifolia I, Ricinus communis I,* wheat germ agglutinin, and Con A.

We use wheat germ lectin affinity chromatography as a simple first-step purification technique for human and porcine thyroid membranes. Immobilized wheat germ agglutinin can be purchased (Pharmacia, Sweden; Vector laboratories, CA) or prepared by using cyanogen bromide (CNBr)-activated agarose (Pharmacia, Sweden) (Adair and Kornfeld, 1974; Lotan et al., 1977). Alternatively, wheat germ lectin can be readily coupled to 1,1'-carbonyldiimadazole (CDI)-activated agarose, (Reacti-Gel (6x), Pierce Chemical Company, IL) (Leedman et al., 1989). The usual ratio of coupled lectin to gel is 1–5 mg/mL. Solubilized thyroid membranes are either batch-adsorbed for 12 h or recycled 3–4 times over the wheat germ lectin-CDI-agarose (lectin-agarose) column at 4°C. Approximately 25 mg of solubilized thyroid membrane glycoproteins can be bound per mL of lectin-agarose. The column is then washed extensively with Tris-NaCl,

0.1% Triton X-100, 0.1 m*M* PMSF (buffer) and eluted with the same buffer containing 0.3*M* *N*-acetyl-D-glucosamine. Samples can contain at least 1.0% Triton X-100 without causing interference to lectin–glycoprotein binding. This approach produces a 5–10-fold purification of human and porcine TSH binding sites, with a recovery of ~30% of human and ~20% of porcine TSH binding activity, respectively.

As outlined in detail by Newman and Harrison (this volume), the concentration of sodium dodecyl sulfate (SDS) used to wash the lectin- agarose columns after elution of receptors can influence the ability of the columns to rebind receptor. It appears that concentrations of 0.01% SDS may not remove all residually bound glycoproteins yet concentrations of 0.05% SDS and greater may promote release of lectin from the gel (Lotan et al., 1977). Thus, we wash our lectin-agarose columns with buffer containing 0.02% SDS as suggested by Hedo (1984), followed by buffer containing 0.3*M* *N*-acetyl-D-glucosamine. Columns regenerated in this way can be reused several times over 6–12 mo.

2.2. TSH Affinity

The demonstration that TSH-agarose conjugates retain the biological activity of the native hormone (Tate et al., 1975; Rickards et al., 1981) provided the basis for ligand-specific affinity-purification of the TSH receptor. Although affinity-purified TSH receptors have been used in crosslinking, immunoprecipitation, and monoclonal antibody studies, very few of these have estimated either the specific binding capacity of the receptor or its degree of purification.

2.2.1. Ligand

All of the available forms of TSH have been coupled to activated agarose and used to affinity-purify TSH receptors. Preparations of partially purified bovine TSH (bTSH) are available from Armour Pharmaceutical Company, Chicago, IL, as Thytropar, 1–3 IU/mg, Sigma Chemical Company, St. Louis, MO, 0.5–1.0 IU/mg, and the NIH, Bethesda, MD, as NIH-TSH-09. Traditionally, highly-purified bTSH (40 IU/mg) was provided by J. G. Pierce (University of California, Los Angeles, CA), although more recently others have made highly-purified bTSH either directly from bovine pituitaries (60–70 IU/mg, Parkes et al., 1985) or from partially-purified bTSH (Sigma Chemical Company) (19.7 IU/mg, Iida et al., 1987). Highly-purified human TSH (hTSH) can be obtained from the National Pituitary Agency (40 IU/mg, Baltimore, MD) or from M. R. Sairam (5 IU/mg, Montreal, Quebec). Sigma Chemical Company manufactures purified hTSH (5–8 IU/mg), although this remains an expensive option.

2.2.2. *TSH-CNBr-Activated-Sepharose*

Receptors from human, porcine, and bovine thyroid, guinea pig fat cells and FRTL-5 cells have been purified on TSH-Sepharose conjugates (Table 1). Bovine TSH can be easily coupled to CNBr-activated-Sepharose 4B (Pharmacia Fine Chemicals, Sweden) (Tate et al., 1975; Koizumi et al., 1982; Iida et al., 1987). The gel (~1 g) is sequentially washed in 1 m*M* HCl, in excess distilled water and, finally, in 0.1*M* $NaHCO_3$, 0.5–1.0*M* NaCl, pH 8.5. TSH (for example, 60 IU bTSH) is dissolved in 0.1*M* $NaHCO_3$ buffer, pH 8.5, added to the gel and mixed at 4°C overnight. Remaining active groups may be blocked with 1*M* ethanolamine, pH 9.0, for 2 h. The gel is then washed with 0.1*N* sodium acetate buffer, pH 4.0, and 0.1*M* $NaHCO_3$, 1.0*M* NaCl, pH 8.5, and finally resuspended in Tris-NaCl, 0.1% Triton X-100. The fractional conjugation of unlabeled bTSH to the gel (estimated from the binding of added tracer ^{125}I-bTSH) varies from 40% (Iida et al., 1983) to 98% (Iida et al., 1987). Receptors have been eluted from bTSH-Sepharose with either low pH (0.1*N* glycine-HCl, pH 3.0; 0.2*M* acetate, pH 2.5) or high salt (2.0–3.0*M* NaCl) (Table 1). Neither appears to offer any special advantage, although marginally higher purification has been reported with high salt elution followed by dialysis against Tris-NaCl buffer (Table 1). The affinity columns can be reused several times over a 6-mo period.

The highest purification of TSH binding sites with TSH-Sepharose chromatography alone is still relatively low (296-fold, Iida et al., 1987) (Table 1), compared to results obtained with similarly constructed affinity columns for other receptors (e.g., 2457-fold for the insulin receptor, Fujita-Yamaguchi et al., 1983). Chan et al. (1987) claim a 12,000-fold purification of TSH binding sites in FRTL-5 cells but this was achieved with a combination of bTSH-Sepharose and TSH receptor monoclonal antibody-Sepharose affinity chromatography (Table 1). In general, the recovery of ^{125}I-TSH binding sites from TSH-Sepharose is usually <10%, although Iida et al. (1987) recovered 23% of the original binding activity present in guinea pig fat cell membranes. This may be the result of the 40% glycerol that was included in their purification procedure in an attempt to improve receptor stability and reduce time and temperature-dependent receptor degradation. Glycerol may also significantly improve recovery of receptor from solubilized thyroid membranes (*see* Section 3.4.).

Recently, insulin coupled to CDI-agarose was used to obtain highly-purified functional insulin receptors (Newman and Harrison, 1985). We therefore adopted a similar approach, employing a simple method for coupling bTSH to CDI-agarose in an attempt to improve the yield and purification of the TSH receptor.

Table 1
Comparison of Purification Techniques for the TSH Receptor

	Elution buffer	^{125}I-bTSH bound cpm/μg	TSH binding sites: Yield, %	TSH binding sites: Purification factor[a]	Subunit structure	Reference
Human thyroid						
hpbTSH-Sepharose	2*M* NaCl	N/D	N/D	25	N/D	Koizumi et al., 1982
hpbTSH-Sepharose	low pH[b]	3168	10	134	150 kDa	Iida et al., 1983
ppbTSH-CDI-agarose	3*M* NaCl[c]	18,950	1	1263	70, 50,[c] 35 kDa 25 kDa + DTT	Leedman et al., 1989
Porcine thyroid						
ppbTSH-Sepharose	2*M* NaCl	N/D	N/D	50	N/D	Rickards et al., 1981
ppbTSH-Sepharose	2*M* NaCl	N/D	N/D	100	50, 25 kDa + DTT	Parkes et al., 1985
ppbTSH-CDI-agarose	3*M* NaCl[c]	929	1.5	794	65, 50,[c] 35 kDa 25, 20 kDa + DTT	Leedman and Harrison (unpublished results)
Bovine thyroid						
hpbTSH-Sepharose	2*M* NaCl	5000	6	65	330, 66, 38 kDa	Koizumi et al., 1982
hpbTSH-Sepharose	low pH[d]	N/D	6.7	17	286, 160, 75 kDa 24 kDa + trypsin	Tate et al., 1975
hpbTSH-Affi-Gel 10	2*M* NaCl	5172	3	808	115,54 kDa + DTT	Kress and Spiro, 1986
Guinea pig fat cells						
hpbTSH-Sepharose	3*M* NaCl	25,620	23	296	50 kDa ± DTT	Iida et al., 1987
FRTL-5 Cells						
bTSH-Sepharose MAb-Sepharose	N/D	N/D	N/D	12,000	280, 70 kDa 50, 25 kDa + DTT	Chan et al., 1987

[a]Compared to Triton-solubilized membranes. [b]0.1*M* glycine-HCl, pH 3.0. [c]Elution with (i) 3*M* NaCl releases M_r 50,000 subunit only (ii) SDS-sample buffer releases all subunits.
[d]0.2*M* acetate, pH 2.5.
hpbTSH = highly-purified bovine thyrotropin.
ppbTSH = partially-purified bovine thyrotropin.
CDI-agarose = 1,1'carbonyldiimidazole-activated agarose.
MAb = monoclonal antibody.
N/D = not determined.
±DTT = in the presence or absence of reducing agent.
+trypsin = in the presence of trypsin.
kDa = kilodalton.

2.2.3. TSH-CDI-Activated-Agarose

TSH can be readily coupled to CDI-agarose. Coupling is performed by mixing the washed gel with bTSH (6–20 IU/mL gel) in 0.1*M* Na borate, pH 8.5, for 30 h at 4°C on a rotator. Shorter coupling times can be used (~18 h) but unreactive groups should then be blocked by mixing for 3 h at room temperature with 1*M* ethanolamine, pH 9.0. The amount of TSH bound, usually 65–80%, can be measured by adding a tracer amount of ^{125}I-labeled TSH. The columns are washed with 20 vol 0.05*M* NaCl, 0.05*M* Tris-HCl, pH 8.0, 20 vol 0.05*M* Na acetate, pH 4.0, 0.5*M* NaCl, and finally with Tris-NaCl, 0.1% Triton X-100, 0.1 m*M* PMSF, and stored at 4°C.

Fractions eluted from wheat germ lectin-CDI-agarose columns and containing ^{125}I-TSH binding activity are routinely batch adsorbed overnight at 4°C to 1 mL lots of TSH-CDI-agarose. Columns are poured, washed with Tris-NaCl, 0.1% Triton X-100, 0.1 m*M* PMSF, and eluted with 3 mL of 3*M* NaCl, 10 m*M* Tris-HCl, pH 7.5, 0.1% Triton X-100. The eluate is dialyzed and concentrated against Tris-NaCl, 2.0*M* sucrose, 0.1% Triton X-100, 0.1 m*M* PMSF, and, finally, against buffer containing 0.75*M* sucrose. This routine scheme of purification is outlined in Fig. 1. A 1 mL TSH-CDI-agarose column is capable of extracting >90% of receptors present in 2 mg of wheat germ lectin purified glycoproteins. In general, ~1–3% of the binding activity present in solubilized membranes is recovered from TSH-CDI-agarose.

We have purified the human TSH receptor ~1270-fold and porcine TSH receptor ~800-fold (compared to Triton solubilized membranes) with sequential wheat germ lectin CDI-agarose and TSH-CDI-agarose chromatography (Table 1). Although the purification of TSH binding sites is high relative to other reports (Table 1), the recovery of binding activity is lower than might be expected. We believe that the low recovery of binding activity is probably owing to a combination of factors that are difficult to separate:

1. Dissociation of the receptor subunits with recovery of only the M_r 50,000 subunit after elution with 3*M* NaCl;
2. Inactivation or degradation of TSH binding sites during the 2–3 d of purification at 4°C, as observed by others (Tate et al., 1975; Dawes et al., 1978; Koizumi et al., 1982; Iida et al., 1983, 1987); and
3. Nonspecific absorption of TSH binding sites when the protein con-centration decreases after TSH affinity chromatography.

Compared to CNBr-activated-agarose, CDI-activated-agarose offers a number of important advantages. In contrast to CNBr-activated gels, in which hydrolysis causes slow leakage of coupled ligand CDI gels are

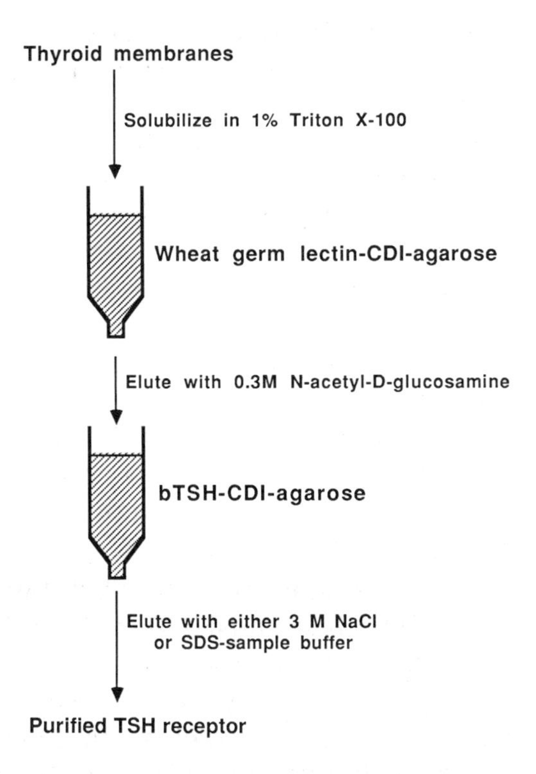

Fig. 1. Schema for TSH receptor purification. CDI = 1,1'-carbonyldiimadazole.

resistant to ligand leakage because of their more stable uncharged *N*-alkyl-carbamate linkage (Bethell et al., 1981). CDI gels also have a higher concentration of activated sites compared to CNBr (Bethell et al., 1981). In addition, CDI gels are easier and faster to prepare than standard CNBr gels. Human and porcine TSH receptors have been purified from four different batches of CDI-agarose and reproducible results obtained for ^{125}I-TSH binding activity and subunit structure.

Analysis by sodium dodecyl sulfate polyacrylamide gel electrophoresis (SDS-PAGE) of nonreduced affinity-purified receptors eluted in SDS-sample buffer from TSH-CDI-agarose demonstates that the receptor is comprised of three noncovalently linked proteins of M_r 70,000, 50,000, and 35,000. When reduced (100 m*M* dithiothreitol, DTT), the major species has a M_r of 25,000 (Leedman et al., 1989). When 3*M* NaCl is used to elute affinity-purified receptors, only the M_r 50,000 nonreduced protein is detected. This M_r 50,000 subunit binds ^{125}I-TSH and is precipitated by Graves' Igs. If, after elution with 3*M* NaCl, the column is eluted with SDS-sample buffer proteins of M_r 70,000 and 35,000 are identified. Modifications to the conventional Western blotting technique enabled putative TSH receptor subunits of M_r 70,000 and 50,000 to be identified in thyroid particulate membranes by Graves' Igs (Fig. 2). Blotting of TSH affinity-purified receptors eluted in SDS-sample buffer revealed either the M_r 70,000 or M_r 50,000 subunit with a minority of Graves' sera. These data lead us to conclude that the nonreduced human TSH receptor is an oligomeric complex composed of three subunits, the two larger of which contain epitopes that bind Graves' Igs by modified Western blotting (Leedman et al., 1989). The reduced receptor appears to consist of only one subunit (M_r 25,000). The stoichiometry of the disulfide-linked subunits that comprise the recep-tor oligomer is yet to be defined.

One might expect that the general use of TSH-agarose chromatography as a purification procedure would produce comparable results for the subunit structure of the TSH receptor. Although a consensus is emerging for the structures derived by affinity-purification (Table 1), there are still significant differences. Presumably, these reflect differences in the sources of receptor, amount and type of degradation of a receptor that is intrinsically labile and the preparative procedures employed prior to TSH-agarose chromatography. Nevertheless, several authors have isolated a protein of M_r ~50,000 from human (Remy et al., 1987; Leedman et al., 1989), porcine (Parkes et al., 1985; Leedman and Harrison, unpublished observation), and bovine (Kress and Spiro, 1986) thyroid membranes, from guinea pig fat cells (Iida et al., 1987) and FRTL-5 cells (Chan et al., 1987). Others have reported the presence of receptor subunits with M_r of ~70,000 and ~35,000 (Tate et al., 1975; Koizumi et al., 1982; Heyma and Harrison, 1984; Chan et al., 1987; Gennick et al., 1987). Tate et al. (1975) identified a M_r 75,000 component of the bovine TSH receptor; upon trypsinization of their solubilized receptor preparation the TSH binding sites were converted into a M_r ~25,000 protein. Affinity-purified proteins of M_r 66,000 and 38,000 were identified by Koizumi et al. (1982) in bovine thyroid. A M_r 70,000 subunit has been identified by us previously (Heyma and Harrison, 1984). TSH receptor subunits of M_r 70,000 and 50,000 have also been identified in FRTL-5 cells (Chan et al., 1987).

Recently we have bulk-purified the porcine thyroid TSH receptor by sequential lectin- and TSH-agarose affinity chromatography. In order to minimize receptor loss we maintain the receptor at 4°C for as short a time as possible. Receptors are solubilized, purified on lectin-agarose, and applied to TSH-agarose within 10 h. After an overnight incubation at 4°C the receptor is eluted in SDS-sample buffer and separated from the SDS on a 1 mL hydroxylapatite column. The reduced and alkylated receptor is finally recovered by electroelution from an SDS-PAGE gel as described by

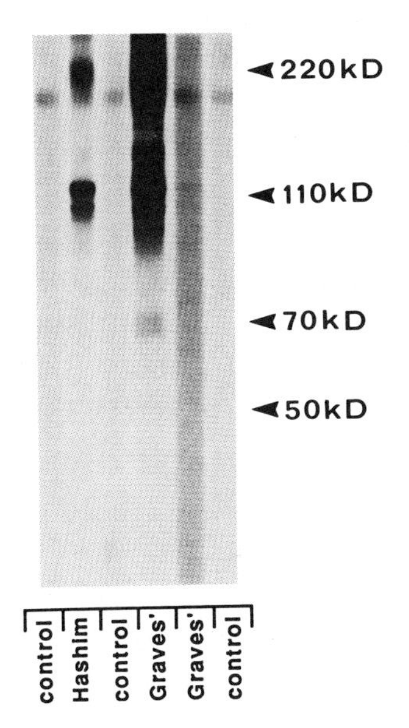

Fig. 2. Modified Western blotting of particulate human thyroid membranes with sera from three control, one Hashimoto's, and two Graves' disease subjects. Membranes were electrophoresed in a SDS-7.5% polyacrylamide gel, the gel was washed 4 × 15 min in 20 m*M* Tris-HCl, pH 8.0, 20% glycerol, and the proteins transferred electrophoretically, in the absence of SDS, to a nylon membrane. The nylon membrane was incubated with 5% skim milk to block nonspecific bind- ing sites and then with serum (1:100), washedwith 5% skim milk, incubated with ^{125}I-protein A, washed again, and visualized by autoradiography. Proteins at 70 and 50 kDa are identical in size to those purified by TSH affinity chromatography; the doublet at 110 kDa corresponds to the microsomal antigen (thyroid peroxidase); the protein at 220 kDa corresponds to a thyroglobulin breakdown product.

Stearne et al. (1985). Using this approach, ~2 µg of purified receptor for microsequencing can be recovered from ~250 g of porcine thyroid.

To avoid the receptor loss encountered during hydroxylapatite chromatography and electroelution from SDS-PAGE gels, we have immunoblotted the reduced receptor onto a nonshrinking, hydrophobic membrane, polyvinylidenedifluoride (PVDF). The immunoblotted proteins can be rapidly stained with Coomassie Blue (Gultekin and Heermann, 1988) and the stained PVDF membrane placed directly into an amino acid sequencer (e.g., Applied Biosystems Sequencer) for *N*-terminal sequencing (Simpson et al., 1989). The advantages of PVDF membrane over other immunoblotting membranes, such as derivatized glass fiber paper, include the ability to directly sequence the Coomassie Blue-stained protein on the membrane and a lower background observed in the sequence analysis (Xu and Shively, 1988).

2.2.4. *TSH-Affi-Gel*

Affi-Gel is an alternative to CNBr-Sepharose and CDI-agarose, but it has not found much application for TSH affinity chromatography. Affi-Gel 10 (Bio-Rad, Richmond, CA) is a *N*-hydroxysuccinimide ester of a derivatized crosslinked agarose gel bead support that efficiently couples neutral to basic proteins with isoelectric points from 6.5–11 (isoelectric point of TSH ~9). It contains a 10-atom spacer arm to prevent steric hindrance. Glycoprotein hormones other than TSH have been successfully coupled to Affi-Gel 10. Dufau et al. (1975) found Affi-Gel 10 superior to CNBr-Sepharose in terms of recovery of functional gonadotropin receptors.

Kress and Spiro (1986) provide the only report of the use of Affi-Gel 10 for the purification of the TSH receptor (Table 1). Highly-purified bTSH was coupled to Affi-Gel 10 by mixing ligand and gel in a small volume of 0.3 *M* HEPES buffer, pH 7.5, for 18 h at 2°C. Remaining active groups were blocked by the addition of 0.1*M* ethanolamine, pH 8.0 for 1 h. A coupling efficiency of ~40% was achieved. The column was washed with 0.3*M* HEPES, pH 7.5, buffer, 2*M* NaCl in HEPES, 50 m*M* NaCl, 0.02% Triton, 0.01 m*M* $CaCl_2$ (HEPES-NaCl-Triton-$CaCl_2$ buffer), and then with HEPES-NaCl-Triton-$CaCl_2$ buffer. Receptors were cycled twice over the column, washed with HEPES-NaCl-Triton-$CaCl_2$ buffer, and eluted with 2*M* NaCl in HEPES buffer. Using sequential TSH-Affi-Gel 10 and lectin-agarose chromatography Kress and Spiro (1986) achieved ~800-fold purification of bovine TSH receptors, with a final yield of TSH binding sites of 3%.

2.3. *Antibody Affinity*

Monoclonal antibodies (MAb) directed against different determinants of the TSH receptor have been produced by Kohn's laboratory, mostly by generating hybridomas from fusions with lymphocytes of patients with Graves' disease (Yavin et al., 1981; Valente et al., 1982; Ealey et al., 1984, 1985; Marshall et al., 1987). TSH receptor MAb have also been produced by hybridization of lymphocytes from a patient with acute-onset type I diabetes mellitus (Tan et al., 1987).

Only one group has utilized these MAb for the purification of the TSH receptor. Chan et al. (1987) claim a 12,000-fold purification of solubilized FRTL-5 cell and bovine TSH receptors after sequential TSH-agarose and antibody-agarose chromatography. Using the MAb 11E8, a thyroid binding inhibitory antibody (Ealey et al., 1985), for the immunopurification step, two TSH receptor proteins of M_r 300,000 and 70,000 were identified.

Unfortunately, no information was provided on either the type of agarose used, the elution buffer or the optimal coupling conditions.

2.4. Other

A number of other chromatographic techniques have been utilized to purify the TSH receptor but none has provided the same specificity and degree of purification as TSH-agarose chromatography. These are briefly discussed below.

2.4.1. Gel Filtration

Several workers have partially purified TSH receptors by gel filtration. Sepharose 6B gel chromatography has been used to characterize the human (Dawes et al., 1978, M_r ~300,000), bovine (Koizumi et al., 1982, M_r ~280,000), and porcine (Dawes et al., 1978, M_r ~300,000) receptors. Separation of high and low affinity binding components of porcine receptors can be achieved with Sepharose 6B chromatography (McQuade et al., 1987). Gel filtration (Sephacryl S-3000) can also be used to partially purify crosslinked ^{125}I-TSH-receptor complexes, which are then more clearly defined by gel electrophoresis (Kajita et al., 1985a). Czarnocka et al. (1979) used sequential gel filtration on Sephadex LH-20 and Ultrogel AcA-44 to identify two bovine TSH binding proteins of M_r 130,000 and 30,000. Interestingly, removal of the detergent from the solubilized membranes by gel filtration on Sephadex LH-20 significantly increased the binding of TSH above that of crude membranes. In general, only miminal purification has been achieved with gel filtration (for example two-fold, Czarnocka et al., 1979) and a poor correlation has been noted between the M_r calculated from SDS-PAGE and the K_{av} of each protein (Koizumi et al., 1982). The apparent discrepancy between the values obtained by gel filtration and SDS-PAGE is probably a result of the association of receptor with Triton micelles during gel filtration (Helenius and Simons, 1975).

2.4.2. Ion-Exchange

Czarnocka et al. (1981) purified bTSH receptors tenfold with DEAE-cellulose. Solubilized receptors (obtained after 1% butanol extraction of plasma membranes) were eluted from DEAE-cellulose with a gradient of ammonium acetate (0.1–0.5*M*) and further purified by chromatography on AcA-54 Ultrogel. Iida et al. (1983) utilized DEAE-Sephadex as a first-step to purify receptors 5–18-fold prior to lectin-agarose and TSH-agarose chromatography. Solubilized human receptors bind to DEAE-Sephadex in 0.05*M* Na acetate, pH 6.3, 0.2% Triton X-100 buffer and can be eluted with a linear gradient from 0.1–1.0*M* ammonium acetate, pH 6.3, 0.2% Triton X-

100. The adsorption and elution profile of TSH receptors with DEAE-Sephadex is similar to that of insulin receptors (Cuatrecasas, 1972).

2.4.3. Sucrose Density Gradient Ultracentrifugation

Several workers have utilized sucrose gradient density centrifugation to characterize the receptor. Amir et al. (1976) found that bovine thyroid membrane proteins extracted from the 30%/40% interface of a discontinuous sucrose gradient were enriched 13-fold in TSH binding. Others have used the technique to separate high and low affinity binding sites in solubilized guinea pig fat cell membranes (Gennick et al., 1986) and to estimate the M_r (~50,000) of a water soluble component of the TSH receptor released from human thyroid membranes treated with DTT (Davies Jones et al., 1985a).

3. Considerations for Optimal Recovery of Functional Receptors

In contrast to the insulin receptor which remains relatively stable during purification (Finn et al., 1982; Newman and Harrison, 1985), the TSH receptor is vulnerable to changes in pH, ionic strength, and temperature, and appears to undergo rapid denaturation or loss of functional integrity. Measures used to prevent receptor loss are outlined below.

3.1. Time/Temperature

The solubilized receptor is extremely heat-labile and several studies have demonstrated its rapid loss of binding activity. There was total loss of binding activity in 5 min at 55°C (Dawes et al., 1978; Koizumi et al., 1982) and 50% loss in 40 min at 37°C (Dawes et al., 1978). During affinity purification, the solubilized receptor is maintained at 4°C for at least 24 h. Inactivation is less rapid at 4°C but, nevertheless, remains an important problem. Iida et al. (1987) demonstrated that 75% of the TSH binding activity present in solubilized guinea pig fat cell membranes was lost in the first 24 h at 4°C and that the loss of binding activity increased with continued exposure at 4°C. Scatchard analysis revealed that reduction in TSH binding was a function of reduced receptor number, without change in receptor affinity. Koizumi et al. (1982) noted that 50% of the TSH binding activity present in solubilized membranes was lost when they were kept at 4°C for 35 h and, in addition, that human receptors were more labile than bovine.

However, solubilized and purified receptors do retain their binding activity for prolonged periods, from 7 d (Iida et al., 1987) to 3 mo (Koizumi et al., 1982), when stored at –70°C.

3.2. pH/Ionic Strength

Optimizing the conditions of binding should maximize receptor extraction by TSH-agarose columns. The binding of solubilized human (Pekonen and Weintraub, 1978; Dawes et al., 1978; Koizumi et al., 1982; Iida et al., 1987), porcine (Dawes et al., 1978), bovine (Koizumi et al., 1982), and guinea pig fat cell (Iida et al., 1987) TSH receptors is optimal at pH 7.4. Nonspecific binding increases at lower pH values.

Binding is also dependent on the nature of the buffers. Increasing the ionic strength of the medium is associated with a decrease in specific binding. A 2.5-fold reduction in binding by solubilized receptors occurs when the NaCl concentration increases from 50–125 m*M* at pH 7.4 (Dawes et al., 1978). In agreement, Iida et al. (1987) documented a significant decrease in the binding capacity of solubilized human receptors, without a change in affinity, when the NaCl concentration increased from 50–150 m*M*.

Magnesium chloride and calcium chloride also inhibit TSH binding by up to 50% at concentrations of 0.2 and 0.4 m*M*, respectively, whereas potassium chloride and potassium iodide inhibit binding by 50% at 20 m*M* (Pekonen and Weintraub, 1978).

The effect of salt on the binding capacity of TSH receptors has important implications for TSH-agarose affinity purification, but the reduction in binding capacity induced by high ionic strength of the elution buffer (2–3*M* NaCl) is potentially reversible. This is evident because even after 24 h exposure to 2–3*M* NaCl, dialysis against Tris-NaCl, pH 7.5, restores TSH binding activity (Koizumi et al., 1982; Iida et al., 1987; Leedman et al., 1989).

3.3. Protease Inhibitors/Sulphydryl Modifiers

Despite the widespread use of protease inhibitors to reduce receptor denaturation owing to proteolysis there is still little evidence that their addition preserves TSH binding activity. Koizumi et al. (1982) could not demonstrate that the addition of either aprotinin (0–250 U/mL) or PMSF (25 µg/mL) prevented loss of binding activity of solubilized bovine and human receptors. More recently, Iida et al. (1987) documented a slight but significant preservation of TSH binding with a combination of PMSF (1 m*M*), aprotinin (500 IU/mL), and leupeptin (10 µg/mL). Despite the lack of concrete data concerning their benefit, we routinely include a cocktail of protease inhibitors comprising 2 m*M* PMSF, 1000 U/mL aprotinin, 100 U/mL bacitracin, and 20 µg/mL leupeptin during purification.

The TSH receptor contains sulphydryl groups and can be easily reduced with either DTT (Leedman et al., 1989) or 2-mercaptoethanol (Kajita et al., 1985a). Davies Jones et al. (1985a) identified a water-soluble fragment of the human thyroid TSH receptor (M_r ~50,000) that was released by concentrations of DTT up to 10 m*M*. Interestingly, they found that even 5 m*M* DTT inactivated the receptor and inhibited binding. Furthermore, they postulated the presence of two types of disulfide bridges within the receptor, one sensitive to DTT, the other to cysteine. We have observed that increasing concentrations of DTT can produce differential reduction of the receptor. With 10 m*M* DTT the higher M_r receptor components are incompletely reduced, but with 100 m*M* DTT all of the receptor components are converted into major subunits of M_r 20-25,000 (Leedman and Harrison, unpublished observations).

These results predict that blockage of free sulphydryl groups and disulphide interchange reactions during purification should help maintain the integrity of the TSH receptor. However, we found that 5 m*M* *N*-ethylmaleimide, added to human thyroid membranes, was without effect (Leedman and Harrison, unpublished observation). Iida et al. (1987) found no effect of 1 m*M* cysteine, added presumably to stabilize sulfydryl groups. Remy et al. (1987) routinely add 10 m*M* iodoacetamide during solubilization, but they do not provide results of purification without iodoacetamide. Further effort should be directed toward understanding the relationship between the thiol redox state and the functional integrity of the receptor.

3.4. Glycerol

Glycerol has been shown to preserve the binding activity of follicle stimulating hormone (FSH) (Dias et al., 1981) and luteinizing hormone (LH) (Ascoli, 1983) receptors. Iida et al. (1987) demonstrated that the addition of glycerol (1–50%) preserved, in a concentration-dependent manner, the binding of ^{125}I-TSH to solubilized guinea pig fat cell receptors that were kept at 4°C for 6 d.

Glycerol appears to act by preferentially hydrating and increasing the hydrophobicity of proteins (Gekko and Timasheff, 1981), thereby preserving their three-dimensional structure and reducing receptor denaturation. Its effects on human thyroid membrane receptors clearly deserve further attention.

4. Future Perspectives

There is a gradually evolving consensus as to the structure of the TSH receptor derived by affinity purification. However, attempts to purify the receptor to homogeneity in reasonable yield are frustrated by receptor denaturation and loss during purification. Modifications to the techniques described above, as well as the availability of cell lines that hyperexpress TSH receptors, will facilitate the purification of sufficient homogeneous receptor for microsequencing in order to synthesize oligonucleotide probes with which to identify clones of the receptor gene in λgt11 cDNA libraries.

Although several groups are at present attempting to clone the receptor by antibody screening of thyroid λgt11 cDNA expression libraries, the difficulty with identifying the receptor by immunoblotting, presumably owing to the strict conformational requirement for antibody binding, mitigates against this method being successful. We have recently used a pool of Graves' Igs, selected on the basis of their ability to detect the affinity-purified receptor with our modified technique of Western blotting, to screen a λgt11 thyroid carcinoma cDNA library. We were unable to identify any positive clones after screening 1.2×10^6 colonies. More recently, a putative "70 kD TSH receptor protein" was cloned by screening a human thyroid carcinoma λgt 11 cDNA expression library with Graves' Igs (Chan et al., 1989). Examination of the amino acid sequence of this kDa protein revealed it to have almost absolute (>99%) sequence identity (Leedman and Harrison, unpublished observation) with another recently cloned autoantigen, the ubiquitous p70 (Ku) nuclear lupus autoantigen (Reeves and Sthoeger, 1989). This unexpected finding of sequence identity serves to illustrate that a variety of autoantibodies may be present in patients with Graves' disease, not all of which are directed at thyroid autoantigens.

Others have used a strategy based on the observation that cyclic AMP is the major intracellular second messenger of TSH, the implication being that the receptor is coupled to a G protein and is likely to share significant amino acid homology with other members of the G protein family of receptors. Libert et al. (1989) utilized, therefore, polymerase chain reaction-amplified DNA fragments derived from homologous regions of other G protein-coupled receptors to screen thyroid cDNA libraries. Although a number of novel G protein-coupled molecules were identified, the TSH receptor remained elusive. The rat (McFarland et al., 1989) and porcine (Loosfelt et al., 1989) lutropin-choriogonadotrophin (LH-hCG) receptors have recently been cloned. In view of the high sequence and structural homology that exists between the glycoprotein hormones TSH, LH-hCG, and follicle-stimulating hormone (FSH), the sequence of the LH-hCG receptor may prove a useful tool for isolating a TSH receptor clone from thyroid cDNA libraries.

Cloning of the TSH receptor will help unravel the uncertainty concerning the structure of this pivotal thyroid membrane protein, as well as clarify our understanding of the pathogenesis of autoimmune thyroid disease, and, most important, provide a basis for the development of specific immunotherapy for the treatment of TSH receptor antibody-mediated diseases.

Acknowledgments

The authors are grateful to Joyce Lygnos for secretarial assistance. Peter J. Leedman is a National Health and Medical Research Council Postgraduate Medical Research scholar.

References

Adair, W. L. and Kornfeld, S. (1974) *J. Cell. Biochem.* **249,** 4696–4704.

Amir, S. M., Goldfine, I. D., and Ingbar, S. H. (1976) *J. Biol. Chem.* **251,** 4693–4699.

Ascoli, M. (1983) *Endocrinology* **113,** 2129–2134.

Bartalena, L., Fenzi, G., Vitti, P., Tombaccini, D., Antonelli, A., Macchia, E., Chiovato, L., Kohn, L. D., and Pinchera, A. (1987) *Biochem. Biophys. Res. Commun.* **143,** 266–272.

Bethell, G. S., Ayers, J. A., Hearn, M. T. W., and Hancock, W. S. (1981) *J. Chromatog.* **219,** 361–372.

Buckland, P. R., Rickards, C. R., Howells, R. D., Davies Jones, E., and Rees Smith, B. (1982) *FEBS Lett.* **145,** 245–249.

Buckland, P. R. and Rees Smith, B. (1984) *FEBS Lett.* **166,** 109–114.

Buckland, P. R., Strickland, T. W., Pierce, J. G., and Rees Smith, B. (1985) *Endocrinology* **116,** 2122–2124.

Buckland, P. R., Rickards, C. R., Howells, R. D., and Rees Smith, B. (1986) *Biochem. J.* **235,** 879–882.

Chan, J., Santisteban, P., De Luca, M., Isozaki, O., Grollman, E., and Kohn, L. (1987) *Acta Endocrinol. (Suppl.) (Copenh.)* **281,** 166–172.

Chan, J. Y. C., Lerman, M. I., Prabhakar, B. S., Isozaki, O., Santisteban, P., Kuppers, R. C., Oates, E. L., Notkins, A. L., and Kohn, L. D. (1989) *J. Biol. Chem.* **264,** 3651–3654.

Cuatrecasas, P. (1972)*Proc. Natl. Acad. Sci. USA* **69,** 1277–1281.

Czarnocka, B., Nauman, J., Adler, G., and Kielczynski, W. (1979) *Acta Endocrinol. (Copenh.)* **92,** 512–521.

Czarnocka, B., Gardas, A., and Nauman, J. (1981) *Acta Endocrinol. (Copenh.)* **96,** 335–341.

Davies Jones, E. and Rees Smith, B. (1984) *J. Endocr.* **100,** 113–118.

Davies Jones, E., Hashim, F. A., Kajita, Y., Creagh, F. M., Buckland, P. R., Petersen, V. B., Howells, R. D., and Rees Smith, B. (1985a) *Biochem. J.* **228,** 111–117.

Davies Jones, E., Hashim, F. A., Creagh, F. M., Williams, S. E., and Rees Smith, B. (1985b) *Mol. Cell. Endocrinol.* **41,** 257–261.

Dawes, P. J. D., Petersen, V. B., Rees Smith, B., and Hall, R. (1978) *J. Endocr.* **78,** 89–102.

Dias, J. A., Huston, J. S., and Reichert, L. E. (1981) *Endocrinology* **109,** 736–742.

Drummond, R. W., McQuade, R., Grunwald, R., Thomas, C. G., Jr., and Nayfeh, S. N. (1982) *Proc. Natl. Acad. Sci. USA* **79,** 2202–2206.

Dufau, M. L., Ryan, D. W., Baukal, A. J., Catt, K. J. (1975) *J. Biol. Chem.* **250,** 4822–4824.

Ealey, P. A., Kohn, L. D., Ekins, R. P., and Marshall, N. J. (1984) *J. Clin. Endocrinol. Metab.* **58,** 909–914.

Ealey, P. A., Valentine, W. A., Ekins, R. P., Kohn, L. D., and Marshall, N. J. (1985) *Endocrinology* **116,** 124–131.

Field, J. B. (1986) in *Werner's The Thyroid. A Fundamental and Clinical Text* (Ingbar, S. H. and Braverman, L. E., eds.), J. B. Lippincott, Philadelphia.
Finn, F. M., Titus, G., Nemoto, H., Noji, T., and Hofmann, K. (1982) *Metabolism* **31,** 691–698.
Fujita-Yamagichi, Y., Choi, S., Sakamoto, Y., and Hakura, K. (1983) *J. Biol. Chem.* **258,** 5045–5049.
Furmaniak, J., Davies Jones, E., Buckland, P. R., Howells, R. D., and Rees Smith, B. (1986) *Mol. Cell. Endocrinol.* **48,** 31–38.
Gardas, A. and Nauman, J. (1981) *Acta Endocrinol. (Copenh.)* **98,** 549–555.
Gekko, K. and Timasheff, S. N. (1981) *Biochemistry* **20,** 4667–4676.
Gennick, S. E., Thomas, C. G., Jr., and Nayfeh, S. N. (1986) *Biochem. Biophys. Res. Commun.* **135,** 208–214.
Gennick, S. E., Thomas, C. G., Jr., and Nayfeh, S. N. (1987) *Endocrinology* **121,** 2119–2130.
Gultekin, H. and Heermann, K. H. (1988) *Anal. Biochem.* **172,** 320–329.
Hedo, J. A. (1984) in *Receptor Biochemistry and Methodology,* vol 2 (Venter, J. C. and Harrison, L. C., eds.), Alan R. Liss, New York.
Helenius, A. and Simons, K. (1975) *Biochem. Biophys. Acta* **415,** 27–79.
Heyma, P. and Harrison, L. C. (1984) *J. Clin. Invest.* **74,** 1090–1097.
Iida, Y., Konishi, J., Kasagi, K., Ikekubo, K., Kuma, K., and Torizuka, K. (1981) *Acta Endocrinol.* **98,** 50–56.
Iida, Y., Konishi, J., Kasagi, K., Endo, K. Misaki, T., Kuma, K., and Torizuka, K. (1983) *Acta Endocrinol. (Copenh.)* **103,** 198–204.
Iida, Y., Amir, S. M. and Ingbar, S. H. (1987) *Endocrinology* **121,** 1627–1636.
Kajita, Y., Rickards, C. R., Buckland, P. R., Howells, R. D., and Rees Smith, B. (1985a) *Biochem. J.* **227,** 413–420.
Kajita, Y., Rickards, C. R., Buckland, P. R., Howells, R. D., and Rees Smith, B. (1985b) *FEBS Lett.* **181,** 218–222.
Koizumi, Y., Zakarija, M., and McKenzie, J. M. (1982) *Endocrinology* **110,** 1381–1391.
Kotani, M., Karinya, T., and Field, J. B. (1975) *Metabolism* **24,** 959–971.
Kress, B. C. and Spiro, R. G. (1986) *Endocrinology* **118,** 974–979.
Leedman, P. J., Newman, J. D., and Harrison, L. C. (1989) *J. Clin. Endocrinol. Metab.* **69,** 134–141.
Libert, F., Parmentier, M., Lefort, A., Dinsart, C., Van Sande, J., Maenhaut, C., Simons, M.-J., Dumont, J. E., and Vassart, G. (1989) *Science* **244,** 569–572.
Loosfelt, H., Misrahi, M., Atger, M., Salesse, R., Thi, M. T. V. H.-L., Jolivet, A., Guiochon-Mantel, A., Sar, S., Jallal, B., Garnier, J., and Milgrom, E. (1989) *Science* **245,** 525–528.
Lotan, R., Beattie, G., Hubbell, W., and Nicolson, G. L. (1977) *Biochemistry* **16,** 1787–1794.
McFarland, K. C., Sprengel, R., Phillips, H. S., Kohler, M., Rosemblit, N., Nikolics, K., Segaloff, D. L., and Seeburg, P. H. (1989) *Science* **245,** 494–499.
Manley, S. W., Bourke, J. R., and Hawker, R. W. (1974) *J. Endocrinol.* **61,** 437–445.
Marshall, N. J., Kohn, L. D., and Ealey, P. A. (1987) *Acta Endocrinol. (Suppl.) (Copenh.)* **281,** 173–180.
McQuade, R., Thomas, C. G., and Nayfeh, S. N. (1986a) *Arch. Biochem. Biophys.* **246,** 52–56.
McQuade, R., Thomas, C. G., and Nayfeh, S. N. (1986b) *Biochem. Biophys. Res. Commun.* **137,** 61–68.
McQuade, R., Thomas, C. G., Jr., and Nayfeh, S. N. (1987) *Arch. Biochem. Biophys.* **252,** 409–417.
Moore, W. and Wolff, J. (1974) *J. Biol. Chem.* **249,** 6255–6263.
Mullin, B. R., Fishman, P. H., Lee, G., Aloj, S. M., Ledley, F. D., Winand, R. J., Kohn, L. D., and Brady, R. O. (1976) *Proc. Natl. Acad. Sci. USA* **73,** 842–846.

Newman, J. D. and Harrison, L. C. (1985) *Biochem. Biophys. Res. Commun.* **132,** 1059–1065.
Nielsen, T. B., Totsuka, Y., Kempner, E. S., and Field, J. B. (1984) *Biochemistry* **23,** 6009–6016.
Parkes, A. B., Kajita, Y., Buckland, P. R., Howells, R. D., Rickards, C. R., Creagh, F. M., and Rees Smith, B. (1985) *Clin. Endocrinol.* **22,** 511–520.
Pekonen, F. and Weintraub, B. D. (1978) *Endocrinology* **103,** 1668–1677.
Pekonen, F. and Weintraub, B. D. (1979) *Endocrinology* **105,** 354–359.
Rapoport, B., Seto, P., Magnusson, R. P. (1984) *Endocrinology* **115,** 2137–2144.
Rees Smith, B., Pyle, B., and Gwyneth, A. (1977) *J. Endocrinol.* **75,** 391–400.
Rees Smith, B., McLachlan, S. M., and Furmaniak, J. (1988) *Endo. Rev.* **9,** 106–121.
Reeves, W. H. and Sthoeger, Z. M. (1989) *J. Biol. Chem.* **264,** 5047–5052.
Remy, J. J., Salamero, J., and Charreire, J. (1987) *Endocrinology* **121,** 1733–1741.
Rickards, C. R., Buckland, P. R., Rees Smith, B., and Hall, R. (1981) *FEBS Lett.* **127,** 17–21.
Simpson, R. J., Moritz, R. L., Begg, G. S., Rubira, M. R., and Nice, E. C. (1989) *Anal. Biochem.,* 177, 221–236.
Stearne, P. A., van Driel, I. R., Grego, B., Simpson, R. J., and Goding, J. W. (1985) *J. Immunol.* **134,** 443–448.
Tan, K. S., Foster, C. S., De Silva, M., Byfield, P. G. H., Medlen, A. R., Wright, J. M., and Marks, V. (1987) *Metabolism* **36,** 327–334.
Tate, R. L., Holmes, J. M., Kohn, L. D., and Winand, R. J. (1975) *J. Biol. Chem.* **250,** 6527–6533.
Teng, C. S., Rees Smith, B., Anderson, J., and Hall, R. (1975) *Biochem. Biophys. Res. Commun.* **66,** 836–841.
Valente, W. A., Vitti, P., Yavin, Z., Yavin E., Rotella, C. M., Grollman, E. F., Toccafondi, R. S., and Kohn, L. D. (1982) *Proc. Natl. Acad. Sci. USA* **79,** 6680–6684.
Xu, Q.-Y. and Shively, J. E. (1988) *Anal. Biochem.* **170,** 19–30.
Yavin, E., Schneider, M. D., and Kohn, L. D. (1981) *Proc. Natl. Acad. Sci. USA* **78,** 3180–3184.

Prolactin Receptors

Purification, Molecular Weight, and Binding Specificity

Arieh Gertler

1. Introduction

Prolactin (PRL), first discovered by Stricker and Grueter (1928), is a protein hormone secreted from the anterior pituitary gland. It is a highly pleiotropic hormone and appears in all vertebrates. Nicoll and Bern (1972) have listed over 85 different biological activities attributable to PRL. The outstanding of these include lactation and reproduction in mammals and regulation of water and ion fluxes in amphibians and fish. PRL receptors have been identified in various tissues of mammals and in the livers and kidneys of fish and amphibians. Nicoll et al. (1986) recently published an extensive review devoted to the analysis of relations beween prolactins, growth hormones, and their receptors. Other recent reviews (Kelly et al., 1984; Akers, 1985; Nicoll and White, 1985; Forsyth, 1986; Rillema et al., 1988) have focused on the mechanism of PRL action and the regulation and evolution of PRL receptors. Since these subjects are beyond the scope of the present review, we shall attempt to summarize our present knowledge on purification of prolactin receptors with emphasis on their molecular weight and specificity.

2. Determination of Receptor Activity

Most researchers adapted the modification of the Cuatrecasas (1973) method introduced by Shiu and Friesen (1974). Briefly, this method was based on incubation of a solubilized receptor with 0.3–0.5 ng of ^{125}I-oPRL or ^{125}I-human growth hormone (hGH) for 15 h at room temperature in 25 m*M* Tris-HCl buffer pH 7.4 containing 10 m*M* $MgCl_2$ and 0.1% bovine serum albumin. Then, the reaction was halted by addition of one volume of cold 0.1*M* sodium phosphate buffer pH 7.5 containing 0.1% (W/V) rabbit γ-globulin and two volumes of 25% (w/v) polyethylene glycol 6000 (PEG) dissolved in the same phosphate buffer. The precipitate containing the hormone–receptor complex was collected by centrifugation and the nonspecific binding was determined in the presence of 0.5–1.0 μg of unlabeled oPRL. ^{125}I-hGH was preferred over ^{125}I-oPRL as the latter is precipitated nonspecifically in the presence of Triton X-100 concentrations exceeding 0.01%. Determination of the binding in the particulate fraction was performed in a similar manner but without PEG. A different method using a polyethyleneimine treated membrane filter was described recently (El-Hamzawy and Costlow, 1988). The method is very rapid, the filters retain 70–90% of the hormone–receptor complex, and the nonspecific binding is low.

3. Purification Procedure

3.1. Preparation of Microsomal Membranes

Most laboratories have been utilizing the original procedure introduced by Shiu et al. (1973). This consists mainly of homogenization of the respective tissue in 0.25–0.3*M* sucrose solution, removal of 1500 and 15000*g* preciptates and recovery of the microsomal fraction by ultacentrifugation at 100,000*g*. This fraction contained over 75% of the prolactin binding activity found in the homogenate. Minor modifications of this method included addition of 0.025*M* Tris-HCl buffer, pH 7.4 containing 10 m*M* $MgCl_2$ and proteinase inhibitors, such as 1 m*M* $PhCH_2SO_2F$ (Djiane et al., 1985) and 0.1 mg/mL Kunitz soybean trypsin inhibitor, 1 m*M* leupeptin, 2 m*M* Nα-*p*-tosyl-L-lysine chloromethyl ketone, 9000 U/mL kallikrein (Bonifacino and Dufau, 1984), 20 m*M* sodium molybdate (Mitani and Dufau, 1986), or 0.1% sodium azide (Amit et al., 1984). A slightly different procedure was used for preparation of rat ovary membranes. The first precipitate (1000*g*) was discarded and the membranes were collected as a 34,000*g* precipitate (Bonifacino and Dufau, 1984). An alternative procedure for microsomal membrane preparation was developed by fractionation of membranes on sucrose gradients (Borst et al., 1985; N'Guema Ename et al., 1986). Using this method, Berthon et al. (1987b) prepared sow mammary gland light membranes by centrifugation of tissue homogenate (in 0.3*M* sucrose) on a 1.7*M* sucrose cushion (100,000*g*, 18 h, 4°C) and collected the membranes located on the interface. This fraction was resuspended in 0.3*M* sucrose and recovered by centrifugation at 200,000*g* for 90 min. The specific binding of PRL was twofold higher as compared to the classical microsomes and recovery was 65%.

In order to increase the numbers of PRL receptors, some research groups pretreated nonlactating rats with β-estradiol or lactating rabbits with bromocryptine (Katoh et al., 1984, Necessary et al., 1984). Others used superovaluated rat ovaries (Koppelman and Dufau, 1982).

3.2. Solubilization of the Receptors

Most published procedures followed the protocol of Shiu and Friesen (1974), in which membrane preparations suspended in 25 m*M* Tris-HCl buffer pH 7.4, containing 10 m*M* $MgCl_2$ were treated with 1% Triton X-100, for 30–60 min at room temperature. Following removal of the insoluble material by centrifugation at 200,000*g* for 2 h at 4°C, the detergent was removed by dialysis or by batch or column (Gertler et al., 1984) treatment with Bio-Beads SM2 (BioRad). It should be noted that removal of the detergent resulted in aggregation of the PRL receptors isolated from bovine mammary gland, as determined by gel filtration studies (Ashkenazi et al., 1987a). Haeuptle et al. (1983) compared the efficiency of several detergents, such as Triton X-100, sodium deoxycholate, β-octylglucoside, CHAPS, or sodium dodecyl sulfate, used at various concentrations for solubilization of PRL receptors from liver and mammary gland of rabbit. Best results were achieved using 0.25–0.5% Triton X-100. In all cases, only 50–80% of the total binding activity was solubilized and the sum of solubilized and nonsolubilized binding sites never exceeded the initial amount found in the membranes.

A two- to fivefold increase in the association constant was observed as a result of solubilization. This finding seems to be a rather general phenomenon and was confirmed by others using membranes from various animals and organs (Shiu and Friesen, 1974; Katoh et al., 1984; Sakai et al., 1985; Ashkenazi et al., 1987a; Berthon et al., 1987a). An alternative procedure was reported by Liscia et al. (1982), who used 0.5% (8 m*M*) solutions of zwitterionic detergent 3-[(3-cholamidopropyl)dimethylammonio]-1-propanesulfonate (CHAPS) or 3-[(3-cholamidopropyl)dimethylammonio]-1-hydroxy-1-propanesulfonate (CHAPSO). The major advantage of these detergents is the lack of absorptivity at 280 n*M* and a low aggregation number compared to Triton X-100; their solubilizing efficiency was, however, slightly lower.

Sakai et al. (1985) and Berthon et al. (1987a) reported modification of this procedure. Microsomal membranes from sow mammary gland were first treated with 1 m*M* CHAPS that solubilized over 30% of proteins but only 10% of receptors and then followed by solubilization with 7.5 m*M* CHAPS. The specific activity of the solubilized receptors was almost three times higher compared to 1% Triton X-100 solubilizate, but the overall yield was less than 50%. A similar zwitterionic detergent Zwittergent 3-12 (*N*-dodecyl-*N*,*N*-dimethyl-3-ammonio-1-propane sulfonate) was used for receptor solubilization from rabbit mammary gland with results similar to those obtained with CHAPS (Church and Ebner, 1982; Necessary et al., 1984).

Water-soluble PRL receptors were identified in the 100,000*g* supernatant of liver and lung microsomes prepared from male rats treated with either PRL or E_2 (Amit et al., 1984). The soluble receptors had the same specificity and affinity as the membrane-bound receptors. Gel chromatography revealed that the soluble receptors from the liver of E_2-treated rats had an M_r of 340,000 and 165,000 kDa. Soluble receptors in female rat livers amounted to 12% of the total binding capacity, whereas the proportion in the E_2-treated male rats increased to 50% (Amit et al., 1984). Soluble PRL receptors were also found in the 200,000*g* supernatant prepared from porcine mammary gland (Berthon et al., 1987b). They constituted 50% of the total binding and were nondistinguishable from the membrane-bound receptors. This was evident from their binding specificity, affinity for lactogenic hormones, and reactivity against three antirabbit mammary gland PRL receptors MAbs. No data concerning the M_r of these receptors were reported.

3.3. Affinity Chromatography Using Immobilized Lectins

Affinity chromatographpy, using either concanavalin A-Sepharose (Mitani and Dufau, 1986) or wheat germ lectin-Sepharose (Dufau and Kusuda, 1987) was performed for further purification of Triton X-100 solubilized PRL receptors from rat ovary. Alpha-methyl D-mannoside or *N*-acetylglucosamine was used, respectively, as the eluting agent. The main advantage of this procedure was the simultaneous purification of human chorionic gonadotropin and PRL receptors. The extent of purification of PRL receptor at this step was 12.8. The K_a of the hormone–receptor complex was unaffected. Concanavalin A-Sepharose also was used for partial purification of PRL receptors from Nb2-1ymphoma cells (Gertler and Elberg, 1988).

3.4. Affinity Chromatography on Immobilized oPRL, hGH, or mAb

The final stage of purification was achieved in most studies by affinity chromatography using the original methodology developed by Shiu and Friesen (1974) or its modification. Most procedures involved application of the solubilized receptor to immobilized oPRL (Liscia and Vonderhaar, 1982; Necessary et al., 1984; Sakai et al., 1985; Berthon et al., 1987a; Katoh et al., 1987) or to immobilized hGH (Shiu and Friesen, 1974; Mitani and Dufau, 1986; Katoh et al., 1984; Dufau and Kusuda, 1987; Ashkenazi et al., 1987a; Dusanter-Fourt et al., 1987). Then, in most cases, the column was washed with the application buffer followed by 4*M* urea and/or 1*M* NaCl. The active fraction was then eluted with 1–2 bed vol of 4–5*M* $MgCl_2$. Inclusion of detergent (0.1% Triton X-100 or 1–10 m*M* CHAPS) was absolutely required to preserve the binding activity (Shiu and Friesen, 1974).

An alternative elution procedure was utilized by Necessary et al. (1984) whereby the receptor was eluted with 0.05*M* ammonium acetate pH 4.2 containing 0.5% Zwittergent 3–12.

Solubilized PRL receptors from sow mammary gland also were partially purified using immobilized mAb rather than lactogenic hormone (Berthon et al., 1987a). The Triton X-100 solubilized fraction was run initially through a precolumn of Affi-gel 10 to which normal mouse γ-globulins were coupled and then through a column of immobilized antiPRL receptor MAb. After washing the column with 1*M* NaCl, 4*M* urea, and 1 m*M* CHAPS (all in Tris buffer pH 7.5), the receptor was eluted with 5*M* $MgCl_2$ in 10 m*M* CHAPS. The authors reported this procedure to be superior to oPRL-affinity chromatography. The specific binding activity and the yield were, respectively, 3.4- and 2.9-fold higher.

A different protocol for purification of PRL receptors from rabbit mammary gland and liver was reported by Haeuptle et al. (1983). These authors prepared a complex of solubilized receptor and biotinilated hGH. Following precipitation with PEG and resolubilization in 1% Triton X-100, the complex was bound to streptavidin-Affi-gel. Then the resin was washed with 0.1% Triton X-100 and 30 m*M* β-octylglucoside and the receptor was eluted subsequently with either 25 m*M* Tris buffer pH 7.6 containing 5*M* $MgCl_2$, 30 m*M* β-octylglucoside, and 0.5% BSA or with 1.0*M* glycine-NaOH buffer pH 10 containing 2% $NaDodSO_4$. No data concerning the specific activity were given, but indirect calculation led to the suggestion that the purified receptor approaches the theoretical maximal activity of 28 nmoles/mg protein.

Compilation of the results achieved by various affinity chromatography procedures (Table 1) provides an extremely heterogenous picture. Although different laboratories used similar procedures, the specific binding activity of the affinity chromatography purified receptor varied from 27 to 28,000 pmoles/mg protein. The purity of the receptor, as estimated by comparison to the theoretical maximal expected value, was sometimes as low as 0.02%. In fact, only two groups reported purity close to 100%: Haeuptle et al. (1983) whose purification procedure was completely different from others (but the calculation was rather speculative) and Dufau and coworkers (Mitani and Dufau, 1986; Dufau and Kusuda, 1987). The latter group used a fairly common procedure but included lectin affinity chromatography prior to the immobilized hGH column. The specific activity of the starting material was one to three orders of magnitude higher than in other cases. Since the extent of purification and the yield of activity were not remarkably different from other groups, the high specific activity of the starting material seems to be of importance. Based on this assumption, enrichment of the starting material by either pretreatment of animals or by additional purifications steps prior to affinity chromatography are recommended. Another procedure in which a relatively high degree of purification was achieved was based on elution at acidic pH rather than 5*M* $MgCl_2$. Surprisingly, this promising procedure was not followed by others. However, in the case of PRL receptor from bovine mammary gland, it was not superior to elution by 5*M* $MgCl_2$ (Ashkenazi, 1986).

As shown in Table 1, recovery of the activity from the affinity column was somewhat low, indicating inactivation or nonspecific absorption. This recovery yield is not sufficient, however, to explain the low specific activity of the purified receptor, which, in most cases, was less than 1% of the value expected for pure receptor. Thus, it seems that with the exception of the work of Mitani and Dufau (1986) and Dufau and Kusuda (1987), the experimental protocol suffers from some basic disadvantages. The hGH or oPRL columns are not specific enough and bind other proteins as well and the elution by 5*M* $MgCl_2$ is not distinctive for lactogenic receptors. Other, more specific procedures for affinity purification must be developed.

4. Cloning of Prolactin Receptor Gene

PRL receptor was purified from livers of estrogen-treated female rats according to Katoh et al. (1987), but using E21 MAb immunoaffinity chromatography instead of oPRL affinity chromatography step (Boutin et al., 1988). No data concerning the specific activity were provided. Further

Table 1
Compilation of Data Concerning the Purification of Prolactin Receptors

Tissue source[a]	Type of affinity column[b]	Binding capacity prior and after chromatography[c]		Purification			Reference
				-fold	yield, %	percent pure[d]	
Rabbit MG	hGH (A)	0.20	28	140	13.9	0.6	Shiu and Friesen (1974)
	STRP (F)	0.09	311,000[e]	20,440	10.0[e]	100.0[e]	Haeuptle et al. (1983)
	hGH (A)	3.47	1181	340	8.6	15.7	Katoh et al. (1984)
	oPRL (E)	0.22	1792	8145	23.6	7.5	Necessary et al. (1984)
	oPRL (C)	0.72	258	352	20.7	0.8	Katoh et al. (1985)
Rat liver	oPRL (D)	0.40	200	500	NR	0.9	Katoh et al. (1987)
Rat ovary	hGH (A)	39.8	20,440	513	15.0	83.8	Mitani and Dufau (1986)
	hGH (A)	95.0	18,500	194	NR	76.0	Dufau and Kusuda (1987)
Mice liver	oPRL (B)	1.33	27	20	1.0	0.1	Liscia and Vonderhaar (1982)
Porcine MG	oPRL (C)	0.04	7	175	6.1	0.02	Sakai et al. (1985)
	oPRL (C)	0.43	36	83	12.4	0.16	Berthon et al. (1987a)
	MAb (D)	0.42	123	294	36.8	0.55	Berthon et al. (1987a)
Bovine MG	hGH (C)	0.06	9	151	15.8	0.03	Ashkenazi et al. (1987a)

[a] MG—mammary gland.

[b] Immobilized hGH, oPRL, or streptavidin (STRP); the letters in the parentheses indicate the elution conditions from the affinity column: (A) 5*M* $MgCl_2$ in 0.1% Triton X-100, pH 7.5; (B) 5*M* $MgCl_2$ in 0.5% CHAPS, pH 7.5; (C) as in B but in 1 m*M* CHAPS; (D) as in B but in 10 m*M* CHAPS; (E) 0.05*M* NH_4Ac in 0.5% Zwittergent 3–12, pH 4.2; (F) 5*M* $MgCl_2$ in 30 m*M* β-octylglucoside, pH 7.6.

[c] pmoles/mg protein.

[d] As a percent of the maximal expected theoretical value, which was calculated assuming the M_r given in Table 2. In the cases when two or more M_r were present, the major species was chosen. In cases when different methodologies resulted in a distinct M_r, a lower value was used for the calculation.

[e] Calculated indirectly.

purification of the reduced and alkylated receptor was achieved by preparative $NaDodSO_4$-PAGE. The mol wt of the pure receptor determined by either immunoblotting or silver staining was 41 kDa. It was found that PRL receptor in rat liver was encoded by a 2.2 kb long messenger RNA, which included an insert that contained a complete reading open frame of 310 amino acids (Fig. 1). The *N*-terminal methionine is followed by 18 hydrophobic amino acids that probably form the signal peptide. Thus, the full sized receptor consists of 291 amino acids with a theroretical M_r of 33,368 D. A strong hydrophobic region of 24 amino acids in positions 230–254 indicates the putative transmembrane region. Accordingly, the cytosolic part of the receptor (Fig. 2) is rather small compared to the hGH receptor. The *N*-terminal extracellular part (20–229) has five Cys residues and three Asn potential glycosylation sites. Two of them were found to be glycosylated by sequence analysis. The F3 clone of the receptor was expressed in *Xenopus laevis* oocytes and in COS-7 or CHO cells. Detailed kinetic analysis of the expressed receptor in these lines revealed expression of ca. 30,000 and 10,000 receptors/cell, and the K_a values were comparable to those found previously in rat liver (Katoh et al., 1987). The binding specificity was preserved as well.

Northern blot analyses of PRL receptor mRNA revealed that the 2.2 kb band was the major species in prostate and ovary. In adrenal gland, a band of slightly higher size was observed. Minor 4 kb bands were also found in these tissues. This is in contrast to the kidney and mammary glands where the major form consisted of 4 kb. Whether these transcripts are encoded by different genes and are translated to other size proteins is not known. A 3.5–4 kb prolactin receptor mRNA was also found in rabbit mammary gland. The nucleotide structure of the respective cDNA was determined (Edery et al., 1989). The purified gene has a reading frame of 592 amino acids and shows high homology to the rat liver receptor in the extracellular and transmembrane parts. The cytosolic part, however, is much longer. In this respect, the rabbit mammary gland receptors resemble the recently described rabbit and hGH receptors (Leung et al., 1987). These results indicate that the diversity in the size of prolactin receptors (*see* next section) exists not only on a protein but also on a genomic level. Neither the PRL nor GH receptors appear to be related to tyrosine kinases or other known receptors; thus, the mechanism of transduction remains obscure. The finding that the liver receptor has a truncated cytosolic domain that is characteristic of receptors that act as transporters (Schneider et al., 1984; Yamamoto et al., 1984; Morgan et al., 1987) points to the possibility that PRL receptors in different organs may have distinct functions.

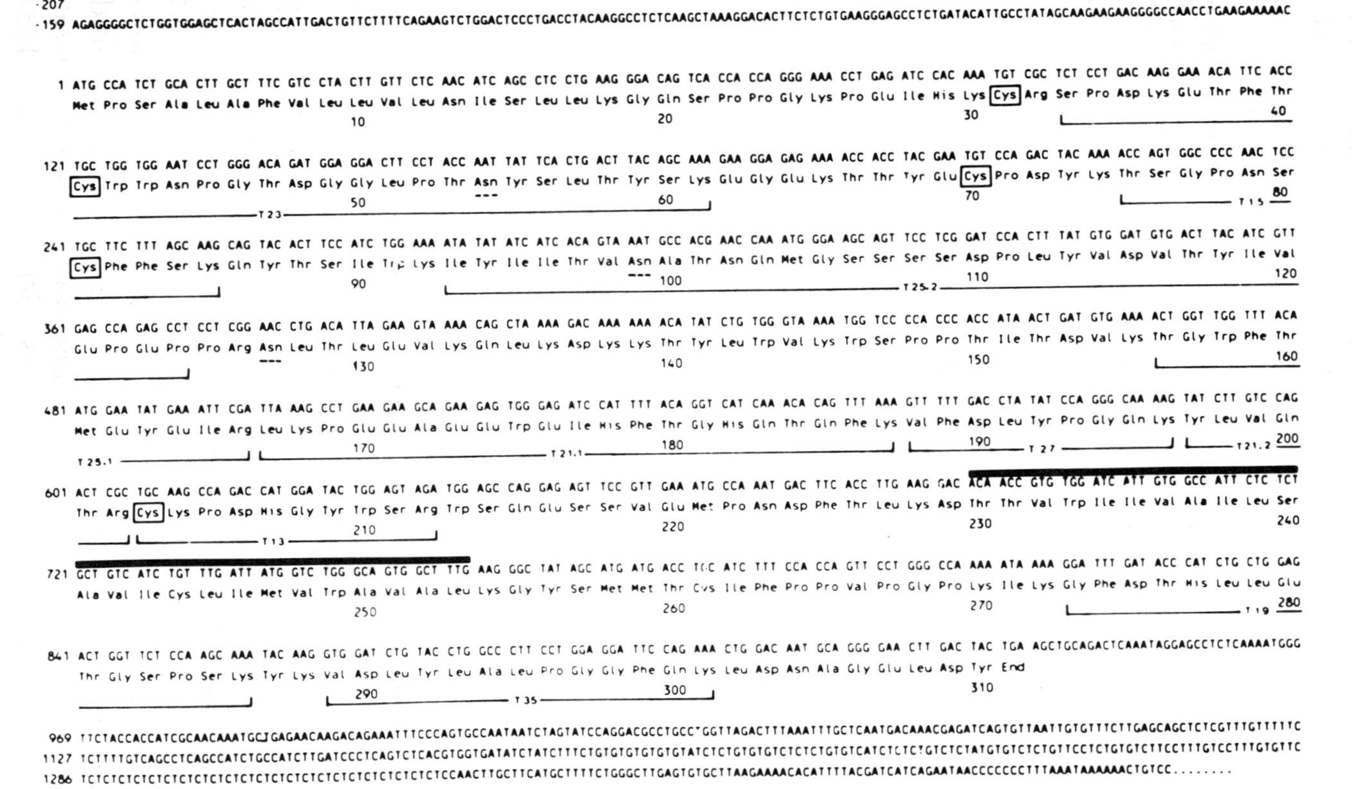

Fig. 1. Nucleic acid and predicted amino acid sequence of the prolactin receptor from rat liver. Nucleotides are numbered at left and amino acid below the sequence, starting with the initiation codon and the corresponding methionine, respectively. Sequences of the tryptic fragments (T) obtained from the purified receptor are underlined and numbered according to their order of elution. The black bar above the nucleotides and the corresponding amino acids represent the putative transmembrane domain. Extracellular cysteines are boxed and potential Asn-linked glycosylation sites are underlined with dashed line (reproduced from Boutin et al., 1988).

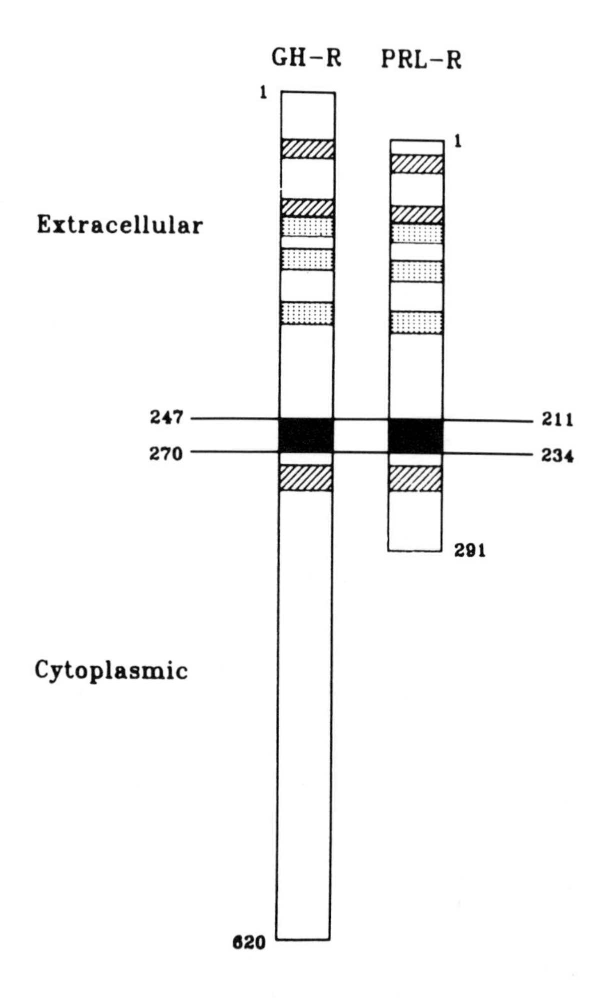

Fig. 2. Schematic comparison of liver GH and PRL receptor structures. Numbers indicate the putative position of the first amino acid, the transmembrane region (black), and the last amino acid of the mature receptor. Regions of high similarity (^{3}67%) are crosshatched and those of moderate similarity (40–60%) are striped (reproduced from Boutin et al., 1988).

5. Molecular Weight of Prolactin Receptors

Several laboratories have attempted to determine the molecular weight of PRL receptor using different species and organs as a receptor source. Intact cells, microsomal fractions, solubilized, or affinity chromatography purified receptors were used. The methodology was mostly based on a chemical crosslinking or photolabeling of the receptor–radioiodinated hormone complex, followed by $NaDodSO_4$-PAGE in the presence or absence of reducing agents.

An alternative approach consisted of gel chromatography or nondenaturing gel electrophoresis of a noncovalent or covalent hormone–receptor complex. The presence of the detergent was absolutely required under these conditions to prevent receptor aggregation (Shiu and Friesen, 1974). The majority of the data published until 1988 are compiled in Table 2. The results are extremely heterogenous with regard to the source of the tissue and the methodology used. Thus, the overall picture is far from clear, and species or tissue difference cannot be excluded.

The major portion of studies on rabbit mammary gland based on affinity labeling led to the conclusion that the M_r of the binding subunit is 35–42 kDa and is not linked to other proteins by disulfide bonds. Only two publications reported a 63 or 70–80 kDa band (Djiane et al., 1985; Katoh et al., 1985). However, gel filtration studies revealed the receptor to appear as a 133–220 kDa species. If these results are corrected for the bound Triton X-100 according to Jaffe (1982), then 100–160 kDa species are obtained. On the other hand, gel filtration in the presence of CHAPS yielded a sole 37 kDa binding subunit. Thus, it would appear that the PRL receptor in rabbit mammary gland exists in an associated state held by noncovalent forces. Whether the receptor was initially composed of one protein chain split by a specific proteolytic cleavage is unclear. This possibility is notably substantiated by the recent finding that the PRL receptor gene in rabbit mammary gland encodes for a protein composed of 592 amino acids (Edery et al., 1989). The M_r of this protein should be at least 60 kDa. A higher molecular weight could be expected because of glycosylation; this would account for the 70–80 species reported by Katoh et al. (1985). It is obvious that the 35–42 kDa binding subunit is most likely a proteolytic product of a higher molecular weight protein. Small variations in proteolytic cleavage could account for the size heterogeneity and occasional occurrence of lower M_r species. Further investigation is required to determine whether this cleavage has any physiological importance or occurs during purification procedures. PRL receptor in the mammary gland of other species (rat, mouse, cow, and sow) has been investigated to a lesser degree, but the general picture appears to be similar.

The M_r reported for the binding subunit in rat liver, using either chemical crosslinking or affinity photolabeling, varies between 35–45 kDa, although an 87 kDa binding subunit was also detected (Haldosen and Gustafsson, 1987,1988). Gel filtration studies revealed 80, 170, and 340 kDa species. If the latter values are corrected for the bound Triton X-100, it is conceivable that high M_r components exist as dimers or trimers of the 87 kDa species. The nature of the 87 kDa binder was studied extensively by Haldosen and Gustafsson (1987, 1988) using various combinations of ex-

Table 2
Compilation of Data Concerning the Molecular Weight of Prolactin Receptors

Tissue source[a]	Stage of purification[b]	Method of determination[c]	M_r, kD	Reference
Rabbit MG	ST or HA	GFT	220[d]	Shiu and Friesen (1974)
	HA	PL, IMP, DGE±	35	Haeuptle et al. (1983)
	MI or HA	CL, DGE+	39	Hughes et al. (1983)
	HA	GFT	133[d]	Katoh et al. (1984)
	HA	GTF	180[d]	Necessary et al. (1984)
	HA	GFZ	37	Necessary et al. (1984)
	HA	DGE	42	Necessary et al. (1984)
	HA	PL, DGE±	32 mj 63 mi	Djiane et al. (1985)
	HA	CL, DGE±	32 mj 70–80	Katoh et al. (1985)
	HA	IMP, DGE±	42	Dusanter-Fourt et al. (1987)
Rabbit liver	HA	PL, IMP	35 mj 67 mi	Haeuptle et al. (1983)
	HA	EB	67	Haeuptle et al. (1983)
	MI or HA	CL, DGE+	50–54	Hughes et al. (1983)
Rat liver	ST	NDE	170	Carr and Jaffe (1982)
	ST	GFT	78[e]	Jaffe (1982)
	WS	GFT	340[e]	Amit et al. (1984)
	MI	PL, DGE±	36	Borst and Sayare (1982)
	MI	CL, DGE±	45	Hughes et al. (1983)
	ST	PL, DGE–	36	Borst et al. (1985)
	HA	IMP, DGE±	44	Katoh et al. (1987)
	HA	EB	42 44	Katoh et al. (1987)
	MI	CL, DGE±	40	Haldosen and Gustafsson
	MI	GFT, DGE±	40 87[e]	(1987, 1988)
	ST	CL, DGE±	35 40 87	Haldosen and Gustafsson
	ST	GFT, DGE±	35 40 87[e]	(1987, 1988)
	MI	CL, DGE±	45–50	Webb and Wallis (1988)
	HA PE	DGE+, IMB, SS	41	Boutin et al. (1988)
Rat hepatocytes	IC	GFT	380[d]	Yamada and Donner (1985)
	IC	DGE±	28 43	Yamada and Donner (1985)
Rat lung	WS	GFT	340 165[e]	Amit et al. (1984)
Rat ovary	MI	CL, DGE±	40	Bonifacino and Dufau (1984)

	ST	CL, DGE±	44 80	Bonifacino and Dufau (1984)
	HA	CL, DGE±	41 88	Dufau and Kusuda (1987)
	HA	EB	50 85	Dufau and Kusuda (1987)
	HA	GFT	150 250[d]	Dufau and Kusuda (1987)
Rat Nb2 lymphoma cells	MI	CL, DGE±	65–75	Ashkenazi et al. (1987b)
Rat Nb2 lymphoma cells	IC	CL, DGE±	65	Gertler and Elberg (1988)
	LA	CL, DGE±	147	Gertler and Elberg (1988)
	LA	GFT	115[e]	Gertler and Elberg (1988)
	LA	EB	123 mj 150 mi	Gertler and Elberg (1988)
	IC	CL, DGE±	72–88	Webb and Wallis (1988)
Mice liver	SC	GFC	37	Liscia et al. (1982)
	HA	GFC, DGE±	37	Liscia and Vonderhaar (1982)
Porcine MG	MI	CL, DGE±	32 mj 68 mi	Sakai et al. (1985)
	MI SC	CL, DGE±	42–45	Berthon et al. (1987a)
	IMA	DGE±	45	Berthon et al. (1987a)
	HA	IMP, DGE±	31 45 mj 66	Berthon et al. (1987a)
Bovine MG	HA	CL, DGE±	37	Ashkenazi et al. (1987a)
	HA	GFT	80–85[e]	Ashkenazi et al. (1987a)
Bovine kidney	ST	GFT	77–80[e]	Elberg et al. (1988)
Tadpole liver	ST	NGE	114	Carr and Jaffe (1982)
Tadpole tail fin	ST	NGE	103	Carr and Jaffe (1982)

[a]MG—mammary gland.

[b]HA—hormone affinity purified; IC—intact cells; MA—immunoaffinity purified; LA—lectin affinity purified; MI—microsomal fraction; PE—purified by preparative PAGE; SC—solubilized with CHAPS; ST—solubilized with Triton X-100; WS—water soluble receptors.

[c]CL—chemical crosslinking; DGE—PAGE under denaturing conditions; EB—electroblotting; GFC—gel filtration in presence of CHAPS; GFT—gel filtration in presence of Triton X-100; GFZ—gel filtration in presence of Zwittergent; IMB—immunoblotting; IMP—immunoprecipitation; NDE—PAGE under nondenaturing conditions; PL—photoaffinity labeling; SS—silver staining; ± with or without reducing agent; mi—minor; mj—major.

[d]Noncorrected for contribution of Triton X-100.

[e]Corrected for contribution of Triton X-100 according to Jaffe (1982).

traction and crosslinking. Four possible models (Fig. 3) were proposed. The first model suggests that the 87 kDa is composed of 40 kDa and 35 kDa subunits. The 40 kDa subunit contains the lactogenic binding site. In the microsomal fraction (Fig. 3a), the 35 kDa subunit that interacts with the 40 kDa subunit cannot form crosslinks with the bound hormone because of a hindrance by surrounding membrane phospholipids. Solubilization by Triton X-100 removes this restriction, and, thus, the 35 kDa subunit also can be labeled (Fig. 3b). Whether the 40 and 35 kDa subunits are linked by disulfide bonds is unclear. Since the two-dimensional diagonal PAGE revealed that the 35 and 40 kDa species may be partially released from the 87 kDa receptor by reducing agents, the possibility that two populations of 87 kDa species exist has been suggested (Haldosen and Gustafsson, 1987). An alternative model proposes that the 35 kDa subunit is a proteolytically modified or less glycosylated form of the 40 kDa subunit. Both subunits are capable of binding the lactogenic hormone and they interact with a ca. 50 kDa unlabeled subunit. Since the 35 kDa subunit could not be detected dur-ing crosslinking in the microsomal membranes, it is possible that the proteolytic cleavage occurred in the solubilization process. Recent findings that the mature prolactin receptor consists of 291 amino acids with a theoretical mol wt of 33,368 kDa and that the purified 41 kDa receptor can be deglycosylated to 36 kDa by endoglycosylate F (Boutin et al., 1988) strongly support the suggested models. Similar evidence of 40 and 88 kDa receptors has been reported in rat ovary (Bonifacino and Dufau, 1984; Mitani and Dufau, 1986; Dufau and Kusuda, 1987), whereas in the mouse liver, only a 37 kDa subunit was detected (Liscia et al., 1982; Liscia and Vonderhaar, 1982).

A higher M_r (65–88 kDa) has been reported for the binding subunit in the microsomal fraction or in intact Nb2 rat lymphoma cells (Ashkenazi et al., 1987b; Shenk et al., 1987; Gertler and Elberg, 1988; Webb and Wallis, 1988). These receptors are extremely sensitive to proteolysis (Shenk et al., 1987) and occasionally, lower M_r species were found. A higher molecular weight species (123–147 kDa) was detected in a partially purified receptor (Gertler and Elberg, 1988), leading to the possibility that binding of lactogenic hormones to intact cells or microsomal fraction leads to a subsequent specific proteolytic event that, in turn, reduces the receptor size.

In summary, the present data suggest that three types of PRL receptors may exist in mammals:

1. The mammary gland type (M_r of ~ 70 kDa), which is proteolyzed to 35–40 kDa subunit;
2. The liver type (M_r ca. 40 kDa), which lacks most of the cytosolic domain existing in the former type; or
3. The Nb2 lymphoma cells type (M_r 65–88 or even 123–150 kDa).

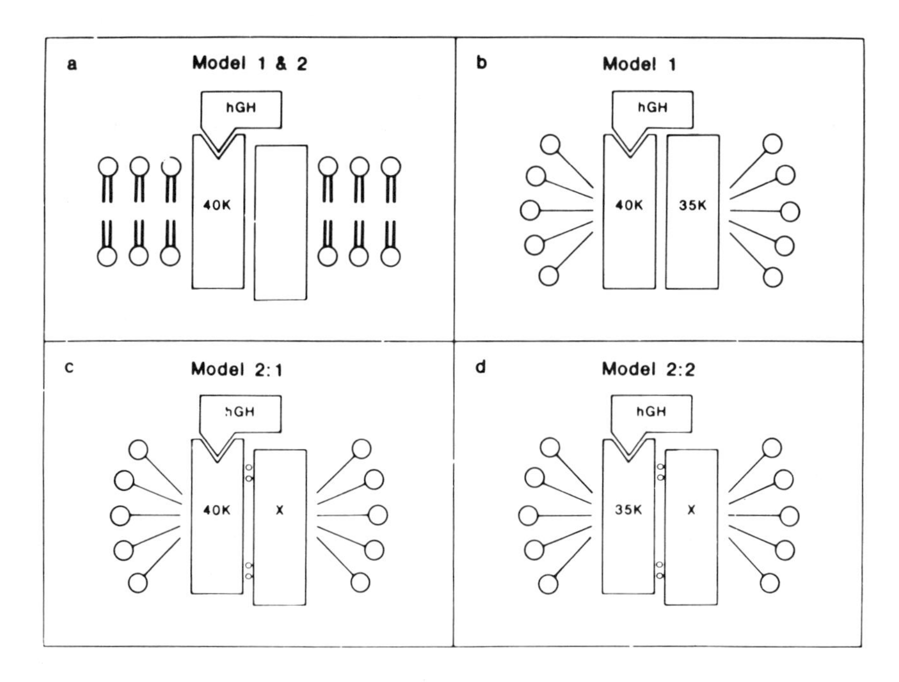

Fig. 3. Proposed models for subunit structure of the female rat liver PRL receptor. Membrane and Triton solubilized situations are pictured in *a* and *b–d*, respectively. o— peptide bonds or noncovalent binding (reproduced from Haldosen and Gustafsson, 1987).

The relationship among these three types is only partially understood and awaits further clarification by purification of the Nb2 and other receptors and/or cloning of their genes.

6. Open Questions Concerning the Binding Specificity of PRL Receptors in Ruminants and Other Species

The availability of oPRL and hGH led many researchers to use these hormones as tracers and/or displacers in determination of the binding activity of PRL receptors. Most of the reports found none or small differences in binding of these hormones to PRL receptors in rabbit, rat, and mouse. In the rabbit, oPRL was as effective as hGH in displacing ^{125}I-hGH bound to mammary gland (Shiu and Friesen, 1974; Posner et al., 1974) and kidney (Roguin et al., 1982), but less effective in liver (Posner et al., 1974; Fix et al., 1981). In the mouse, oPRL and hGH were almost equally effective in displacing ^{125}I-hGH from receptors in mammary gland and liver (Bhattacharya and Vonderhaar, 1981) or ^{125}I-oPRL from dissociated mammary epithelial cells (Sakai et al., 1978). Similar results were reported for displacement of ^{125}I-hGH from rat liver receptors by Posner et al. (1974), Borst and Sayare (1982), and Sasaki et al. (1982).

Equal binding potency of hGH and oPRL was also reported for binding to porcine mammary gland receptors (Sakcai et al., 1985) and to particulate and water-soluble receptors from rat liver (Amit et al., 1984). In the latter case, hPRL was equally active. Data from other laboratories using membranes from different organs of various animals have shown that hGH was less potent than oPRL in inhibiting the binding of radiolabeled oPRL to lactogenic receptors (Nicoll et al., 1980; Carr and Jaffe, 1982; Buntin et al., 1984; Edery et al., 1984; Haro and Talamantes, 1985). Similar results were also recently reported for amphibians and reptiles (Nicoll et al., 1986, Tarpey and Nicoll, 1987). The difference in all cases were rather small (ca. two- to threefold), and could be attributed to impurity of the hormone preparation.

Remarkably different results were observed in binding experiments using membrane-bound and solublized receptors from bovine liver and mammary gland (Gertler et al., 1984). In contrast to ^{125}I-hGH, the ^{125}I-oPRL binding was very low. Excess of unlabeled ovine or bovine PRLs were capable of inhibiting the binding of ^{125}I-hGH (Fig. 4), but the ED_{50} values were 40–80-fold higher. Association constants for bPRL or oPRL binding were calculated by Dixon plot and found to be 40–90-fold lower than for binding of hGH. A 110-fold lower association constant for oPRL in solubilized receptors from bovine mammary gland was also found using homologous binding experiments whereby the free and bound hormones were separated by gel filtration (Ashkenazi et al., 1987a). Mutual competition experiments led to the conclusion that both hormones bind to the same receptor but the affinity is different. We have found also that this distinct affinity is not restricted to bovine mammary gland and liver but also exists in ovary, adrenal gland, and kidney (Cohen et al., 1987; Elberg et al., 1989). It occurs in lactating and in virgin animals and is independent of the degree of purification (Ashkenazi, 1986).

Other laboratories using either ovine or bovine tissues reported similar discrepancies in hGH and ovine placental lactogen (Servely et al., 1983), or in hGH and oPRL binding (Akers and Keys, 1984; Jammes et al., 1985; Kazmer et al., 1986; Bramley et al., 1987a; Keys and Djiane, 1988). Our initial suggestion that this preferential binding of hGH is specific for ruminants (Gertler et al., 1984) requires, however, modification as similar phenomena have been since reported for other homologous and heterologous systems. In rat Leyding cells, lactogenic receptors have ca. 35-fold higher affinity for hGH than oPRL (Bonifacino and Dufau, 1985). In rat

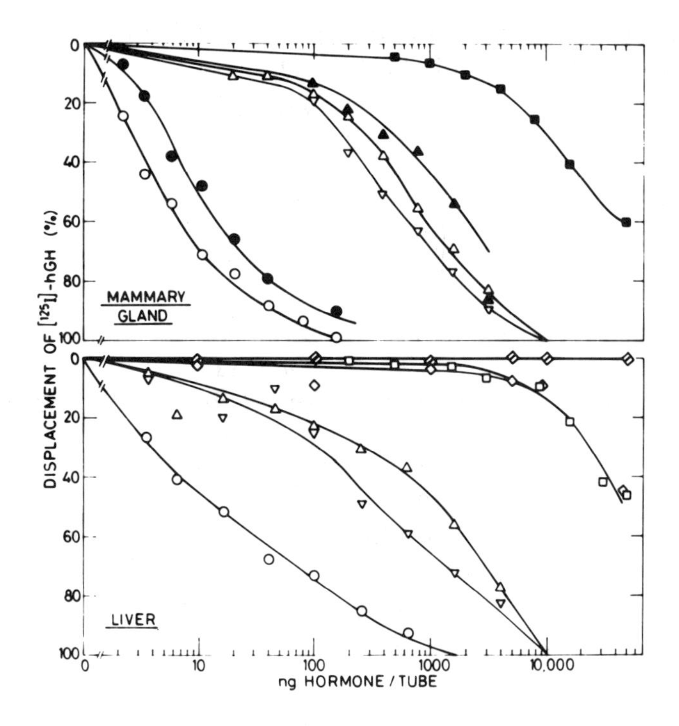

Fig. 4. Inhibition of the ^{125}I-hGH binding to bovine membrane bound and solubilized mammary gland receptors or membrane bound liver receptors by increased concentrations of unlabeled hormones. Closed symbols—solubilized receptors, open symbols—membrane bound receptors; (○—●)—hGH; (∇)—bPRL; (△,▲)—oPRL; (□,■)—pGH; (◊)—bGH; (◆) — recombinant bGH (reproduced from Gertler et al., 1984).

lung and liver, hGH and oPRL were equipotent, although rat PRL exhibited a 10–20-fold lower affinity (Amit et al., 1984; Boutin et al., 1988). Ovine prolactin and hGH were also equipotent in rabbit mammary gland, whereas rabbit prolactin was less active (Berthon et al., 1987b). Ovine PRL was ca. 10 times more potent than hGH in PRL receptors purified from porcine mammary gland (Berthon et al., 1987b) or ovaries (Bramley and Menzies, 1987), whereas rabbit and porcine PRLs were almost completely inactive. Identical results were obtained using either microsomal or water soluble receptors (Berthon et al., 1987b). These results are paradoxical, since hGH and oPRL are almost equally active in their effects on galactopoietic activities in explants of lactating bovine mammary gland (Gertler et al., 1983), whereas porcine and ovine prolactins are equally active in various bioassay in porcine tissue (Berthon et al., 1987b).

To date, the preferential binding for heterologous hormones remains unexplained, although three possible hypotheses can be offered:

1. The binding experiments using crude, solubilized, or partially purified receptors do not reflect the in vivo situation. They occur as an experimental artifact caused by desensitization of the receptors toward homologousPRL (but not toward hGH or other heterologous PRL) during cell disruption;
2. PRL used in the assays is not equivalent to the secreted and circulating hormone. It has been shown that the circulating prolactin is phosphoryl ated (Oetting et al., 1986) and/or glycosylated (Markoff and Lee, 1987);
3. The affinity for PRL is indeed low, but the majority of biological processes stimulated by PRL are already fully activated when the occupancy of the receptor is very low, similar to insulin-stimulated activities.

An alternative explanation has been proposed recently that suggests that selective inactivation of oPRL but not hGH occurs in ovine corpus luteum homogenates and membrane fraction (Bramley et al., 1987b). The nature of this inactivation remains unknown. It could not be prevented by proteinase inhibitors of different specificities and did not affect the gel filtration profile of the radioiodinated hormone.

Acknowledgments

I wish to thank P. A. Kelly of McGill University and L. A. Haldosen and J. A. Gustafsson of Karolinska Institute for the permission to use figures published in their papers, and J. Djiane of the Institute National Recherche Agronomique for the information concerning the cloning of prolactin receptors from rabbit mammary gland. The editorial assistance of Marsha Rockitter is highly appreciated.

References

Akers, R. M. (1985) *J. Dairy Sci.* **68,** 501–519.
Akers, R. M. and Keys, J. E. (1984) *J. Dairy Sci.* **67,** 2224– 2235.
Amit, T., Barkey, R. J., Gavish, M., and Youdin, M. B. H. (1984) *Endocrinology* **114,** 545–552.
Ashkenazi, A. (1986) PhD Thesis, The Hebrew University of Jerusalem, Israel.
Ashkenazi, A., Madar, Z., and Gertler, A. (1987a) *Mol. Cell. Endocrinol.* **50,** 79–87.
Ashkenazi, A., Cohen, R., and Gertler A. (1987b) *FEBS Lett.* **210,** 51–55.
Berthon, P., Katoh, M., Dusanter-Fourt, I., Kelly, P.A., and Djiane, J. (1987a) *Mol. Cell. Endocrinol.* **51,** 71–81.
Berthon, P., Kelly, P. A., and Djiane, J. (1987b) *Proc. Soc. Exp. Biol. Med.* **184,** 300–306.
Bhattacharya, A. and Vonderhaar, B. K. (1981) *Proc. Natl. Acad. Sci. USA* **78,** 5704–5707.
Bonifacino, J. S. and Dufau, M. L. (1984) *J. Biol. Chem.* **259,** 4542–4549.
Bonifacino, J. and Dufau, M. L. (1985) *Endocrinology* **116,** 1610–1614.
Borst, D. W. and Sayare, M. (1982) *Biochem. Biophys. Res. Commun.* **105,** 194–201.

Borst, D. W., Sayare, M., and Posner, B. I. (1985) *Mol. Cell. Endocrinol.* **39,** 125–130.
Boutin, J. M., Jolicoeur, C., Okamura, M., Ganon, J., Edery, M., Shirota, M., Banville, D., Dusanter-Fourt, I., Djiane, J., and Kelly, P. A. (1988) *Cell* **53,** 69–77.
Bramley, T. A. and Menzies, G. S. (1987) *J. Endocrinol.* **113,** 355– 364.
Bramley, T. A., Menzies, G. S., McNeilly, A. S., and Friesen, H. G. (1987a) *J. Endocrinol.* **113,** 365–374.
Bramley, T. A., Menzies, G. S., McNeilly, A. S., and Friesen, H. G. (1987b) *J. Endocrinol.* **113,** 375–381.
Buntin, J. D., Keskey, T. S., and Hanik, D. S. (1984) *Gen. Comp. Endocrinol.* **55,** 418–428.
Carr, F. E. and Jaffe, R. C. (1982) *Endocrinology* **109,** 943–949.
Church, W. R. and Ebner, K. E. (1982) *Experientia* **38,** 434–435.
Cohen, R., Ashkenazi, A., Elberg, G., and Gertler, A. (1987) *J. Recept. Res.* **7,** 931–936.
Cuatrecasas, P. (1973) *J. Biol. Chem.* **254,** 3528–3534.
Djiane, J., Kelly P. A., Katoh, M., and Dusanter-Fourt, I. (1985) *Hormone Res.* **22,** 179–188.
Dufau, M. L. and Kusuda, S. (1987) *J. Recept. Res.* **7,** 167–193.
Dusanter-Fourt, I., Kelly, P. A., and Djiane, J. (1987) *Biochimie* **69,** 639–646.
Edery, M., Young, G., Bern, H. A., and Steiney, S. (1984) *Gen. Comp. Endocrinol.* **56,** 19–23.
Edery, M., Jolicoeur, C., Levi-Meyrueis, C., Dusanter-Fourt, I., Petridov, B., Bovtin, J.-M., Lesueur, L., Kelly, P. A., Djiane, J. (1989) *Proc. Natl. Acad. Sci. USA* **86,** 2112–2116.
Elberg, G., Ashkenzi, A., and Gertler, A. (1989) *Mol. Cell. Endocrinol.* **61,** 77–85.
El-Hamzawy, M. A. and Costlow, M. E. (1988) *Anal. Biochem.* **171,** 300–304.
Fix, J. A., Leppert, P., and Moore, W. V. (1981) *Horm. Metab. Res.* **13,** 508–515.
Forsyth, I. A. (1986) *J. Dairy Sci.* **69,** 886–903.
Gertler, A., Cohen N., and Maoz, A. (1983) *Mol. Cell. Endocrinol.* **33,** 169–182.
Gertler, A., Madar, Z., and Ashkenazi, A. (1984) *Mol. Cell. Endocrinol.* **34,** 51–57.
Gertler, A. and Elberg, G. (1988) *Proc. Vth International Congress on Prolactin,* P-65. Kyoto, Japan.
Haeuptle, M. T., Aubert, M. L., Djiane, J., and Kraehenbuhl, J. P. (1983) *J. Biol. Chem.* **258,** 305–314.
Haldosen, L. A. and Gustafsson, J. A. (1987) *J. Biol. Chem.* **262,** 7404– 7411.
Haldosen, L.-A. and Gustafsson, J. A. (1988) *Biochem. J.* **252,** 509–514.
Haro, L. S. and Talamantes, F. J. (1985) *Mol. Cell. Endocrinol.* **41,** 93–104.
Hughes, J. P., Simpson, J. S. A., and Friesen, H. G. (1983) *Endocrinology* **112,** 1980–1985.
Jaffe, R. C. (1982) *Biochemistry* **21,** 2936–2945.
Jammes, H., Schirar, A., and Djiane, J. (1985) *J. Reprod. Fert.* **73,** 27–35.
Katoh, M., Djiane, J., Leblanc, G., and Kelly, P. A. (1984) *Mol. Cell. Endocrinol.* **34,** 191–200.
Katoh, M., Djiane, J., and Kelly, P. A. (1985) *Endocrinology* **116,** 2612–2620.
Katoh, M., Raguet, S., Zachwieja, J., Djiane, J., and Kelly, P. A. (1987) *Endo-crinology* **120,** 739–749.
Kazmer, G. W., Barnes, M. A., Akers, R. M., and Whittier, W. D. (1986) *J. Endocrinol.* **109,** 175–180.
Kelly, P., Djiane, J., Katoh, M., Ferland, L. H., Houdebine, L. M., Teyssot, B., and Dusanter-Fourt, I. (1984) *Recent Prog. Horm. Res.* **40,** 379–436.
Keys, J. E. and Djiane, J. (1988) *J. Recept. Res.* **8,** 731–750.
Koppelman, M. and Dufau, M. L. (1982) *Endocrinology* **111,** 1350–1356.
Leung, D. W., Spencer, S. A., Cachianes, G., Hammonds, R. G., Collins, C., Henzel, W. J., Barnard, R., Waters, M. J., and Wood, W. I. (1987) *Nature* **330,** 537–543.

Liscia, D. S., Alhadi, T., and Vonderhaar, B. K. (1982) *J. Biol. Chem.* **257,** 9401–9405.
Liscia, D. S. and Vonderhaar, B. K. (1982) *Proc. Natl. Acad. Sci. USA* **79,** 5930–5934.
Markoff, E. and Lee, D. W. (1987) *J. Clin. Endocrinol. Metab.* **65,** 1102–1108.
Mitani, M. and Dufau, M. L. (1986) *J. Biol. Chem.* **261,** 1309–1315.
Morgan, D. O., Edman, J. C., Standring, F. N., Fried, V. A., Smith, M. C., Roth, R. A., and Rutter, W. J. (1987) *Nature* **320,** 301–307.
Necessary, P. C., Humphrey, P. A., Mahajan, P. B., and Ebner, K. E. (1984) *J. Biol. Chem.* **259,** 6942–6946.
N'Guema Ename, N. M., Delouis, C., Kelly, P. A., and Djiane, J. (1986) *Endocrinology* **118,** 695–700.
Nicoll, C. S. and Bern, H. A. (1972) in *Ciba Foundation on Lactogenic Hormones* (Wolstenholme, G.E.W. and Knight, J., eds.), Chruchill Livingstone, London, pp. 299–324.
Nicoll, C. S., White, B. A., and Leung, F. C. (1980) in *Central and Peripheral Regualtion of Prolactin* (MacLeod, R.M. and Scapagnini, U., eds.), Raven, New York, pp. 11–25.
Nicoll, C. S. and White, B. A. (1985) in *Current Trends in Comparative Endocrinology* (Lofts, B. and Holmes, W. H., eds.), Hong Kong University Press, Hong Kong, pp. 781–784.
Nicoll, C. S., Tarpey, J. F., Mayer, G. M., and Russell, S. M. (1986) *Am. Zool.* **26,** 965–984.
Oetting, W. S., Tuazon, P. T., Traugh, J. A., and Walker, A. M. (1986) *J. Biol. Chem.* **261,** 1649–1652.
Posner, B. I., Kelly, P. A., Shiu, R. P. C., and Friesen, H. G. (1974) *Endocrinology* **96,** 521–531.
Rillema, J. A., Etindi, R. N., Ofenstein, J. P., and Waters, S. B. (1988) in *The Physiology of Reproduction* (Knobil, E. and Neill, J. D., Ewing, L. L., Greenwald, G. S., Market, C. L., and Pfaff, D. W., eds.), Raven, New York, pp. 2217–2234.
Roguin, L. P., Bonifacino, J. S., and Paladini, A. C. (1982) *Biochim. Biophys. Acta* **715,** 222–229.
Sakai, S., Enami, J., Nandi, S., and Banerjee, M. R. (1978) *Mol. Cell. Endocrinol.* **12,** 285–298.
Sakai, S., Katoh, M., Berthon, P., and Kelly, P. A. (1985) *Biochem. J.* **224,** 911–922.
Sasaki, M., Tanaki, Y., Imai, Y., Tsushima, T., and Matsuzaki, F. (1982) *Biochem. J.* **203,** 653–662.
Schneider, C., Owen, M. J., Banville, D., and Williams, J. G. (1984) *Nature* **311,** 675– 678.
Servely, J. L., N'Guema Emane, M., Houdebine, L.-M., Djiane, J., Delouis, C. and Kelly, P. A. (1983) *Gen. Comp. Endocrinol.* **51,** 255–262.
Shenk, B. A., Laherty, F., and Hughes, J. P. (1987) *Proc. 69th Ann. Meet. Endocr. Soc. Indianapolis,* Abstract no. 818.
Shiu, R. P., Kelly, P. A., and Friesen, H. G. (1973) *Science* **180,** 968–971.
Shiu, R. P. C. and Friesen, H. G. (1974) *J. Biol. Chem.* **249,** 7902–7911.
Stricker, P. and Grueter, R. (1928) *C.R. Soc. Biol. (Paris)* **99,** 1978–1980.
Tarpey, J. F. and Nicoll, C. S. (1987) *J. Exp. Zool.* **241,** 317–325.
Webb, C. F. and Wallis, M. (1988) *Biochem. J.* **250,** 215–219.
Yamada, K. and Donner, D. B. (1985) *Biochem. J.* **220,** 361–369.
Yamamoto, T., Davis, C. G., Brown, M. S., Schneider, W. J., Casey, M. L., Goldstein, J. L., and Russell, D. W. (1984) *Cell* **39,** 27–38.

Ligand- and Antibody-Affinity Purification of the Epidermal Growth Factor Receptor

George Panayotou, J. Justin Hsuan, and Michael D. Waterfield

1. Introduction

The study of the mechanism by which growth factors transmit a mitogenic signal through the plasma membrane has been greatly facilitated by the use of the EGF receptor as a model system. Binding of EGF to its receptor initiates a cascade of cellular events, presumably through stimulation of its intrinsic tyrosine-specific protein kinase activity, which culminate in DNA synthesis and cell division (*reviewed by* Carpenter, 1987). Studies using purified EGF receptor have helped to determine the mechanism of its activation and elucidate its primary structure (Downward et al., 1984a). The EGF receptor has a molecular weight of 170 kDa. A single, membrane-spanning stretch of 23 amino acids links the extracellular, EGF binding domain to the cytoplasmic kinase domain (Ullrich et al., 1984). A C-terminal 15 kDa domain carries the major autophosphorylation sites of the receptor and is connected to the cytoplasmic domain through a putatively flexible, protease-sensitive link (Downward et al., 1984a).

Purification of the EGF receptor was first reported by Cohen et al. (1980) using EGF coupled to an immobilized support in a specific affinity step. A variety of other affinity-purification methods have since been reported. Table 1 presents them in a brief form and details can be obtained from the relevant references. In this paper, the EGF-affinity method and two immunoaffinity methods are presented in detail.

2. Sources of EGF Receptor

The major source of EGF receptor in most purification studies has been the human epidermoid carcinoma cell line A431, which expresses the EGF receptor at very high levels, approximately 2×10^6/cell (Fabricant et al., 1977). Human placenta has also been used in some studies.

In the case of the A431 cell line, whole cells can be used to obtain a maximum amount of receptor at a satisfactory purity level. If a higher purity is required, plasma membranes can be prepared from the cells using one of two published protocols. The method of Thom et al. (1977) provides a relatively high yield of plasma membranes and involves hypotonic lysis of cells and centrifugation steps. The exact protocol described in the original paper should be followed with the important exception that Ca^{2+} and Mg^{2+} salts must be omitted from the buffers in order to avoid proteolysis of the EGF receptor to a 150 kDa form. The method of Cohen et al. (1982b) involves swelling cells in monolayers in a hypotonic buffer, under conditions where plasma membrane vesicles are shed from the cell surface and can be collected from the medium by centrifugation. High purity plasma membranes can be obtained by this method but, in our experience, the yield is low.

Syncytiotrophoblastic plasma membrane vesicles can be prepared from human placentas using the method described by Smith et al. (1977). Because of the lower level of EGF receptors in placenta compared to A431 cells, a first purification step involving wheatgerm agglutinin chromatography should be performed before the ligand- or antibody-affinity step to obtain highly purified receptor (*see below* for details).

3. Culture of A431 Cells

A431 cells grow optimally in Dulbecco's Modified Eagle's Medium (DMEM) containing 4.5 g/L glucose, 10% fetal calf serum, penicillin (30 mg/mL), and streptomycin (75 mg/L). For bulk culture, the concentration of serum may be reduced to 5% or substituted by newborn calf serum. The cells are kept at 37°C in an atmosphere of 5% CO_2–95% air in a humidified incubator. A431 cells form confluent monolayers and then continue

Table 1
Methods of Purification of the EGF Receptor

Source	Affinity matrix	Elution	Detergent	Reference
A431 membranes	EGF-Affigel	pH 9.7	0.2% Triton	Cohen et al. (1980); Cohen (1983)
Mouse liver	EGF-Affigel	pH 9.7	0.2% Triton	Cohen et al. (1982a)
A431 cells	EGF-Affigel	pH 9.7	0.5% Triton or deoxycholate	Gregoriou and Rees (1984)
A431 cells/human placenta	Ab R1-Affigel	pH 3.0	0.05% NP40	Downward et al. (1984b)
A431 cells	Ab 9A-Affigel	0.25*M* GlcNAc	0.1% Triton	Parker et al. (1984)
A431 cells	Ab 528-Sepharose	EGF, pH 7.4	0.05% Triton	Weber et al. (1984)
A431 cells	Ab 29.1-Sepharose	pH 2.5 or 11	0.1% Triton	Yarden et al. (1985)
A431 membranes	Tyr-Sepharose	0.25*M* Amm. sulfate	0.05% Triton	Akiyama et al. (1985)

dividing with excess cells shedding into the culture medium. Adhesion to the tissue culture plastic is very strong and removal for routine subculturing usually requires 5 ×-concentrated trypsin-EDTA solution. The cells should be fed with fresh medium every 3–4 d, until they reach confluence.

For receptor purification, the cultures should be expanded in roller-bottles (glass or tissue culture plastic). Approximately 2×10^8 cells can be obtained from each confluent roller bottle. The following procedure should be used to harvest the cells, since trypsinization may result in digestion of the EGF receptor: The culture medium is discarded and the cells washed twice with 50 mL of 5 m*M* EDTA in Ca^{2+}- and Mg^{2+}-free phosphate-buffered saline (PBSA). The roller bottles are returned to the incubators at 37°C with a fresh 50 mL of the above solution for 10–15 min. By the end of this period, cell monolayers begin to detach and should be collected in suitable centrifuge tubes, where the cells are pelleted at 1000*g* for 10 min at 4°C. The pellet is then washed once in PBSA. It should be noted that harvesting the cells in the presence of EDTA has the benefit of protecting the EGF receptor from Ca^{2+}-activated proteases. The collected cells can either be used directly for EGF receptor extraction and purification or used to prepare plasma membranes.

4. Solubilization of the EGF Receptor

Triton X-100 is the detergent most commonly used to extract the receptor from A431 cells or membranes. The cell pellet (or plasma membranes) is resuspended in solubilization buffer (20 m*M* Tris, pH 7.5, 1% (v/v) Triton X-100, 1 m*M* EDTA, 150 m*M* NaCl, 25 m*M* benzamidine, 0.2 m*M* PMSF, 10% (v/v) glycerol). Approximately 1 mL of buffer should be used for 2×10^6 cells. Solubilization is allowed to proceed for 30 min at 4°C with occasional stirring. Cell debris is removed by centrifugation at 3,000*g* for 10 min at 4°C and the supernatant is further centrifuged at 100,000*g* for 1 h at 2°C. A cloudy material may be obtained at the top of the centrifuge tube that does not interfere with following procedures. The pellet is discarded and the supernatant stored at –70°C. No loss of activity occurs for several months at this temperature.

5. Wheatgerm Agglutinin Affinity Chromatography

A lectin-affinity step is not usually required for EGF receptor purification from A431 cells but should be performed when placental membranes are used as a source of receptor.

The EGF receptor-containing lysate is incubated with agarose-bound wheatgerm agglutinin (10 mL, Vector Laboratories) for 1 h at 4°C. The matrix is then washed by filtration through a scintered-glass funnel with 200 mL of PBSA containing 0.5*M* NaCl and 0.1% Triton X-100 at 4°C, followed by 200 mL of PBSA/0.1% Triton X-100 and then 200 mL of 20 m*M* Tris, pH 7.5, 150 m*M* NaCl, 0.1% Triton at 4°C. Bound glycoproteins are eluted by tumbling the lectin matrix with 15 mL of 0.25*M* *N*-acetylglucosamine (Sigma) in 20 m*M* Tris, pH 7.5, 0.1% Triton X-100 for 15 min at room temperature and the supernatant collected by filtration. This procedure is repeated twice and the eluates are combined.

6. Ligand-Affinity Chromatography

6.1. Coupling of EGF to Affigel

A variety of bead supports are available for coupling proteins and peptides through their free amino groups. In our laboratory, as in most published purification procedures, Affigel-10 (Bio-Rad), a *N*-hydroxysuccinimide ester of a crosslinked agarose gel bead support, has been found to be the most convenient matrix for coupling EGF, both regarding ease of preparation and low nonspecific protein binding during purification.

The EGF solution should be carefully prepared as follows to ensure maximal coupling:

1. Approximately 10 mg of EGF, prepared as described by Savage and Cohen (1972), are dissolved in 4 mL of 10 m*M* HCl and the pH adjusted to 8.0 with 1*M* $NaHCO_3$. It is essential at this point to dialyze the solution against 3× 1 lt changes of 0.1*M* $NaHCO_3$, pH 8.2 to ensure that no contaminating low molecular weight compounds with free amino groups remain. After dialysis, the amount of EGF may be calculated b amino acid analysis or from the absorbance of the solution at 280 nm (extinction coefficient $E_{1\%} = 3.09$).
2. Affigel 10 is supplied swollen in isopropyl alcohol and should be washed in a scintered glass funnel under vacuum with three bed volumes of this solvent, followed by three volumes of cold distilled water, and ten volumes of 0.1*M* $NaHCO_3$, pH 8.2. This procedure should be carried out at 4°C and completed in less than 20 min to avoid hydrolysis of the active sites of the support. It is also essential to avoid drying the gel during washing.
3. Approximately 5 g of washed gel are mixed with the EGF solution in a tightly closed tube which is then rotated end-over-end for 16 h at 4°C.
4. Under optimal conditions, 70–90% of the EGF should react with the matrix. Uncoupled EGF can be collected by filtering the mixture through the scintered glass funnel and washing twice with a small vol of 0.2*M* glycine-HCl, pH 2.5. The percentage of uncoupled EGF can be estimated from the absorption at 280 nm. It is important, however, to keep the pH at 2.5, since by products of the coupling reaction absorb strongly at higher pH.
5. The gel is then washed with 100 mL each of 0.5*M* NaCl, 5 m*M* ethanolamine, pH 9.7, and 0.1*M* $NaHCO_3$, pH 8.2 and then incubated with 0.1*M* ethanolamine, pH 8.2 with shaking for 2 h at room temperature. This step is essential for blocking excess, unreacted sites on the Affigel, especially if the matrix is going to be used immediately for receptor purification.
6. Finally, the gel should be washed extensively with 2 L of 50 m*M* Tris, pH 7.5, 150 m*M* NaCl and stored at 4°C. Addition of sodium azide to 0.05% (w/v) is essential to prevent microbial growth.

6.2. EGF Receptor Purification

1. The prepared EGF-Affigel is added to the cell or membrane extract or to the wheatgerm lectin column eluate and mixed end-over-end for 1–2 h at room temperature.
2. Unbound material is removed by filtration through a scintered-glass funnel and the gel is washed with approx 100 bed volumes of ice-cold 20 m*M* Tris, pH 7.5, 1% Triton X-100, 2 m*M* EDTA, 25 m*M* benzamidine, 10% glycerol, followed by a wash with 20 bed volumes of 0.2% Triton, 10% glycerol, pH 7.5 (adjusted with NaOH).
3. The gel is transferred to a 10 mL column (e.g., Bio-Rad Econocolumn) and

the elution buffer, consisting of 5 m*M* ethanolamine, 0.2% Triton, 10% glycerol, pH 9.7, is applied at 4°C. After one column vol has passed through, the flow is stopped for 10–15 min and the elution is continued with six column volumes of elution buffer.

4. The high pH during elution may result in substantial loss of kinase activity of the receptor and therefore the eluate should be brought immediately to pH 8 by addition of 2*M* Tris-HCl, pH 6.8 to a final concentration of 20 m*M* Tris. Benzamidine is also added to 25 m*M*. The purified receptor should preferably be used immediately after preparation.
5. The EGF-Affigel matrix can be used for multiple receptor purifications. After each use it should be washed with at least 50 volumes of PBSA and stored at 4°C in PBSA/0.05% sodium azide.

For some applications, it may be necessary to obtain the receptor in a detergent other than Triton X-100. In this case, anion-exchange chromatography can be used both for detergent exchange and sample concentration. The eluate is applied to a small column (approx 0.5 mL) of DE52, which is then washed with at least five bed volumes of 20 m*M* Tris, pH 7.5, 10% glycerol containing the required detergent. Bound receptor is eluted with 0.4*M* NaCl in the same buffer.

In order to avoid exposure of the receptor to high pH during elution, an alternative method may be employed that utilizes EGF to specifically elute the receptor from the affinity matrix. The gel is incubated with one bed vol of 0.2% Triton, 10% glycerol, pH 7.5 containing 1 mg/mL EGF for 2 h at 4°C with gentle shaking. The eluted material is collected as above. The major disadvantage of this method is that the basal level of receptor kinase activity and the EGF-binding capacity of the receptor cannot be determined. Moreover, large amounts of EGF are required that are difficult to recover.

7. Antibody-Affinity Chromatography

As with the purification of receptor using ligand-affinity chromatography, methods based on antibody-affinity are frequently restricted by the need to preserve receptor activity. Thus, high affinity antibodies may give cleaner receptor preparations and greater yields when compared with low affinity antibodies, but bound receptor may be eluted from the latter without substantial loss of activity. In our laboratory we have developed two methods for the purification of EGF receptor, one using a high affinity antibody and the other using a lower affinity antibody from which the receptor can be eluted by competition with carbohydrate.

In our hands, methods using antipeptide antibody affinity and elution by competing peptide give poor yields and require large amounts of synthetic peptides. On the other hand, the eluted receptor is active and a variety of receptor sources are possible.

Preparation of immobilized MAb essentially follows the protocol described for immobilizing EGF. For the preparative scales described below, approximately 10 mg MAb in solution are dialyzed against 0.1*M* $NaHCO_3$, pH 8.2 at 4°C. The protein concentration is estimated using the Bradford reagent, before coupling to 25 mL washed gel.

7.1. MAb 9A-Affinity

This antibody was generated using whole A431 cells as immunogen, followed by a simple screening assay involving binding to immobilized cells (Parker et al., 1984). MAb 9a is an IgG3 that recognizes the human blood group A antigen that is found on the EGF receptor as well as glycolipids of A431 cells. No EGF receptor from any other human cell line has yet been found to display this determinant. The nature of the determinant allows elution by competition with a simple, dialyzable sugar that leaves the receptor fully active.

1. The pH of the A431 cell lysate is adjusted to 8.5 with 1*M* NaOH, before centrifugation at 100,000*g*.
2. The supernatant is decanted carefully, added to the 9A matrix, and tumbled at 4°C for 2 h.
3. The supernatant is then removed from the matrix in a scintered-glass funnel. Any residual matrix is collected by washing out the flask used for tumbling with 50 mL PBSA containing 0.5*M* NaCl, 1 m*M* EDTA, 1 m*M* DTT, 0.1% Triton X-100 (Buffer A) at 4°C and this is then added to the matrix.
4. Nonspecifically bound material is removed by washing with 100 mL Buffer A containing 0.25*M* D-glucose followed by 500 mL Buffer A also at 4°C.
5. The matrix is eluted with 15 mL 1 m*M* DTT, 0.3*M* *N*-acetylgalactosamine, 1 m*M* EDTA, 0.15*M* NaCl 0.05% Triton X-100, 50 m*M* HEPES, pH 7.5 at 4°C for 30 min and the eluate collected by filtration as above. This elution step is repeated and the filtrates are pooled.
6. The pooled filtrates are dialyzed at 4°C for 2 h against 50% glycerol, 1 m*M* DTT, 1 m*M* EDTA, 50 m*M* HEPES, pH 7.5 for storage at –20°C. This step also removes free sugar and concentrates the preparation.
7. The matrix is washed with PBSA and stored at 4°C in this solution containing 0.05% azide. Further purification can be achieved using ion exchange chromatography that does not destroy receptor activity. This can also be used to concentrate a preparation. All steps are carried out at 4°C.

8. The receptor in 50% glycerol is diluted with three volumes of 50 m*M* Tris/HCl pH 7.4, 1 m*M* b-mercaptoethanol, 0.1% Triton X-100 (running buffer) and applied to a MonoQ column (HR5/5, Pharmacia), previously equilibrated with running buffer at 0.1 mL/min. After washing with running buffer, receptor is eluted by a step to Buffer A containing 0.5*M* NaCl.
9. The elution can be monitored by the absorbance at 280 nm if a reference cell is available that can be filled with running buffer in order to offset the absorbance of the detergent. The clearest profile is, however, obtained using a Bradford analysis of each fraction.
10. This preparation can be stored at –20°C following dialysis into 50% glycerol buffer as above.

7.2. MAb R1-Affinity

As for MAb 9A, this antibody was generated following immunization with whole A431 cells, but the subsequent screening used inhibition of EGF binding to immobilized A431 cells (Waterfield et al., 1982). MAb R1 is an IgG2b that recognizes a conformation-sensistive epitope of the human EGF receptor external domain. It has an extremely high affinity for the receptor and accordingly a low pH is required for elution, which destroys receptor kinase activity as well as most of ligand binding, but high yields of denatured receptor can be obtained (Downward et al., 1984b).

1. Initial steps are performed as above but EGFR1 matrix is added in place of the 9A matrix.
2. The matrix is filtered as above and then washed with 500 mL PBSA containing 0.65*M* NaCl, 0.1% NP40 followed by 500 mL PBSA containing 0.1% NP40 at 4°C.
3. Elution is carried out with 10 mL 50 m*M* sodium citrate pH 3.0 containing 0.05% NP40 at 4°C for 10 min. This step is repeated and the eluates are pooled. Finally, the pH of the eluates and the matrix are adjusted to 7.0.
4. The receptor purity can be determined by SDS-PAGE. If necessary, further purification can be achieved using gel permeation chromatography following disulfide reduction and alkylation.
5. The EGF receptor preparation is lyophilized, resuspended in 0.5*M* Tris/HCl pH 8.5 containing 6*M* guanidine hydrochloride at 0.5–1 mg/mL, and incubated overnight at 37°C with 10 m*M* DTT under nitrogen.
6. Alkylation is performed with excess iodoacetamide for 60 min at 20°C.
7. The preparation is applied to a TSK 4000 column (0.7× 60 cm, LKB or equivalent), previously equilibrated with 0.1*M* KH_2PO_4, pH 4.5 containing 6*M* guanidine hydrochloride at a flowrate of 0.5 mL/min while monitoring the absorbance at 280 nm.

Figure 1 shows representative examples of EGF receptor purified by ligand- or antibody-affinity chromatography.

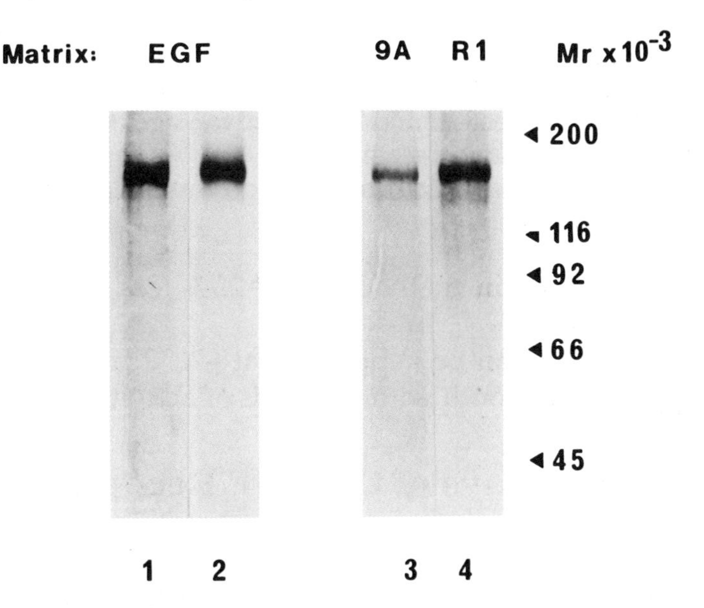

Fig. 1. Purification of EGF receptor. EGF receptor purified from A431 cells by EGF-affinity (1,2) and MAb 9A-affinity (3) or from placental membranes by WGA- and MAb R1-affinity chromatography (4) was run on 7.5% SDS-polyacrylamide gels and stained with Coomassie blue, with the exception of lane 2, which shows an autoradiograph of material purified from [^{35}S]methionine-labeled A431 cells using EGF-affinity.

8. Secreted, Truncated Receptor

Apart from producing large amounts of EGF receptor, A431 cells also secrete a truncated form of the receptor that lacks the kinase and transmembrane domains (Weber et al., 1984). The methods described above for purification of intact EGF receptor are appropriate for purification of the secreted EGF-binding form from conditioned A431 media. A first wheat-germ agglutinin-affinity step is desirable because of the presence of large amounts of serum albumin in the conditioned medium. As this 110 kDa form is soluble, detergents may be avoided during purification.

9. Concentration Methods and Storage of the Purified Material

Depending on the use intended for the purified material, it may be necessary to concentrate the eluted receptor to small volumes. The following methods are recommended:

1. Ultrafiltration: The eluted material is concentrated using an Amicon ultrafiltration unit with XR50 filters. If possible, the procedure should be carried out at 4°C.

2. Dialysis: The material is dialyzed against 20 m*M* Tris, pH 7.5, 50% glycerol, 2 m*M* EDTA, 1 m*M* DTT. The concentration obtained by this method is usually around threefold and has the advantage that the material can be kept at –20°C without freezing.
3. Centrifugation using Centricon microconcentrators (Amicon), which can concentrate about tenfold.

For all concentration methods, a 10–20% loss of receptor protein should be expected.

Purified receptor can be kept either at –20°C in 50% glycerol or at –70°C for approx one month with little loss of activity.

10. Assays of the EGF Receptor

10.1. Binding Assay

The presence of detergents as well as the relatively low amounts of EGF receptor obtained, make the usual protein determination assays (e.g., Lowry, Bradford) unsuitable. The amount of receptor prepared by EGF- and 9A-affinity chromatography can be determined by ligand-binding assays. There are two ways of separating receptor-EGF complexes from unbound factor. The first was described by Carpenter (1979) using polyethylene glycol precipitation. The second is the radioimmunoassay method described by Gullick et al. (1984), in which the receptor is immunoprecipitated with antibody R1 coupled to Protein A Sepharose beads. Scatchard analysis of the data gives estimates of both the amount of receptor present at various purification stages and, if needed, its affinity for EGF (expressed with the dissociation constant, K_d). The radioimmunoassay is performed as follows.

The EGF receptor solution is made up to a final vol of 0.2 mL in 20 m*M* Tris, pH 7.5, 0.2% Triton X-100, 150 m*M* NaCl, 10% glycerol, 1 mg/mL BSA. ^{125}I-EGF is added to a final concentration of 200 n*M* (or a range of 0.2–100 n*M* if estimation of the dissociation constant is desired) and monoclonal antibody R1 to a final 67 n*M*. The solution is incubated at room temperature for 1 h followed by addition of 20 µL of a 1:1 suspension of protein A-Sepharose (Pharmacia) and tumbling for 30 min at room temperature. The Sepharose beads are then washed rapidly 3 times by centrifugation, aspiration of the supernatant, and resuspension of the pellet in 1 mL of ice-cold assay buffer. The pellets are then counted for bound radioactivity in a γ-counter. Assays are performed in triplicates. The nonspecific background binding is determined by the use of 5 µL of a 1:100 dilution of normal mouse

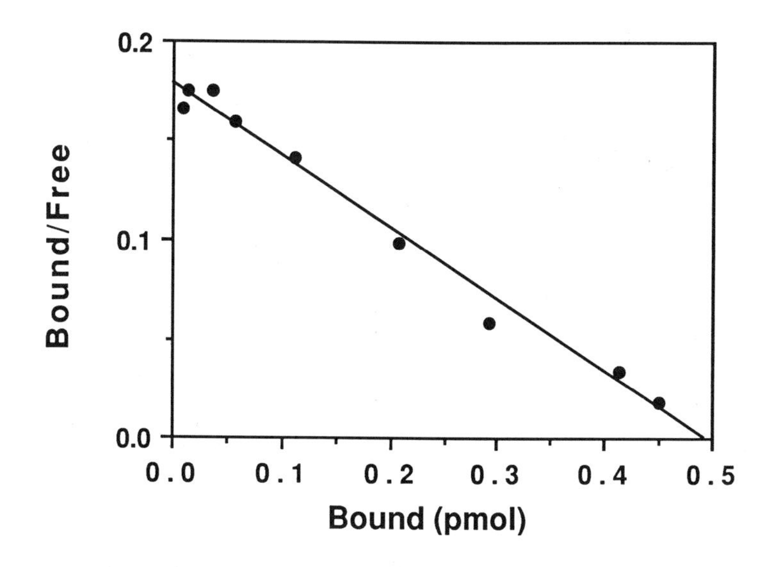

Fig. 2. Scatchard analysis of ^{125}I-EGF binding on purified EGF receptor. The binding of ^{125}I-EGF to EGF-affinity purified receptor was measured by radioimmunoassay. The specific activity of ligand was 1.1×10^5 cpm/ng. Background values were obtained for each point by incubating with a 100-fold excess of unlabeled EGF.

serum or in the presence of a 200-fold excess of unlabeled EGF. Figure 2 shows a Scatchard analysis of ^{125}I-EGF binding data obtained with a typical preparation of EGF-affinity purified receptor.

10.2. Kinase Assays

The protein tyrosine kinase activity of the purified receptor can be determined both toward an exogenous substrate and toward itself (autophosphorylation). A variety of tyrosine-containing peptides may be used, including angiotensin II and src-peptide (both available from Sigma).

Purified receptor is made up to a final concentration of 50 m*M* HEPES, pH 7.4, 150 m*M* NaCl, 2 m*M* $MnCl_2$, 12 m*M* $MgCl_2$, 0.2% Triton X-100, 10% glycerol, 100 μ*M* sodium orthovanadate, and exogenous substrate peptide added to 1 m*M* (or no addition in the assay of autophosphorylation). The mixture is incubated with or without 200 n*M* EGF for 30 min at room temperature. [γ-^{32}P]ATP is then added to a final concentration of 5 μ*M* (for autophosphorylation) or 20 μ*M* (for exogenous substrates) at 500 cpm/pmol and the incubation continued for 10 min at 0°C. The final incu-

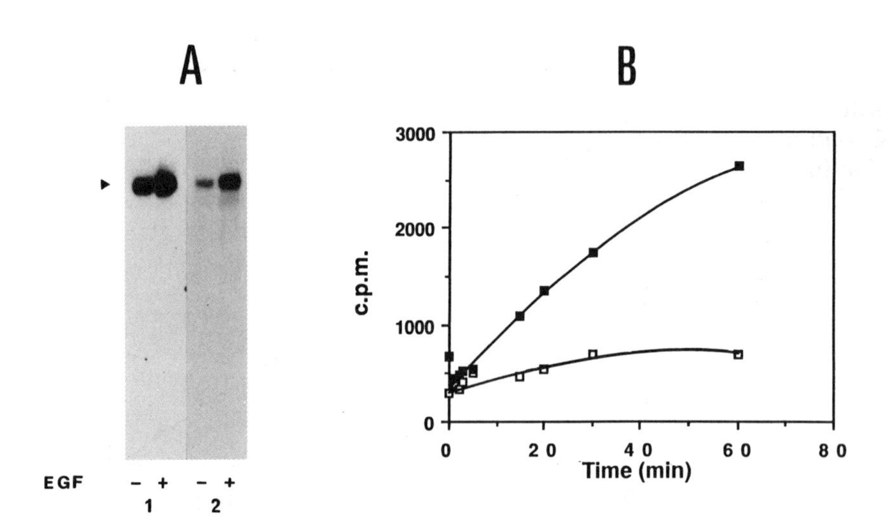

Fig. 3. Kinase activity of purified EGF receptor. **A.** Autophosphorylation: EGF receptor purified by EGF-affinity (1) or MAb 9A-affinity (2) was assayed for self-phosphorylation as described in the text in the absence or presence of EGF as indicated. Autoradiographs of 7.5% SDS-polyacrylamide gels are shown. Approximately 10 times more material was used in (1). Note the difference in the level of stimulation by EGF in the two preparations. Material purified by R1-affinity showed no activity in this assay. **B.** Phosphorylation of exogenous peptide: EGF receptor was purified by 9A-affinity and assayed for kinase activity against angiotensin II as described in the text in the absence (open symbols) or presence (filled symbols) of EGF.

bation volume is adjusted to 50 µL. The reaction may be stopped in two different ways depending on the type of assay. For autophosphorylation, concentrated Laemmli SDS gel sample buffer is added and the mixture heated to 100°C for 2 min, followed by SDS-PAGE in a 7.5% gel. Following autoradiography, the EGF receptor band is excised and counted for Cerencov radiation. For peptide substrate phosphorylation, the reaction is stopped by the addition of an equal volume of 5% trichloroacetic acid and then bovine serum albumin to 1 mg/mL. After standing on ice for 20 min, the mixture is centrifuged at 10,000*g* for 10 min and the supernatants spotted onto Whatman P11 phosphocellulose paper squares, which are then washed 3 times for 10 min with 30% acetic acid, dried, and counted for radioactivity.

Figure 3 shows typical examples of autophosphorylation and peptide phosphorylation by purified EGF receptor.

Table 2
Purification of the EGF Receptor[a]

	A431 cells	Placental membranes
Lysate	100%	100%
WGA-affinity	—	62%
EGF-affinity	8%	5%
9A-affinity	6%	—
R1-affinity	33%	21%

[a]Typical yields of purified EGF receptor by the methods described in the text are shown. The amount of EGF receptor in the 100,000 g lysate and in the column eluates was estimated by radioimmunoassay except for antibody R1-affinity where amino acid analysis was used. The lysate from approximately 2×10^8 A431 cells (obtained from one roller-bottle) contained 80–100 µg EGF receptor. Placental syncytiotrophoblastic membranes contained 120 µg EGF receptor per 100 mg total protein.

11. Comparison of the Three Methods

The first consideration for choosing one of the above described methods is the source from which the EGF receptor has to be purified. The monoclonal antibody 9A recognizes a carbohydrate epitope and is specific to the receptor from A431 cells. Antibody R1 reacts with the human receptor from any source, as it recognizes the protein core. An EGF-affinity column, however, is suitable for obtaining receptor from a wide range of species, with the exception of chicken.

The need for an active receptor kinase should also be taken into account. Purification using antibody R1 results in complete loss of activity and this method is therefore preferable for primary structural studies. The EGF-affinity method gives variable results, but usually a very small stimulation of the kinase activity by EGF is observed. Purification by antibody 9A, however, allows considerable stimulation by EGF, because of the neutral pH used during elution.

The best yield of receptor protein is obtained by the R1-affinity method, whereas EGF-affinity and 9A-affinity give roughly similar yields (Table 2). Purity depends on the source of the receptor and the inclusion of a wheatgerm agglutinin-affinity step. EGF-affinity and R1-affinity result in an essentially homogenous product, whereas the 9A-affinity purified receptor is usually less pure.

Finally, it should be noted that the 9A-Affigel column is not very stable and can only be used a few times. EGF- and R1-Affigel, on the other hand, if stored properly, remain effective for over a year and for multiple receptor purifications.

References

Akiyama, T., Kaddooka, T., and Ogawara, H. (1985) *Biochem. Biophys. Res. Commum.* **131,** 442–448.

Carpenter, G. (1979) *Life Sci.* **24,** 1691–1698.

Carpenter, G. (1987) *Ann. Rev. Biochem.* **56,** 881–914.

Cohen, S. (1983) *Meth. Enzymol.* **99,** 379–387.

Cohen, S., Carpenter, G., and King, L. (1980) *J. Biol. Chem.* **255,** 4834–4842.

Cohen, S., Fava, R. A., and Sawyer, S. T. (1982a) *Proc. Natl. Acad. Sci. USA* **79,** 6237–6241.

Cohen, S., Ushiro, H., Stoscheck, C., and Chinkers, M. (1982b) *J. Biol. Chem.* **257,** 1523–1531.

Downward, J., Parker, P., and Waterfield, M. D. (1984a) *Nature* **311,** 483–485.

Downward, J., Yarden, Y., Mayes, E., Scrace, G., Totty, N., Stockwell, P., Ullrich, A., Schlessinger, J., and Waterfield, M. D. (1984b) *Nature* **307,** 521–527.

Fabricant, R. N., DeLarco, J. E., and Todaro, G. J. (1977) *Proc. Natl. Acad. Sci. USA* **74,** 565–569.

Gregoriou, M. and Rees, A. R. (1984) *Biochem. Soc. Trans.* **12,** 160–165.

Gullick, W. J., Downward, D. J. H., Marsden, J. J., and Waterfield, M. D. (1984) *Anal. Biochem.* **141,** 253–261.

Parker, P. J., Young, S., Gullick, W. J., Mayes, E. L. V., Bennett, P., and Waterfield, M. D. (1984) *J. Biol. Chem.* **259,** 9906–9912.

Savage, C. R. and Cohen, S. (1972) *J. Biol. Chem.* **247,** 7609–7611.

Smith, C. H., Nelson, D. M., King, B. F., Donohue, T. M., Ruzycki, S. M., and Kelley, L. K. (1977) *Am. J. Obstet. Gynecol.* **128,** 190–196.

Thom, D., Powell, A. J., Lloyd, C. W., and Rees, D. A. (1977) *Biochem. J.* **168,** 187–194.

Ullrich, A., Coussens, L., Hayflick, J. S., Dull, T. J., Gray, A., Tam, A. W., Lee, J., Yarden, Y., Libermann, T. A., Schlessinger, J., Downward, J., Mayers, E. L. V., Whittle, N., Waterfield, M. D., and Seeburg, P. H. (1984) *Nature* **309,** 418–425.

Waterfield, M. D., Mayes, E. L. V., Stroobant, P., Bennet, P. L. P., Young, S., Goodfellow, P. N., Banting, G. S., and Ozanne, B. (1982) *J. Cell. Biochem.* **20,** 149–161.

Weber, W., Bertics, P. J., and Gill, G. N. (1984) *J. Biol. Chem.* **259,** 14631–14636.

Yarden, Y., Harari, I., and Schlessinger, J. (1985) *J. Biol. Chem.* **260,** 315–319.

Platelet-Derived Growth Factor B Type Receptor

Carl-Henrik Heldin and Lars Rönnstrand

1. Introduction

Platelet-derived growth factor (PDGF) is a major mitogen in serum for connective tissue cells (for reviews, *see* Heldin et al., 1985; Ross et al., 1986). The in vivo function of PDGF is not known. However, the fact that PDGF is released from platelets in conjunction with the blood coagulation and stimulates not only proliferation, but also chemotaxis and synthesis of matrix proteins of connective tissue cells, is compatible with a function for PDGF in wound healing (Heldin et al., 1985; Ross et al., 1986). The fact that PDGF is expressed in the placenta also suggests a role for PDGF during the development (Goustin et al., 1985). In addition, PDGF has been found to be produced by type-1 astrocytes and to act on O-2A progenitor cells that differentiate to oligodendrocytes and type-2 astrocytes, suggesting a role for PDGF in the developing rat brain (Richardson et al., 1988; Noble et al., 1988). PDGF may also have adverse effects in several conditions involving cell proliferation, such as atherosclerosis, various fibrotic conditions, as well as malignant diseases (Heldin et al., 1985; Ross et al., 1986). The latter possibility is illustrated by the fact that PDGF is homologous to $p28^{sis}$, the transforming protein of simian sarcoma virus (Waterfield et al., 1983; Doolittle et al., 1983).

Receptor Purification, vol. 1

Structurally, PDGF is a 30 kDa disulphide-bonded dimer of A chains and B chains (Johnsson et al., 1982). The two PDGF chains are 60% similar in their amino acid sequences (Josephs et al., 1984; Betsholtz et al., 1986), and all three dimeric combinations (AA, AB, and BB) have been identified and purified from platelets and transformed cells (Stroobant and Waterfield, 1984; Heldin et al., 1986; Hammacher et al., 1988). The various isoforms of PDGF have been found to have different functional activities; thus, in contrast to PDGF-AB, PDGF-AA was found to have low mitogenic activity for human fibroblasts, and no chemotactic activity or ability to stimulate actin reorganization (Nistér et al., 1988a). The dissimilarities in function are most likely owing to the fact that the isoforms of PDGF bind with different affinities to two distinct receptor classes, denoted type A and type B (also called α and β) (Heldin et al., 1988; Hart et al., 1988).

The A type PDGF receptor binds all three isoforms of PDGF with high affinities (K_d:s about 0.5 n*M*), whereas the B type PDGF receptor binds only PDGF-BB with high affinity (K_d about 0.5 n*M*), PDGF-AB with lower affinity (K_d about 6 n*M*), and not PDGF-AA (Östman et al., 1989; Severinsson et al., 1989). The B type PDGF receptor has been purified (Daniel et al., 1985; Bishayee et al., 1986; Rönnstrand et al., 1987), its cDNA cloned (Yarden et al., 1986; Gronwald et al., 1988; Claesson-Welsh et al., 1988), and monclonal (Hart et al., 1987; Rönnstrand et al., 1988) as well as polyclonal (Rönnstrand et al., 1987; Keating and Williams, 1987) antisera are available. It is a 180 kDa transmembrane glycoprotein with a tyrosine-specific protein kinase activity associated with its cytoplasmic domain (Ek et al., 1982; Nishimura et al., 1982). The A type receptor is a 170 kDa component (Claesson-Welsh et al., 1989), which remains to be further characterized.

The intracellular messenger system that is activated in response to binding of a growth factor to its receptor seems to be conserved between different cell types. Therefore, the responsiveness to a certain growth factor is most likely determined by the expression of the corresponding growth factor receptor. Investigations of the structure, mechanism of activation, and the relation between structure and function of growth factor receptors, are dependent on the availability of pure receptor. In this communication we describe a method to purify large quantities of native B type PDGF receptor from porcine uterus. The purification method takes advantage of the fact that the receptor is a glycoprotein and binds to certain lectins, that the receptor is rather acidic and therefore purified several-fold by ion exchange chromatography, and, finally, that the receptor after autophosphorylation on tyrosine residues has affinity for antiphosphotyrosine antibodies.

2. Purification of the B Type PDGF Receptor

2.1. Assay for Solubilized PDGF Receptor

Several attempts have been made by us and others to establish an assay for binding of ^{125}I-PDGF to solubilized receptor. This, however, has been difficult since:

1. PDGF is a very "sticky" molecule, being both highly charged and hydrophobic,
2. The ligand is of relatively large size, which makes it difficult to separate it from the ligand-receptor complex by precipitation assays,
3. The solubilized B type PDGF receptor has relatively low affinity for its ligand, with half-maximal stimulation of autophosphorylation at about 5 n*M* (Heldin et al., 1989).

Daniel et al. (1985) incorporated the soluble PDGF receptor within liposomes, and showed that the purified receptor could bind PDGF. The percentage of nonspecific binding was, however, very high (45%). In our laboratory, we have employed the ligand-stimulated autophosphorylation of the B type receptor as an assay in the purification procedure (Rönnstrand et al., 1987).

2.2. Preparation of Membranes

In order to localize a source of B type PDGF receptor that would allow the purification of nanomole quantities of receptor, we explored the possibility of using a tissue as starting material rather than cultured cells that are expensive and time consuming to grow in sufficiently large scale (>1000 roller bottles of fibroblasts would be required for the purification of one nanomole of receptor). Membranes were, therefore, prepared from about 20 different porcine organs and screened for B type PDGF receptor by the autophosphorylation assay. Uterus was found to be the best source of receptor and was chosen as starting material in the purification.

Membranes were prepared by differential centrifugations of homogenized porcine uteri (Rönnstrand et al., 1987). Five kg of tissue was processed at a time, which gave about 1.5 g of membranes. In order to prevent proteolysis during the preparation, phenylmethylsulfonyl fluoride and EDTA was included in the homogenization buffer. Without protease inhibitors, the B type PDGF is degraded to a 130 kDa fragment by a calcium-dependent SH-protease (Ek and Heldin, 1986).

The prevention of oxidation of the B type PDGF receptor was also found to be very important. Therefore, the homogenization of tissue was done under an atmosphere of nitrogen. As a further protection against oxidation, dithiothreitol (1 m*M*) was included in the buffers throughout the purification. Similar sensitivity to oxidation has also been found for the insulin receptor (Petruzzelli et al., 1984).

2.3. Chromatography on Wheatgerm Agglutinin (WGA)-Sepharose

The first step in the purification (Table 1) was chromatography on WGA-Sepharose. Approximately 1.5 g of uterus membranes were solubilized in 400 mL of 2.5% Triton X-100, 10% glycerol, 20 m*M* HEPES, 4 m*M* EGTA, 1 m*M* dithiothreitol, pH 7.4. After 30 min of incubation on ice, the solubilizate was centrifuged at 100,000*g* for 30 min. The resulting supernatant was pumped through a 20-mL WGA-Sepharose column (20 mg of lectin/mL of gel), at a flowrate of approx 30 mL/h. The column was washed with 100 mL of 0.2% Triton X-100, 10% glycerol, 0.15*M* NaCl, 20 m*M* HEPES, 1 m*M* EGTA, 1 m*M* dithiothreitol, pH 7.4, at a flowrate of 50 mL/h. After a second wash with the same buffer, but lacking NaCl, bound material was eluted with 50 mL of 0.3*M* *N*-acetylglucosamine, 0.2% Triton X-100, 10% glycerol, 20 m*M* HEPES, 1 m*M* EGTA, 1 m*M* dithiothreitol, pH 7.4, at a flowrate of 20 mL/h.

2.4. Chromatography on an FPLC Mono-Q Column

The eluate of the lectin column was applied directly onto an FPLC Mono-Q column (bed vol 2 mL) and eluted using a gradient of 0–0.5*M* NaCl in 0.2% Triton X-100, 10% glycerol, 20 m*M* HEPES, 1 m*M* EGTA, 1 m*M* dithiothreitol, pH 7.4, at a flowrate of 2 mL/min. Individual fractions were analyzed for PDGF receptor content by the autophosphorylation assay.

2.5. Chromatography on Antiphosphotyrosine-Sepharose

In order to make an antiphosphotyrosine immunoglobulin column, the immunoglobulin fraction of a rabbit antiserum against phosphotyrosine (Ek and Heldin, 1984) was affinity purified on an L-phosphotyrosine-Sepharose Cl-4B column. After elution with 40 m*M* phenylphosphate, the immunoglobulins were further purified by chromatography on Protein A-Sepharose (Pharmacia P-L Biochemicals). The antibodies were then coupled to cyanogen bromide-activated Sepharose (2 mg of immunoglobulin/mL of gel).

Table 1
Purification of the B Type PDGF Receptor from Membranes of Porcine Uterus (modified from Rönnstrand et al., 1987)

	Total amount of protein, mg	Specific activity, pmol phosphate/min/mg	Purification, -fold	Yield, %
Membranes solubilized with Triton X-100	1280	0.023	1	100
Eluate from WGA-Sepharose	14.2	2.0	90	99
Pool from the Mono-Q chromatogram	2.5	5.3	233	46
Eluate from anti-phosphotyrosine-Sepharose after dephosphorylation	0.16	14.3	630	8

Active fractions from the Mono-Q chromatogram were pooled and incubated overnight on ice in the presence of 3 m*M* $MnCl_2$ and 15 μ*M* ATP. During this long incubation time, the background activity of the receptor kinase leads to autophosphorylation on tyrosine residues of the receptor, even in the absence of ligand. A small quantity of ^{32}P-ATP was also included to monitor the purification. The phosphorylated Mono-Q fractions were applied to a 2-mL antiphosphotyrosine antibody column (2 mg/mL) at a flowrate of 10 mL/h. The column was subsequently washed with 10 mL of Triton X-100, 10% glycerol, 0.5*M* NaCl, 20 m*M* HEPES, 1 m*M* EGTA, 1 m*M* dithiothreitol, pH 7.4, and eluted with 10 mL of the same buffer containing 40 m*M* phenylphosphate, at a flowrate of 10 mL/h.

2.6. Dephosphorylation of Phosphorylated Receptor by Alkaline Phosphatase

The eluate from the immunoaffinity column was dialyzed against 0.2% Triton X-100, 10% glycerol, 0.15*M* NaCl, 5 m*M* benzamidine, 1 m*M* $MgCl_2$, 1 m*M* EGTA, 20 m*M* HEPES, 1 m*M* dithiothreitol, pH 7.4, at 4°C for 6 h with one change of dialysis buffer. Thereafter, 500 U of alkaline phosphatase (from calf intestine, Grade I, cat no 108 146, Boehringer Mannheim) was added, and the dialysis was continued for 12 h against the same dialysis buffer. In order to separate the receptor from the phosphatase, the dialyzate was applied to a 200-μL DEAE-Sepharose Cl-4B column, equilibrated with the dialysis buffer, and washed with 5 mL of dialysis buffer. The phosphatase did not bind to the column under these conditions, and the receptor was eluted with 2 mL of 0.2% Triton X-100, 10% glycerol, 0.5*M* NaCl, 20 m*M* HEPES, 1 m*M* EGTA, 1 m*M* dithiothreitol, pH 7.4. The recovery of receptor during this procedure was approx 70%. Starting from 1.5 g of membranes about 160 μg of pure receptor was obtained, at an overall recovery of 8% (Table 1).

2.7. Characteristics of the Purified B Type PDGF Receptor

The purified B type PDGF receptor was shown to be essentially homogeneous by analysis by SDS-gel electrophoresis and silver-staining (Fig. 1). The product was also functionally active and responded to PDGF by autophosphorylation. The different dimeric forms of PDGF showed different degree of agonist activity; PDGF-BB was a potent stimulator of the receptor kinase, PDGF-AB had a moderate effect, whereas PDGF-AA had no stimulatory effect on the B type PDGF receptor (Fig. 1). The autophosphorylation of the PDGF receptor was shown, by dilution experiments, to be an intramolecular event (Rönnstrand et al., unpublished observation).

3. Properties of the B Type PDGF Receptor

3.1. Biosynthesis and Processing

The availability of large quantities of pure PDGF B type receptor has made it possible to make polyclonal (Rönnstrand et al., 1987) as well as monoclonal (Rönnstrand et al., 1988) antibodies against this protein. With these reagents, the biosynthesis and processing of the receptor in human fibroblasts were studied. The receptor is synthesized as a 160 kDa precursor that carries immature *N*-linked carbohydrate complexes. The receptor is then processed to a 180 kDa form by maturation of the *N*-linked carbohydrate complexes and possibly other posttranslational modifications (Claesson-Welsh et al., 1987,1988b). The receptor at the cell surface is turned over fairly rapidly; the degradation rate is increased further by addition of ligand. Similar observations on the B type receptor have been made in other cell systems (Hart et al., 1987; Keating and Williams, 1987).

3.2. In Vivo Distribution

The two monoclonal antibodies obtained against the porcine PDGF B type receptor were both found to crossreact with the human receptor and

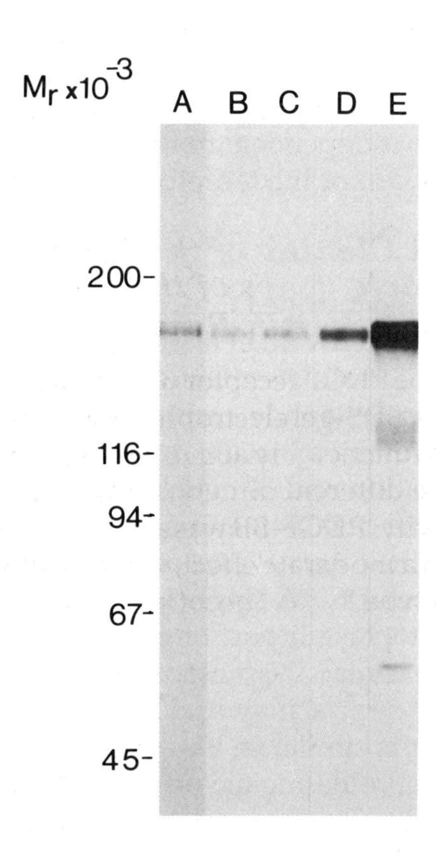

Fig. 1. Characterization of the purified B type PDGF receptor. Aliquots of the purified receptor were analyzed by SDS-gel electrophoresis and silver staining (A), and by autophosphorylation in the absence of ligand (B) or after stimulation with 100 ng of either PDGF-AA (C), PDGF-AB (D), or PDGF-BB (E). The autophosphorylation assay was performed as described (Ek and Heldin, 1982) and a 5–10% gradient polyacrylamide gel was used for SDS-gel electrophoresis. An autoradigram of the gel is shown (Lanes B–E). The 130 kDa component visible in lane E is a proteolytic degradation product of the receptor (Ek and Heldin, 1986).

to recognize the external domain of the receptor (Rönnstrand et al., 1988). Use of the monoclonal antibodies in immunohistochemical stainings revealed that the B type PDGF receptor was present on connective tissue cells of the endometrium of the uterus, but was not present on smooth muscle cells of the myometrium (Terracio et al., 1988). This finding was unexpected since smooth muscle cells in vitro have PDGF receptors. Analysis of porcine myometrial cells explanted into tissue culture revealed that B type PDGF receptors appeared on the surface of the smooth muscle cells after 2–3 d in vitro (Terracio et al., 1988). Analysis by immunohistochemistry of biopsies of other human tissues revealed that PDGF B type receptors are generally not expressed on connective tissue cells of normal tissues. However, receptors become expressed, e.g., in conjunction with inflammatory processes (Rubin et al., 1988a,b). These observations indicate that the action of PDGF in vivo is dependent not only on the availability of ligand, but also on the expression of the receptor on the responder cells.

3.3. cDNA Cloning of the B Type Receptor

cDNA clones for the murine (Yarden et al., 1986) and human (Claesson-Welsh et al., 1988; Gronwald et al., 1988) PDGF B type receptor have been obtained. The nucleotide sequence of the human cDNA predicts a 1106 amino acid long polypeptide, including a cleavable signal sequence. The extracellular part of the molecule contains 11 potential acceptor sites for *N*-linked glycosylation and 10 evenly spaced cysteine residues. The spacing of the cysteine residues and some other properties of the sequence suggests that the external part of the receptor consists of five immunoglobulin-like domains. A single hydrophobic transmembrane domain is found. The cytoplasmic portion of the receptor contains a tyrosine kinase domain that, in contrast to most other tyrosine kinases, contains an insert sequence of 104 amino acids without homology to other kinases. The B type PDGF receptor is similar in its general organization to two other receptors, the colony stimulating factor-1 receptor (Coussens et al., 1986) and the c-*kit* product, the ligand of which remains unknown (Yarden et al., 1987). The structure of the B type PDGF receptor is schematically illustrated in Fig. 2.

3.4. Mechanism of Activation of the PDGF B Type Receptor Kinase

Incubation of PDGF-BB with purified PDGF B type receptor was found to induce dimerization of the receptor in a dose-dependent fashion; the dimerization was maximal at about 0.5 μg/mL of PDGF-BB and decreased at higher concentrations of ligand (Heldin et al., 1989). This indicates that the dimeric complex contains one molecule PDGF-BB and two receptor molecules; most likely each of the two subunits in the PDGF-BB molecules bind one receptor molecule. The dimerization was found to be closely associated with activation of the receptor kinase. Phosphorylation reactions are not a prerequisite for dimerization, however, since dimerization occurred also in the absence of ATP. Further support the notion that dimerization is connected with kinase activation comes from the observation that the monoclonal receptor antibodies, but not Fab' fragments

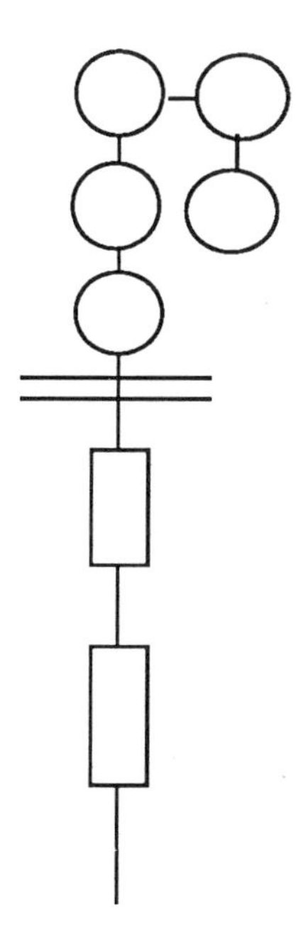

Fig. 2. Schematic illustration of the structure of the B type PDGF receptor. The extracellular part of the receptor consists of five immunoglobulin-like domains. The cytoplasmic part of the receptor contains a protein tyrosine ki-nase domain with an insert sequence without homology to other kinases.

thereof, increase the kinase activity of the receptor (Rönnstrand et al., 1988). In conclusion, the data indicate that ligand binding to the external domain of the B type PDGF receptor induces dimerization of the receptor. It is conceivable that this makes possible an interaction between the tyrosine kinase domains of the cytoplasmic parts of the two receptors, leading to their activation.

4. Future Perspectives

The availability of large quantities of pure and functionally active B type PDGF receptor, together with specific antibodies and cDNA clones, will now make it possible to address several important questions related to the mechanism of action and in vivo function of the PDGF B type receptor. One of the most important tasks is of course to identify the physiologically relevant substrates for the receptor kinase. Several components in fibroblasts, that are phosphorylated on tyrosine residues after PDGF stimulation, have been found (Cooper et al., 1982; Ek and Heldin, 1984; Frackelton et al., 1984), but no substrate has yet been identified with a proven role in the mitogenic pathways. It will also be important to identify which tyrosine residues in the molecule are autophosphorylated, and, furthermore, to determine the functional role of the autophosphorylation.

Transfection of PDGF B type receptor cDNA in Chinese hamster ovary cells has yielded expression of functionally active receptors at the cell surface (Escobedo et al., 1988; Severinsson et al., 1989). Thus, it is now possible to use site directed mutagenesis to elucidate the structure–function relationship of the receptor. Studies are on the way to determine the functional consequences of mutation of the ATP-binding lysine residue, thereby knocking out the receptor kinase, mutation of autophosphorylation sites, and deletion of the insert in the tyrosine kinase domain. It will be of interest to determine which structures in the external part of the receptor are involved in ligand binding, and which domains of the receptor are necessary for the stimulatory effects of PDGF on, e.g., phosphatidylinositol turnover, Ca^{2+}-mobilization, and gene expression.

The availability of high yield expression systems and methods for receptor purification may also eventually make it possible to elucidate the three-dimensional structure of the PDGF B type receptor.

PDGF occurs as three isoforms that bind to A type and B type receptors with different affinities. The available data indicate that both receptor types transduce a mitogenic signal, but there are also functional differences (Nistér et al., 1988; Heldin et al., 1988; Hart et al., 1988). An important aim for future studies will be to characterize the signals transduced by the two receptor types. It will, furthermore, be important to elucidate whether dimerization also occur of A type receptors. If so, an interesting possibility would be that heterodimeric receptor complexes are formed after binding of PDGF-AB or PDGF-BB that bind to both receptor types; one ligand could in such a case induce the activation of two different signals via simultaneous binding to two different receptors.

Acknowledgment

We thank Linda Baltell for valuable help in the preparation of this manuscript.

References

Betsholtz, C., Johnsson, A., Heldin, C.-H., Westermark, B., Lind, P., Urdea, M. S., Eddy, R., Shows, T. B., Philpott, K., Mellor, A., Knott, T. J., and Scott, J. (1986) *Nature* **320,** 695–699.

Bishayee, S., Ross, A. H., Womer, R., and Scher, C. D. (1986) *Proc. Natl. Acad. Sci. USA* **83,** 6756–6760.

Claesson-Welsh, L., Rönnstrand, L., and Heldin, C.-H. (1987) *Proc. Natl. Acad. Sci. USA* **84,** 8796–8800.

Claesson-Welsh, L., Eriksson, A., Morén, A., Severinsson, L., Ek, B., Östman, A., Betsholtz, C., and Heldin, C.-H. (1988) *Mol. Cell. Biol.* **8,** 3476–3486.

Claesson-Welsh, L., Hammacher, A., Westermark, B., Heldin, C.-H., and Nistér, M. (1989) *J. Biol. Chem.* **264,** 1742–1747.

Cooper, J. A., Bowen-Pope, D. F., Raines, E., Ross, R., and Hunter, T. (1982) *Cell* **31,** 263–273.

Coussens, L., van Beveren, C., Smith, D., Chen, E., Mitchell, R. L., Isacke, C. M., Verma, I., and Ullrich, A. (1986) *Nature* **320,** 277–280.

Daniel, T. O., Tremble, P. M., Frackelton, A. R., Jr., and Williams, L. T. (1985) *Proc. Natl. Acad. Sci. USA* **82,** 2684–2687.

Doolittle, T. F., Hunkapiller, M. W., Hood, L. E., Devare, S. G., Robbins, K. C., Aaronson, S. A., and Antoniades, H. N. (1983) *Science* **221,** 275–277.

Ek, B., Westermark, B., Wasteson, Å., and Heldin, C.-H. (1982) *Nature* **295,** 419, 420.

Ek, B. and Heldin, C.-H. (1984) *J. Biol. Chem.* **259,** 11145–11152.

Ek, B. and Heldin, C.-H. (1986) *Eur. J. Biochem.* **155,** 409–413.

Escobedo, J. A., Keating, M. T., Ives, H. E., and Williams, L. T. (1988) *J. Biol. Chem.* **263,** 1482–1487.

Frackelton, A. R., Jr., Tremble, P. M., and Williams, L. T. (1984) *J. Biol. Chem.* **259,** 7909–7915.

Goustin, A. S., Betsholtz, C., Pfeiffer-Ohlsson, S., Persson, H., Rydnert, J., Bywater, M., Holmgren, G., Heldin, C.-H., Westermark, B., and Ohlsson, R. (1985) *Cell* **41,** 301–312.

Gronwald, R. G. K., Grant, F. J., Haldeman, B. A., Hart, C. E., O'Hara, P. J., Hagen, F. S., Ross, R., Bowen-Pope, D. F., and Murray, M. J. (1988) *Proc. Natl. Acad. Sci. USA* **85,** 3435–3439.

Hammacher, A., Hellman, U., Johnsson, A., Gunnarsson, K., Östman, A., Westermark, B., Wasteson, Å., and Heldin, C.-H. (1988) *J. Biol. Chem.* **263,** 16493–16498.

Hart, C. E., Seifert, R. A., Ross, R., and Bowen-Pope, D. F. (1987) *J. Biol. Chem.* **262,** 10780–10785.

Hart, C. E., Forstrom, J. W., Kelly, J. D., Seifert, R. A., Smith, R. A., Ross, R., Murray, M., and Bowen-Pope, D. F. (1988) *Science* **240,** 1529–1531.

Heldin, C. -H., Wasteson, Å., and Westermark, B. (1985) *Mol. Cell. Endocrinol.* **39,** 169–187.

Heldin, C. -H., Johnsson, A., Wennergren, S., Wernstedt, C., Betsholtz, C., and Westermark, B. (1986) *Nature* **319,** 511–514.

Heldin, C.-H., Bäckström, G., Östman, A., Hammacher, A., Rönnstrand, L., Rubin, K., Nistér, M., and Westermark, B. (1988) *EMBO J.* **7,** 1387– 1394.

Heldin, C.-H., Ernlund, A., Rorsman, C., and Rönnstrand, L. (1989) *J. Biol. Chem.* **264,** 8905–8912.

Johnsson, A., Heldin, C.-H., Westermark, B., and Wasteson, Å. (1982) *Biochem. Biophys. Res. Comm.* **104,** 66–74.

Josephs, S. F., Guo, C., Ratner, L., and Wong-Staal, F. (1984) *Science* **223,** 487–491.

Keating, M. T. and Williams, L. T. (1987) *J. Biol. Chem.* **262,** 7932–7937.

Nishimura, J., Huang, J. S., and Deuel, T. F. (1982) *Proc. Natl. Acad. Sci. USA* **79,** 4303–4307.

Nistér, M., Hammacher, A., Mellström, K., Siegbahn, A., Rönnstrand, L., Westermark, B., and Heldin, C.-H. (1988a) *Cell* **52,** 791–799.

Noble, M., Murray, K., Stroobant, P., Waterfield, M., and Riddle, P. (1988) *Nature* **333,** 560–562.

Östman, A., Bäckström, G., Fong, N., Betsholtz, C., Wernstedt, C., Hellman, U., Westermark, B., Valenzuela, P., and Heldin, C.-H. (1989) *Growth Factors* **1,** 271–281.
Petruzzelli, L., Herrera, R., and Rosen, O. M. (1984) *Proc. Natl. Acad. Sci. USA* **82,** 3327–3331.
Richardson, W. D., Pringle, N., Mosley, M. J., Westermark, B., and Dubois-Dalcq, M. (1988) *Cell* **53,** 309–319.
Rönnstrand, L., Beckmann, M. P., Faulders, B., Östman, A., Ek, B., and Heldin, C.-H. (1987) *J. Biol. Chem.* **262,** 2929–2932.
Rönnstrand, L., Terracio, L., Claesson-Welsh, L., Heldin, C.-H., and Rubin, K. (1988) *J. Biol. Chem.* **263,** 10429–10435.
Ross, R., Raines, E. W., and Bowen-Pope, D. F. (1986) *Cell* **46,** 155–169.
Rubin, K., Terracio, L., Rönnstrand, L., Heldin, C.-H., and Klareskog, L. (1988a) *Scand. J. Immunol.* **27,** 285–294.
Rubin, K., Tingström, A., Hansson, G. K., Larsson, E., Rönnstrand, L., Klareskog, L., Claesson-Welsh, L., Heldin, C.-H., Fellström, B., and Terracio, L. (1988b) *Lancet* **1,** 1353–1356.
Severinsson, L., Claesson-Welsh, L., and Heldin, C.-H. (1989) *Eur. J. Biochem.* **182,** 679–686.
Stroobant, P. and Waterfield, M. D. (1984) *EMBO J.* **3,** 2963–2967.
Terracio, L., Rönnstrand, L., Tingström, A., Rubin, K., Claesson-Welsh, L., Funa, K., and Heldin, C.-H. (1988) *J. Cell Biol.* **107,** 1947–1957.
Waterfield, M. D., Scrace, T., Whittle, N., Stroobant, P., Johnsson, A., Wasteson, Å., Westermark, B., Heldin, C.-H., Huang, J. S., and Deuel, T. F. (1983) *Nature* **304,** 35–39.
Yarden, Y., Escobedo, J. A., Kuang, W.-J., Yang-Feng, T. L., Daniel, T. O., Tremble, P. M., Chen, E. Y., Ando, M. E., Harkins, R. N., Francke, U., Fried, V. A., Ullrich, A., and Williams, L. T. (1986) *Nature* **323,** 226–232.
Yarden, Y., Kuang, W.- J., Yang-Feng, T., Coussens, L., Munemitsu, S., Dull, T. J., Chen, E., Schlessinger, J., Francke, U., and Ullrich, A. (1987) *EMBO J.* **6,** 3341–3351.

Colony-Stimulating Factor 1 Receptor

Yee-Guide Yeung and E. Richard Stanley

1. Introduction

Colony stimulating factors (CSFs) regulate the development of granulocytes and macrophages from hemopoietic precursor cells (Stanley and Jubinsky, 1984; Metcalf, 1986; Clark and Kamen, 1987). One of these growth factors, CSF-1, is specifically required for the in vitro survival, proliferation, and differentiation of mononuclear phagocytes (precursor cell→monoblast→promonocyte→monocyte→macrophage). The response to CSF-1 is pleiotropic and varies with mononuclear phagocyte cell type (Stanley et al., 1983). The actions of CSF-1 are mediated via a specific receptor that is selectively expressed on all mononuclear phagocytes and their precursors (Guilbert and Stanley, 1980; Byrne et al., 1981) and trophoblastic cells (Rettenmier et al., 1986). The CSF-1 receptor has been shown immunologically and functionally identical to the c-fms proto-oncogene product (Sherr et al., 1985; Sacca et al., 1986; Roussel et al., 1987) that belongs, with the platelet-derived growth factor receptor, to a subclass of tyrosine receptor kinases (Coussens et al., 1986; Yarden and Ullrich, 1988). Chemical crosslinking studies (Morgan and Stanley, 1984) indicated that the cell surface CSF-1 receptor is a macromolecule of approximately 165 kDa and is not disulfide bonded to any other macromolecule in the macrophage membrane. In this chapter we describe a two-step method that was initially developed in 1983 for the purification of the CSF-1 receptor from a CSF-1 independent, receptor bearing cell line.

2. Source of Receptor

2.1. Selection of Cell Source

A previous CSF-1 binding study (Byrne et al., 1981) showed that a wide variety of mononuclear phagocytes, including bone marrow derived macrophages, peritoneal, and pulmonary alveolar macrophages, possess CSF-1 receptors. The low concentration of these cells in tissues suggested that the best source for receptor purification would be cultured cells. However, these primary cells and cells of the mouse macrophage cell line, BAC1.2F5 (Morgan et al., 1987), which express a high receptor number (approx. 1.2×10^5/cell), have two major disadvantages as sources for receptor purification. They require CSF-1 for proliferation, necessitating the production of large amounts of growth factor, and they all attach strongly to the substratum during cell growth, requiring large and/or sophisticated culture devices compared with those used for suspension culture. Cells of the mouse macrophage cell line, J774.2 (Ralph et al., 1975; Diamond et al., 1978) possess approximately 6 times fewer receptors than BAC1.2F5 cells, but they can be rapidly grown to high cell density (10^6 cells/mL) in suspension culture in the absence of CSF-1. We have, therefore, used these cells exclusively as a source for the purification of the receptor.

2.2. Large Scale Cell Culture

The J774.2 cells,which double every 12 h, are maintained in log phase growth in 30 mL MEM-alpha medium (Gibco Laboratory) containing 10% heat inactivated horse serum (56°C, 30 min) in a T75 tissue culture flask in a humidified atmosphere of 10% CO_2 in air at 37°C.

One day prior to the onset of mass culture, 10 L of alpha medium containing 0.1 mg/mL penicillin and 0.1 mg/mL streptomycin is Millipore filtered directly into a sterile 15 L spinner bottle. The bottle is incubated overnight at 37°C to check sterility and equilibrate temperature.

The next day the cells from two T75 cultures (30 mL each) at a cell density between $0.5–1 \times 10^6$/mL are collected by centrifugation (400*g*, 4°C, 5 min). The cell pellet is suspended in 5 mL culture medium, and a small vol diluted in Trypan Blue solution {0.2%, w/v, Trypan Blue in phosphate-buffered saline (PBS, 0.14 m*M* NaCl/3 m*M* KCl/8 m*M* Na_2HPO_4/1.5 m*M* KH_2PO_4, pH 7.4)} and the cells counted by hemocytometer. Cells are only used if viability by Trypan Blue exclusion exceeds 90%.

The 15 L spinner bottle containing the prewarmed medium is placed in sterile hood and warm, heat inactivated horse serum is added to a final concentration of 10%. The required number of cells (4×10^7 cells for a 4-d culture period, or 1.4×10^8 cells for 3-d culture period) is seeded. The bottle is gassed for 5 min with a 10% CO_2 in air mixture, the caps are covered with aluminum foil, flamed, and then sealed with Parafilm. The spinner is finally placed in 37°C warm room with constant gentle stirring.

The cell density at harvest should be between $0.6–1 \times 10^6$/mL. Cells are collected by centrifugation (400*g*, 4°C, 10 min) in 1 L bottles. The pellet in each bottle is resuspended in 10 mL culture medium and pooled into one 125 mL glass conical centrifuge bottle at 4°C. Of the pooled suspension, 0.5 mL is removed for cell counting in Trypan Blue. Cells are only used if viability exceeds 90% by Trypan Blue exclusion. The bottle is centrifuged (400*g*, 10 min, 4°C) and the supernatant aspirated off. Usually, 10^{10} cells yields a pellet of approx 10 mL.

The cell pellet (8–10 mL) is washed briefly by resuspending in 5 pellet vol of ice-cold PBS containing 4 m*M* sodium iodoacetate (pH 7.4), 1 m*M* sodium EGTA (pH 7.4). Addition of iodoacetate and EGTA will inhibit cell surface sulfhydryl- and calcium-activated proteases. One hundred µL of the suspension is removed and stored on ice for receptor assay. The remainder is centrifuged (400*g*, 5 min, 4°C) to pellet the cells for homogenization.

3. Assay Methods for Receptor

Successful assay of solubilized receptors requires the efficient separation of the receptor–ligand complex from the free ligand. Experiments with the Triton X-100 solubilized membrane fraction from cells prebound with ^{125}I-CSF-1 (Yeung et al., 1986) indicated that separation of the receptor-CSF-1 complex from free CSF-1 can be effected by slight modification of the method of Cuatrecasas (1972), in which the complex is precipitated with polyethylene glycol and the precipitate is separated from free growth factor by membrane filtration. The filtration procedure has also been used to separate free CSF-1 from membrane bound CSF-1 receptor complex in the determination of membrane associated receptor (Yeung et al., 1986).

L-cell CSF-1 is purified and iodinated as previously described (*reviewed in* Stanley, 1985). To reduce retention of free ^{125}I-CSF-1 by filters used in the assay of both membrane associated and solubilized receptor, the ^{125}I-CSF-1 is brought to 10% (v/v) with respect to normal rabbit serum, to 10% (w/v) with respect to polyethylene glycol (PEG 8000; 8000 mol wt, Sigma), centrifuged (2,000*g*, 10 min, 4°C), filtered (0.45 µm cellulose acetate membrane), dialyzed (assay buffer, *see below*, 18 h, 4°C), and refiltered (0.22 µm cellulose acetate membrane) prior to use (or storage at –20°C).

3.1. Assay for Membrane-Associated Receptor

Standard incubation mixtures consist of 10 µL of sample, 40 µL assay buffer (MEM-alpha medium containing 25 m*M* HEPES in lieu of bicarbonate (alpha-HEPES), 0.2% (w/v) bovine serum abumin (BSA; A4378, Sigma) and 0.02% (w/v) NaN_3, pH 7.35), 5 µL of either purified CSF-1 (120,000 U/mL in PBS) or 5 µL PBS, and 5 µL of ^{125}I-CSF-1 (100,000 counts/min; specific activity = 3×10^5 counts/min/ng protein). The competing unlabeled CSF-1 is added 30 min prior to the ^{125}I-CSF. Incubation with ^{125}I-CSF-1 is carried out at 0°C (ice bath) for 1 h. Then 50 µL of the mixture is filtered by suction through a 0.22 µm membrane filter (13 mm diameter, cellulose acetate) that has been presoaked in 0.2% (w/v) BSA in PBS at 0°C. The filter is washed three times with 0.4 mL aliquots of the same BSA solution at 0°C prior to counting in a gamma counter. The filter retains 90% of the counts of the postnuclear fraction of cells to which ^{125}I-CSF-1 has been prebound. In contrast, it retains 0.02% of the ^{125}I-CSF-1 in assay buffer in the absence of the postnuclear fraction (Yeung et al., 1986).

3.2. Assay for Solubilized Receptor

Incubation buffer and conditions are the same as for the assay of membrane-bound receptor except that the assay buffer contains 0.1% (v/v) Triton X-100. At the end of the incubation, 10 µL of 50% (v/v) normal rabbit serum in PBS at 0°C is added as carrier protein, followed by 70 µL of 20% (w/v) polyethylene glycol (PEG 8000) in 10 m*M* HEPES, pH 7.4 at 0°C. The mixture is allowed to stand for 5 min in ice. A 100 µL aliquot is filtered through a 0.45 µm membrane (13 mm diameter, cellulose acetate) which has been soaked in 10% (w/v) PEG 8000 in 10 m*M* HEPES, pH 7.4 at 0°C. The precipitate retained by the filter is washed three times with 0.4 mL aliquots of 10 % PEG 8000 in 10 m*M* HEPES, pH 7.4 at 0°C prior to counting in a gamma counter. All data are corrected for the binding of ^{125}I-CSF-1 to the membrane filters by subtracting the counts retained from incubation mixtures that are incubated with the competing unlabeled CSF-1 for 30 min, followed by the addition of the ^{125}I-CSF-1, the rabbit serum carrier, PEG 8000, and immediate filtration. The retention of ^{125}I-CSF-1 by the filters is not prevented by presoaking them in 0.2% (w/v) BSA or 1000 U/mL CSF-1 and can amount to as much as 6% of the total counts added to the incubation mixture. Since this retention varies with different filter batches it is wise to test several different batches prior to selecting one for use in the assay. The overall efficiency of the assay is 19% (Yeung et al., 1986) and this should be taken into account when calculating the receptor concentration in assay samples.

4. Receptor Purification

4.1. Preparation of Crude Receptor Extract

The cell pellet (e.g., 8 mL) is resuspended in 5 times its vol of ice-cold, freshly prepared hypotonic buffer, composed of 5 m*M* Tris-HCl, 75 m*M* sucrose, 1 m*M* sodium iodoacetate, 0.5 m*M* EGTA, 10 μg/mL leupeptin (Sigma), 0.2 TIU/mL aprotinin (Sigma, TIU = trypsin inhibitor U), and 1000 U/mL soybean trypsin inhibitor (Sigma), pH 8.0. The cells are allowed to swell in the hypotonic buffer on ice for 15–20 min. Swelling is confirmed by examining the cells under phase contrast microscope immediately before and 15 min after addition of the hypotonic buffer.

Swollen cells are homogenized in a 40 mL Dounce homogenizer in ice with 15–20 fast strokes of a tight fitting pestle. Cells breakage is checked under phase contrast microscope. Further homogenization may be necessary. When more than 90% of cells are broken, 10 mL (i.e., 0.25 volumes with respect to the total volume of hypotonic buffer used) of compensating buffer consisting of 20 m*M* Tris-HCl, 0.95*M* sucrose, 25 m*M* $MgCl_2$, 0.15*M* NaCl, 0.15*M* KCl, and 5 m*M* EGTA, pH 7.4 is added. This renders the mixture isotonic and stabilizes the nuclei.

The homogenate is transferred to a 125 mL glass conical centrifuge tube and centrifuged at 1000*g*, 4°C for 45 s and the supernatant is saved. The nuclei pellet is suspended in 1/3 of the homogenate volume of reconstituted hypotonic buffer (1 part hypotonic buffer: 0.25 parts compensating buffer) and centrifuged as above. The postnuclear supernates are combined and 50 μL saved for receptor and protein assay.

Nine mL of a 15% sucrose solution containing 0.1*M* Tris-HCl, 2 μg/mL leupeptin, 0.04 TIU/mL aprotinin, and 200 U/mL soy-bean trypsin inhibitor, pH 7.4 is pipeted into each of four Beckman SW27 rotor thick wall polycarbonate tubes (capacity 30 mL each). The postnuclear supernatant is carefully layered on top and the tubes are centrifuged at 22,000 rpm, 2°C, for 30 min. The resulting supernatant is aspirated, and the pellets are combined and resuspended in 1/3 of the original homogenate vol of ice-cold membrane resuspension buffer (0.1 M Tris-HCl pH 7.4, 10% sucrose, 50 m*M* NaCl, 10 μg/mL leupeptin, 0.2 TIU/mL aprotinin, and 1000 U/mL soybean trypsin inhibitor) by Dounce homogenization with a loose fitting pestle in an ice bath. Forty μL is then removed for receptor assay and protein determination. At this stage, the membrane suspension can be quickly frozen in polyethylene tube(s) in liquid nitrogen and stored at –80°C. The CSF-1 receptor in the membrane suspension is stable under this storage condition for at least 3 mo.

To solubilize the receptor, the membrane suspension is adjusted to 2% (v/v) Triton X-100 with 20% (v/v) Triton X-100 and mixed by intermittent Dounce homogenization in an ice bath with a loose fitting pestle for 5 min. The resultant suspension is centrifuged in a Beckman SW27 rotor at 22,000 rpm, 30 min, 2°C, and the supernatant after centrifugation is saved as the solubilized crude receptor preparation.

4.2. Affinity Chromatography on CSF-l Sepharose-4B

4.2.1. Preparation of CSF-1 Sepharose 4B

CSF-1 is coupled through its tyrosine residue via an azo linkage on a short (benzamidoethyl-) spacer arm to Sepharose 4B.

4.2.1.1. Preparation of p-amino-benzamido-ethyl-Sepharose. The method of March et al. (1974) is adopted with substantial modification for the cyanogen bromide (CNBr) activation step. Fifty mL (packed volume) of Sepharose 4B (Pharmacia) is washed thoroughly with double glass distilled water (DDW) by decantation to remove fines and preservative, then twice by vacuum filtration on a coarse fritted glass filter funnel with 0.75*M* Na_2CO_3. The washed Sepharose is suspended in an equal vol of 0.75*M* Na_2CO_3 in a 250 mL Erlenmeyer flask and chilled to 0°C on ice. Two mL of acetonitrile (CH_3CN) containing 3 g CNBr (Sigma) (freshly prepared) at room temperature is added dropwise over a period of 30 s to the Sepharose suspension with constant vigorous shaking in an ice water bath (bath of crushed ice and water mixture to ensure efficient cooling). The final mixture is shaken continuously for a further 1.5 min at 0°C, vacuum filtered, and washed rapidly with 1 L of ice-cold DDW in a coarse fritted glass filter funnel (150–300 mL capacity) by suction. Finally, the gel is filtered until just moist.

The major modifications to the method of March et al. (1974) are the following:

1. Reaction at 0–4°C;
2. Reduction of the Na_2CO_3 concentration to prevent its crystallization at low temperature;
3. Shortening of the reaction time;
4. Lowering the concentration of CNBr used.

Gel that has been activated by the original method of March et al. (1974), or by the conventional titration method results in a substantial increase in contaminating protein in receptor preparation, probably because of the introduction of charged groups.

Subsequent synthesis of the *p*-amino-benzamido-ethyl (pABAE) arm follows the method described by Assimeh et al. (1974) with some minor changes. The CNBr activated gel cake is transferred immediately after the last washing to 50 mL ice-cold freshly prepared mixture containing 16% (v/v) ethylene diamine (Sigma) and 25 m*M* $NaHCO_3$, pH 10.0 (adjust by addition of 6*M* HCl). The gel suspension in the ethylene diamine solution is mixed gently (by rotating end to end in a sealed 150 mL plastic centrifuge bottle) overnight (18 h) at 4°C to form the aminoethyl derivative of Sepharose. The amino-ethyl-Sepharose is washed in a coarse fritted glass filter funnel with DDW using suction at room temperature until the washings are neutral. The gel is filtered until just moist.

The resulting gel cake is resuspended in 50 mL of *p*-nitro-benzyl azide (pNBA, Eastman Kodak™) reaction mixture that contains 20 mL dimethylformamide (DMF, Sigma), 30 mL 0.2*M* sodium borate buffer pH 9.3, and 400 mg pNBA. The pNBA solution must be freshly prepared by dissolving the pNBA first in DMF. Ice-cold borate buffer is then added with constant shaking, and the mixture cooled to room temperature if necessary. A slight turbidity will develop shortly after the addition of the borate buffer. The gel and pNBA suspension is mixed gently at room temperature (24°C) for 4 h. The resulting *p*-nitro-benzamido-ethyl Sepharose is washed free from excess pNBA with 50% (v/v) aqueous DMF that has been prechilled to 15°C, followed by DDW on a coarse fritted glass filter funnel, and finally vacuum filtered until just moist.

The gel cake is then transferred to 50 mL of freshly prepared 0.2*M* dithionite in 0.5*M* $NaHCO_3$, pH 8.5 and mixed. The *p*-nitrobenzamido-ethyl-Sepharose is reduced in this solution to its *p*-amino-derivative for 2 h at 40°C. The gel is then washed twice with 0.5*M* pH 8.5 bicarbonate buffer and stored in the same buffer for 1 wk at 4°C. This storage further reduces nonspecific absorption, probably by promoting hydrolysis of some of the more labile nitrogen containing groups from the gel, as suggested by Tesser et al. (1974). Finally, the gel is washed with DDW by filtration, and stored in 20% (v/v) ethanol at 4°C.

Each mL (packed volume) of the gel synthesized by this method is able to covalently bind approximately 10 µg CSF-1. The low capacity is beneficial since a relatively small amount of the growth factor is sufficient to saturate all of the reactive sites, leaving little or no unoccupied sites available for nonspecific interaction with proteins in the receptor preparation.

4.2.1.2. Coupling of CSF-1 to pABAE-Sepharose. The method of Assimeh et al. (1974) is essentially followed and all operations are carried out in a 4°C cold room.

One packed vol of pABAE-Sepharose-4B is washed 4 times, each with 5 volumes of 0.5*M* HCl by filtration in a coarse fritted glass filter funnel. The gel is then suspended in one vol 0.5*M* HCl and cooled in ice bath to below 4°C.

One vol of freshly prepared 0.2*M* $NaNO_2$ at 0°C is added and the mixture is incubated in an ice bath for 9 min with occasional mixing. The suspension is vacuum filtered through a coarse fritted glass filter funnel. The gel on the filter is washed rapidly 2 times, each with 5 volumes of ice-cold DDW, filtered until just moist, and the gel cake transferred immediately into one vol of an ice-cold solution of pure CSF-1 (15 µg protein/mL) that had been dialyzed overnight in at least 200 volumes of a 0.2*M* sodium borate buffer pH 8.0. The mixture is rocked gently in a sealed plastic tube for 18–24 h at 4°C. The gel is sedimented by centrifugation (400*g*, 2 min, 4°C), the volumes of gel and supernatant noted, and the supernatant saved for assay of unbound CSF-1.

The CSF-1-Sepharose is then packed into a column (1.2 cm diameter × 10 cm in length for 5 mL of gel) in 0.2*M* borate buffer, pH 8.0. The column is washed at a slow flowrate (8–10 mL/h) for 2 h with 1% (v/v) triethanolamine (or ethanolamine) containing 50 m*M* Tris (pH adjusted to 8.5 with HCl) to block any remaining reactive groups. The column is then subjected to 2 cycles of washing, each with 6 column volumes of freshly prepared 6*M* urea in 0.1*M* Tris-HCl, pH 7.4 followed by 6 column volumes of 0.4*M* $NaHCO_3$, pH 8.6 to remove any nonspecifically adsorbed protein. Finally the column is equilibrated in the binding buffer composed of 0.1*M* Tris-HCl pH 7.4, 10% (w/v) sucrose, 50 m*M* NaCl, 0.1% (v/v) Triton-X 100, 0.02% (w/v) NaN_3 and can be stored in this buffer at 4°C.

4.2.2. Binding of Receptor to CSF-1 Sepharose

The solubilized membrane extract is diluted by addition of 1/2 of its vol of binding buffer and passed through the column twice at a flowrate of 30–50 mL/h. The flowthrough fraction is saved for assay of residual receptor (if a significant amount of receptor remains in the flowthrough, it should be rechromatographed on a regenerated column). The column is then washed at least 5 times, each with 3 column volumes of binding buffer, or until no protein is found in the washing. Five mL of the CSF-1 gel prepared as above is enough to process the solubilized membrane extract from 10 mL of packed J. 774.2 cells. More than 80% of receptor in the extract can be bound after two passages of the membrane extract through the column, and a third passage will not increase the final yield significantly.

4.2.3. Elution of CSF-1 Receptor

The receptor is eluted by passing 8–10 aliquots, each of one-half column vol, of a 1.2*M* NaCl solution containing 0.1% (v/v) Triton X-100, 10% (w/v) sucrose and 0.1*M* Tris-HCl, pH 7.4 through the column. The 1.2*M* NaCl solution used is of sufficiently high ionic strength to elute all the bound receptor and elution with 0.2*M* acetic acid as described earlier (Yeung et al., 1987) is not necessary. Each aliquot of eluate is collected in separate plastic tubes at 4°C and assayed for receptor activity. The fractions (total vol about 10 mL) containing high receptor binding activity are pooled, concentrated and desalted by vacuum dialysis to approx 0.5 mL against 200 mL of a buffer containing 50 m*M* Tris-HCl pH 7.4, 5% (w/v) sucrose, 25 m*M* NaCl, 0.5% (v/v) Triton X-100, and 0.02% (w/v) NaN_3.

4.2.4. Regeneration of the CSF-1 Sepharose Column

The column is regenerated by washing using the following sequence: 1 column vol 0.2*M* acetic acid in 0.1% (v/v) Triton X-100, 1 column vol binding buffer, 3 column volumes freshly prepared 6*M* urea in 0.1*M* Tris-HCl pH 7.4, and finally 4 volumes of binding buffer. Acetic acid must be washed from the column as soon as possible and not allowed to stay in the column for longer than necessary to avoid denaturation of the coupled CSF-1. The column may be used repeatedly for many cycles of purification.

4.3. Sepharose 6B Chromatography

The molecular sieving chromatography is a necessary step to remove traces of contaminating, mostly low mol wt, proteins. A column of Sepharose 6B (1.6 × 95 cm) is packed under standard procedure at 4°C in a buffer composed of 0.1*M* Tris-HCl pH 7.4, 0.5*M* NaCl, 5% (w/v) sucrose, 0.1% (v/v) Triton X-100, and 0.02% (w/v) NaN_3. The affinity purified receptor, concentrated to 0.5–1 mL, is applied. The flowrate is adjusted to about 10 mL/h and 2 mL fractions are collected. The protein profile on the eluate is monitored continuously with a UV-monitor at 280 nm. The fractions having high CSF-1 binding activity (receptor) are pooled and concentrated to about 200 µL by vacuum dialysis against 20 volumes of a buffer consisting of 50 m*M* Tris-HCl pH 7.4, 0.5% (v/v) Triton X-100, 10% (w/v) sucrose, and 0.02% (w/v) NaN_3. The concentrated receptor solution can be frozen quickly in liquid nitrogen and stored at –80°C.

4.4. Protein Determination

Protein determination of the receptor fractions are carried out according to the method of Bradford (1976). In order to eliminate the interference of Triton X-100, protein samples are pretreated prior to the assay. One mL of 5% (w/v) trichloroacetic acid (TCA) is added to 50 or 100 µL of sample, and the mixture incubated at 4°C for 15 min. The precipitate is collected by centrifugation (2000*g*, 10 min, 4°C) and extracted twice with acetone at room temperature. The final pellet is dried in a gentle stream of nitrogen. The dried pellet is dissolved in 150 µL of 0.2*M* NaOH and 50 µL of 0.5*M* HCl is then added to reduce the alkalinity. Ten to 50 µL of the resulting protein solution is used with 2.5 mL of the Coomassie Blue reagent.

5. Properties of the Purified Receptor

5.1. Stability and Ligand Binding

The purified receptor retains full CSF-1 binding activity for more than 3 mo on storage at –80°C. The binding activity of the receptor is increased by approximately twofold by heating at 37°C for 2 h in the storage buffer, but is lost if 50 m*M* dithiothreitol is included. Incubation with 0.1% sodium dodecyl sulfate (SDS) at 4°C for 2 min also completely abolishes the CSF-1 binding activity (Yeung et al., 1987).

5.2. Protein Kinase Activity

The purified receptor possesses CSF-1 stimulated autophosphorylation activity at tyrosine residue(s) (Yeung et al., 1987). The receptor kinase is also able to phosphorylate casein but not histone (Yeung et al., 1987). The CSF-1-stimulated increase in autophosphorylation by the purified receptor is small (about twofold) compared with CSF-1 stimulated receptor phosphorylation at tyrosine residues in membrane preparations from the same cell line (Jubinsky et al., 1988). Solubilization of the receptor probably relieves steric or other restrictions on its tyrosine kinase activity. The CSF-1 stimulated kinase activity is lost after storage at –80°C or by incubation at 37°C for 30 min, although endogenous activity remains.

5.3. Molecular Weight

Sepharose 6B chromatography of the receptor-detergent complex yields an apparent molecular weight of approximately 600,000. The purified receptor usually behaved as a single band of approx 165 kDa on SDS polyacrylamide gel electrophoresis (SDS-PAGE) (Yeung et al., 1987), corresponding to the mol wt of the membrane protein to which specifically bound ^{125}I-CSF-1 was chemically crosslinked (Morgan and Stanley, 1984). The high molecular weight observed on Sepharose 6B chromatography may result from detergent binding with or without dimerization.

6. Production of Antibody

Specific antiserum was successfully raised in a goat that received nine consecutive injections of purified receptor (1.5 μg each) emulsified in 50% (v/v) complete Freund's adjuvant that were administered intradermally at multiple sites and at weekly intervals. The antiserum is monospecific for the CSF-1 receptor, is able to block the binding of ligand and to bind cell surface receptor whether it has bound CSF-1 or not. However it is unable to elicit CSF-1 stimulated effects on CSF-1 dependent cells (Ferrante, A., Li, W., Yeung, Y. G., and Stanley E. R., unpublished observations).

7. Results and Conclusions

The results of a typical purification are summarized in Table 1. A previous study showed that 76% of ^{125}I-CSF-1 prebound to J774.2 cells could be recovered in the postnuclear particulate fraction (Yeung et al., 1986). The apparent yield of CSF-1 binding activity (receptor) of the post-nuclear supernatant from whole cells is about 20–40%. Homogenization probably creates some inside-out vesicles in which the CSF-1 binding domain of the receptors will be masked. Consistent with this possibility, subsequent solubilization of the membrane preparation resulted in a 1.4-fold increase in the binding activity of the postnuclear supernatant. This phenomenon was also observed after the solubilization of the insulin receptor (Deutsch et al., 1982).

Affinity chromatography is the major step in the present purification, resulting in high purity (more than 800-fold from solubilized membrane and 3000-fold from the postnuclear supernatant), a very small amount of contaminating protein and a high yield (better than 80% from solubilized membrane) (Yeung et al., 1987). This is achieved by using modification of the published methods for the preparation of the affinity gel. Preliminary experiments, using commercially available affinity matrix, or other methods for the preparation of the immobilized ligand, resulted in either no receptor binding or the coelution of an extremely large quantity of contaminating protein with the receptor. The adoption of the present method was based on two major considerations. The first is that CSF-1 can be iodinated to saturation without loss of its receptor binding activity (Stanley and Guilbert, 1981), suggesting that most, if not all, of its tyrosine residues are

Table 1
Purification of CSF-1 Receptor from J774.2 Cells[a]

Fraction	Volume, mL	Protein, mg	^{125}I-CSF-1 bound, counts/min × 10^{-6}	Specific ^{125}I-CSF-1 binding activity, counts/min/mg × 10^{-6}	Yield, %	Purification, –fold
Whole cells	7.5 × 10^9	cells	1000			
Post nuclear supernatant	63.5	235	190	0.81	100	1.0
Concentrated membrane	11.0	86	150	1.70	79	2.1
Solubilized membrane	12.5	75	266[a]	3.54	140	4.4
CSF-1 affinity column	0.5	0.082	229[a]	2813	121	3473
Sepharose 6B column	0.75	0.026	146[a]	5627	77	6947

[a]Corrected for 19% efficiency of the solubilized receptor assay (Yeung et al., 1986). This table is reproduced from Yeung et al., 1987, with additions and permission of the authors.

not essential for ligand–receptor interaction and indicating that the tyrosine residues could be used to link CSF-1 to a gel matrix without impairing receptor binding to any significant extent. The coupling of the tyrosine residue of CSF-1 to the pABAE-Sepharose (Yeung et al., 1987) via an azo linkage (Cohen, 1974) resulted in an immobilized CSF-1 preparation with high ligand binding activity. The second consideration is that existing methods of gel preparation would have to be modified to reduce nonspecific binding of contaminating proteins to the gel. In fact, the modifications to the CNBr activation method of March et al. (1974) are critical and result in a dramatic reduction in nonspecific binding.

The protein eluted from the CSF-1 affinity column consists of essentially a single silver stained 165 kDa band in 7.5% SDS-PAGE with a negligible degree of contamination (Yeung et al., 1987). The Sepharose 6B step removes the remaining contaminating protein. Approximately 25 µg of purified receptor protein can be obtained from membranes prepared from about 10^{10} cells. The receptor is purified approximately 7000-fold with respect to the protein present in the postnuclear supernatant.

Acknowledgment

We thank Anne Palestroni and Susan Bassett for technical assistance. This work was supported by NIH Grants CA26504, CA32551 (to ERS), P30-CA1330, and 5P60DK20541-11 (Diabetes Research and Training Centre).

References

Assimeh, S. N., Bing, D. H., and Painter, R. H. (1974) *J. Immunol.* **113,** 225–234.

Bradford, M. M. (1976) *Anal. Biochem.* **72,** 248–254.

Byrne, P. V., Guilbert, L. J., and Stanley, E. R. (1981) *J. Cell Biol.* **91,** 848– 853.

Clark, S. C. and Kamen, R. (1987) *Science* **236,** 1229–1237.

Cohen, L. (1974) in *Methods in Enzymology—Affinity Techniques (Enzyme Purification: Part B),* (Colowick, S. P. and Kaplan, N. O., eds.), vol. 34, Academic, New York, pp. 102–108.

Coussens, L., Van Beveren, C., Smith, D., Chen, E., Mitchell, R. L., Isacke, C. M., Verma, I. M., and Ullrich, A. (1986) *Nature* **320,** 227–280.

Cuatrecasas, P. (1972) *Proc. Natl. Acad. Sci. USA* **69,** 318–322.

Deutsch, P. J., Rosen, O. M., and Rubin, C. S. (1982) *J. Biol. Chem.* **257,** 5350– 5358.

Diamond, B., Bloom, B. R., and Scharff, M. D. (1978) *J. Immunol.* **121,** 1329–1333.

Guilbert, L. J. and Stanley, E. R. (1980) *J. Cell Biol.* **85,** 153–159.

Jubinsky, P. T., Yeung, Y. G., Sacca, R., Li, W., and Stanley, E. R. (1988) in *Biology of Growth Factor: Molecular Biology, Oncogenes, Signal Transduction, and Clinical Application* (Kudlow, J. E., MacLennan, D., Bernstein, A., and Gotlieb, A. I., eds.), Plenum, New York, pp. 75–90.

March, S. C., Parikh, I., and Cuatrecasas, P. (1974) *Anal. Biochem.* **60,** 149– 152.
Metcalf, D. (1986) *Blood* **67,** 257–267.
Morgan, C. J. and Stanley, E. R. (1984) *Biochem. Biophys. Res. Commun.* **119,** 35–41.
Morgan, C., Pollard, J. W., and Stanley, E. R. (1987) *J. Cell. Physiol.* **130,** 420–427.
Ralph, P., Prichard, J., and Cohn, M. (1975) *J. Immunol.* **114,** 898–905.
Rettenmier, C. W., Sacca, R., Furman, W. L., Roussel, M. F., Holt, J., Nienhuis, A. W., Stanley, E. R., and Sherr, C. J. (1986) *J. Clin. Invest.* **77,** 1740–1746.
Roussel, M. F., Dull, T. J., Rettenmier, C. W., Ralph, P., Ullrich, A., and Sherr, C. J. (1987) *Nature* **325,** 549–552.
Sacca, R., Stanley, E. R., Sherr, C. J., and Rettenmier, C. W. (1986) *Proc. Natl. Acad. Sci. USA* **83,** 3331–3335.
Sherr, C. J., Rettenmier, C. W., Sacca, R., Roussel, M. F., Look, A. T., and Stanley, E. R. (1985) *Cell* **41,** 665–676.
Stanley, E. R. (1985) in *Method in Enzymology—Immunochemical Techniques,* (Colowick, S. P. and Kaplan, N. O., eds.), vol. 116, Academic, New York, pp. 564–568.
Stanley, E. R. and Guilbert, L. J. (1981) *J. Immunol. Methods* **42,** 253–284.
Stanley, E. R., Guilbert, L. J., Tushinski, R. J., and Bartelmez, S. H. (1983) *J. Cell Biochem.* **21,** 151–159.
Stanley, E. R. and Heard, P. M. (1977) *J. Biol. Chem.* **252,** 4305–4321.
Stanley, E. R. and Jubinsky, P. T. (1984) *Clin. Haematol.* **13,** 329–348.
Tesser, G. I., Fisch, H .U., and Schwyzer, R. (1974) *Helv. Chim. Acta* **57,** 1718– 1730.
Yarden, Y. and Ullrich, A. (1988) *Biochemistry* **27,** 3113–3119.
Yeung, Y. G., Jubinsky, P. T., and Stanley, E. R. (1986) *J. Cell. Biochem.* **31,** 259–269.
Yeung, Y. G., Jubinsky, P. T., Sengupta, A., Yeung, D. C. Y., and Stanley, E. R. (1987) *Proc. Natl. Acad. Sci. USA* **84,** 1268–1271.

Insulin-Like Growth Factor-II Receptors

Robert C. Baxter and Carolyn D. Scott

1. Introduction

In the past few years there has been an explosive increase in interest in peptides of the insulin-like growth factor (IGF) family. This follows a period of over two decades, dating back to the discovery of "somatomedins," as the IGFs were originally known, in the late 1950s. Independent research groups isolated and characterized peptides with three apparently unrelated groups of biological activities: "sulfation factor," peptides that stimulate proteoglycan synthesis in cartilage; "nonsuppressible insulin-like activity," peptides with the ability to mimic the various actions of insulin, such as the stimulation of glucose oxidation by fat cells; and "multiplication-stimulating activity" or MSA, peptides that promote cell replication (Baxter, 1986). In time it was realized that all of these properties were in fact possessed by all of the peptides in the three groups; these, in turn, were eventually categorized, on the basis of their primary sequences, into two peptide groups, IGF-I and IGF-II. For example, MSA is now regarded as rat IGF-II (Marquardt et al., 1981). Although there have been suggestions that different IGF classes (such as "fetal somatomedins") might exist, it can be concluded from recent protein and cDNA sequencing studies that, notwithstanding the many proven or predicted structural variants of each peptide type, all known IGFs may be classified as either IGF-I or IGF-II.

1.1. Chemistry and Biology of IGF-II

Since IGF-I and its receptor will be discussed elsewhere in this book, this chapter will concentrate on IGF-II and the receptor with primary specificity for IGF-II. IGF-II is a slightly acidic protein of molecular mass 7471 Da (Rinderknecht and Humbel, 1978). Highly homologous with proinsulin, it can be considered (together with IGF-I) to consist of four domains, termed the B, C, A, and D regions. The first and third correspond to, and are homologous with, the B and A chains of insulin, respectively, whereas the C region is analogous to the C-peptide of proinsulin, although there is no primary homology. Both IGFs differ from proinsulin in having a carboxy-terminal extension, termed the D-peptide, which in the case of IGF-II is five amino acids long.

In humans, the gene for IGF-II is found on chromosome 11, contiguous with the insulin gene (Tricoli et al., 1984). Transcription of the IGF-II gene yields multiple mRNA forms, including species of approx 1.2, 1.7, 2.2, 3.8–4.0, and 6.0 kb in the rat (Brown et al., 1986), and 1.8–2.0, 2.8–3.0, 4.0, 4.8, and 6.2 kb in the human fetus (Han et al., 1988). Sequencing of cloned cDNAs has led to the prediction that pro-IGF-II should contain a fifth peptide domain, a carboxy-terminal extension of the D-region (Bell et al., 1984; Whitfield et al., 1984). Antibodies raised against a synthetic peptide representing this putative carboxy-terminal extension (termed the E-domain) are, in fact, able to recognize this sequence in serum and conditioned cell culture medium, indicating that newly synthesized IGF-II does indeed contain an E-peptide that is, presumably, cleaved off under normal circumstances (Hylka et al., 1987).

Whereas IGF-I is considered an important growth factor in childhood and pubertal growth, IGF-II is thought to be more important in fetal growth. This view is based on many studies showing highest levels of IGF-II mRNA, IGF-II synthesis, and circulating IGF-II levels, in fetal animals (Underwood and D'Ercole, 1984; Brown et al., 1986; Han et al., 1988). For example, fetal rat hepatocytes in primary culture secrete an IGF that appears to be MSA or rat IGF-II (Richman et al., 1985), whereas adult hepatocytes produce IGF-I (Scott et al., 1985). Similarly, fibroblast cultures from fetal tissues predominantly release IGF-II, whereas fibroblasts from neonatal and adult tissues switch to IGF-I secretion (Adams et al., 1983). Nevertheless, the IGF-II concentration in the adult human circulation, approx 600 ng/mL or 80 nmol/L, is 3–4 times higher than that of IGF-I (Baxter, 1986), and, in the rat, although circulating IGF-II levels are very low, IGF-II receptors have been detected at very high levels in a wide variety of tissues (Taylor et al., 1987). Thus, it appears that IGF-II might also play a role in postnatal life.

What this role of IGF-II might be is, as yet, unknown. IGF-II mRNA levels are much higher in the brain than in any other adult tissue, suggesting a possible function in the central nervous system (Murphy et al., 1987). One such function might be in the regulation of food intake (Lauterio et al., 1987). In some cell types, IGF-II has been shown to stimulate DNA synthesis or cell proliferation in an apparently specific way (Tally et al., 1987; Rosenfeld et al., 1987; Nishimoto et al., 1987), and in rat hepatocytes, but not hepatoma cells, cells cultured at high density have greatly decreased expression of both cell surface and intracellular IGF-II receptors, a regulatory pattern consistent with the involvement of the receptor in growth-related processes (Scott and Baxter, 1987b; Scott et al., 1988b). However, the mitogenic effects of IGF-II are not always mediated by specific IGF-II receptors (*see* Section 1.2.) (Mottola and Czech, 1984; Furlanetto et al., 1987).

Recent studies also implicate IGF-II (acting through its specific receptor) in the regulation of calcium entry into cells via a voltage-independent calcium channel (Nishimoto et al., 1987; Kojima et al., 1988). However, it has never been demonstrated that normal growth or metabolism in vivo is IGF-II-dependent. It is also unclear what part IGF-II might play in the growth of neoplastic cells. Various reports describe increased concentrations of IGF-II, or of its mRNA, in tumor types such as Wilm's tumor, pheochromocytoma, colon carcinoma, and liposarcoma (Reeve et al., 1985; Tricoli et al., 1986; Haselbacher et al., 1987), but whether these observations represent casual occurrences, or are indicative of a specific function of IGF-II in neoplastic growth, is unclear.

1.2. The Type II IGF Receptor: Comparison with the Type I Receptor

1.2.1. Structure

Although IGF-II is structurally very similar to IGF-I, the receptors for the two peptides show few features in common. The IGF-I receptor, also known as the type I IGF receptor, is homologous to the insulin receptor; both are heterotetrameric glycoproteins with two disulfide-linked extracellular α-chains of approx 130 kDa containing IGF-I or insulin binding sites, joined by further disulfide bonds to two transmembrane β-chains of approx 90 kDa that have tyrosine kinase activity (Massague and Czech, 1982; Sasaki et al., 1985; Yu et al., 1986; Ullrich et al., 1986). These tyrosine kinases can phosphorylate both their own intracellular domains and other tyrosine-containing substrates (Sasaki et al., 1985; Yu et al., 1986). The full primary sequence of the IGF-I receptor precursor, deduced from the cloned cDNA sequence, indicates that the α and β chains are derived from a single gene product (Ullrich et al., 1986).

In contrast to the tetrameric structure of the IGF-I receptor, the IGF-II receptor is a single chain glycoprotein with an apparent molecular mass, estimated from sodium dodecyl sulfate polyacrylamide gel electrophoresis (SDS-PAGE) studies, of approx 220 kDa nonreduced, or 250 kDa reduced (Massague and Czech, 1982). Also termed the type II IGF receptor, to distinguish it from the structurally distinct type I or IGF-I receptor, most of the primary sequence of this protein from rat and humans has recently been deduced following cDNA cloning (Morgan et al., 1987; MacDonald et al., 1988). The human receptor has a deduced length of 2451 amino acids, consisting of a long amino-terminal extracellular domain of 2264 residues, a hydrophobic region of 23 residues, thought to be the transmembrane domain, and an intracellular region of only 164 residues (Morgan et al., 1987). The receptor cDNA also codes for a 40 amino acid putative signal peptide at the amino-terminus of the extracellular domain, which is presumably removed in the mature receptor. There are 19 potential *N*-glycosylation sites in the extracellular domain, to which 20–30 kDa of carbohydrate is thought to be attached; thus, the molecular mass of the mature receptor is probably approx 300 kDa, somewhat larger than the estimates based on SDS-PAGE. The IGF-II receptor apparently lacks intrinsic kinase activity, although it can itself be phosphorylated (Corvera et al., 1986). The phosphorylation state has been implicated in the regulation of receptor translocation between plasma membrane and intracellular compartments (Corvera and Czech, 1985).

1.2.2. Specificity

Many studies indicate that IGF-II can bind with high affinity to both type I and type II IGF receptors. Type I receptors, which are regarded as having primary specificity for IGF-I, in fact bind IGF-II with 10–50% of the affinity of IGF-I, and generally also show crossreactivity of about 1% toward insulin (Nissley and Rechler, 1984; Ballard et al., 1988). Thus, in tissues that are rich in type I receptors, it is not uncommon to detect IGF-II binding to a protein of approx 130 kDa, i.e., the α-subunit of the type I receptor (Casella et al., 1986; Misra et al., 1986). Type II receptors show much greater selectivity, with an affinity for IGF-II that is frequently 100 times that seen for IGF-I (Baxter and De Mellow, 1986; Taylor et al., 1987; Scott et al., 1988b). In addition, it is characteristic of type II receptors that they show no reactivity toward insulin (Massague and Czech, 1982; Taylor et al., 1987).

The unusually high specificity of IGF-II receptors in various tissues makes them ideally suitable for use in radioreceptor assay, particularly since they tend to bind IGF-II with very high affinity. For example, a radioreceptor assay method using the ovine placental IGF-II receptor (Baxter and De Mellow, 1986) can detect 100 pg of human IGF-II (equivalent to less than 1 μL of human serum), with crossreactivity toward recombinant human IGF-I estimated at only 0.03% (Baxter et al., 1987). All rat tissues tested have IGF-II receptors with even higher affinity (association constants greater than 10 L/nmol) (Bryson and Baxter, 1987; Taylor et al., 1987); in a rat liver IGF-II RRA, the limit of detection is only 10 pg of human IGF-II (Baxter et al., 1987). Such RRA methods, in fact, surpass most published IGF-II radioimmunoassays in their combination of sensitivity and specificity.

1.3. Circulating Form of the IGF-II Receptor

In the past few years, several laboratories have identified a high mol wt circulating IGF binding protein with selective affinity for IGF-II. Immunological and affinity crosslinking experiments indicate that this protein is closely related, or identical, to the IGF-II receptor, with an apparent mol mass on SDS-PAGE of 210 kDa (Kiess et al., 1987a). In the rat, circulating levels are high in fetal and neonatal serum, but greatly reduced in the serum of mature animals. A similar protein also exists in fetal sheep serum (Hey et al., 1987), monkey serum (Gelato et al., 1988), and human serum (Causin et al., 1988). Since over 90% of the plasma membrane IGF-II receptor is extracellular (Morgan et al., 1987), it seems most likely that the circulating receptor results from cleavage at or near the transmembrane region; however, the mechanism and significance of this phenomenon remain to be determined.

1.4. Identity Between Receptors for IGF-II and Mannose-6-Phosphate

As mentioned above, the primary structure of the IGF-II receptor has been deduced from the sequence of its cloned cDNA (Morgan et al., 1987). This study revealed a striking homology between this receptor and another recently characterized protein, the cation-independent mannose-6-phosphate receptor (Lobel et al., 1987). This receptor has been known for several years to be involved in the targeting of lysosomal enzymes to lysosomes (Von Figura and Hasilik, 1986). The functional identity between these two proteins is supported by a variety of ligand binding studies. For example, mannose-6-phosphate can stimulate the binding of labeled IGF-II to the receptor by twofold (Roth et al., 1987; MacDonald et al., 1988), although this effect is not always observed (Tong et al., 1988). In addition, antisera against purified mannose-6-phosphate receptor react with purified IGF-II receptor (Roth et al., 1987); and a monoclonal antibody against mannose-6-phosphate receptor partially inhibits IGF-II binding to fibroblast monolayers (Braulke et al., 1988). Finally, as detailed in Section 2.5., IGF-II receptors can be purified by affinity chromatography on columns of immobilized pentamannosyl-6-phosphate and other mannose-6-phosphate derivatives (MacDonald et al., 1988).

2. Receptor Purification by Affinity Chromatography

2.1. General Considerations

Because the IGF-II receptor, like other peptide hormone receptors, is a transmembrane protein, it must first be solubilized before it can be purified (this step is obviously unnecessary when purifying the circulating form of the receptor). This is achieved by stirring membranes rich in receptors with a detergent such as Triton X-100 or CHAPS (3-[(3-cholamidopropyl)-dimethylammonio]-1-propane sulfonate). A low concentration of detergent must also be added to all buffers to maintain the receptor in a soluble form. The continual presence of detergents places some restriction on the range of separation methods that may be applied. For example, hydrophobic interaction or reverse phase chromatography columns may become saturated with detergent and lose their binding capacity for proteins. However, many other fractionation techniques are relatively unaffected by the presence of detergents (especially uncharged detergents, such as Triton X-100 and Nonidet P-40); these include gel permeation, many ion-exchange methods, and affinity chromatography.

Affinity chromatography is the key to most successful purifications of receptors and other binding proteins, although for proteins whose ligand is not readily available in large quantities, this method may only be applied in a limited number of centers. In the case of the IGF-II receptor, ligand availability has been a major limiting factor in the past; however, since this peptide is now being produced by recombinant DNA technology (Furman et al., 1987), it is likely that it will become readily available over the next few years, as IGF-I has in recent years. It should also be stressed that the purification of IGF-II from human plasma (or Cohn Fraction IV of plasma), as performed in the authors' laboratory (Baxter and De Mellow, 1986), is not difficult (though rather time-consuming), and readily yields sufficient IGF-II to prepare affinity columns.

Successful affinity chromatography depends not only on the availability of sufficient ligand to prepare a column of adequate capacity, but also on the preparation of the receptor in a form in which it will readily bind to the immobilized ligand, and the development of an elution protocol that causes efficient dissociation of the receptor from the column, without resulting in irreversible loss of receptor activity. Both stages can be optimized for maximal yields.

2.1.1. Optimization of Loading Conditions

As mentioned above, the membrane-bound IGF-II receptor must be detergent-solubilized before affinity chromatography. This is conveniently achieved with Triton X-100, which must be present at a concentration of at least 1% for efficient solubilization. However, at high concentrations this detergent inhibits the interaction between IGF-II and its receptor; thus, care must be taken to dilute out the detergent before the sample is applied to the affinity column.

Another consideration when preparing receptor solutions for affinity chromatography is whether the receptor binding site is likely to be occupied by its ligand, thus preventing interaction with the immobilized ligand. In the case of the IGF-II receptor from adult rat liver, described below, it might be expected that receptor occupancy would not be a problem, since circulating IGF-II levels in the rat are extremely low. This appears to be the case, since almost no receptor flows through the affinity column in this preparation. In contrast, experience in the authors' laboratory with IGF-II receptors from various fetal and adult human tissues suggests that prior occupancy could indeed influence the binding of these receptors to the affinity column, as might be predicted in view of the high IGF-II concentrations in fetal and adult circulation. Receptor occupancy is indicated by the fact that various putative desaturation techniques are able to increase IGF-II binding to the receptor. These techniques include transient exposure of the receptor, highly diluted, to low pH, high ionic strength, or calcium chelators (unpublished data).

Effectors that can increase the affinity of IGF-II binding to its receptor might also be expected to improve the efficiency of the adsorption step. The affinity of this interaction is known to be increased two- to threefold by 10 m*M* calcium ions (Bryson and Baxter, 1987). Other divalent cations such as Mg^{2+} and Mn^{2+} similarly increase IGF-II binding to solubilized receptor, although to a smaller degree. For this reason, calcium is included during the loading of the affinity column. As described above, mannose-

6-phosphate is also reported to increase the affinity of the interaction between IGF-II and its receptor twofold (MacDonald et al., 1988). Therefore, it might be predicted that the efficiency of the adsorption step could be improved by adding mannose-6-phosphate to the loading buffer, although this has not yet been demonstrated in practice.

2.1.2. Optimization of Eluting Conditions

The elution of specifically bound protein from an affinity column appears most efficient when the available binding sites on the column are substantially saturated. This is particularly true when, as in the case of the IGF-II receptor, the protein is relatively unstable under the eluting conditions, so that the mildest possible conditions must be chosen. If the protein is bound at the top of a largely unoccupied column, then the passage of the protein down the column during the elution step will be retarded as it continually adsorbs (albeit with lowered affinity) and desorbs to free binding sites. This will result in a broad peak of eluted activity, unless harsh eluting conditions are chosen.

Conversely, the use of a column that can be saturated by the available amount of soluble receptor allows the use of more gentle eluting conditions. Even though there might be some tendency for dissociated receptor to rebind to the column, there are few available binding sites, resulting in a sharp peak of eluted protein. The optimization of elution conditions is, thus, clearly dependent on both the size of the affinity column and the stability of the receptor, with the general strategy being to use the harshest conditions that will still allow an acceptable recovery of activity. In the case of the rat liver IGF-II receptor, exposure to buffer at pH 4 is sufficient to elute the protein from a saturated column, but the exposure must be transient to prevent the irreversible loss of receptor activity. Therefore each eluted fraction must be reneutralized as soon as it is collected.

2.2. Assay for IGF-II Receptors

An essential feature of any purification method is the ability to determine activity throughout the procedure. Although the authors have reported the use of a radioimmunoassay for the IGF-II receptor (Scott and Baxter, 1987a; *see* Section 3.2.1.), the usual way to monitor activity is by radioligand binding. Radioiodinated IGF-II typically is prepared by iodinating 5 μg of peptide in 50 μL 0.5*M* sodium phosphate, pH 7.5, with 1 or 2 mCi Na ^{125}I for 20 s in the presence of 10 μL of a 1 mg/mL solution of chloramine-T, and then terminating the reaction by adding 10 μL of a 5 mg/mL solution of sodium metabisulfite (Scott and Baxter, 1987a). Unreacted iodide is removed by chromatography on a 1 × 20 cm column of Sephadex G-50 (Pharmacia, Uppsala, Sweden) equilibrated and eluted in 25 m*M* sodium HEPES, pH 7.4, containing 2.5 g/L bovine albumin. This method yields [^{125}I]iodo-IGF-II of specific activity 50–100 Ci/g if 1 mCi of Na ^{125}I is used, or up to 200 Ci/g if 2 mCi is used.

To determine IGF-II binding to receptors in particulate membrane fractions, membranes are incubated with IGF-II tracer; then bound and free tracer are separated by sedimenting the membranes by centrifugation (Bryson and Baxter, 1987). However, when receptors are solubilized as part of the purification procedure, a precipitant must be added to sediment the receptor with its bound radioligand. The following method is suitable for use in the rat liver IGF-II receptor purification (Scott and Baxter, 1987a). Soluble receptor is diluted to 400 µL in buffer containing 25 m*M* sodium HEPES, 10 m*M* calcium chloride, 0.1% Triton X-100, and 0.2% bovine albumin, pH 7.4. IGF-II tracer (approx 10,000 cpm) is added in 100 µL of the same buffer, and the mixture incubated 2 h at 22°C. To facilitate precipitation of the soluble receptor, 20 µL of bovine gamma globulin solution (40 mg/mL) and 1 mL cold polyethylene glycol solution (180 g/L) in 25 m*M* sodium HEPES, 0.15*M* NaCl, pH 7.4, are added, and the mixture is centrifuged 20 min at 4000 × g. The supernatant is removed by aspiration and the pellet is counted in a gamma counter. Nonspecific binding is determined in tubes to which excess unlabeled IGF-II is added in the first incubation. For the rat liver receptor, 100 ng IGF-II is sufficient excess; however, for receptors of lower affinity, higher concentrations of unlabeled IGF-II will be required. Specific binding is calculated as the difference between total and nonspecific binding.

2.3. Purification of the Rat Liver IGF-II Receptor

The following method has been used in the authors' laboratory for several years (Scott and Baxter, 1987a). It will serve, therefore, to illustrate the approach required for a successful IGF-II receptor purification. However, as the following sections will indicate, many alternative methods are also capable of yielding pure receptor preparations.

2.3.1. Preparation and Solubilization of Microsomal Membranes

The starting material is a preparation of microsomal membranes from fresh rat liver, or liver thawed after being stored at –70°C. To minimize contamination by blood, livers can be perfused before dissection with 50 mL cold saline via the vena cava. The tissue is homogenized with 4 vol of 0.25*M* sucrose (containing 0.5 m*M* bacitracin, 50 mg/L soybean trypsin inhibitor, and 0.1 m*M* L-1-tosyl-2-phenyl-ethylchloromethyl ketone as protease inhibitors) at 2°C in a motor-driven Teflon™-glass homogenizer, followed by 30 s with the large probe of a Polytron (Kinematica, Switzerland) or Ultra-Turrax (Junke and Kunkel, Staufen, FRG) homogenizer on full power. The homogenate is centrifuged at 2°C for 20 min at 12,000 × *g*, and the supernatant is decanted and again centrifuged. The supernatant is then ultracentrifuged at 2°C for 60 min at 100,000 × *g*, and, after discarding the supernatant and removing any lipid adhering to the tube, the pellet is resuspended by gentle homogenization in the original volume of 0.25*M* sucrose (plus inhibitors). The resuspended pellet is again sedimented by centrifuging 45 min at 100,000 × *g* and, if the supernatant appears colored by hemoglobin, the pellet is once again resuspended and centrifuged as above. The supernatant is discarded, and the membrane pellet is resuspended by homogenization in 25 m*M* sodium HEPES, pH 7.4 (plus protease inhibitors), frozen in liquid air, and stored at –70°C until used.

The above procedure typically yields 1 g microsomal membrane protein from 50 g liver. This is solubilized by thawing rapidly, adding 0.01 vol of Triton X-100 (i.e., final concentration 1%) and mixing end-over-end for 30 min at 22°C. It appears that transient exposure to concentrations higher than 1% might, in fact, be required, since the solubilization is less efficient when the detergent is added as a diluted solution than when 0.01 vol of undiluted detergent is used. The solubilized membrane is diluted tenfold in 25 m*M* sodium HEPES, 10 m*M* calcium chloride, pH 7.4, containing protease inhibitors as above, and centrifuged at 22°C for 20 min at 12,000 × *g* to remove insoluble material.

2.3.2. IGF-II Affinity Chromatography

The agarose-IGF-II column is prepared by coupling human IGF-II to Affi-Gel 15 (Bio-Rad, Richmond, CA), an *N*-hydroxysuccinimide ester of crosslinked agarose (Martin and Baxter, 1986). Human IGF-II (3 mg) is dissolved in 10 mL of 0.1*M* sodium HEPES, pH 7.5, and Affi-Gel 15 (8 g moist wt), washed according to the manufacturer's instructions, is added. After mixing for 1 h at 22°C, 1 mL of 1*M* Tris-HCl, pH 8.0, is added to block any remaining reactive sites, and the mixing continued for 1 h. The gel is poured into a 1 cm diameter glass column, giving a bed height of approx 12 cm, and washed with 1 L of coupling buffer. By monitoring the IGF-II content of the wash fractions, the efficiency of coupling can be determined; this is frequently almost 100% under the above conditions. The column is cycled through high salt and low pH washes prior to use; typically, these might be 100 mL of 0.5*M* NaCl and 100 mL of 0.5*M* sodium acetate, pH 3.0, before final equilibration at neutral pH.

To protect the affinity column from any sediment that might develop in the solubilized membrane preparation, a guard column of Sephadex G-10 (Pharmacia), 2.5 cm diameter × 5 cm high, is placed in-line before the affinity column. The two columns are equilibrated in 25 m*M* HEPES, 10 m*M* $CaCl_2$, 0.1% Triton X-100, pH 7.4, at 22°C, and the solubilized membrane preparation is passed through the columns at 22°C, pumping at 1 mL/min. When all of the sample has been applied, the columns are washed with 100 mL of equilibration buffer plus 0.15*M* NaCl to minimize nonspecifically adsorbed proteins.

The IGF-II receptor is then eluted from the affinity column by lowering the pH to a point where the receptor no longer binds IGF-II. This is achieved by applying 50 mL of wash buffer that has been adjusted to pH 4.0 with glacial acetic acid. This is pumped at 1 mL/min, collecting 1 mL fractions that are immediately neutralized with 2*M* Tris base. The purified IGF-II receptor emerges from the column as a sharp peak in the few fractions following the drop in pH. In five such preparations, the mean recovery of active IGF-II receptor was 28%, yielding up to 1 mg of receptor with 2000-fold overall purification (Scott and Baxter, 1987a).

2.4. Other Purifications Using IGF-II Columns

The purification method described above depends for its success on the high affinity binding of the rat liver IGF-II receptor to the affinity column, but also on the fact that no other receptor in rat liver microsomal membranes can adsorb to the column. As mentioned previously, IGF-II also interacts strongly with type I IGF receptors; however, this is not a problem in adult rat liver, since this tissue contains little or no type I receptors. Other rat cells or tissues have also yielded high-purity IGF-II receptor preparations by a similar approach, although on a smaller scale. For example, in a preparation from rat placental membranes, 20 mg total membrane protein, purified on a rat IGF-II column, yielded 13 μg of 1100-fold purified receptor protein, representing a 37% recovery (Oppenheimer and Czech, 1983). In this study, the elution buffer, at pH 5.0, contained 1.5 *M* NaCl to facilitate receptor dissociation from the column, a technique that has also been successfully applied to the purification of receptors from rat 18,54-SF cells (Rosenfeld et al., 1986). Similarly, 41 mg of solubilized membranes from rat chondrosarcoma cells, loaded onto a rat IGF-II affinity column and eluted with a pH 6.0 buffer containing 4.5*M* urea, yielded 46 μg of purified receptor (August et al., 1983). Since, in a survey of IGF-II binding to a wide variety of rat tissues, only binding to type II receptors could be detected by affinity labeling and SDS-PAGE (Taylor et al., 1987), it seems likely that interference by type I receptors will not present a problem in IGF-II receptor purification from this species.

In contrast, it is the experience of the authors and others that human tissues in which IGF-II receptors can be demonstrated frequently also contain type I receptors, with the potential to interfere in the affinity chromatography procedure. A typical example is human placenta, a convenient tissue to use in large scale preparations because of its ready availability. Competitive binding studies with IGF-II tracer and human placental microsomal membranes indicate a component of binding displaceable readily by IGF-I and insulin as well as by IGF-II, and another component that is entirely unaffected by high insulin concentrations. The latter component (approx 40% of total IGF-II tracer binding) presumably represents type II receptors, and the former, type I receptors (Baxter, 1986). Since IGF-II binding to the type I receptors is inhibited by high concentrations of insulin, whereas that to type II receptors is unaffected by insulin, the specificity of the affinity chromatography step for type II receptors can be ensured by including a high concentration (e.g., 50 μg/mL) of insulin in the purification buffers.

2.5. Use of Other Affinity Ligands

With the recent discovery that receptors for IGF-II and mannose-6-phosphate are identical, purification methods based on affinity chromatography using mannose-6-phosphate derivatives can now be included in a survey of methods for purifying IGF-II receptors. The affinity ligand of choice appears to be a high mol wt core fragment derived by mild acid hydrolysis of a yeast phosphomannan (Bretthauer et al., 1973) that, presumably, terminates in a mannose-6-phosphate residue. When covalently coupled to agarose beads, this polysaccharide efficiently binds the receptor, which can then be specifically eluted with a buffer containing 5 or 10 m*M* mannose-6-phosphate (Sahagian et al., 1982; Steiner and Rome, 1982). The yield obtained by this method from rat chondrosarcoma cells, 700 μg receptor from 1 g membrane protein, or 0.07% by weight (Steiner and Rome, 1982), compares closely to the yield of 0.11% by weight obtained from the same cell line by IGF-II affinity chromatography (August et al., 1983). Glycoproteins containing mannose-6-phosphate have also been used as affinity ligands. For example, rat liver receptor has been isolated on a column of immobilized *Dictyostelium discoideum* glycoprotein (Geuze et al., 1984), and bovine liver receptor on a column of β-galactosidase-agarose (Sahagian et al., 1981). In both cases, elution was with buffer containing mannose-6-phosphate.

3. IGF-II Receptor Antibodies

3.1. Preparation and Characterization of Antibodies

One of the reasons for purifying receptors is to enable production of antireceptor antibodies. As discussed in detail below, these antibodies can be useful in measuring receptor concentration, and in determining the subcellular localization and biological roles of receptors. The purified rat liver IGF-II receptor, described above, has proved to be a particularly good antigen, leading to the production of high titer antisera by standard immunological methods. In the authors' laboratory, antisera were raised by emulsifying 400 µg pure receptor in 4 mL of isotonic buffer with an equal volume of complete Freund's adjuvant, and immunizing three 12-wk female New Zealand White rabbits with one-third of the mixture (130 µg receptor) each at multiple subcutaneous sites on the back. After 2 wk, the procedure was repeated using 100 µg receptor per rabbit, in incomplete adjuvant. Rabbits were boosted 2 wk later with 100 µg receptor administered into the hind leg muscles in isotonic buffer without adjuvant, and bled after another week.

Antisera can be tested for immunity in either (or both) of two ways (Scott and Baxter, 1987a). Antibodies that inhibit IGF-II binding to the receptor (and are, therefore, presumed to bind at or near the IGF-II binding site) are detected by preincubating membranes containing receptor (e.g., rat liver microsomal membranes) with various dilutions of test sera or control nonimmune rabbit serum, before measuring IGF-II tracer binding under normal conditions, centrifuging the membranes to separate bound from free tracer. Antibodies that precipitate prelabeled receptor (and, therefore, presumably react with an epitope distal to the IGF-II binding site) are detected by incubating detergent-solubilized membranes with IGF-II tracer to equilibrium, then adding various dilutions of test sera, and precipitating antibody-receptor-IGF-II complexes with second antibody, i.e., anti-rabbit immunoglobulin, plus 40 g/L (final concentration) polyethylene glycol (Scott and Baxter, 1987a).

Immunoblotting can be used to determine that antibodies react only with the receptor, and not with any other membrane-derived proteins (Geuze et al., 1984; Valentino et al., 1988; Hartshorn et al., 1989). Crude membranes are detergent-solubilized, subjected to SDS-PAGE, electroblotted onto nitrocellulose or another appropriate medium, and probed with solutions of test antisera or control nonimmune rabbit serum. Receptor-bound antibody is detected by reacting with [^{125}I]iodo-protein A followed by autoradiography, or with an enzyme-labeled second antibody and an appropriate chromogenic substrate. Different antibodies might react preferentially with the nonreduced or reduced receptor; therefore, the SDS-PAGE should be carried out under both conditions. No immunoreactive bands other than the receptor itself (i.e., 220 kDa nonreduced, 250 kDa reduced) should be detectable.

3.2. Applications of Receptor Antibodies

3.2.1. Radioimmunoassay

This technique is relatively rarely used to quantitate receptors. It can be useful, however, when it is necessary to determine total receptor concentration independently of the concentration of available receptor binding sites, for example, when the state of occupancy by ligand is unknown. In the case of the rat liver IGF-II receptor, a conventional radioimmunoassay was established using antireceptor antiserum at 1:1,000,000 final dilution, [^{125}I]iodo-receptor as radioligand, and increasing concentrations of pure receptor, over the range 0.01–1 ng/tube, as standard. With this assay it was possible to confirm by an immunological method the degree of receptor purification previously estimated by radioligand binding (Scott and Baxter, 1987a).

3.2.2. Immunohistochemistry

Receptor antibodies are an invaluable tool for determining tissue and subcellular distribution of receptors. Antibodies raised against purified IGF-II receptors have been used to localize these receptors by immunoperoxidase staining within a number of rat tissues, and to demonstrate by immunofluorescence their predominant intracellular location in juxtanuclear elements probably corresponding to Golgi bodies (Valentino et al., 1988; Hartshorn et al., 1989). This confirmed earlier immunofluorescence studies using antibodies raised against mannose-6-phosphate receptor (Brown and Farquhar, 1984; Geuze et al., 1984). The latter antibodies have also been used in immunoelectron microscopy studies that localized the receptor to coated vesicles, endosomes, lysosomes, and *cis* Golgi cisternae (Brown and Farquhar, 1984; Geuze et al., 1984).

3.2.3. Use of Receptor Antibodies in Functional Studies

The use of IGF-II receptor antibodies has greatly facilitated investigations into the function of the type II receptor. Demonstration of inhibitory (or agonistic) actions of antibodies on various biological processes is assumed to provide the most definitive proof of the involvement of type II receptors in those processes. As an example of this application, Kiess et al. (1987b) demonstrated that the stimulation by IGF-II of sugar and amino acid uptake in L6 skeletal muscle cells was not affected by antibodies known to inhibit IGF-II binding to the type II receptor, providing convincing evidence against the involvement of this receptor in these IGF-II actions (Kiess et al., 1987b). This contradicted the conclusion from earlier ligand binding studies (Beguinot et al., 1985) that the type II receptor did in fact mediate these IGF-II actions. Conversely, the demonstration that thymidine incorporation and calcium uptake in primed 3T3 fibroblasts is stimulated by IGF-II receptor antibodies that block receptor binding argues in favor of a role for the type II receptor in calcium gating in these cells (Kojima et al., 1988). Similarly, an antibody against human IGF-II receptor has been shown to mimic IGF-II in stimulating glycogen synthesis in HepG2 cells, indicating that this function is mediated by the IGF-II receptor (Hari et al., 1987). The wider application of receptor antibodies with both inhibitory and stimulatory actions should help to further delineate biological functions that are coupled to the type II receptor.

4. Concluding Comments

This review has attempted to put into perspective the many recent advances made in the area of IGF-II/mannose-6-phosphate receptor biochemistry, while providing practical advice for investigators who wish to purify and study this receptor. Although some of the reagents required to prepare the necessary affinity columns may not be readily available through commercial sources at present, there are sufficient published methods available to allow the techniques to be carried out successfully.

Many important questions concerning the biochemistry and physiology of IGF-II and its receptors remain to be answered. Among these could be included the following:

1. What is the full range of biological functions mediated via the type II receptor?
2. What is the origin and significance of the soluble IGF-II receptor in serum?
3. How do the growth-related and metabolic functions attributed to IGF-II acting via the type II receptor relate to the function of lysosomal enzyme transport attributed to the same (i.e., mannose-6-phosphate) receptor?

With the increased availability of receptor antibodies, IGF-II, and other reagents and techniques that might be anticipated in the near future, it should be possible to find answers to these questions over the next few years.

References

Adams, S. O., Nissley, S. P., Handwerger, S., and Rechler, M. M. (1983) *Nature* **302**, 150–153.
August, G. P., Nissley, S. P., Kasuga, M., Lee, L., Greenstein, L., and Rechler, M. M. (1983) *J. Biol. Chem.* **258**, 9033–9036.
Ballard, F. J., Ross, M., Upton, F. M., and Francis, G. L. (1988) *Biochem. J.* **249**, 721–726.
Baxter, R. C. (1986) *Adv. Clin. Chem.* **25**, 49–115.
Baxter, R. C. and De Mellow, J. S. M. (1986) *Clin. Endocrinol.* **24**, 267–278.
Baxter, R. C., De Mellow, J. S., and Burleigh, B. D. (1987) *Clin. Chem.* **33**, 544–548.
Beguinot, F., Kahn, C. R., Moses, A. C., and Smith, R. J. (1985) *J. Biol. Chem.* **260**, 15892–15898.
Bell, G. I., Merryweather, J. P., Sanchez-Pescador, R., Stempien, M. M., Priestley, L., Scott, J., and Rall, L. B. (1984) *Nature* **310**, 775–777.
Braulke, T., Causin, C., Waheed, A., Junghans, U., Hasilik, A., Maly, P., Humbel, R. E., and von Figura, K. (1988) *Biochem. Biophys. Res. Commun.* **150**, 1287–1293.
Bretthauer, R. K., Kaczorowski, G. J., and Weise, M. J. (1973) *Biochemistry* **12**, 1251–1256.
Brown, A. L., Graham, D. E., Nissley, S. P., Hill, D. J., Strain, A. J., and Rechler, M. M. (1986) *J. Biol. Chem.* **261**, 13144–13150.
Brown, W. J. and Farquhar, M. G. (1984) *Cell* **36**, 295–307.
Bryson, J. M. and Baxter, R. C. (1987) *J. Endocrinol.* **113**, 27–35.
Casella, S. J., Han, V. K., D'Ercole, A. J., Svoboda, M. E., and Van Wyk, J. J. (1986) *J. Biol. Chem.* **261**, 9268–9273.
Causin, C., Waheed, A., Braulke, T., Junghans, U., Maly, P., Humbel, R. E., and von Figura, K. (1988) *Biochem. J.* **252**, 795–799.
Corvera, S. and Czech, M. P. (1985) *Proc. Natl. Acad. Sci. USA* **82**, 7314–7318.
Corvera, S., Whitehead, R. E., Mottola, C., and Czech, M. P. (1986) *J. Biol. Chem.* **261**, 7675–7679.
Furlanetto, R. W., DiCarlo, J. N., and Wisehart, C. (1987) *J. Clin. Endocrinol. Metab.* **64**, 1142–1149.
Furman, T. C., Epp, J., Hsiung, H. M., Hoskins, J., Long, G. L., Mendelsohn, L. G., Schoner, B., Smith, D. P., and Smith, M. C. (1987) in *Escherichia coli. Bio/Technology* **5**, 1047–1051.
Gelato, M. C., Kiess, W., Lee, L., Malozowski, S., Rechler, M. M., and Nissley, P. (1988) *J. Clin. Endocinol. Metab.* **67**, 669–675.
Geuze, H. J., Slot, J. W., Strous, G. J. A. M., Hasilik, A., and von Figura, K. (1984) *J. Cell Biol.* **98**, 2047–2054.
Han, V. K. M., Lund, P. K., Lee, D. C., and D'Ercole, A. J. (1988) *J. Clin. Endocrinol. Metab.* **66**, 422–429.
Hari, J., Pierce, S. B., Morgan, D. O., Sara, V., Smith, M. C., and Roth, R. A. (1987) *EMBO J.* **6**, 3367–3371.
Hartshorn, M. A., Scott, C. D., and Baxter, R. C. (1989) *J. Endocrinol.* **121**, 221–227.
Haselbacher, G. K., Irminger, J. C., Zapf, J., Ziegler, W. H., and Humbel, R. E. (1987) *Proc. Natl. Acad. Sci. USA* **84**, 1104–1106
Hey, A. W., Browne, C. A., and Thorburn, G. D. (1987) *Endocrinology* **121**, 1975–1984.
Hylka, V. W., Kent, S. B. H., and Straus, D. S. (1987) *Endocrinology* **120**, 2050–2058.
Kiess, W., Greenstein, L. A., White, R. M., Lee, L., Rechler, M. M., and Nissley, S. P. (1987a) *Proc. Natl. Acad. Sci. USA* **84**, 7720–7724.
Kiess, W., Haskell, J. F., Lee, L., Greenstein, L. A., Miller, B. E., Aarons, A. L., Rechler, M. M., and Nissley, S. P. (1987b) *J. Biol. Chem.* **262**, 12745–12751.

Kojima, I., Nishimoto, I., Iiri, T., Ogata, E., and Rosenfeld, R. (1988) *Biochem. Biophys. Res. Commun.* **154**, 9–19.
Lauterio, T. J., Marson, L., Daughaday, W. H., and Baile, C. A. (1987) *Physiol. Behav.* **40**, 755– 758.
Lobel, P., Dahms, N. M., Breitmeyer, J., Chirgwin, J. M., and Kornfeld, S. (1987) *Proc. Natl. Acad. Sci. USA* **84**, 2233–2237.
MacDonald, R. G., Pfeffer, S. R., Coussens, L., Tepper, M. A., Brocklebank, C. M., Mole, J. E., Anderson, J. K., Chen, E., Czech, M. P., and Ullrich, A. (1988) *Science* **239**, 1134–1137.
Marquardt, H., Todaro, G. J., Henderson, L. E., and Oroszlan, S. (1981) *J. Biol. Chem.* **256**, 6859–6865.
Martin, J. L. and Baxter, R. C. (1986) *J. Biol. Chem.* **261**, 8754–8760.
Massague, J. and Czech, M. P. (1982) *J. Biol. Chem.* **257**, 5038–5045.
Misra, P., Hintz, R. L., and Rosenfeld, R. G. (1986) *J. Clin. Endocrinol. Metab.* **63**, 1400–1405.
Morgan, D. O., Edman, J. C., Standring, D. N., Fried, V. A., Smith, M. C., Roth, R. A., and Rutter, W. J. (1987) *Nature* **329**, 301–307.
Mottola, C., and Czech, M. P. (1984) *J. Biol. Chem.* **259**, 12705–12713.
Murphy, L. J., Bell, G. I., and Friesen, H. G. (1987) *Endocrinology* **120**, 1279– 1282.
Nishimoto, I., Ohkuni, Y., Ogata, E., and Kojima, I. (1987) *Biochem. Biophys. Res. Commun.* **142**, 275–286.
Nissley, S. P. and Rechler, M. M. (1984) *Clinics Endocrinol. Metab.* **13**, 43–67.
Oppenheimer, C. L. and Czech, M. P. (1983) *J. Biol. Chem.* **258**, 8539–8542.
Reeve, A. E., Eccles, M. R., Wilkins, R. J., Bell, G. I., and Millow, L. J. (1985) *Nature* **317**, 258–260.
Richman, R. A., Benedict, M. R., Florini, J. R., and Toly, B. A. (1985) *Endocrinology* **116**, 180–188.
Rinderknecht, E. and Humbel, R. E. (1978) *FEBS Lett.* **89**, 283–286.
Rosenfeld, R. G., Hodges, D., Pham, H., Lee, P. D. K., and Powell, D. R. (1986) *Biochem. Biophys. Res. Commun.* **138**, 304–311.
Rosenfeld, R. G., Pham, H., James, P., Shah, R., Diaz, G., and Wyche, J. (1987) *Abstracts, 69th Annual Meeting, The Endocrine Society, Indianapolis, IN, June 10-12*, p. 43.
Roth, R. A., Stover, C., Hari, J., Morgan, D. O., Smith, M. C., Sara, V., and Fried, V. A. (1987) *Biochem. Biophys. Res. Commun.* **149**, 600–606.
Sahagian, G. G., Distler, J., and Jourdian, G. W. (1981) *Proc. Natl. Acad. Sci. USA* **78**, 4289–4293.
Sahagian, G. G., Distler, J. J., and Jourdian, G. W. (1982) *Methods Enzymol.* **83**, 392–396.
Sasaki, N., Rees-Jones, R. W., Zick, Y., Nissley, S. P., and Rechler, M. M. (1985) *J. Biol. Chem.* **260**, 9793–9804.
Scott, C. D. and Baxter, R. C. (1987a) *Endocrinology* **120**, 1–9.
Scott, C. D. and Baxter, R. C. (1987b) *J. Cell. Physiol.* **133**, 532–538.
Scott, C. D., Martin, J. L., and Baxter, R. C. (1985) *Endocrinology* **116**, 1094–1101.
Scott, C. D., Taylor, J. E., and Baxter, R. C. (1988b) *Biochem. Biophys. Res. Commun.* **151**, 815–821.
Steiner, A. W. and Rome, L. H. (1982) *Arch. Biochem. Biophys.* **214**, 681–687.
Tally, M., Li, C. H., and Hall, K. (1987) *Biochem. Biophys. Res. Commun.* **148**, 811–816.
Taylor, J. E., Scott, C. D., and Baxter, R. C. (1987) *J. Endocrinol.* **115**, 35–41.
Tong, P. Y., Tollefsen, S. E., and Kornfeld, S. (1988) *J. Biol. Chem.* **263**, 2585–2588.
Tricoli, J. V., Rall, L. B., Scott, J., Bell, G. I., and Shows, T. B. (1984) *Nature* **310**, 784–786.
Tricoli, J. V., Rall, L. B., Karakousis, C. P., Herrera, L., Petrelli, N. J., Bell, G. I., and Shows, T. B. (1986) *Cancer Res.* **46**, 6169–6173.

Ullrich, A., Gray, A., Tam, A. W., Yang-Feng, T., Tsubokawa, M., Collins, C., Henzel, W. et al. (1986) *EMBO J.* **5**, 2503–2512.
Underwood L. E. and D'Ercole, A. J. (1984) *Clinics Endocrinol. Metab.* **13**, 69–89.
Valentino, K. L., Pham, H., Ocrant, I., and Rosenfeld, R. G. (1988) *Endocrinology* **122**, 2753–2763.
von Figura, K. and Hasilik, A. (1986) *Ann. Rev. Biochem.* **55**, 167–193.
Whitfield, H. J., Bruni, C. B., Frunzio, R., Terrell, J. E., Nissley, S. P., and Rechler, M. M. (1984) *Nature* **312**, 277–280.
Yu, K. T., Peters, M. A., and Czech, M. P. (1986) *J. Biol. Chem.* **261**, 11341–11349.

Purification of Insulin and Insulin-Like Growth Factor (IGF)-I Receptor

Yoko Fujita-Yamaguchi and Thomas R. Le Bon

1. Introduction

Peptide hormone/growth factor receptors are membrane glycoproteins. The receptor specifically binds its ligand, which approaches the cell from the outside, transduces the signal to the inside, and, consequently initiates biological programs specific to the receptor–ligand complex. In the case of insulin and IGF-I, the ligands not only share amino acid sequence homology (Rinderknecht and Humbel, 1978; Blundell et al., 1978), but also the receptors are structurally and functionally similar (Jacobs et al., 1983; Rechler and Nissley, 1985; Fujita-Yamaguchi et al., 1986). Recent progress made in purification and subsequent cloning of both receptors revealed ~60% identity in their overall amino acid sequence (Ullrich et al., 1985; Ebina et al., 1985; Ullrich et al., 1986). The receptors for insulin and IGF-I consist of α and β subunits, linked in a β–α–α–β form by disulfide bonds. The α subunit of M_r = ~125,000 is extracellular and responsible for ligand binding, whereas the β subunit of M_r = ~90,000 is a transmembrane protein carrying a tyrosine-specific protein kinase in its cytoplasmic domain. When the specific ligand binds to the extracellular domain, the β subunit kinase is activated, and consequently, phosphorylates itself (autophosphorylation) as well as intracellular components (Rees-Jones and Taylor, 1985; White et al., 1985; Bernier et al., 1987). The ability of the kinase to phosphorylate other proteins can be measured in vitro by the phosphorylation of exogenously added substrates such as histone H2B, certain types of Tyr-containing synthetic polymers and some tyrosine-containing peptides (Kasuga et al., 1983b; Braun et al., 1984; Sasaki et al., 1985; Sahal et al., 1988).

Since biological membranes are very complex structures, the mechanism of the receptor and ligand binding event could best be studied with completely purified fully active receptors. Complete purification of fully functional receptors are essential not only for gaining the amino acid sequence information, but also for determining the properties intrinsic to the receptors, as well as for studying effects of the membranes and its components on receptor function. We planned to isolate functional receptors for insulin and IGF-I so as first, to compare their structural homology and, second, to determine the subtle functional differences between the insulin and IGF-I receptors, toward understanding the signal transduction mechanism that is characteristic of each receptor. Thus, we focused on purification of sufficient amounts of biologically active receptors. We purified the insulin receptor to apparent homogeneity from human placenta, a rich and readily available source for human insulin receptor. The purified receptor retained full insulin binding activity (Fujita-Yamaguchi et al., 1983) as well as tyrosine-specific protein kinase activity (Kasuga et al., 1983a,b), indicating that our purified receptor retained the basic functions of the native insulin receptor. We have purified the IGF-I receptor from human placenta to apparent homogeneity (LeBon et al., 1986). Both IGF-I binding and kinase activities were copurified in our IGF-I receptor preparation, which confirmed that the receptor is a tyrosine-specific protein kinase. Our purified IGF-I receptor was partially sequenced, and its cDNA was cloned using oligonucleotide probes designed on the basis of the sequence data (Fujita-Yamaguchi et al., 1986; Ullrich et al., 1986).

This chapter summarizes our studies on purification of insulin and IGF-I receptors from human placenta (Table 1) (Fujita-Yamaguchi et al., 1983; LeBon et al., 1986). Protein sequences and activities of the two purified receptors are also discussed in a comparative manner (Fujita-Yamaguchi et al., 1986; Ullrich et al., 1986).

1.1. The Insulin Receptor

The basis of using ligand-affinity chromatography for receptor purification lies in taking advantage of the high affinity and specificity of the receptor for its hormone. This approach was pioneered by Cuatrecasas

Table 1
Purification of Insulin and IGF-I Receptors from Human Placenta

A. Preparation of **Placental Membranes** (Section 3).
Fresh placenta —> differential centrifugation —> placental membranes.

B. **Solubilization** of the receptors from placental membranes followed by **WGA-Sepharose chromatography** (Section 4).
Triton X-100 extracts —> WGA-Sepharose chromatography —> glycoprotein fractions with insulin and IGF-I binding activities.

C1. **Purification of the insulin receptor** from the WGA-Sepharose eluates (Section 5).
The glycoprotein fractions with insulin and IGF-I binding activities —> insulin-Sepharose chromatography —> purified insulin receptor.

C2. **Purification of the IGF-I receptor** from the WGA-Sepharose eluates (Section 6).
The glycoprotein fractions with insulin and IGF-I binding activities —> αIR-3-Sepharose chromatography —> purified IGF-I receptor.

(1970,1972a; Cuatrecasas and Parikh, 1972) and used by his colleagues and others to purify the insulin receptor (Jacobs et al., 1977; Siegel et al., 1981). Using insulin-Sepharose affinity chromatography, Jacobs et al. (1977) first purified the insulin receptor from rat liver. Later, Siegel et al. (1981) reported purification of human placental insulin receptor. These purification protocols, however, resulted in receptor preparations with low specific insulin binding activity. In addition, there was no kinase activity, as a result of either the lack of an intact β subunit or inactivation of the kinase. We have modified the elution conditions for insulin-Sepharose chromatography, which resulted in purified insulin receptor with full binding activity and insulin-stimulated kinase activity (Fujita-Yamaguchi et al., 1983; Kasuga et al., 1983 a,b). Although some additional modifications have been reported since (Petruzzelli et al., 1984; Pike et al., 1986), we basically use our original procedures. Other purification procedures different from ours have also been reported. These include use of anti-insulin receptor monoclonal antibodies (Roth et al., 1986) and polyclonal antibodies (Harrison and Itin, 1980) and of avidin-Sepharose coupled with biotinyl insulin (Finn et al., 1984; Kohanski and Lane, 1985). However, it is more difficult to prepare these columns than our insulin-Sepharose, since antibodies to the insulin receptor are not readily available and synthesis of biotinyl insulin is not simple.

1.2. The IGF-I Receptor

In contrast to insulin that is commercially available and inexpensive, IGF-I had not been readily available until recombinant IGF-I was marketed 4–5 y ago. Although ligand-affinity chromatography has been used for IGF-I receptor purification (Maly and Lüthi, 1986; Tollefsen et al., 1987), the method remains impractical because of the relative scarcity and expense of the recombinant IGF-I. Thus, we have chosen immunoaffinity chromatography to specifically purify the IGF-I receptor (LeBon et al., 1986). A monoclonal antibody to the IGF-I receptor, αIR-3, (Jacobs et al., 1983; Kull et al., 1983) was covalently linked to Sepharose. The IGF-I receptor was eluted from an αIR-3-Sepharose column under different elution conditions. Alkaline elution resulted in purified receptor preparations with IGF-I binding and kinase activities. Acid elution, although it resulted in inactive receptors, was effective in purifying a large amount of the receptor that was used for protein sequencing and molecular cloning studies.

2. Setting Up Assay Systems and Affinity Columns

2.1. Materials

Wheat germ agglutinin (WGA) and Sepharose 4B were purchased from E-Y Laboratories, San Mateo, CA., and Pharmacia, Pleasant Hill, CA. Crystalline porcine insulin was provided by Eli Lilly Co., Indianapolis, IN. IGF-I was purchased from Amgen, Thousand Oaks, CA. Phenylmethylsulfonylfluoride (PMSF) and N^{α}-benzoyl-l-arginine ethyl ester (BAEE) were from Sigma, St. Louis, MO. IgG class of monoclonal antibody (αIR-3), which is directed against the IGF-I receptor, was supplied by Steven Jacobs, the Wellcome Research Laboratories, Research Triangle Park, NC. A synthetic peptide resembling the tyrosyl phosphorylation site of $pp60^{src}$ (Arg-Arg-Leu-Ile-Glu-Asp-Ala-Glu-Tyr-Ala-Ala-Arg-Gly) was purchased from Peninsula Laboratories, Belmont, CA. ^{125}I-labeled IGF-I was from Amersham, Arlingon Heights, IL. (^{125}I-Tyr-A14)-insulin and [γ-^{32}P]ATP were purchased from New England Nuclear, Boston, MA. Molecular weight markers for sodium dodecyl sulfate (SDS)-polyacrylamide gels were from Bio-Rad, Richmond, VA.

2.2. Insulin and IGF-I Binding Assay

Activity of the insulin receptor was monitored at each purification step by insulin binding assay. Samples were incubated at 4°C for 16 h with ^{125}I-labeled porcine insulin (20,000 cpm, 0.2 ng) in a final vol of 0.4

mL of 50 mM Tris-HCl buffer, pH 7.4, containing 0.1% Triton X-100 and 0.1% bovine serum albumin. The receptor-[^{125}I]insulin complex was separated from free [^{125}I]insulin by adding 0.1 mL of 0.4% bovine γ-globulin and 0.5 mL of 20% polyethylene glycol 6000 and allowing to precipitate for 20 min at 4°C. The proteins were precipitated by centrifuging at 1500*g* for 20 min at 4°C. The supernatants were aspirated, and the radioactivity in the pellets was counted with a gamma counter. The radioactivity precipitated in the presence of excess unlabeled insulin (20 µg/mL) was considered "nonspecific" binding.

The nonspecific binding was always less than 7% (3% in the case of purified receptor) of the total radioactivity added to the reaction mixture. Competition binding studies are performed by varying the amount of unlabeled insulin (0.1–200 ng/mL) in the assay tube, and the data were analyzed by the method of Scatchard (1949) in order to calculate the maximum number of insulin molecules bound to crude or purified receptor. Specific activity was expressed as µg of insulin bound per mg of protein.

IGF-I binding activity was measured using ^{125}I-labeled IGF-I (20,000 cpm, 0.2 ng) under the same assay conditions, except that Triton X-100 was omitted (*see* Section 7.2. for explanation). For "nonspecific" binding controls, unlabeled bovine insulin (250 µg/mL) was used in place of IGF-I.

2.3. Tyrosine-Specific Protein Kinase Assay

2.3.1. Phosphorylation of an Exogenous Substrate (Kinase Assay)

Phosphorylation assays were carried out at 25°C for 40 min in 30 µL of 50 mM Tris-HCl buffer, pH 7.4, containing 1 mM of the src-related peptide, 2 mM $MnCl_2$, 15 mM $MgCl_2$, 0.1% Triton X-100, and 40 µM [γ-^{32}P]ATP (~12,000 cpm/pmole) and the purified insulin or IGF-I receptor (~0.1 µg), which was incubated in the absence or presence of insulin or IGF-I at 25°C for 1 h. The reaction was terminated by the addition of 50 µL of 5% trichloroacetic acid (TCA) and 20 µL of bovine serum albumin (10 mg/mL). After incubating this solution at 0°C for 30 min, the proteins were precipitated by centrifugation. Duplicate 35 µL aliquots of the supernatant were spotted on pieces (1.5 × 1.5 cm^2) of phosphocellulose paper (Whatman, P-81). The papers were extensively washed in 75 mM phosphoric acid. Incorporation of ^{32}p into the peptide was quantitated by liquid scintillation counting.

Alternatively, synthetic tyrosine-containing polymers, which are much less expensive than src-related peptide, such as (Glu:Tyr, 4:1)n (available from Sigma) can be used as substrates. The phosphorylation reaction is performed using 1 m*M* tyrosine-containing polymer. After the reaction, aliquots of reaction mixtures were spotted in duplicate on Whatman P-81 papers (1.5 × 1.5 cm^2) and washed in cold 10% TCA twice followed by 5% TCA-wash twice at 25°C, as described (Sahal and Fujita-Yamaguchi, 1987). The paper disks were transferred to scintillation vials, and ^{32}p incorporation was determined by liquid scintillation counting.

2.3.2. Phosphorylation of the β-Subunit (Autophosphorylation Assay)

Phosphorylation of the purified receptor was carried out in the same manner as described for the exogenous substrate phosphorylation. The reaction was stopped by the addition of 3 ×-concentrated Laemmli sample buffer (Laemmli, 1970), followed by boiling for 5 min. SDS-PAGE was performed according to Laemmli under reducing conditions. The gels were stained with silver, dried, and autoradiographed.

2.4. Preparation of Wheat Germ Agglutinin-Sepharose

Wheat germ agglutinin (WGA)-Sepharose was prepared by coupling WGA to cyanogen bromide-activated Sepharose 4B following the method of Porath et al. (1967). The amount of WGA bound to Sepharose 4B was approx 6 mg/mL of gel. WGA-Sepharose is also commercially available.

2.5. Preparation of Insulin-Sepharose

Insulin-Sepharose was prepared by a modification of the methods of Cuatrecasas (1970,1972; Cuatrecasas and Parikh, 1972) as summarized in Figure 1.

Step 1 Diaminodipropylamine-agarose was prepared by coupling 3,3'- diaminodipropylamine (60 nmol; the diamine solution adjusted to pH 10) to 30 mL of cyanogen bromide-activated Sepharose 4B at 4°C overnight with gentle stirring using a magnetic stirrer.

Step 2 The diaminodipropylamine-agarose was washed with 3–5 vol of distilled water and suspended in 30 mL of distilled water. Succinic anhydride (2.8 g) was added, the suspension was raised to, and maintained at, a pH of 6.0 by titrating with 5*N* NaOH for 10 min, and the mixture was stirred at 4°C overnight. The resin was washed with 3–5 vol of distilled water and then with 0.1*N* NaOH for 30–40 min, followed by washing with approx 5 vol of distilled water at room temperature on a sintered glass filter under vacuum.

PREPARATION OF INSULIN-SEPHAROSE

STEP

BrCN-activated Sepharose 4B

1 → coupled with 3,3'-diaminopropylamine

O-$NHCH_2CH_2CH_2NHCH_2CH_2CH_2NH_2$: Diaminodipropylamine-agarose

2 → coupled with succinic anhydride

O-$NHCH_2CH_2CH_2NHCH_2CH_2CH_2NHCOCH_2CH_2COOH$: Succinyldiaminodipropylamino-agarose

3 → coupled with N-hydroxysuccinimide in dioxane in the presence of DCC (N,N'-dicyclohexylcarbodiimide)

O-$NHCH_2CH_2CH_2NHCH_2CH_2CH_2NHCOCH_2CH_2\overset{O}{\overset{\|}{C}}$-O-N(succinimide)

4 → coupled with porcine insulin

O-$NHCH_2CH_2CH_2NHCH_2CH_2CH_2NHCOCH_2CH_2\overset{O}{\overset{\|}{C}}$-NH-Gly(A1) and/or -Phe(B1) and/or -Lys(B29)-insulin:

"INSULIN-SEPHAROSE"

(Insulin-succinyldiaminodipropylamino-agarose)

Fig. 1. Summary of preparation of insulin-Sepharose.

Step 3 The resulting succinyldiamino-dipropylamino-agarose was extensively washed with dioxane (previously treated with Molecular Sieve type 4A, mesh 8012) to obtain anhydrous conditions and resuspended in 60 mL of dioxane. Solid *N*-hydroxysuccinimide and solid *N,N*[1]-dicyclohexylcarbodiimide were added to the suspension to obtain a concentration of 0.1*M* each, and the suspension was gently stirred for 70 min at room temperature. The activated agarose was washed with 8 vol of dioxane over a 10-min period, followed by further washing with 4 vol of methanol (for 5 min to remove dicyclohexylurea), 3 vol of dioxane, and distilled water.

Step 4 This activated agarose was immediately mixed with 30 mL of 0.1*M* Na-phosphate buffer, pH 7.4, containing 6*M* urea, 30 mg of crystalline porcine insulin, and [^{125}I]insulin as a tracer. The coupling reaction was performed at 4°C overnight with gentle shaking, followed by 6 washes with 50 mL of 50 m*M* Tris-HCl buffer, pH 7.4. After transferring to a column, the gel was thoroughly washed with 50 m*M* Tris-HCl buffer (pH 7.4) containing 1*M* NaCl and 0.1*M* Na-phosphate buffer (pH 7.4) containing 6*M* urea, followed by equilibration in 50 m*M* Tris-HCl buffer (pH 7.4) containing 0.1% Triton X-100 and 0.1 m*M* PMSF (phenylmethlysulfonyl fluoride). The amount of insulin bound was 0.26 mg/mL of gel.

Alternatively, one can use a commercially available activated agarose, such as Affi-Gel 10 (Bio-Rad) or activated CH Sepharose 4B (Pharmacia). The use of these resins would avoid steps 1, 2, and 3. Although we were not able to utilize commercial gels, later other groups successfully used those gels (Sweet et al., 1985; Pike et al., 1986).

Preparation of insulin-Sepharose is one of the most crucial steps in the purification of insulin receptor. However, if one can synthesize one functional insulin-Sepharose preparation, our experience has shown that the same insulin-Sepharose column can be used without loss of function more than 200 times over a period of 5 y. A larger affinity column, ~25 mL, of insulin-Sepharose is required to purify 50–100 μg of the receptor, since the affinity of bound insulin seems to be very low. It is important to wash out free insulin completely from the column, or during application the receptor will bind to trace amounts of insulin rather than to the insulin-Sepharose and pass through the column as an insulin receptor complex.

2.6. Preparation of IGF-I Receptor Monoclonal Antibody (αIR-3)-Sepharose

αIR-3, a monoclonal antibody to the IGF-I receptor, is purified from ascites fluid by precipitation with 50% ammonium sulfate, followed by chromatography on DEAE-cellulose equilibrated with 10 m*M* potassium phosphate, pH 8.0, and then coupled to cyanogen bromide-activated Sepharose, as described (Porath et al., 1967).

3. Preparation of Placental Membranes

Fresh normal human placentas were obtained within 1 h after delivery, kept on ice, trimmed of amnion and chorion, washed with 0.25 *M* sucrose, and cut into small pieces. The pieces were transferred to 1 vol of 50 m*M* Tris-HCl buffer, pH 7.4, containing 0.25 *M* sucrose, 0.1 m*M* PMSF, and 2 m*M* BAEE, homogenized 5× for 1 min each with a Tekmar™ tissuemizer at a setting of 30, and centrifuged at 15,000 g for 20 min at 4°C. The supernatant was then centrifuged at 100,000 g for 1 h at 4°C. The pellet was suspended in 10 vol of 50 m*M* Tris-HCl buffer, pH 7.4, containing 0.1 m*M* PMSF and 2 m*M* BAEE, homogenized with a Tekmar™ tissuemizer at a setting of 10, and centrifuged at 100,000 g for 60 min at 4°C. The sedimented membranes were resuspended in 2 vol of the same buffer and stored at –75°C. Membranes that have been stored for over 2 y at –75°C are still usable for purification of functional receptors.

4. Solubilization of Insulin Receptor and IGF-I Receptor from Placental Membranes with Triton X-100, Followed by WGA-Sepharose Chromatography

Placental membranes (0.8–1.2 g of protein) prepared from two placentas (wet wt, 800–1200 g) were usually used for purification of the receptors for insulin and IGF-I. Insulin or IGF-I binding activity was monitored throughout the purification.

Step 1 Placental membrane proteins (25–30 mg/mL) were solubilized in 50 m*M* Tris-HCl buffer, pH 7.4, containing 2% Triton X-100, 0.1 m*M* PMSF, and 2 m*M* BAEE for 45 min at room temperature with stirring. A clear supernatant (Triton X-100 extracts) was obtained by centrifugation at 100,000 g for 90 min at 4°C. This temperature was maintained throughout the rest of the purification.

Step 2 The supernatant was diluted with 3 vol of 50 m*M* Tris-HCl buffer, pH 7.4, containing 0.1% Triton X-100, 10 m*M* $MgCl_2$, 0.1 m*M* PMSF and 2 m*M* BAEE, and applied to a column of WGA-Sepharose (2.5 × 8 cm), that was previously equilibrated with the same buffer. The column was washed with the buffer, then eluted with 50 m*M* Tris-HCl buffer, pH 7.4, containing 0.3*M* *N*-acetylglucosamine, 0.1% Triton X-100, 0.1 m*M* PMSF, and 2 m*M* BAEE.

Figure 2 shows typical elution profiles of both receptors on WGA-Sepharose chromatography. Although both receptors started eluting at the same position, the elution profile of IGF-I receptor was always broader than the insulin receptor. "Frac I" indicated in Fig. 2B contained both insulin and IGF-I binding activities and was usually used to purify both receptors. When IGF-I receptor was purified from "Frac II" by αIR-3-Sepharose chromatography and compared with the receptor purified from "Frac I," we found that the receptor purified from "Frac II" had no kinase activity, although it exhibited good IGF-I binding activity (LeBon and Fujita-Yamaguchi, unpublished data). During purification of the insulin receptor, we noted that the major insulin binding activity peak was often followed by an additional binding activity peak (Fujita-Yamaguchi and Kathuria, 1985b). The late eluting peak had insulin binding activity comparable to that of the early peak. However, the autophosphorylation activity and the ability to phosphorylate exogenous substrates by the late eluting peak was greatly reduced (Fujita-Yamaguchi and Kathuria, 1985b).

IGF-I receptor kinase activity appears more susceptible to proteolysis, and is, therefore, more easily lost during purification as compared to

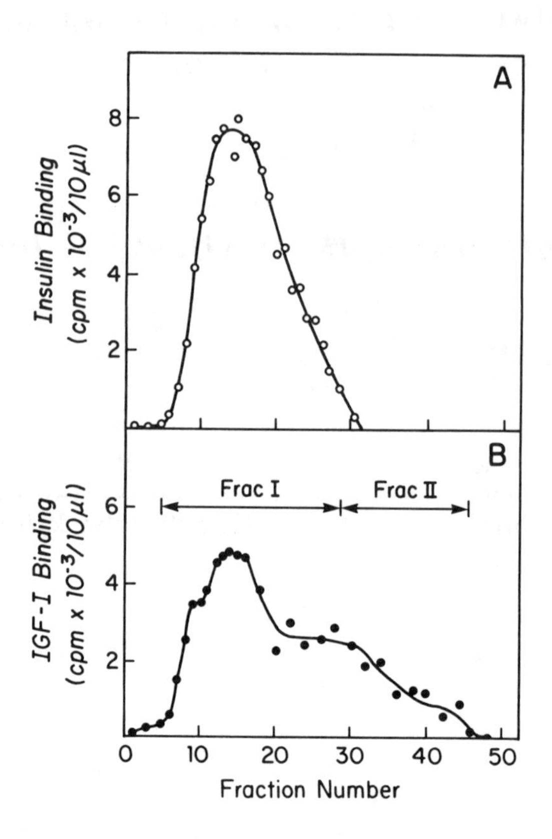

Fig. 2. WGA-Sepharose chromatography of Triton X-100 extracts of human placental membranes ~ 1 g). The sample was applied and eluted as described in Section 4. The profiles show the elution part in which 10 μL from each fraction (3 mL) was assayed for insulin binding (A) or IGF-I binding (B).

insulin receptor kinase. In the case of both receptors, kinase-active receptors appear to elute earlier than kinase-inactive receptors from a WGA-Sepharose column.

5. Purification of the Insulin Receptor from Glycoprotein Fractions (WGA-Sepharose Eluates)

5.1. Insulin-Sepharose Affinity Chromatography

WGA-Sepharose eluates with insulin binding activity were pooled and applied to a column (2 × 8 cm) of insulin-Sepharose that had been equilibrated with 50 m*M* Tris-HCl buffer, pH 7.4, containing 0.1% Triton X-100, 0.1 m*M* PMSF, and 2 m*M* BAEE. The column was thoroughly washed with 50 m*M* Tris-HCl buffer, pH 7.4, containing 1*M* NaCl, 0.1% Triton X-100, 0.1 m*M* PMSF, and 2 m*M* BAEE, then eluted with 50 m*M* acetate buffer, pH 5.0, containing 1*M* NaCl and 0.1% Triton X-100. To neutralize the eluates, fractions (2 mL) were collected in tubes containing 1 mL of 0.5*M* Tris-HCl buffer, pH 7.4.

5.2. Concentration of the Purified Insulin Receptor

The fractions containing insulin binding activity were pooled and concentrated by pressure dialysis using a PM-10 Diaflo ultrafiltration membrane (Amicon Corp.), replacing the elution buffer with 50 m*M* Tris-HCl buffer, pH 7.4. Glycerol (final 10%) was added to the concentrated receptor. The purified receptor can be stored for 6 mo at –75°C. This method was useful and practical in concentrating the receptor to ~0.5 mg/mL without loss of activity. One drawback, however, was that Triton X-100 was also concentrated. Alternatively, the insulin receptor can be concentrated by the use of a small column (0.2 mL) packed with DEAE-cellulose (Whatman DE-52) (Pike et al., 1986). The column method may be more suitable to concentrate the receptor in the appropriate buffer.

5.3. Purity and Activity of the Purified Insulin Receptor

Purification of the insulin receptor from Triton X-100 extracts is summarized in Table 2 (Fujita-Yamaguchi et al., 1983). Insulin binding activity was purified 2500-fold to apparent homogeneity. At the time we established the purification procedure in 1982, insulin binding activity was only an intrinsic activity of the insulin receptor. When assayed for autophosphorylation and kinase activities, our purified receptor was found to have tyrosine-specific protein kinase activity (Kasuga et al., 1983a,b). Later, others reported a modification of our procedures in an effort to recover better kinase activity (Pike et al., 1986): The specific activity of 300 nmol/min/mg of protein reported is ~30-fold higher than that of our preparation (Kasuga et al., 1983b). However, we did not find as large a variation as they observed. The difference could be owing to different assay conditions between theirs (+DTT, 30°C), and ours (no DTT, 22°C). Since mild DTT treatment increases kinase activity 5–10-fold(Fujita-Yamaguchi and Kathuria, 1985a), inclusion of DTT and changing the reaction temperature from 22 to 30°C may easily increase the specific kinase activity of our receptor preparation 10–20-fold.

Table 2
Purification of Human Insulin Receptor from Two Placentas

Step	Total protein, mg	Insulin-binding (specific activity)[a] μg/mg	Total activity, μg	Purification -fold	Yield %
Triton X-100 extracts	323[b]	0.0116	3.75	1	100
WGA-Sepharose	23.1[b]	0.087	2.01	7.5	54
Insulin-Sepharose	0.052[c]	28.5	1.48	2457	40

[a]Insulin-binding activity was calculated from the Scatchard plots and expressed as μg of insulin bound/mg of protein.
[b]Determined by a modified Lowry's method (Bensadoun and Weinstein, 1976).
[c]Determined by amino acid analysis (Del Valle and Shively, 1979).

Figure 3 shows SDS-PAGE analysis of the purified insulin receptor; silver-stained insulin receptor α and β subunits and autoradiography of phosphorylated β subunit in the absence or presence of insulin.

6. Purification of the IGF-I Receptor from Glycoprotein Fractions (WGA-Sepharose Eluates)

6.1. Affinity Chromatography on Sepharose Coupled with Monoclonal Antibody Against IGF-I Receptor (αIR-3)

IGF-I receptor could be eluted from the αIR-3 column either with acidic or alkaline buffers. Optimal elution conditions vary, depending on the level of coupling of αIR-3 and the amount of receptor loaded on the column. Highly substituted columns not only have a higher capacity for receptor, but require more stringent elution conditions. Applying a larger load to the column causes the receptor to elute over a broader range of elution conditions, which, in fact, give apparently homogeneous receptor preparations with high binding and kinase activities (*see* Fraction 1 in Table 3). Carbonate buffer with pH as high as 11 is used to elute receptor without loss of binding or kinase activity. However, in some experiments where highly substituted columns are used, a fraction of the retained receptor cannot be completely eluted at pH 11, but can be eluted with pH 2.2 glycine buffer. Of the buffers tested, pH 2.2 glycine

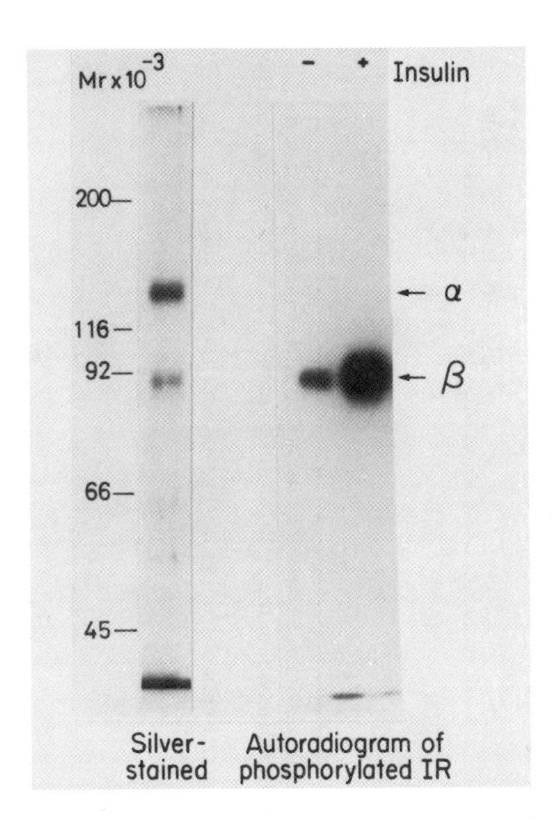

Fig. 3. SDS-PAGE analysis of purified insulin receptor. Insulin receptor (~0.5 μg) purified as described in the text was autophosphorylated in the absence or presence of 1 μ*M* insulin and analyzed by SDS-PAGE (7.5%) under reducing conditions, followed by silver-staining and autoradiography.

buffer is the most effective for eluting the receptor. However, receptor eluted at that pH loses tyrosine kinase activity and has considerably reduced binding activity. The pH 5.0 buffer elutes only a fraction of the receptor, but the receptor binds IGF-I and has tyrosine kinase activity.

6.1.1. Large-Scale Purification of IGF-Receptor on aIR-3-Sepharose: Acid Elution

WGA-Sepharose eluate prepared from a total of 21 placentas was passed two times through an insulin-Sepharose column and then applied to an αIR-3-Sepharose column (2 × 6.4 cm, ~0.5 mg of antibody/mL of packed gel) that had been equilibrated with 50 m*M* Tris-HCl buffer, pH 7.4, containing, 0.1% Triton X-100, 0.1 m*M* PMSF and 2 m*M* BAEE.

Table 3
Large-Scale Purification of IGF-I Receptor on αIR-3-Sepharose: Acid Elution[a]

	Total protein	Binding Activity				IGF-I-Stimulated kinase activity[b]		
		IGF-I Binding	Purification	Yield	Insulin binding	Specific activity	Purification	Yield
	mg	cpn	–fold	%	cpm	pmol/min/mg	–fold	%
WGA-Sepharose	680[c]	1.76×10^8	1	100	(6.85×10^7)	1.89	1	100
αIR-3-Sepharose								
Fraction I	0.094[d]	0.52×10^8	2765	30	(0)	5264	2785	39
Fraction II	0.168[d]	0.02×10^8	60	1.2	(0)	0	—	—

[a]The IGF-I receptor was purified from glycoprotein fractions of 21 placentas as described in Section 6.1.1. Fractions containing IGF-I binding activity appeared in the initial wash and the pH 5 elution (Fraction I) as well as the pH 2.2 elution (Fraction II). Modified after Table 2 (LeBon et al., 1986).

[b]Receptor preparations were assayed for kinase activity in the presence of IGF-I (1 μg/μL) as described in Section 2.3.1.

[c]Determined by a modified Lowry's method (Bensadoun and Weinstein, 1976).

[d]Determined by amino acid analysis (Del Valle and Shively, 1979).

The column was washed with the same buffer containing 1*M* NaCl, eluted with 50 m*M* acetate buffer, pH 5, containing 1*M* NaCl, 0.1% Triton X-100, 0.1 m*M* PMSF, and 2 m*M* BAEE, and then with 1*M* glycine buffer, pH 2.2, containing 1*M* NaCl, 0.1% Triton X-100, 0.1 m*M* PMSF, and 2 m*M* BAEE. Fractions (2 mL) were collected in tubes containing 3 mL of 0.5 *M* Tris-HCl, pH7.4, or 1*M* Tris-base to neutralize the eluates.

6.1.2. Purification of IGF-I Receptor with High Binding and Kinase Activities: Alkaline Elution

WGA-Sepharose eluate (125 mL) prepared from ~4 g of human placental membranes was applied to a 20-mL αIR-3-Sepharose column (~0.1 mg of antibody/mL of gel) that had been equilibrated with 50 m*M* Tris-HCl buffer, pH 7.4, containing 1*M* NaCl, 0.1% Triton X-100, 0.1 m*M* PMSF, and 2 m*M* BAEE. The column is washed with the equilibration buffer, then eluted with 50 m*M* carbonate buffer, pH 11, containing 1*M* NaCl, 0.1% Triton X-100, 0.1 m*M* PMSF, and 2 m*M* BAEE.

6.2. Concentration of the Purified IGF-I Receptor

Immediately after the elution, fractions containing IGF-I binding activity were pooled and concentrated using a PM-10 Diaflo ultrafiltration membrane (Amicon Corp.) and replacing the elution buffer with 50 m*M* Tris-HCl buffer, pH 7.4. Glycerol (final 10%) was added to the concentrated receptor. Aliquots of the receptor were stored at –75°C.

6.3. Purity and Activity of the Purified IGF-I Receptor

Purification of the IGF-I receptor from glycoprotein fractions of 21 placentas as described in Section 6.1.1. is summarized in Table 3 (LeBon et al., 1986). Since the acid elution causes inactivation of both insulin binding and kinase activities, the receptors eluted at neutral pH (because of overloading) and pH 5 (Fraction I) were used to calculate specific activities. Both IGF-I binding and kinase activities were purified ~3000-fold from the WGA-Sepharose eluates. Although pH 2.2 eluates (Fraction II) lost binding and kinase activities, this fraction was pure and especially useful for protein chemical studies.

Figure 4 shows SDS-PAGE analysis of the purified IGF-I receptor; silver-stained IGF-I receptor α and β subunits and autoradiography of phosphorylated β subunit in the absence or presence of IGF-I.

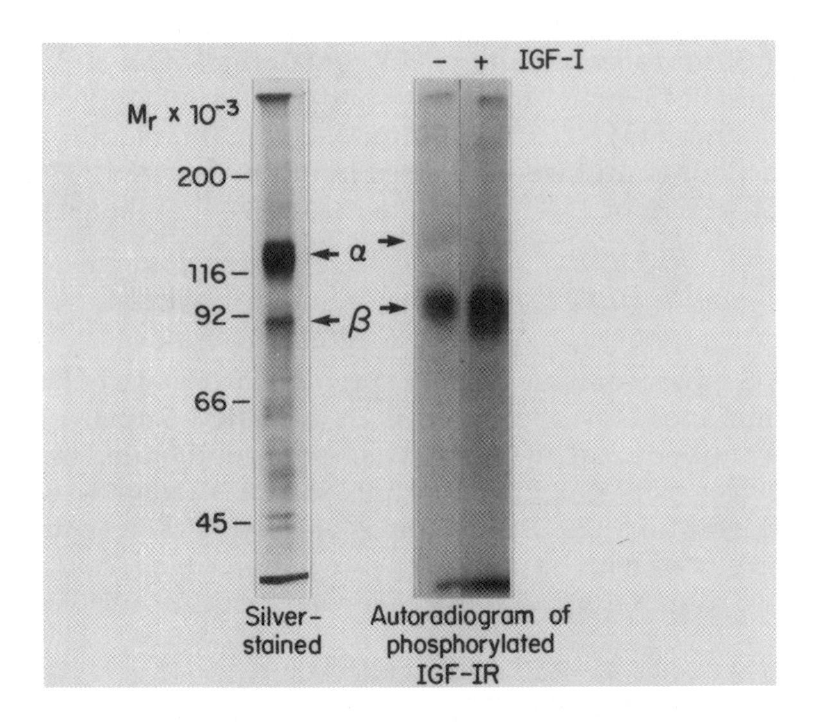

Fig. 4. SDS-PAGE analysis of purified IGF-I receptor. IGF-I receptor (2 µg) of Fraction II shown in Table 3 was analyzed by SDS-PAGE (7.5%) under reducing conditions and silver-stained. The receptor (~0.5 µg) eluted by alkaline conditions as described in Section 6.1.2., was autophosphorylated in the absence or presence of IGF(~1 µg/mL) and analyzed by SDS-PAGE as above, followed by autoradiography.

7. Comparison of the Two Purified Receptors

7.1. Primary Structure Deduced from cDNAs Encoding Insulin and IGF-I Receptors

Amino acid sequences of IGF-I and insulin receptors are shown in Fig. 5A (Ullrich et al., 1986). Hydropathy analysis (Kyte and Doolittle, 1982) of both receptors and receptor domains are presented in Fig. 5B. The numbers written in the domains show percent identity. The kinase domains are the most homologous, showing 84% identity.

7.2. Binding Activity

Competition of unlabeled IGF-I and insulin for binding of ^{125}I-IGF-I to purified IGF-I receptor and for ^{125}I-insulin to purified insulin receptor are shown in Fig. 6A and B, respectively (Fujita-Yamaguchi et al., 1986).

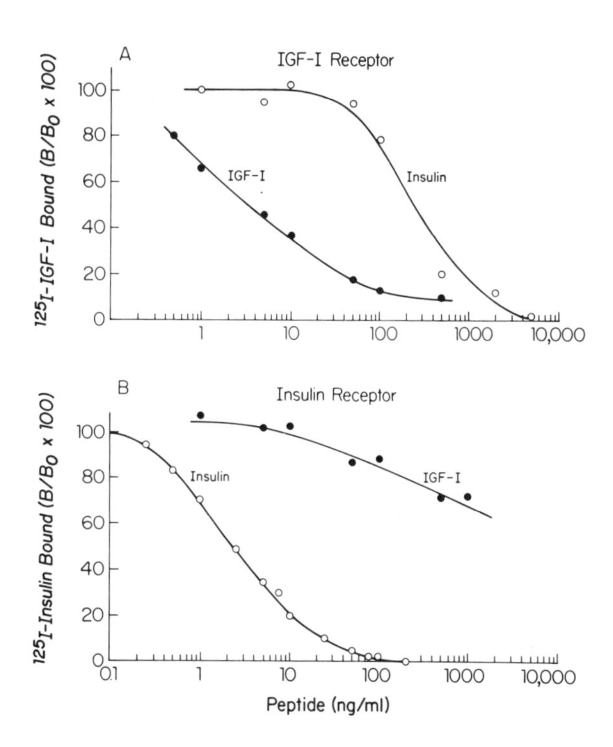

Fig. 6. Competition binding studies on purified IGF-I and insulin receptors (Fig. 5 from Fujita-Yamaguchi et al., 1986; permission granted). ^{125}I-labeled IGF-I (~20,000 cpm) was incubated with purified IGF-I receptor (~30 ng) in the presence of different concentrations of unlabeled IGF(●) or insulin (o) for 16 h at 4°C **(A)**. ^{125}I-insulin (~20,000 cpm) was incubated with purified insulin receptor (~20 ng) in the presence of different concentrations of unlabeled insulin (o) and IGF-I (●) for 16 h at 4°C **(B)**. The percentage of maximal specific binding of ^{125}I-labeled IGF-I or ^{125}I-insulin is plotted against ng/mL of peptide.

The difference in affinity between insulin and IGF-I appeared in both cases to be at least 100-fold, based on estimates of the concentration required to displace 50% of the bound radioactive ligand.

A striking difference between the two receptors is seen in the Triton X-100 effect on their binding activities (Fig. 7). When the concentration of Triton X-100 in the assay was reduced from 0.1 to 0.015%, binding activity of the insulin receptor increased up to 2.5-fold, whereas that of the IGF-I was not significantly affected.

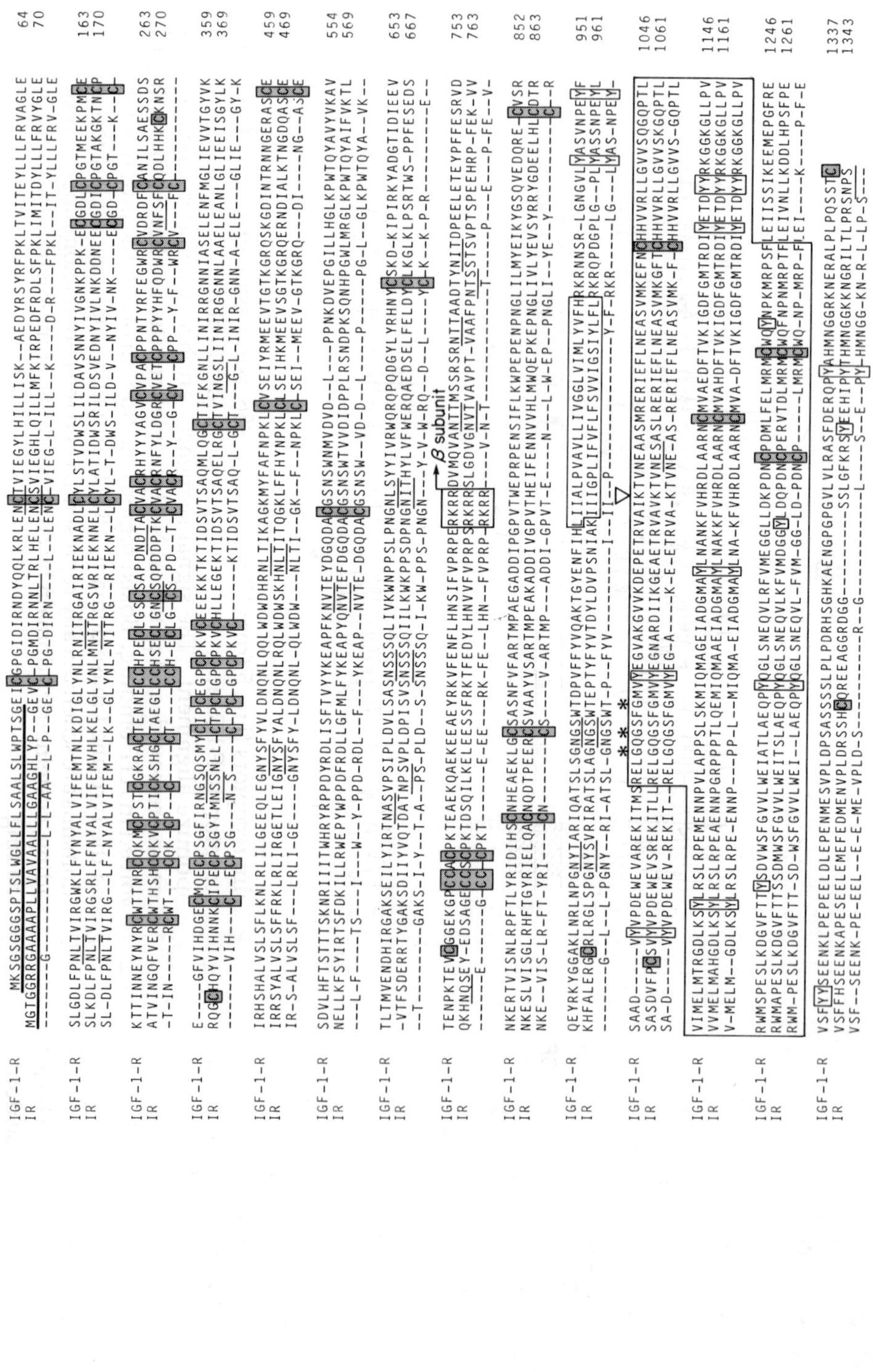

A

```
IGF-1-R    MKSGSGGGSPTSLWGLLFLSAALSLWPTSGEICGPGIDIRNDYQQLKRLENCTVIEGYLHILLISK--AEDYRSYRFPKLTVITEYLLLFRVAGLE     64
IR       MGTGGRRGAAAAPLLVAVAALLLGAAGHLYP--GEVC-PGMDIRNNLTRLHELENCSVIEGHLQILLMFKTRPEDFRDLSFPKLIMITDYLLLFRVYGLE     70
         -------G-----------L-L-AA--L-P--GE-C-PG-DIRN----L--LENC-VIEG-L-ILL--K----D-R---FPKL--IT-YLLLFRV-GLE

IGF-1-R  SLGDLFPNLTVIRGWKLFYNYALVIFEMTNLKDIGLYNLRNITRGAIRIEKNADLCYLSTVDWSLILDAVSNNYIVGNKPPK-ECGDLCPGTMEEKPMCE    163
IR       SLKDLFPNLTVIRGSRLFFNYALVIFEMVHLKELGLYNLMNITRGSVRIEKNNELCYLATIDWSRILDSVEDNYIVLNKDDNEECGDICPGTAKGKTNCP    170
         SL-DLFPNLTVIRG--LF-NYALVIFEM--LK--GLYNL-NITRG--RIEKN--LCYL-T-DWS-ILD-V--NYIV-NK----ECGD-CPGT---K--C-

IGF-1-R  KTTINNEYNYRCWTTNRCQKMCPSTCGKRACTENNECCHPECLGSCSAPDNDTACVACRHYYYAGVCVPACPPNTYRFEGWRCVDRDFCANILSAESSDS    263
IR       ATVINGQFVERCWTHSHCQKVCPTICKSHGCTAEGLCCHSECLGNCSQPDDPTKCVACRNFYLDGRCVETCPPPYYHFQDWRCVNFSFCQDLHHKCKNSR    270
         -T-IN-----RCWT---CQK-CP--C----CT----CCH-ECLG-CS-PD--T-CVACR--Y--G-CV--CPP--Y-F--WRCV---FC----------

IGF-1-R  E----GFVIHDGECMQECPSGFIRNGSQSMYCIPCEGPCPKVCEEEKKTKTIDSVTSAQMLQGCTIFKGNLLINIRRGNNIASELENFMGLIEVVTGYVK    359
IR       RQGCHQYVIHNNKCIPECPSGYTMNSSNLL-CTPCLGPCPKVCHLLEGEKTIDSVTSAQELRGCTVINGSLIINIRGGNNLAAELEANLGLIEEISGYLK    369
         -------VIH---C--ECPSG---N-S----C-PC-GPCPKVC------KTIDSVTSAQ-L-GCT---G-L-INIR-GNN-A-ELE---GLIE---GY-K

IGF-1-R  IRHSHALVSLSFLKNLRLILGEEQLEGNYSFYVLDNQNLQQLWDWDHRNLTIKAGKMYFAFNPKLCVSEIYRMEEVTGTKGRQSKGDINTRNNGERASCE    459
IR       IRRSYALVSLSFFRKLRLIRGETLEIGNYSFYALDNQNLRQLWDWSKHNLTITQGKLFFHYNPKLCLSEIHKMEEVSGTKGRQERNDIALKTNGDQASCE    469
         IR-S-ALVSLSF---LRLI-GE----GNYSFY-LDNQNL-QLWDW---NLTI--GK--F--NPKLC-SEI--MEEV-GTKGRQ---DI----NG--ASCE

IGF-1-R  SDVLHFTSTTTSKNRIIITWHRYRPPDYRDLISFTVYYKEAPFKNVTEYDGQDACGSNSWNMVDVD--L---PPNKDVEPGILLHGLKPWTQYAVYVKAV    554
IR       NELLKFSYIRTSFDKILLRWEPYWPPDFRDLLGFMLFYKEAPYQNVTEFDGQDACGSNSWTVVDIDPPLRSNDPKSQNHPGWLMRGLKPWTQYAIFVKTL    569
         ---L-F----TS---I---W--Y-PPD-RDL--F---YKEAP--NVTE-DGQDACGSNSW--VD-D--L----P-----PG-L--GLKPWTQYA--VK--

IGF-1-R  TLTMVENDHIRGAKSEILYIRTNASVPSIPLDVLSASNSSSQLIVKWNPPSLPNGNLSYYIVRWQRQPQDGYLYRHNYCSKD-KIPIRKYADGTIDIEEV    653
IR       -VTFSDERRTYGAKSDIIYVQTDATNPSVPLDPISVSNSSSQIILKWKPPSDPNGNITHYLVFWERQAEDSELFELDYCLKGLKLPSRTWS-PPFESEDS    667
         --T--------GAKS-I-Y--T-A--PS-PLD--S-SNSSSQ-I-KW-PPS-PNGN---Y-V-W-RQ--D--L----YC-K--K-P-R---------E--

                                                          β subunit
IGF-1-R  TENPKTEVCGGEKGPCCACPKTEAEKQAEKEEAEYRKVFENFLHNSIFVPRPERKRRDVMQVANTTMSSRSRNTTAADTYNITDPEELETEYPFFESRVD    753
IR       QKHNQSEY-EDSAGECCSCPKTDSQILKELEESSFRKTFEDYLHNVVFVPRPSRKRRSLGDVGNVTVAVPT-VAAFPNTSSTSVPTSPEEHRP-FEK-VV    763
         ------E------G-CC-CPKT------E-EE---RK-FE--LHN--FVPRP-RKRR----V-N-T-----------T-----P---E---P-FE--V-

IGF-1-R  NKERTVISNLRPFTLYRIDIHSCNHEAEKLGCSASNFVFARTMPAEGADDIPGPVTWEPRPENSIFLKWPEPENPNGLILMYEIKYGSQVEDQRE-CVSR    852
IR       NKESLVISGLRHFTGYRIELQACNQDTPEERCSVAAYVSARTMPEAKADDIVGPVTHEIFENNVVHLMWQEPKEPNGLIVLYEVSYRRYGDEELHLCDTR    863
         NKE--VIS-LR-FT-YRI----CN-------CS----V-ARTMP---ADDI-GPVT-E----N---L-W-EP--PNGLI--YE--Y----------C--R

IGF-1-R  QEYRKYGGAKLNRLNPGNYTARIQATSLSGNGSWTDPVFFYVQAKTGYENFIHLIIALPVAVLLIVGGLVIMLYVFHRKRNNSR-LGNGVLYASVNPEYF    951
IR       KHFALERGCRLRGLSPGNYSVRIRATSLAGNGSWTEPTYFYVTDYLDVPSNIAKIIIGPLIFVFLFSVVIGSIYLFLRKRQPDGPLG--PLYASSNPEYL    961
         -------G--L--L-PGNY--RI-ATSL-GNGSWT-P--FYV---------I--II--P--------------Y-F-RKR-----LG---LYAS-NPEY-
                                     * *  *

IGF-1-R  SAAD-----VYVPDEWEVAREKITMSRELGQGSFGMVYEGVAKGVVKDEPETRVAIKTVNEAASMRERIEFLNEASVMKEFNCHHVVRLLGVVSQGQPTL   1046
IR       SASDVFPCSVYVPDEWEVSREKITLLRELGQGSFGMVYEGNARDIIKGEAETRVAVKTVNESASLRERIEFLNEASVMKGFTCHHVVRLLGVVSKGQPTL   1061
         SA-D-----VYVPDEWEV-REKIT--RELGQGSFGMVYEG-A----K-E-ETRVA-KTVNE-AS-RERIEFLNEASVMK-F-CHHVVRLLGVVS-GQPTL

IGF-1-R  VIMELMTRGDLKSYLRSLRPEMENNPVLAPPSLSKMIQMAGEIADGMAYLNANKFVHRDLAARNCMVAEDFTVKIGDFGMTRDIYETDYYRKGGKGLLPV   1146
IR       VVMELMAHGDLKSYLRSLRPEAENNPGRPPPTLQEMIQMAAEIADGMAYLNAKKFVHRDLAARNCMVAHDFTVKIGDFGMTRDIYETDYYRKGGKGLLPV   1161
         V-MELM--GDLKSYLRSLRPE-ENNP---PP-L--MIQMA-EIADGMAYLNA-KFVHRDLAARNCMVA-DFTVKIGDFGMTRDIYETDYYRKGGKGLLPV

IGF-1-R  RWMSPESLKDGVFTTYSDVWSFGVVLWEIATLAEQPYQGLSNEQVLRFVMEGGLLDKPDNCPDMLFELMRMCWQYNPKMRPSFLEIISSIKEEMEPGFRE   1246
IR       RWMAPESLKDGVFTTSSDMWSFGVVLWEITSLAEQPYQGLSNEQVLKFVMDGGYLDQPDNCPERVTDLMRMCWQFNPNMRPTFLEIVNLLKDDLHPSFPE   1261
         RWM-PESLKDGVFTT-SD-WSFGVVLWEI--LAEQPYQGLSNEQVL-FVM-GG-LD-PDNCP-----LMRMCWQ-NP-MRP-FLEI----K----P-F-E

IGF-1-R  VSFYYSEENKLPEPEELDLEPENMESVPLDPSASSSSLPLPDRHSGHKAENGPGPGVLVLRASFDERQPYAHMNGGRKNERALPLPQSSTC   1337
IR       VSFFHSEENKAPESEELEMEFEDMENVPLDRSSHCQREEAGGRDGG---------SSLGFKRSYEEHIPYTHMNGGKKNGRILTLPRSNPS   1343
         VSF--SEENK-PE-EEL--E-E-ME-VPLD-S----------R--G----------L----S--E--PY-HMNGG-KN-R-L-LP-S---
```

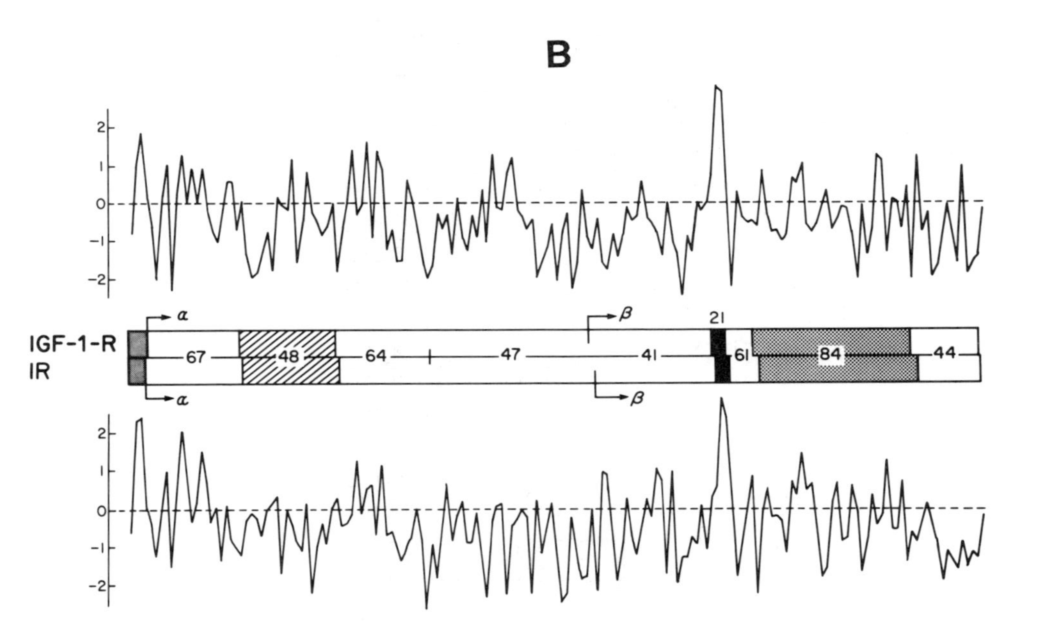

Fig. 5. Comparison of IGF-I and insulin receptors (Fig. 5 from Ullrich et al., 1986; permission granted). **(A)** Comparison of primary amino acid sequences of IGF-I receptor (IGF-1-R) and insulin receptor (IR). Identical residues are indicated on the third line, gaps having been introduced to optimize alignment. The signal sequences are heavily underlined, cysteine residues shaded, tyrosines in the cytoplasmic domain are boxed, and potential N-glycosylation sites are lightly underlined. Asterisks and open triangles indicate residues involved in ATP binding. Boxed regions include, in order, the putative precursor processing site, the transmembrane domain and the tyrosine kinase domain. **(B)** Hydropathy analysis of IGF-I receptor and insulin receptor precursor sequences. Receptor domains are schematically represented and the percent homologous residues indicated. The signal sequence is shown by fine shading, cysteine-rich domain by cross hatching, trans-membrane domain by a black bar, and tyrosine kinase domain by coarse shading.

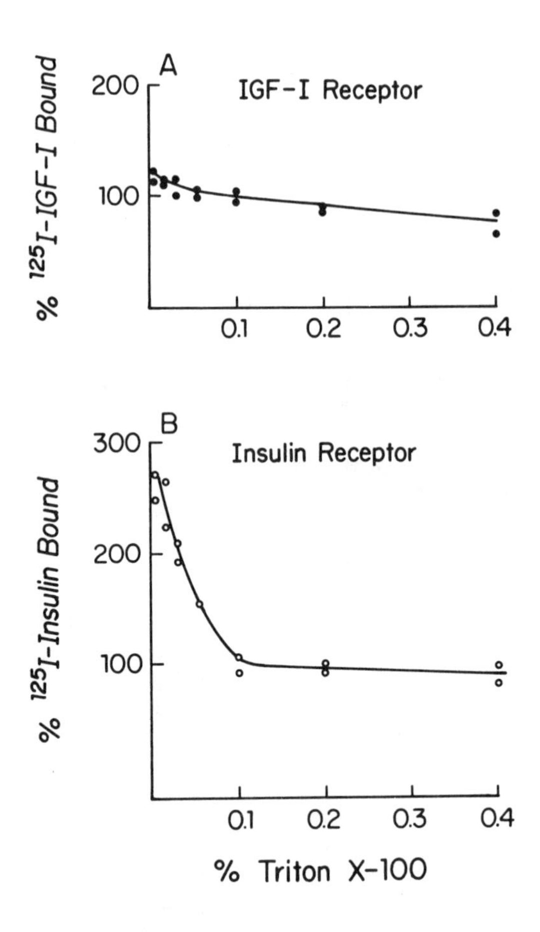

Fig. 7. Effects of Triton X-100 on ligand binding activity on purified IGF-I receptor **(A)** and insulin receptor **(B)**. Purified receptors (~30 ng) were assayed for their corresponding ligand binding activity as described in Section 2.2 in the presence of different concentrations of Triton X-100. The binding activity is expressed as a percent of that obtained in the presence of 0.1% Triton X-100.

7.3. Tyrosine-Specific Protein Kinase Activity

Both purified receptors exhibited tyrosine-specific protein kinase activity, as judged by ^{32}P incorporation into the src-related peptide. Table 4 shows examples of kinase activity of different receptor preparations. The kinase activity of the receptors for insulin and IGF-I was usually stimulated by corresponding ligand ~5- and ~2-fold, respectively (A and D). It is to be noted that kinase activation by ligand was reduced when the receptor preparations were not fresh (B and C), and that it was

Table 4
Phosphorylation of the Src-Related Peptide by Purified Insulin and IGF-I Receptors

Receptor preparation[a]	Specific Activity (pmol/min/mg)[b] Basal	Hormone-stimulated[c]	Fold stimulation
Insulin receptor			
A	1230	6340	5.2
B	1710	2310	1.4
IGF-I receptor			
C	970	1550	1.6
D	2130	5260	2.5

[a]Receptor preparations A and C were purified from the same source of fresh placental membranes. IGF-I receptor preparations C and D were the fractions eluted from αIR-3-Sepharose with pH 11 buffer and pH 5 buffer, respectively. Insulin receptor preparation A represents our typical results. Insulin receptor preparation B represents purified receptors that had been partially inactivated (Kathuria et al., 1986).

[b]The tyrosine kinase activity was measured using 1 m*M* Src-related peptide. The amount of protein was determined by amino acid analysis.

[c]Kinase activity of insulin receptor and IGF-I receptor was measured in the presence of 1 μ*M* insulin and IGF-1 (1 μg/mL), respectively.

Table 5
Substrate Preference for Insulin Receptor vs IGF-I Receptor

	Km for $(Glu:Tyr,4:1)_n$[a]	Ratio of ^{32}P Incorporation into $(Glu:Tyr,4:1)_n/(Glu:Ala:Tyr,6:3:1)_n$
Insulin receptor	0.6 m*M*	1.1 + 0.2 (6) [b]
IGF-I receptor	10 m*M*	4.0 + 0.6 (7)

[a]Phosphorylation of (Glu:Tyr,4:1)*n* or (Glu:Ala:Tyr,6:3:1)*n* were performed using a substrate concentration of 1 m*M*.

[b]Numbers in parentheses indicate numbers of independent experiments from which mean ratios ± SD were calculated.

more difficult to prepare the IGF-I receptor kinase than the insulin receptor kinase from placental membranes (A and C).

Tyrosine-containing polymers, such as (Glu:Tyr,4:1)*n* and (Glu:Ala:Tyr,6:3:1)*n* have also been used as substrates for tyrosine-specific protein kinases (Sahal et al., 1988). Km values and substrate preference for the insulin and IGF-I receptors are summarized in Table 5. Although the kinase domains shared 84% identity, significant difference had also been found in their substrate specificity as well as in the effects of inhibitors (Sahal et al., 1988; Fujita-Yamaguchi and Kathuria, 1988).

8. Closing Remarks

Six years have passed since we established our purification procedure for the human placental insulin receptor. We still use the same procedure that was optimized for a large-scale purification. The original insulin-Sepharose column that has been used more than 200 times remains in good condition. This may be attributed to the condition we keep the column (in the elution buffer at 4°C). As mentioned earlier, it is not easy to purify the intact IGF-I receptor, since the β subunit seems more susceptible to proteolysis than the insulin receptor β subunit. We did not feel that it was necessary to use a combination of protease inhibitors since our receptor preparations retained full binding and kinase activity, as seen in Fig. 4. However, inclusion of leupeptin (1μg/mL), and pepstatin (1 μg/mL) during preparation and solubilization of placental membranes may improve the recovery of the intact IGF-I receptor β subunit.

Our purified receptor preparations as described in this chapter have proven to be consistently useful for structural and functional studies of the two receptors over the past 6 y. It was our insulin receptor preparation with an intact β subunit that was initially characterized as purified insulin receptor kinase (Fujita-Yamaguchi et al., 1983; Fujita-Yamaguchi, 1984; Kasuga et al., 1983a,b). We obtained amino acid sequence data from our purified insulin receptor (Fujita-Yamaguchi et al., 1987) and IGF-I receptor (Fujita-Yamaguchi et al., 1986). Molecular cloning based on our amino acid sequence data led to the isolation of the cDNAs encoding the IGF-I receptor (Ullrich et al., 1986). The cDNA sequence information revealed that the overall primary structure of the IGF-I receptor is very similar to that of the insulin receptor (Fig. 5). There are also some significant differences that may reflect subtle and yet important functional differences between the two receptors. Our studies of the structure–function relationships using the purified receptors have been performed toward gaining an understanding of binding characteristics of the insulin receptor (Fujita-Yamaguchi and Harmon, 1988; Wang et al., 1988) as well as kinase activity of both receptors (Fujita-Yamaguchi and Kathuria, 1985a,b; Fujita-Yamaguchi et al., 1986; Kathuria et al., 1986; Sahal et al., 1988; Fujita-Yamaguchi and Kathuria, 1988). Although X-ray crystallography analysis should ultimately lead to the exact structure of these receptors and the understanding of the important functional domains, protein chemical studies, such as chemical modification and substructural domain analysis, are essential to fill the gap that exists between the primary and three-dimensional structures. These studies are now in progress in our laboratory.

In addition, using our purified insulin and IGF-I receptors, sensitive receptor immunoprecipitation assays have been developed. These have allowed us to screen a new type of autoantibody in human diabetes that cannot be identified by conventional binding inhibition assays (Boden et al., 1988; Tappy et al., 1988).

Acknowledgments

We are grateful for encouragement and support by Keiichi Itakura and Rachmiel Levine. We thank Izumi Hayashi and Pamela J. Smith for valuable comments on this chapter, Song Choi and Satish Kathuria for their excellent work in purifying the receptors, and Steven Hartman, Martha Burgeson, Kirk Kanzelberger, and Dipali Koyal for preparing placental membranes. Secretarial assistance given by Faith Sorensen is greatly appreciated. This work was supported by NIH grants DK29770, DK34427, and CA33572.

References

Bensadoun, A. and Weinstein, D. (1976) *Anal. Biochem.* **70,** 241–250.

Bernier, M., Laird, D. M., and Lane, M. D. (1987) *Proc. Natl. Acad. Sci., USA* **84,** 1844–1848.

Blundell, T. L., Bedarkar, S., Rinderknecht, E., and Humbel, R. E. (1978) *Proc. Natl. Acad. Sci., USA* **75,** 180–184.

Boden, G., Fujita-Yamaguchi, Y., Shimoyama, R., Shelmet, J. J., Tappy, L., Rezvani, I., and Owen, O. E. (1988) *J. Clin. Invest.* **81,** 1971–1978.

Braun, S., Raymond, W. E., and Racker, E. (1984) *J. Biol. Chem.* **259,** 2051–2054.

Cuatrecasas, P. (1970) *J. Biol. Chem.* **245,** 3059–3065.

Cuatrecasas, P. (1972) *Proc. Natl. Acad. Sci., USA* **69,** 1277–1281.

Cuatrecasas, P. and Parikh, I. (1972) *Biochemistry* **11,** 2291–2299.

Del Valle, U. and Shively, J. E. (1979) *Anal. Biochem.* **96,** 77–83.

Ebina, Y., Ellis, L., Jarnagin, K., Edery, M., Graf, L., Clauser, E., Ou, J.-H., Masiarz, F., Kan, Y. W., Goldfine, I. D., Roth, R. A., and Rutter, W. J. (1985) *Cell* **40,** 747–758.

Finn, F. M., Titus, G., Horstman, D., and Hofmann, K. (1984) *Proc. Natl. Acad. Sci., USA* **81,** 7328–7332.

Fujita-Yamaguchi, Y. (1984) *J. Biol. Chem.* **259,** 1206–1211.

Fujita-Yamaguchi, Y., Choi, S., Sakamoto, Y., and Itakura, K. (1983) *J. Biol. Chem.* **258,** 5045–5049.

Fujita-Yamaguchi, Y. and Harmon, J. T. (1988) *Biochemistry* **27,** 3252–3260.

Fujita-Yamaguchi, Y., Hawke, D. H., Shively, J. E., and Choi, S. (1987) *Protein Sequences and Data Analysis* **1,** 3–6.
Fujita-Yamaguchi, Y. and Kathuria, S. (1985a) *Proc. Natl. Acad. Sci., USA* **82,** 6095–6099.
Fujita-Yamaguchi, Y. and Kathuria, S. (1985b) in *Cancer Cells: Growth Factors and Transformation,* vol 3 (Seramisco, J., Ozanne, B., and Stiles, C., eds.), Cold Spring Harbor, NY, pp. 123–129.
Fujita-Yamaguchi, Y. and Kathuria, S. (1988). *Biochem. Biophys. Res. Commun.* **157,** 955–962.
Fujita-Yamaguchi, Y., LeBon, T.R., Tsubokawa, M., Henzel, W., Kathuria, S., Koyal, D., and Ramachandran, J. (1986) *J. Biol. Chem.* **261,** 16727–16731.
Harrison, L. C. and Itin, A. (1980) *J. Biol. Chem.* **255,** 12066–12072.
Jacobs, S., Kull, F. C., Earp, H. S., Svoboda, M., VanWyk, J. J., and Cuatrecasas, P. (1983) *J. Biol. Chem.* **258,** 9581–9584.
Jacobs, S., Shechter, Y., Bissel, K., and Cuatrecasas, P. (1977) *Biochem. Biophys. Res. Commun.* **77,** 981–988.
Kasuga, M., Fujita-Yamaguchi, Y., Blithe, D. L., and Kahn, C. R. (1983a) *Proc. Natl. Acad. Sci., USA* **80,** 2137–2141.
Kasuga, M., Fujita-Yamaguchi, Y., Blithe, D.L., White, M. F., and Kahn, C. R. (1983b) *J. Biol. Chem.* **258,** 10973–10980.
Kathuria, S., Hartman, S., Grunfeld, C., Ramachandran, J., and Fujita-Yamaguchi, Y. (1986) *Proc. Natl. Acad. Sci., USA* **83,** 8570–8574.
Kohanski, R. A. and Lane, M. D. (1985) *J. Biol. Chem.* **260,** 5014–5025.
Kull, F. C., Jr., Jacobs, S., Su, Y.-F., Svoboda, M. E., VanWyk, J. J., and Cuatrecasas, P. (1983) *J. Biol. Chem.* **258,** 6561–6566.
Kyte, J. and Doolittle, R. F. (1982) *J. Mol. Biol.* **157,** 105–132.
Laemmli, U. K. (1970) *Nature* (London) **227,** 680–685.
LeBon, T. R., Jacobs, S., Cuatrecasas, P., Kathuria, S., and Fujita-Yamaguchi, Y. (1986) *J. Biol. Chem.* **261,** 7685–7689.
Maly, P. and Lüthi, C. (1986) *Biochem. Biophys. Res. Commun.* **137,** 695–701.
Petruzzelli, L., Herrera, R., and Rosen, O. M. (1984) *Proc. Natl. Acad. Sci., USA* **81,** 3327–3331.
Pike, L. J., Eakes, A. T., and Krebs, E. G. (1986) *J. Biol. Chem.* **261,** 3782–3789.
Porath, J., Axen, R., and Ernback, S. (1967)*Nature* (London) **215,** 1491–1492.
Rechler, M. M. and Nissley, S. P. (1985) *Ann. Rev. Physiol.* **47,** 425–442.
Rinderknecht, E. and Humbel, R. E. (1978) *J. Biol. Chem.* **253,** 2769–2776.
Rees-Jones, R. W. and Taylor, S. I. (1985) *J. Biol. Chem.* **260,** 4461–4467.
Roth, R. A., Morgan, D. O., Beandoin, J., and Sara, V. (1986) *J. Biol. Chem.* **261,** 3753–3757.
Sahal, D. and Fujita-Yamaguchi, Y. (1987) *Anal. Biochem.* **167,** 23–30.
Sahal, D., Ramachandran, J., and Fujita-Yamaguchi, Y. (1988) *Arch. Biochem. Biophys.* **260,** 416–426.
Sasaki, N., Rees-Jones, R. W., Zick, Y., Nissley, S. P., and Rechler, M. M. (1985) *J. Biol. Chem.* **260,** 9793–9804.
Scatchard, G. (1949) *Ann. NY. Acad. Sci.* **51,** 660–672.
Siegel, T. W., Ganguly, S., Jacobs, S., Rosen, O. M., and Rubin, C. S. (1981) *J. Biol. Chem.* **256,** 9266–9273.
Sweet, L. J., Wilden, P. A., Spector, A. A., and Pessin, J. E. (1985) *Biochemistry* **24,** 6571–6580.
Tappy, L., Fujita-Yamaguchi, Y., LeBon, T. R., and Boden, G. (1988) *Diabetes* **37,** 1708–1714.
Tollefsen, S. E., Thompson, K., and Petersen, D. J. (1987) *J. Biol. Chem.* **262,** 16461–16469.

Ullrich, A., Bell, J. R., Chen, E. Y., Herrera, R., Petruzzelli, L. M., Dull, T. J., Gray, A., Coussens, L., Liao, Y.-C., Tsubokawa, M., Mason, A., Seeburg, P. H., Grunfeld, C., Rosen, O. M., and Ramachandran, J. (1985) *Nature* **313,** 756–761.
Ullrich, A., Gray, A., Tam, A. W., Yang-Feng, T., Tsubokawa, M., Collins, C., Henzel, W., LeBon, T. R., Kathuria, S., Chen, E., Jacobs, S., Francke, U., Ramachandran, J., and Fujita-Yamaguchi, Y. (1986) *EMBO J.* **5,** 2503–2512.
Wang, C.-C., Goldfine, I. D., Fujita-Yamaguchi, Y., Gattners, H. G., Brandenburg, D., and De Meyts, P. (1988) *Proc. Natl. Acad. Sci. USA* **85,** 8400–8404.
White, M. F., Maron, R. and Kahn, C. R. (1985) *Nature* **318,** 183–186.
Williams, P. F. and Turtle, J. R. (1979) *Biochim. Biophys. Acta* **579,** 367–374.

Insulin Receptor Isolation Through Avidin-Biotin Technology

Frances M. Finn and Klaus Hofmann

1. Introduction

The ability to quantitate hormone receptors through binding of their radiolabeled ligands has significantly enlarged our view of endocrine disorders. For example, the finding that the plasma membrane concentration of receptors is a dynamic property of the cell, i.e., that receptor number can decrease in the presence of chronically elevated concentrations of hormones and, thus, dampen the cellular response to the hormone, has given us a new insight into the basis for endocrine dysfunction. A better understanding of the receptors themselves should help us to identify abnormal receptors and lead to an appreciation of the function of the hormone–receptor complex.

For this reason we became interested, a few years ago, in developing generally applicable techniques for the isolation of peptide hormone receptors. It seemed to us that the method of affinity chromatography would be essential to any purification scheme, owing to the low concentration of receptors even in their target tissues. One of the primary difficulties encountered in constructing affinity resins using peptide ligands is that the methods for attaching the ligand to the resin are nonspecific. Coupling usually occurs either through amino or carboxyl groups on the peptide and often results in a "family" of ligands attached to the resin through one or more reactive groups along the peptide chain. These immobilized ligands can have widely differing affinities (or even no affinity) for the receptor. The insulin molecule illustrates this point clearly. Reagents that couple to amino groups can react with insulin at three sites (Fig. 1) the α-amino groups on the *N*-terminal amino acid residues of the A and B chains and the ε-amino group of the lysine residue at position B^{29}. When insulin is reacted with the *N*-hydroxysuccinimide ester of biotin, all three possible monobiotinylinsulins, all three possible dibiotinylinsulins and tribiotinylinsulin are produced. The results of a large body of structure-function studies with insulin indicate that modification of the *N*-terminus of the A chain results in loss of biological activity. Reaction at the ε-amino group is less detrimental to activity but May et al. (1978) have shown that when $N^{\varepsilon,B29}$-biotinyl insulin is allowed to form a complex with avidin the complex has very low biological activity. Only the derivative with biotin attached to the α-amino group of the B chain *N*-terminal amino acid residue has high biological activity even when complexed with avidin. Unfortunately, both the α-amino group at Iysine^{29} and the α-amino group on the A chain are considerably more reactive than the α-amino group on the B chain. Thus, if insulin is coupled to a resin via its amino groups, the majority of the products will be unable to bind receptor specifically and may bind significantly to other proteins in the applied solution. Our goal was to find a method of attaching the ligand to the resin in a targeted manner such that it would retain its capacity to interact with receptor. If the hormone ligand could be prepared by well-defined solution chemistry and tested for its receptor binding capacity, we could ascertain whether the derivative would be suitable for affinity chromatography.

The unusually strong noncovalent interaction between the egg-white protein avidin and biotin (the $t_{1/2}$ for dissociation of the complex is 200 d, for a review, *see* Green, 1975) afforded a unique and generally applicable solution to the problem. We reasoned that if avidin were immobilized on Sepharose, then an affinity resin could be formed by simply mixing a suitably biotinylated hormone with this resin. This approach would eliminate the need for performing chemical manipulations on the hormone, such as coupling it to the resin or deprotecting the coupled product on the resin. Furthermore, the amount of ligand on the column could be varied to achieve optimal receptor binding. In theory, once this technique is worked out for a particular receptor, it could easily be scaled up for production of large amounts, since it should be highly reproducible. The only problem remaining was to find a method to release the bound receptor. In the case of insulin receptor, a degree of success was achieved by using a

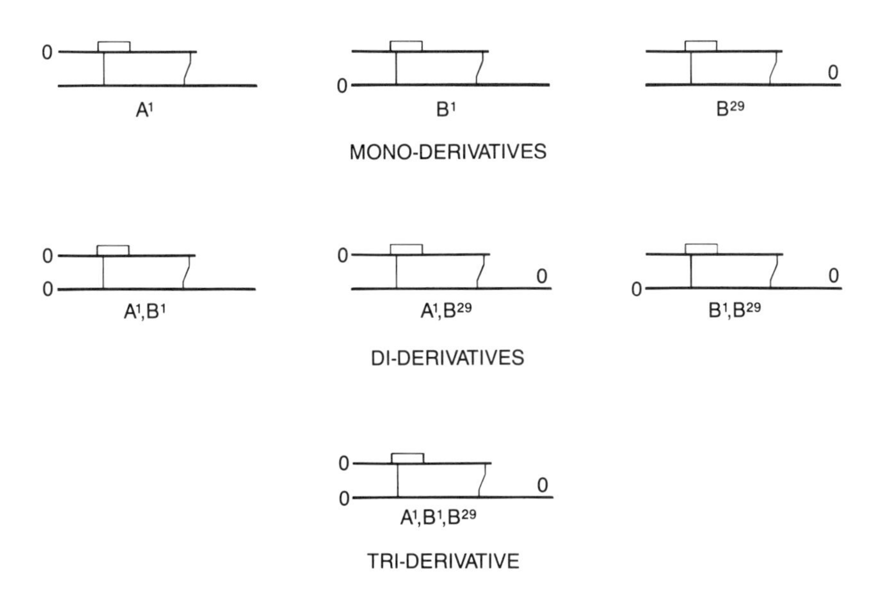

Fig. 1. Schematic representation of all possible derivatives obtained by derivatizing amino groups on native insulin.

combination of pH 5 and 1*M* salt (Fujita-Yamaguchi et al., 1983). Although the receptor isolated by this procedure has full insulin binding capacity, the receptor eluted at pH 5 has poor tyrosine kinase activity. In any case, we were interested in a more general solution to the problem of receptor retrieval, and this has been achieved by eluting the affinity resin with a detergent in which hormone–receptor binding is weak.

1.1. Principles of the Technique

The principles of the technique are illustrated in Fig. 2. In method A, soluble receptor (R) is percolated through an affinity column to form the complex shown in the center. The column is then exhaustively washed to remove contaminating materials. For this reason, it is desirable to have a hormone ligand that retains strong affinity for the receptor. Alternatively, the biotinylated hormone ligand (B–H) can be added to a solution of solubilized receptor to form the soluble complex BHR. Percolating a solution containing this complex through a column of immobilized suc-avidin (SA) as in B will result in the formation of the same complex obtained by method A. Both these schemes have been used for the isolation of insulin receptors from human placenta (Finn et al., 1984). Removal of functional, purified

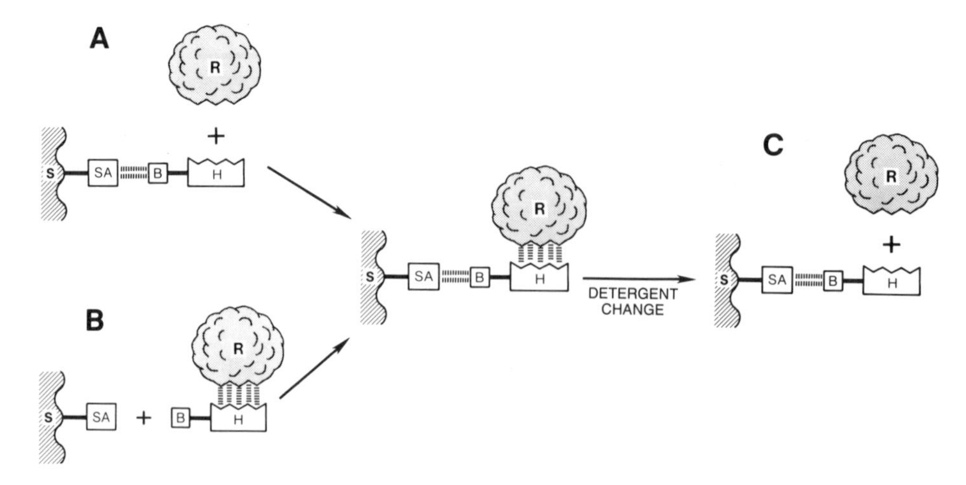

Fig. 2. General scheme for the application of the avidin-biotin method to the isolation of hormone receptors. Dashed lines represent noncovalent bonds.

insulin receptor from affinity resins is achieved by eluting the column with a buffer solution containing a detergent in which hormone–receptor binding is weakened. In the case of insulin receptor octyl β-glucoside is used.

2. The Tools

In order to exploit the avidin-biotin interaction for receptor purification, certain design principles must be observed in preparing the biotinylated hormone. Essentially, three criteria must be met:

1. Derivatization of the ligand with biotin must not substantially weaken its ability to interact with its receptor;
2. The biotinylated hormone must retain strong affinity for avidin; and, most important;
3. The biotinylated hormone must be able to interact simultaneously with avidin and its receptor.

2.1. The Biotinylated Probe

As previously stated, derivatization at the $N^{\alpha,B1}$ position has little effect on the biological activity of insulin (Pullen et al., 1976). A series of insulins derivatized at this position with biotin or dethiobiotin (attached either directly or via a spacer arm) were synthesized and simplified structures of these potential ligands are shown in Fig. 3. The effect of derivatization on their biological activity was assessed using stimulation of glucose oxida-

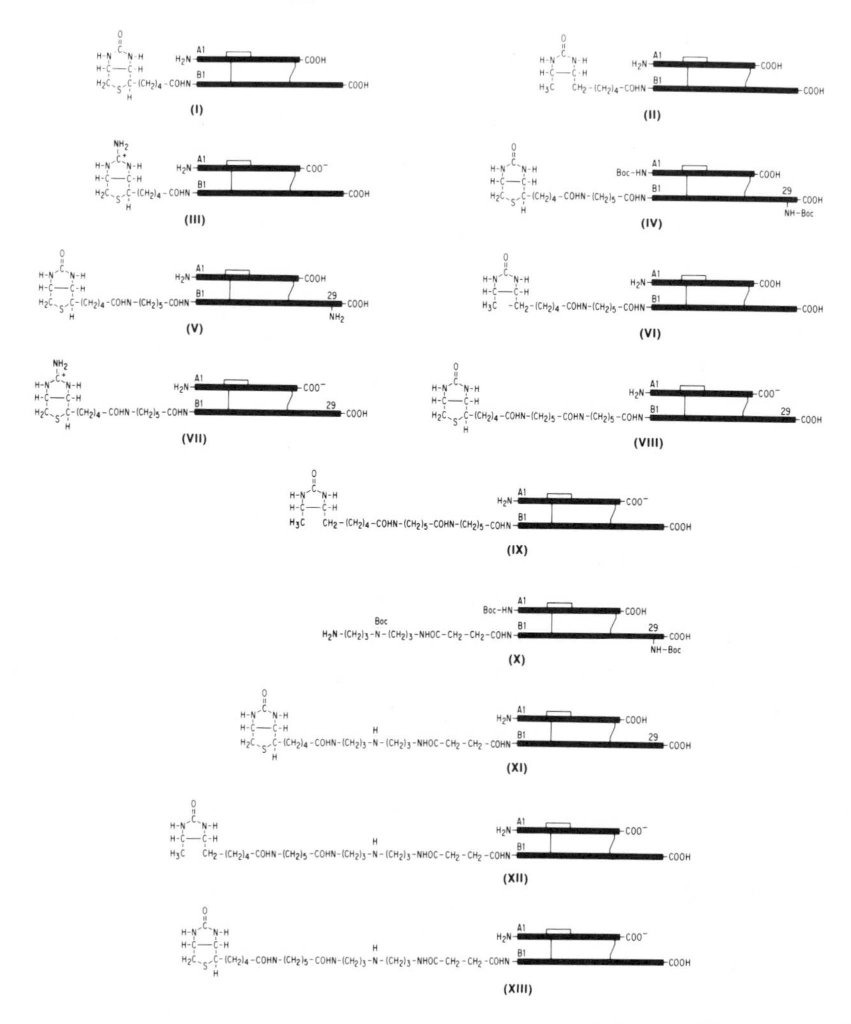

Fig. 3. Simplified structures of ligands synthesized for insulin receptor isolation.

tion in isolated rat adipocytes. Ligand V, in which the biotin is separated from insulin by an amino hexanoic acid spacer arm, was one of several ligands whose dose response curves did not differ significantly from that of insulin (Finn et al., 1984). Furthermore, it retained its affinity for succinoyl-avidin (suc-avidin, *see below*). The $t_{1/2}$ for dissociation of the ligand V:suc-avidin complex was not significantly different from the value for the biotin:suc-avidin complex. The same was true for a number of other analogs, but from the synthetic point of view ligand V is the simplest to prepare and was, therefore, chosen for constructing affinity resins. In fact, we have recently used the blocked precursor of this molecule, ligand IV, for construction of affinity columns and have obtained better yield of receptor using this ligand.

2.2. The Avidin Molecule

Avidin is a highly basic glycoprotein composed of 4 identical subunits each having a biotin binding site (Green, 1975). The use of this protein for the construction of affinity columns has the serious disadvantage that avidin binds nonspecifically to many components, especially negatively charged membrane proteins. This property can be eliminated by exhaustively succinoylating avidin with succinic anhydride to form suc-avidin. There are 9 lysine and 8 arginine residues/avidin subunit. Succinoylation changes the direction of migration of avidin from the cathode to the anode on electrophoresis without significantly altering its biotin binding characteristics. The dissociation constant for the avidin:biotin$_4$ complex is 4×10^{-8} s^{-1}, for suc-avidin:biotin$_4$, it is 6.3×10^{-8} s^{-1} (Finn et al., 1984). An alternative solution to this problem would be to use streptavidin (so named because it is secreted by *Streptomyces avidinii*), which is not a basic protein. Although streptavidin is employed widely in histocytochemistry, we prefer suc-avidin for the reason that the physical and chemical properties of the avidin molecule have been analyzed exhaustively and its affinity for biotin as well as for many biotin analogs has been determined by Green and his coworkers (reviewed in Green, 1975). This is not the case for streptavidin.

The avidin molecule, because of its size (M_r 68,000), has the potential to interfere with the interaction between a biotinylated hormone and its receptor. The biologically inactive avidin: $N^{\varepsilon,B29}$–biotinylinsulin complex, referred to earlier, illustrates this problem. Once it has been established that the biotinylated hormone forms a strong complex with avidin, so that activity resulting from complex dissociation can be ruled out, the biological activity of the avidin:biotinylated hormone complex must be measured. If the avidin:biotinylated hormone complex is active, it is safe to assume that the biotinylated hormone interacts simultaneously with avidin and the receptor.

In the case of the biotinylated derivatives of insulin, biological activity was measured using glucose oxidation in isolated adipocytes (Finn et al., 1984). To ensure complex formation, excess suc-avidin was added to the derivatives prior to exposure to the cells. Ligand V, as well as other ligands containing a spacer arm between the biotin (or dethio biotin) and the insulin portions of the molecule were able to interact with avidin and receptor simultaneously. The biological activity of the suc-avidin complexes of all of these derivatives was 20–30% that of the uncomplexed derivatives. Thus, ligand V meets all three criteria and could be predicted to function successfully as a ligand for insulin receptor affinity chromatography.

3. Materials

3.1. Preparation of the Ligand

3.1.1. $N^{\alpha,A1}$, $N^{\epsilon,B29}$-(Boc)$_2$ Insulin

$N^{\alpha,A1}$, $N^{\epsilon,B29}$-(Boc)$_2$ insulin is prepared by procedures developed by Geiger et al., (1971). Boc azide (Carpino, 1957) (2.3 mL, 16 mmol) is added with stirring to a solution of insulin (1.2 g, 0.2 mmol) in DMSO (45 mL), water (11.5 mL), and 1*N* sodium bicarbonate (3.3 mL) and the mixture is stirred at 35°C for 5 h. The solution is cooled to room temperature, acidified with 50% aqueous acetic acid (4 mL), and the solvents are removed *in vacuo* at a bath temperature of 35°C. The residue is triturated with diethyl ether (40 mL), dried, and rubbed twice with 1% aqueous acetic acid and redried; yield 1.08 g.

SP-Sephadex C 25 (250 g) is swelled in a buffer composed of glacial acetic acid, 0.45 L/*i*-propanol, 2.0 L/water, 2.55 L, and a portion of $N^{\alpha,A1}$, $N^{\epsilon,B29}$ (Boc)$_2$ insulin (580 mg) is dissolved in 20 mL of this buffer to which a small volume of 10% aqueous acetic acid is added if solution is not complete. A column of SP-Sephadex (5 × 56 cm) is equilibrated with the buffer, peptide solution is applied, and the column is developed with a linear gradient containing 1.0 L each of the acetic acid/*i*-propanol/water buffer and the same buffer to which NaCl (14.5 g) is added. Fractions corresponding to the major protein peak (280 nm) are pooled, evaporated to dryness, and the residue is washed with water at 4°C and filtered. Yield, 350 mg.

3.1.2. Ligand IV

The synthesis of ligand IV has been described by Hofmann et al. (1977). *N*-Hydroxysuccinimido 6-(biotinylamido)hexanoate (268 mg, 0.59 mmol) is added at room temperature to a stirred solution of $N^{\alpha,A1}$, $N^{\epsilon,B29}$-(Boc)$_2$ insulin (1.0 g, 0.16 mmol) and imidazole (200 mg, 2.94 mmol) in DMSO (30 mL) and the solution is stirred for 6 h. The solution is cooled in an ice bath, and ice water (approx 2 vol) is added. The solution is desalted on a Sephadex G-25 column (5 × 35 cm) equilibrated with 0.05*M* ammonium bicarbonate, and fractions containing protein are pooled and lyophilized.

3.1.3. Ligand V

Ligand V is prepared from the diBoc derivative (Hofmann et al., 1977). Ligand IV is deprotected with anhydrous trifluoroacetic acid (TFA) (20 mL) for 30 min at room temperature and the TFA is removed *in vacuo* at 25°C.

3.2. Wheat Germ Agglutinin Sepharose 4B

For preparation of this material CNBr-activated Sepharose 4B (15 g) (Pharmacia) is rapidly washed with I m*M* HCl (3.0 L) at room temperature on a Buchner funnel. The resin is sucked dry, immediately transferred to a plastic bottle (250 mL), and a solution of wheat germ agglutinin (500 mg) (E-Y Laboratories, San Mateo, CA) in 0.1*M* $NaHCO_3$ (100 mL) containing 0.5*M* NaCl is added at 4°C. The suspension is rotated at 4°C for 24 h and is filtered on a Buchner funnel; the filtrate normally contains negligible amounts of protein. The filter cake is resuspended in 50 mL of 1*M* ethanolamine • HCl (pH 8.0) and the suspension is rotated for 3 h at 4°C. The resin is collected by filtration on a sintered glass filter and subjected to three washing cycles, each consisting of a wash with 0.1*M* $NaHCO_3$/1.0*M* NaCl (pH 8.0 buffer, 1.0 L) and 0.1*M* NaOAc/1.0*M* NaCl (pH 4.0 buffer, 1.0 L). The washed resin is stored at 4°C in pH 4.0 buffer to which sodium azide (0.05%) is added.

3.3. Sepharose 4B-Immobilized Suc-Avidin

CNBr-activated Sepharose 4B (10 g) is washed at room temperature with 1 m*M* HCl (2.0 L) and the washed resin, in a plastic bottle, is rotated at 4°C for 22 h with a solution of avidin (100 mg) dissolved in 0.1*M* $NaHCO_3$ (50 mL) containing 0.5 *M* NaCl. The resin is collected by filtration and resuspended in 1*M* ethanolamine • HCl (25 mL, pH 8.0). The filtrate, which can be assayed for protein (Lowry et al., 1951), normally contains only negligible amounts of avidin. The suspension is rotated at 4°C for 3 h, the resin is collected and washed with three cycles (600 mL each) of pH 8.0 and pH 4.0 buffers and then with 100 mL of 0.5*M* Na_2CO_3, pH 9.0. The washed resin is slurried in the sodium carbonate buffer (75 mL), the suspension is cooled in an ice bath, and succinic anhydride (59 mg) in dioxane (1.8 mL) is added with vigorous overhead stirring. Stirring is continued for 1 h at ice-bath temperature and for 4 h at room temperature. The resin is filtered, washed with three cycles (600 mL each) of pH 8.0 and pH 4.0 buffers, and stored in pH 4.0 buffer to which sodium azide (0.05%) is added.

The ^{14}C-biotin binding capacity of the resin can be determined in the following manner. An aliquot (0.2 mL) of a 1:1 slurry of the resin in pH 4.0 buffer is layered on 0.3 mL of Sephadex G-5O in a Pasteur pipet and the resin is washed with 50 mL of water. A solution of *d*-[carbonyl-^{14}C] biotin (Amersham, sp. act. 50 mCi/mmol, 1.85 GBq) is diluted with water to 1 m*M*, applied to the resin, and allowed to diffuse into the resin for 30 min at room temperature. The biotin solution is recycled 3 times through the resin and then fractions (2 mL each) are collected. The column is washed with water (16 mL) and portions (0.1 mL) of the eluate fractions are counted. Binding capacity is the difference between applied and recovered biotin.

3.4. Insulin Affinity Resin (Human Placental Receptor)

Suc-avidin Sepharose 4B (5.0 mL of settled resin) is poured into a column for equilibration with buffer. The resin is washed with 15 volumes of 50 m*M* NH_4HCO_3. If the resin has been stored for several weeks or longer, it is first washed by cycling with the pH 8.0 and 4.0 buffers (15 vol × 3 of each buffer) prior to equilibration with the NH_4HCO_3 buffer. The equilibrated resin is suspended in 6 vol of NH_4HCO_3 buffer and transferred to a beaker (100 mL). The probe, ligand IV (700 µg dissolved in 28 mL of NH_4HCO_3 buffer) is added dropwise with overhead stirring over a period of 1 h at room temperature. The resin is stirred for an additional 30 min, poured into a column (0.9 × 7 cm), and washed with 1.0 L of 50 m*M* HEPES, pH 7.6/1*M* NaCl/0.1% Triton X-100 to remove nonspecifically bound probe. The washed resin is stored in the HEPES buffer to which sodium azide (0.05%w/v) is added.

3.5. Insulin Affinity Resin (Rat Liver Receptor)

For attachment of ligand X to the solid support, CNBr-activated Sepharose 4B (2 g) is washed with 1 m*M* HCl (400 mL), and the washed resin is rotated for 16 h at 4°C with a solution of the ligand (20 mg) in 5 mL of 0.1*M* $NaHCO_3$, pH 8.0, containing 0.5*M* NaCl and 10^6 counts/min of ^{125}I labeled ligand as a tracer. The resin is collected and rotated at 4°C for 3 h with 1*M* ethanolamine • HCl, pH 8.0 (5 mL). Once the resin has been deactivated, it is collected, washed with three cycles each of 100 mL of pH 8.0 and 4.0 buffers (*see* Section 3.2.), then washed with 1 L of 50 m*M* HEPES/1*M* NaCl/0.1% Triton X-100 and stored in the HEPES buffer containing 0.05% sodium azide. Ligand loading is usually 1.25 mg/mL of settled resin.

3.6. ^{125}I-Insulin

Labeling insulin is patterned after the procedure described by De Meyts (1976). The changes that have been introduced include omission of sodium metabisulfite reduction of unreacted Chloramine T, and chromatography to remove unreacted iodine is performed on Sephadex G-25 instead of cellulose. Briefly, to a 1.5 mL microfuge tube are added in the order given, 0.3*M* sodium phosphate buffer, pH 7.4 (35 μL), insulin (1 mg/mL in 0.01*N* HCl, 10 μL) and ^{125}I (Amersham, 1 mCi, 37 mGBq, 10 μL). A solution of Chloramine T is prepared immediately before use by dissolving 64.3 mg in 10 mL 0.3*M* sodium phosphate buffer, pH 7.4 (stock solution). The stock solution is diluted 1 to 100 in the same buffer and an aliquot (10 μL) is added to the reaction solution. After 5 min, a portion of the solution (2 μL) is added to 2 mL of a solution of bovine serum albumin (BSA, 1 mg/mL in 0.03*M* sodium phosphate buffer, pH 7.4), and this solution is used to determine the total radioactivity in the reaction solution. An aliquot of this solution is diluted again with the BSA-containing buffer (2 μL/1.0 mL) and 10% (w/v) trichloroacetate (TCA) is added to precipitate the labeled insulin. A portion (200 μL) of the TCA precipitated mixture is centrifuged for 1 min in a microfuge, the supernatant is removed, transferred to a counting vial and the tip of the centrifuge tube is excised and transferred to a second vial. The percent of incorporation of ^{125}I is determined from the ratio of radioactivity in the pellet to the radioactivity in this sample. The level of incorporation should be less than 0.3 at of ^{125}I per mol of insulin for optimum binding results. Once the desired level of incorporation has been reached, the remainder of the reaction solution is applied to a Sephadex G-25 column (0.8 × 16 cm) equilibrated with 50 m*M* HEPES, pH 7.6/0.1% Triton X-100. This column can be conveniently prepared in a Disposaflex column (Kontes). The column is eluted with the same buffer, and 0.5 mL fractions are collected. ^{125}I-insulin should appear at 3–3.5 mL of eluate. The peak fractions (usually 2 fractions) are combined, diluted with an equal portion of column buffer, and stored as 50–100 μL aliquots at –15°C.

If too little radioactivity is incorporated, a second portion of the Chloramine T stock solution can be diluted and added to the reaction solution. The efficiency of Chloramine T oxidation of the iodide is sensitive to the purity of reagents, especially the water. The presence of contaminents that can be oxidized will consume the Chloramine T before it is added to the reaction. For this reason it may be necessary to adjust the level of Chloramine T to achieve the proper level of incorporation.

4. Methods

4.1. Insulin Binding

4.1.1. Charcoal Method (Williams and Turtle, 1979)

4.1.1.1. Solutions. TMA buffer: 50 m*M* Tris/10 m*M* $MgSO_4$/1% w/v BSA, adjusted to pH 7.6 with HCl. Charcoal suspension: 2.5 g Norit A in 100 mL TMA buffer. Unlabeled insulin: a solution containing 1 mg/mL in 0.01*N* HCl is diluted to 1 nanomol/mL with TMA buffer. ^{125}I Insulin: 0.1 picomol/250 µL TMA. Chase solution: 0.1 picomol ^{125}I insulin plus 33 picomol insulin/250 µL TMA.

4.1.1.2. Assay. For each assay, one tube (12 × 75 mm polystyrene) for measuring total binding (250 µL ^{125}I insulin) and one tube for measuring nonspecific binding (250 µL chase solution) are prepared. Receptor (10 or 20 µL) is added to each set of tubes and the contents are mixed by vortexing. Reagent additions are made at ice-bath temperature and the assay tubes are incubated for 50 min at 24°C, then they are returned to the ice-bath and charcoal suspension (1 mL) is added to each. To keep the charcoal suspended, the suspension is stirred with a magnetic stirrer. The tubes are centrifuged for 15 min at 1900*g*, the supernatant decanted into counting vials and counted. Standards (250 µL of ^{125}I insulin) are also counted.

4.1.2. Polyethylene Glycol (PEG) Method (Cuatrecasas, 1972)

4.1.2.1. Solutions. Tris-albumin: 50 m*M* Tris/0.1% w/v BSA adjusted to pH7.4 with HCl. Insulin: a stock solution of 1 mg/mL insulin in 0.01*N* HCl is diluted to 2 nanomol/mL with Tris-albumin before use. ^{125}I Insulin: specific activity 0.1 µCi/picomol (3.7 µGBq/picomol). Various dilutions of this stock solution are made with Tris-albumin to produce solutions containing 0.05, 0.1, 0.3, 0.6, 1.0, 1.5, 3.0, 5.0, and 10.0 picomol/150 µL. Gamma globulin: 0.1% w/v bovine gamma globulin (Miles Scientific) in 0.1*M* sodium phosphate buffer, pH 7.4. PEG (A): 25% w/v PEG (Sigma Chemical Co., M_r 8,000) in water. PEG (B): 8% w/v PEG in 0.1*M* Tris • HCl, pH 7.4.

4.1.2.2. Assay. For each point two tubes to measure total and two tubes to measure nonspecific binding are prepared. Microfuge tubes (1.5 mL) are placed in an ice bath. To each tube are added 100 µL insulin (nonspecific binding) or the same volume of Tris-albumin (total binding), 150 µL of ^{125}I-insulin dilutions, receptor, and Tris-albumin to bring the volume to 300 µL. Receptor is added last and the tubes are mixed by vortexing before and after the addition of receptor. The assay tubes are incubated at 4°C for 16 h. Gamma globulin (0.5 mL) and PEG A (0.5 mL) are added and the contents of the tubes are mixed by vortexing. After 15 min at ice-bath temperature, the tubes are centrifuged in a microfuge (Beckman Instruments) for 5 min and the supernatant is aspirated. The pellet is washed with 1 mL of PEG B that is also aspirated. The tubes (without covers) are placed in counting vials and counted in a gamma counter. Duplicate standards for each concentration of ^{125}I insulin are pipetted directly into counting vials.

4.2. Insulin Receptor Isolation (Human Placenta)

4.2.1. Preparation of a 40,000g Pellet

All operations are performed at 4°C unless otherwise noted. A fresh term placenta is dissected to remove the chorionic and amnionic membranes. The tissue is cut into pieces (ca. 5 × 5 cm) and these are washed several times, to remove blood, with 0.25*M* sucrose (adjusted to pH 7.5 with $NaHCO_3$). The sucrose is decanted, the tissue is weighed, and 0.25*M* sucrose/5 m*M* Tris • HCl, pH 7.4/0.1 m*M* phenylmethylsulfonyl fluoride (PMSF) is added (2 mL/g). The suspension is homogenized using a Polytron homogenizer (Brinkmann) at a setting of 5 for 1 min. The suspension is centrifuged at 650*g* for 10 min. The centrifugate often contains three layers, a spongy material, an infranatant liquid, and a pellet. The infranatant is saved and centrifuged for 10 min at 12,000*g*. Sodium chloride and $MgSO_4$ are added to a final concentration of 0.1*M* and 0.2 m*M*, respectively, the vol of the supernatant is adjusted to 1.0 L, and the suspension is centrifuged at 40,000*g* for 40 min. The pellet is resuspended with the aid of a Dounce homogenizer in 50 m*M* Tris • HCl, pH 7.4/0.1 m*M* PMSF/2 m*M* ethylenediamine tetraacetic acid (EDTA) (1.2 L) and washed by centrifugation at 40,000*g* for 40 min. The pellet is resuspended in 50 m*M* HEPES, pH 7.6/ 0.1 m*M* PMSF/2 m*M* ethyleneglycolbis-(β-aminoethyl ether) *N,N,N',N',*-tetraacetic acid (EGTA)/25 m*M* benzamidine • HCl to a final vol of 85.5 mL (protein concentration is ca. 15 mg/mL). The suspension can be stored frozen (flash freeze in a dry-ice acetone bath and store at –70°C) or solubilized immediately.

4.2.2. Solubilization

To the resuspended pellet (85.5 mL) is added the HEPES/PMSF/ EGTA/benzamidine • HCl buffer containing 20% Triton X-100 (4.5 mL). The sample is stirred for 30 min at 4°C. In earlier procedures, solubilization was performed at room temperature. The same degree of solubilization is achieved at 4°C, but with less proteolytic destruction. Insoluble material is removed by centrifugation at 100,000*g* for 2 h. The supernatant from this step can either be flash frozen and stored at –70°C or immediately applied to a wheat germ affinity column. A sample (1.0 mL) is removed for protein (Udenfriend et al., 1972) and binding activity (Scatchard, 1949) determinations.

4.2.3. Wheat Germ Agglutinin Affinity Chromatography

The wheat germ agglutinin Sepharose 4B resin (20 mL) is washed in a column with 4 column vol of 50 m*M* HEPES, pH 7.6/10 m*M* $MgCl_2$/0.1 m*M* PMSF/0.1% Triton X-100 and 4 column vol of 50 m*M* HEPES pH 7.6/

10 m*M* $MgCl_2$/0.1 m*M* PMSF/0.1% Triton X-100/2 m*M* EGTA/25 m*M* benzamidine • HCl. Magnesium chloride is added to the receptor solution to a concentration of 10 m*M*. A portion (1/4) of the soluble receptor is slurried with the washed resin, the slurry is transferred quantitatively with the aid of the rest of the receptor solution to a 250 mL plastic bottle, and the slurry is rotated for 16 h. The rotated mixture is transferred to a chromatography column (3 cm in diameter) that is washed with 1.0 L of the HEPES/$MgCl_2$/PMSF/Triton buffer and eluted with 0.3*M* *N*-acetyl-D-glucosamine/50 m*M* HEPES, pH 7.6/0.1 m*M* PMSF/0.1% Triton X-100 (ca. 80 mL). A portion of the elution buffer (0.5 column vol) is allowed to diffuse into the column and the column is kept for 30 min before elution is started. Fractions (4 mL each, 15–20 fractions) are collected and assayed for binding activity by the charcoal assay (Williams and Turtle, 1979). Fractions containing receptor are pooled, flash frozen, and stored at –70°C. A sample (1.0 mL) is reserved for protein (Udenfriend et al., 1972) and binding activity (Scatchard, 1949) determinations. For reuse, the resin is washed with 4 column vol of 50 m*M* HEPES, pH 7.6/10 m*M* $MgCl_2$/0.1 m*M* PMSF/0.1% Triton X-100 and it is stored in the same buffer to which sodium azide (0.05%) is added.

4.2.4. Insulin Affinity Chromatography

Ligand IV is used as the affinity ligand. The affinity resin (5 mL) is washed with 500 mL of 50 m*M* HEPES, pH 7.6/1*M* NaCl/0.1 m*M* PMSF/0.1% Triton X-100. The wheat germ agglutinin eluate is adjusted to 1*M* in NaCl by adding solid NaCl. The receptor solution (800 picomole of binding activity) is cycled over the column for 12 h at a flowrate of 0.2 mL/min. This flowrate is sufficient to recycle the receptor approx 8 times over the resin. The flow through is retained for assay and reapplication, and can be stored frozen at –70°C until used. The affinity resin is washed with 50 column vol of 50 m*M* HEPES, pH 7.6/1*M* NaCl/0.1% Triton X-100/0.1 m*M* PMSF at a flowrate of 50 mL/h. Triton X100 is exchanged for octyl β– glucoside by washing with one column vol of 50 m*M* HEPES, pH 7.6/1*M* NaCl/0.6% octyl β-glucoside buffer containing bacitracin (2 mg/mL), antipain (1.5 μg/mL), and PMSF (0.1 m*M*). The column is equilibrated to room temperature and eluted with the same buffer. Fractions (20, 2 mL each) are collected into 1 mL of 50 m*M* HEPES, pH 7.6 (chilled to 4°C) and the pooled fractions are concentrated in an Amicon stirred cell (PM-10 filter) to approximately 300–500 pmol binding activity/mL.

4.2.5. Detergent Exchange

The Amicon concentrate is diluted with 50 mL of 50 m*M* HEPES, pH 7.6/0.1% Triton X-100 and concentrated as before. The dilution and concentration processes are repeated once more. Evidence that the receptor is in a Triton X-100 rather than octyl β-glucoside environment is obtained by measuring insulin-stimulated autophosphorylation as a function of insulin concentration.

4.3. Insulin Receptor Isolation (Rat Liver)

Having available a reliable affinity resin for the isolation of insulin receptors from human placenta, we reasoned that it should be a simple matter to isolate insulin receptor from rat liver by the same procedure (Hofmann et al., 1987). This, however, proved not to be the case. We were surprised to find that wheat germ lectin purified rat insulin receptor failed to bind to the affinity resin employed to isolate the placental receptor. This observation suggested that human placenta and rat liver receptors are different and prompted a systematic comparison of the adsorption-desorption characteristics of the two receptors with a series of affinity resins containing different ligands. Affinity resins were constructed using ligands IV, V, XI, and X at a ligand loading of approximately 15 nmol/mL (low loading). In addition, a resin loaded with ligand X at a concentration of approx 190 nmol/mL (high loading) was prepared. Results with these resins are shown in Table 1.

4.3.1. Preparation of a 40,000g Pellet

All operations are performed at 4°C unless otherwise noted. In a typical preparation, livers from 10–12 rats (ca. 100 g) are collected into 0.25*M* sucrose (adjusted to pH 7.5 with $NaHCO_3$). The sucrose is decanted, replaced with 0.25*M* sucrose/5 m*M* Tris, pH 7.4/0.1 m*M* PMSF (2 mL/g of tissue) and the tissue homogenized in a Polytron (setting 5) for 1 min. The suspension is centrifuged at 12,000*g* for 10 min and the supernatant is collected. Sodium chloride and $MgSO_4$ are added to a concentration of 0.1*M* and 0.2 m*M*, respectively, and the suspension is centrifuged at 40,000*g* for 40 min. The pellet is resuspended in 50 m*M* Tris, pH 7.4/0.1 m*M* PMSF/2 m*M* EDTA (60 mL) with the aid of a Dounce homogenizer (tight pestle) and stored frozen at –70°C. A sample is reserved for protein determination (Lowry et al., 1951). Protein concentration at this step is approx. 15 mg/g wet wt of tissue. Assays of binding activity with the 40,000*g* pellet as well as with the crude solubilized receptor are beset with error because of contamination with proteolytic enzymes that destroy the insulin. Addition of *N*-ethylmaleimide, which inhibits the insulinase described by Duckworth and Kitabchi (1981), only partially overcomes the problem.

Table 1
Affinity Behavior of Liver and Placental Receptor

Ligand	Receptor	Bound, pmol	Eluted, pmol
IV	Liver	—	—
IV	Placenta	438	285 (65%)
V	Liver	—	—
V	Placenta	367	131 (36%)
XI	Liver	410	41 (10%)
XI	Placenta	550	328 (60%)
X	Liver	293	—
X	Placenta	543	178 (33%)
X[a]	Liver	577	351 (61%)
X[a]	Placenta	710	168 (24%)

[a]Resin with high loading of ligand.

4.3.2. Solubilization

The crude membrane fraction (40,000*g* pellet) is washed by resuspension (Dounce homogenizer, loose pestle) and centrifugation at 40,000*g* for 40 min twice in 50 m*M* HEPES, pH 7.6/0.1 m*M* PMSF/2 m*M* EGTA/25 m*M* benzamidine • HCl (2 mL/g wet wt). The pellet is resuspended at a concentration of 15–30 mg of protein/mL in the HEPES buffer, and Triton X-100, in the same buffer (20 % v/v) is added to a final concentration of 1%. The mixture is stirred at 4°C for 30 min and centrifuged at 100,000*g* for 2 h. The supernatant is collected and can be flash frozen at –70°C. A sample (1 mL) is taken for protein determination (Udenfriend et al., 1972). The protein concentration at this stage is 10–20 mg/mL. Because of the considerable proteolytic activity associated with liver preparations it is preferable to chromatograph the solution on a wheat germ agglutinin affinity column as soon as possible after solubilization.

4.3.3. Wheat Germ Agglutinin Affinity Chromatography

The crude receptor solution is made 10 m*M* in $MgCl_2$ and added to 40 mL of wheat germ agglutinin Sepharose 4B resin washed as previously described (*see* Section 4.2.3.). A portion (1/4) of the soluble receptor is slurried with the washed resin, the slurry is transferred quantitatively with the aid of the rest of the receptor solution to a 250 mL plastic bottle, and the slurry is rotated for 16 h. The rotated mixture is transferred to a chromatography column (3 cm in diameter) that is washed with 2.0 L of 50 m*M* HEPES, pH 7.6/10 m*M* $MgCl_2$/0.1 m*M* PMSF/0.1% Triton X-100, and eluted with 0.3*M* *N*-acetyl-D-glucosamine/50 m*M* HEPES, pH 7.6/0.1 m*M*

PMSF/0.1% Triton X-100 (20 fractions, 4 mL each). Fractions are screened for binding activity by the charcoal assay of Williams and Turtle (1979). Fractions containing activity are pooled and flash frozen at -70°C. A sample is removed for protein determination (Udenfriend et al., 1972) and binding activity analysis (Scatchard, 1949).

4.3.4. Insulin Affinity Chromatography

The affinity column (2 mL of the resin containing ligand X, high loading) is washed with 500 mL of 50 m*M* HEPES, pH 7.6/1*M* NaCl/0.1% Triton X-100/0.1 m*M* PMSF. The receptor solution (800 pmol of binding activity) is made 1*M* in NaCl by adding solid NaCl and cycled through the affinity column at a flowrate of 0.2 mL/min for 12–16 h. The flow through is retained for assay and reapplication and can be stored frozen at -70°C until used. The resin is washed with 50 column vol of 50 m*M* HEPES, pH 7.6/1*M* NaCl/0.1% Triton X-100/0.1 m*M* PMSF at a flowrate of 50 mL/h and then equilibrated with 20 mL of the same buffer containing (per mL of buffer) aprotinin, pepstatin, antipain, and leupeptin (1.5 μg each), benzamidine • HCl (15 μg), and bacitracin (2 mg). The resin is eluted with 50 m*M* sodium acetate, pH 5.0/1*M* NaCl/0.1% Triton X-100/0.1 m*M* PMSF containing the same additives. Owing to the detrimental effects on autophosphorylation of exposing the receptor to pH 5.0, the eluate fractions (25 fractions, 2 mL each) are collected into 0.5*M* HEPES, pH 7.6/0.1 m*M* PMSF (1.0 mL, containing the inhibitors) to neutralize them as soon as possible. Fractions containing ^{125}I-insulin binding activity (as measured by the charcoal assay) are pooled and subjected to Scatchard (1949) analysis. The pooled fractions are concentrated in an Amicon stirred cell (PM-10 filter) to a volume of 1–2 mL.

5. Properties of the Human Placental Receptor

5.1. Binding and Molecular Weight

Insulin receptor from human placenta (M_r 350,000, as judged by sodiumdodecyl-sulfate/polyacrylamide gel electrophoresis, SDS/PAGE) is composed of two α-(M_r 125,000) and two β-subunits (M_r 90,000). The mol wt of the receptor (based on the sum of its constituent amino acids) as determined from its amino acid sequence (Ullrich et al., 1985; Ebina et al., 1985) is 306 kDa. Using this value, we estimate that the purified receptor is capable of binding 1.5 mol of ^{125}I insulin per mol of receptor or 29 μg insulin per mg receptor. From the subunit composition one would predict that the receptor should bind two mol of ligand per mol and this may indeed be the case, however, the binding isotherms are curvilinear and for this reason it is difficult to make reliable measurements of binding capacity. Suffice it to say that this pH 7.6 receptor preparation has binding activity as high as receptor prepared by any other method (Fujita-Yamaguchi et al., 1983).

Figure 4 compares the yields of ^{125}I-insulin binding activity of human placental receptor obtained by eluting the affinity column with a pH 7.6 buffer containing octyl β-glucoside (Ridge et al., 1988) or with a pH 5.0 buffer containing Triton X-100 (Fujita-Yamaguchi et al., 1983). The yields of receptor retrieved at 4°C using Triton X-100 at pH 5.0 (pH 5.0 receptor) or at room temperature using octyl β-glucoside at pH 7.6 (pH 7.6 receptor) are comparable. Generally, 50–60% of the bound receptor is eluted. The autoradiogram of the SDS/PAGE of pH 7.6 receptor under reducing conditions (insert in Fig. 4) indicates the purity of this preparation and documents its tyrosine kinase activity.

The nature of the detergent employed exerts a marked influence on the strength of the binding of insulin to its receptor. This is illustrated in Fig. 5 which compares the dose response curves for autophosphorylation of receptor in Triton X-100 and octyl β-glucoside. Autophosphorylating activity of the pH 5.0 receptor in Triton X-100 reaches a maximum at an insulin concentration of 100 n*M* (Kasuga et al., 1983), whereas the pH 7.6 receptor in octyl β-glucoside reaches maximum activity only at insulin concen-trations of 250 n*M*; however, if the octyl β-glucoside is exchanged for Triton X-100 (detergent exchanged receptor), the dose response curve returns to the position characteristic for Triton X-100-solubilized material (Fig. 5).

5.2. Autophosphorylation Activity

A comparison of the autophosphorylation activity of a single receptor preparation purified by different methods is shown in Fig. 6. After wheat germ affinity chromatography, the eluate was divided into two portions that were applied to identical insulin affinity columns. One of these columns was eluted at pH 7.6 in octyl β-glucoside (pH 7.6 receptor) as described in Section 4.2.4., the other was eluted at pH 5.0 according to the procedure of Fujita-Yamaguchi et al. (1983).

The pH 7.6 receptor exhibits autophosphorylating activity (Fig. 6) that is significantly higher than that obtained with the pH 5.0 receptor. Autophosphorylation activity is comparable to that obtained with the wheat germ extracts. In addition to the excellent tyrosine kinase activity of the pH 7.6 receptor, the autophosphorylation activity is linear as a function of a

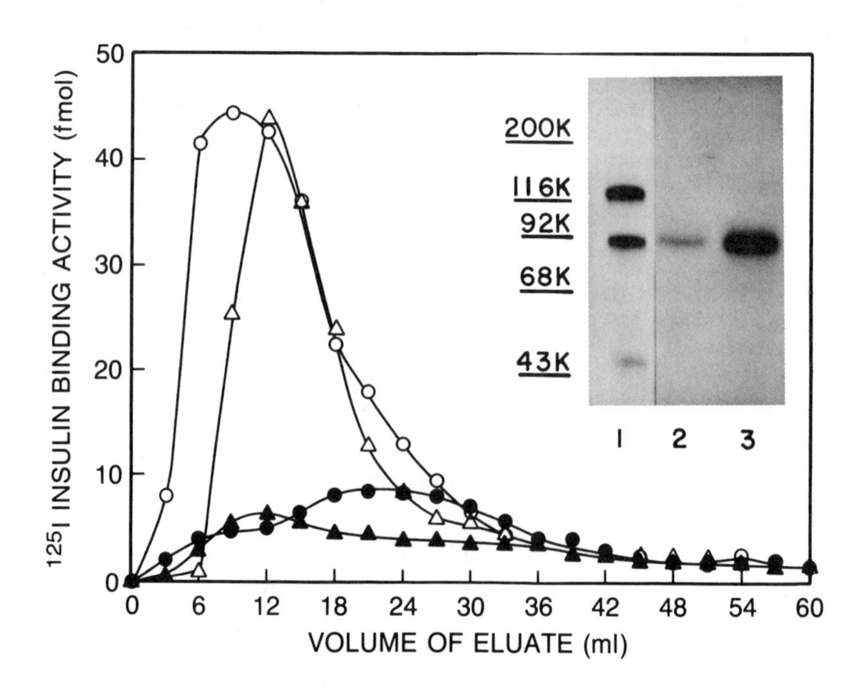

Fig. 4. Elution of human placental insulin receptor from insulin affinity columns by various methods: (O) pH 7.6 buffer (25°C) containing octyl β-glucoside, (●) pH 7.6 buffer (25°C) containing Triton X-100, (Δ) pH 5.0 buffer (4°C) containing Triton X-100, and (▲) pH 7.6 buffer (4°C) containing octyl β-glucoside. Inset: Autoradiogram of SDS/PAGE of pH 7.6 receptor labeled with ^{125}I (lane 1) and basal (lane 2) and 250 n*M* insulin-stimulated (lane 3) autophosphorylation. Standards were myosin (M_r 200,000), β-galactosidase (M_r 116,000), phosphorylase b (M_r 92,000), bovine serum albumin (M_r 68,000), and ovalbumin (M_r 43,000).

wide range of receptor concentrations (Fig. 7). In previous reports (Kasuga et al., 1983; Petruzzelli et al., 1984) autophosphorylation activity was found to be linear both for the wheat germ as well as the pH 5.0 receptor at receptor concentrations considerably below 1 pmol. In this study, where a broader range of receptor concentrations was examined, linearity was obtained only with the pH 7.6 receptor.

A study of the time course for receptor autophosphorylation revealed that ^{32}P incorporation was complete by 10 min and no phosphotyrosine phosphatase activity could be demonstrated. This was true both at ATP concentrations at or below the K_m. Furthermore, addition of a 100-fold excess of unlabeled ATP did not reduce ^{32}P incorporation when incubation was continued for as long as 35 min (Fig. 8).

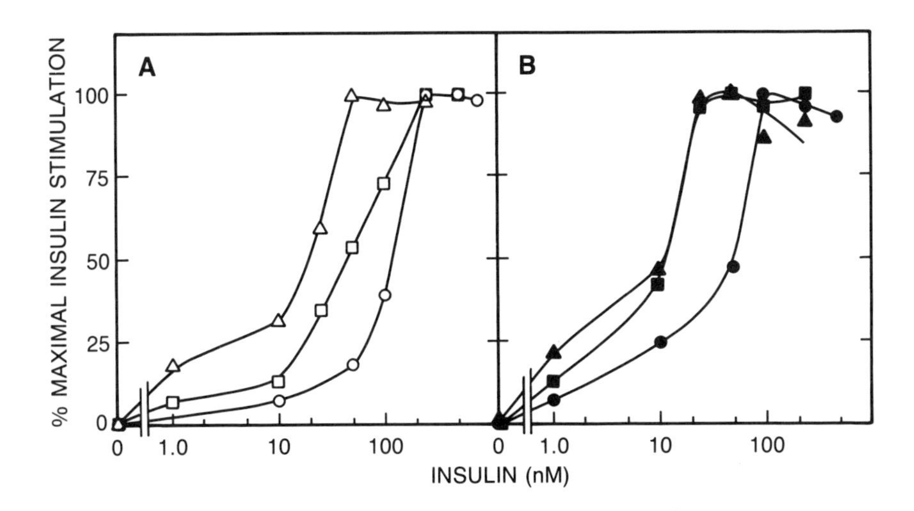

Fig. 5. Insulin dose response curves for autophosphorylation of pH 7.6 receptor in octyl β-glucoside (A) in the presence of 50 μ*M* ATP (O), 50 μ*M* ATP, and 250 μ*M* AMP-PNP (□), 1 m*M* ATP (Δ); detergent exchanged pH 7.6 receptor (B) in the presence of 50 μ*M* ATP (●), 50 μ*M* ATP, and 250 μ*M* AMP-PNP (■), 1 m*M* ATP (▲).

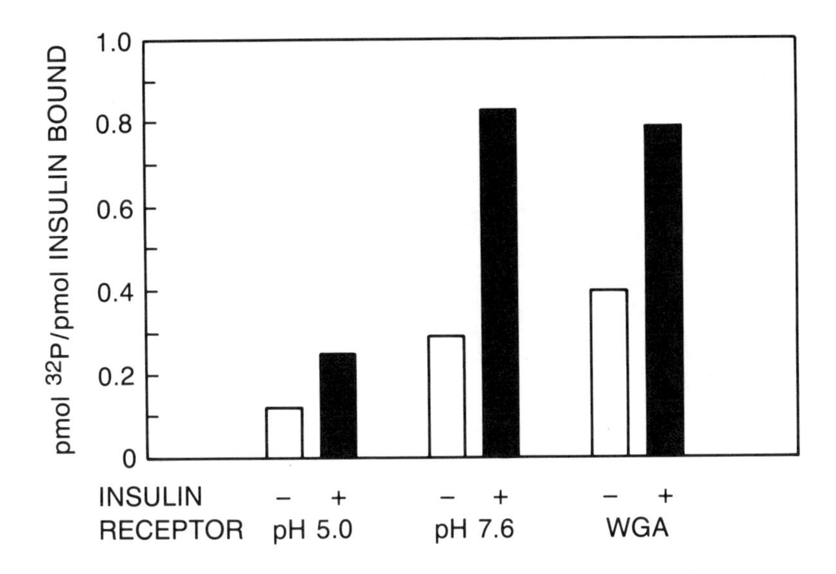

Fig. 6. Comparison of the autophosphorylating activities of an insulin receptor preparation after wheat germ agglutinin affinity chromatography (WGA) and after elution from an insulin affinity column at pH 5 or 7.6.

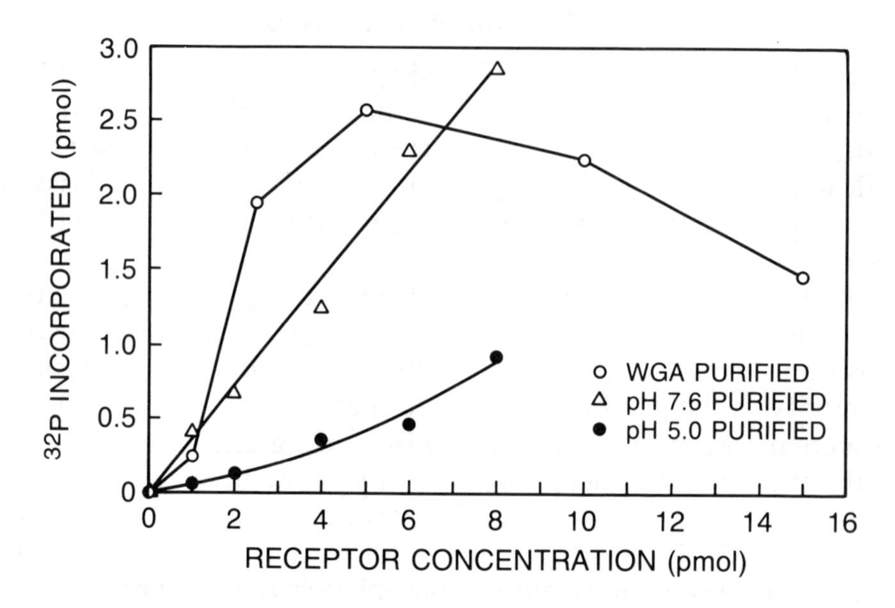

Fig. 7. Autophosphorylation activity as a function of receptor concentration.

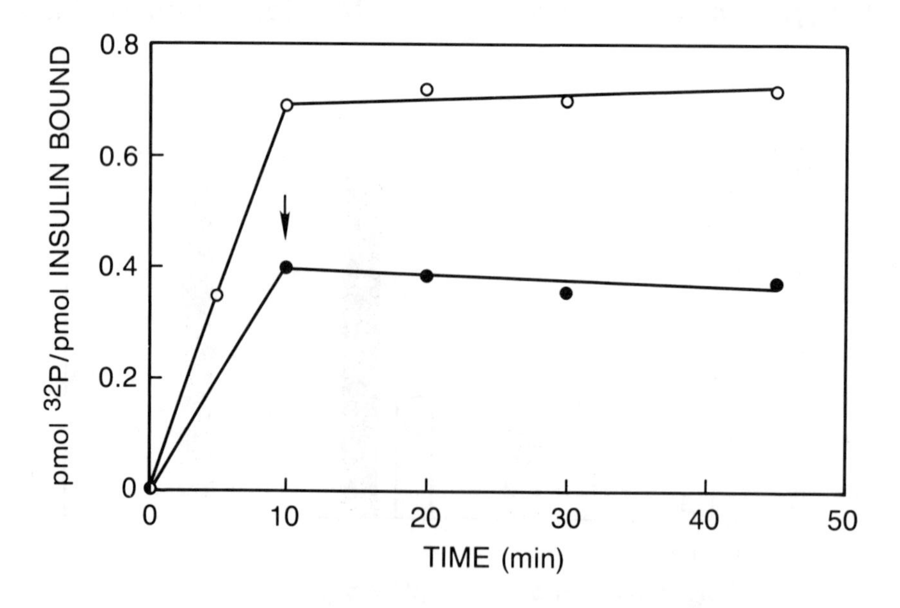

Fig. 8. pH 7.6 Receptor autophosphorylation as a function of time. (●) 50 μM ATP; (O) 200 μM ATP. The arrow denotes the point at which a 100-fold excess of unlabeled ATP was added to measure possible contamination of the receptor with phosphatase activity.

5.3. NEM Alkylation

NEM alkylation of the pH 7.6 receptor inhibited autophosphorylation, as shown in Fig. 9. The linearity of the 1/v vs inhibitor plot (Fig. 9, inset) suggests that the decrease in tyrosine kinase activity results from modification of a single SH group although modification of a class of sites with similar inhibition constants could provide the same result (Brooker and Slayman, 1982).

When the [^{3}H]NEM-alkylated pH 7.6 receptor was subjected to SDS/PAGE under reducing conditions, radioactivity was found exclusively at the position corresponding to the β subunit. Quantitative analysis of the radioactivity in the β subunit indicated that 1.13 ± 0.37 mol (SD, n = 13) NEM were incorporated per mol insulin binding activity.

Inhibition of autophosphorylation by NEM has been observed by several investigators (Shia et al. 1983; Petruzzelli et al., 1984; Fujita-Yamaguchi et al., 1985). Interestingly, however, Wilden and Pessin (1987) have shown that autophosphorylation is not inhibited when iodoacetamide is used to alkylate the receptor and they suggest that inhibition by NEM is a result of steric interference.

5.4. Effect of Adenosine 5'-[b, g-Imido] Triphosphate (AMP-PNP) on Autophosphorylation

We attempted to examine alkylation in the presence of a nonhydrolyzable substrate analog, AMP-PNP, with the expectation that occupation of the ATP site by the analog might prevent alkylation. At AMP-PNP concentrations of 1 m*M*, a modest inhibition of alkylation (33%) was observed. To determine the concentration of AMP-PNP necessary to fill the ATP binding site, we measured autophosphorylation as a function of AMP-PNP concentration (Table 2) and found, surprisingly, that concentrations of AMP-PNP up to 500 μ*M* not only did not inhibit ^{32}P incorporation but, on the contrary, stimulated autophosphorylation. This role for AMP-PNP could not be subserved by GMP-PNP, GTP, or ADP.

Two possible reasons for the effect of AMP-PNP were considered. AMP-PNP stimulates the autophosphorylation in an "indirect" manner, e.g., by inhibiting ATPase activity present in the pH 7.6 receptor, thereby increasing the concentration of ATP available for the autophosphorylation reaction; or AMP-PNP has a "direct" stimulatory effect on the receptor.

The possibility that ATPase activity copurified with the receptor could not be overlooked, especially in light of the hypothesis that insulin activation of the membrane bound (Na^+, K^+) and Mg^{2+}-ATPases is re-

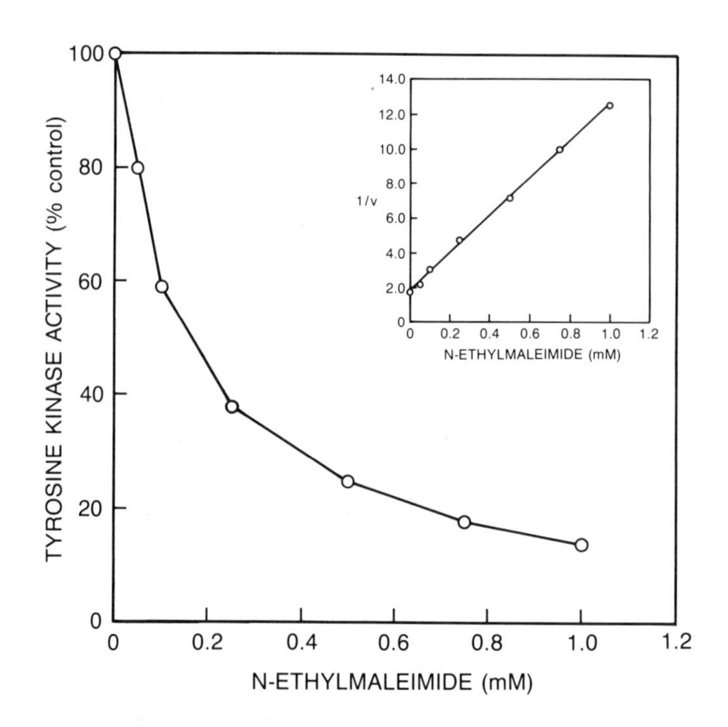

Fig. 9. NEM inhibition of insulin-stimulated autophosphorylation by the pH 7.6 receptor. Inset shows a 1/v vs inhibitor plot of the same data.

sponsible for some of its biological effects (Resh, 1985). This consideration prompted an investigation of the effects of known ATPase inhibitors on insulin-stimulated autophosphorylation. Neither ouabain, at 1 m*M*, a concentration capable of inhibiting 90% of (Na^+, K^+) ATPase activity (Stekhoven and Bonting, 1981), nor sodium orthovanadate, at concentrations ranging from 1 μ*M* to 1 m*M* had any effect on the level of ^{32}P incorporation into tyrosine residues on the β subunit of the purified receptor (data not shown).

Finally, the disappearance of ATP was measured directly (Fig. 10). The production of $^{32}P_i$ from radiolabeled ATP was compared for the wheat germ and pH 7.6 receptors in the presence of vanadate or AMP-PNP. A profound decrease in ATP was observed when wheat germ purified receptor was incubated with ATP for periods up to 30 min. Vanadate diminished, but did not abolish the formation of $^{32}P_i$ from ATP. In contrast, hydrolysis of ATP was barely detectable with the pH 7.6 receptor. Furthermore, AMP-PNP had no effect on this reaction. Thus, AMP-PNP stimulation of autophosphorylation of the pH 7.6 receptor is not a result of inhibition of ATPase activity.

Table 2
Effect of AMP-PNP on Phosphotyrosine Kinase Activity[a]

AMP-PNP, μ*M*	^{32}P Incorporated pmol/pmol insulin bound
0	0.21
100	0.32
250	0.51
500	0.52
1000	0.45
2500	0.21
5000	0.20

[a]Insulin concentration was 250 n*M* and ATP was 50μ*M*.

AMP-PNP, at concentrations of ATP below and at the K_m stimulated ^{32}P incorporation into tyrosine residues in the β subunit. The lower the substrate concentration, the more pronounced was the AMP-PNP stimulation. At 1 m*M* ATP, where substrate is in excess, the effect was no longer evident.

The alternative explanation, namely that AMP-PNP stimulates autophosphorylation by acting directly on the receptor, was explored by measuring its influence on the velocity of the reaction as a function of substrate concentration. Addition of AMP-PNP increases maximum reaction velocity and lowers the K_m for ATP. In the absence of AMP-PNP, a V_{max} of 0.13 pmol ^{32}P incorporated/pmol insulin binding activity/min and a K_m of 208 μ*M* were obtained; with AMP-PNP added, the V_{max} was 0.25 and the K_m was 129 μ*M*.

Since binding of insulin to the receptor is reported to increase the V_{max} for autophosphorylation (White et al., 1983), we hypothesized that the effect of AMP-PNP on these kinetic parameters might actually be caused by a change in insulin binding. For this reason, we examined the effect of AMP-PNP on the binding of ^{125}I insulin to the receptor (Fig. 11). Scatchard (1949) analysis of the binding revealed that AMP-PNP markedly increased binding affinity with no change in the number of sites. The same was true when ATP was substituted for AMP-PNP. The studies shown were performed with detergent exchanged receptor but the same behavior was observed with receptor in octyl β-glucoside demonstrating that the effect is not peculiar to the type of detergent surrounding the receptor.

The consequence of increasing the affinity of the receptor for insulin on the autophosphorylation reaction was predictable, viz, autophosphorylation should be stimulated by much lower insulin concentrations in the presence of AMP-PNP. Analysis of the effect of AMP-PNP on the dose response curve for insulin stimulated autophosphorylation confirmed this

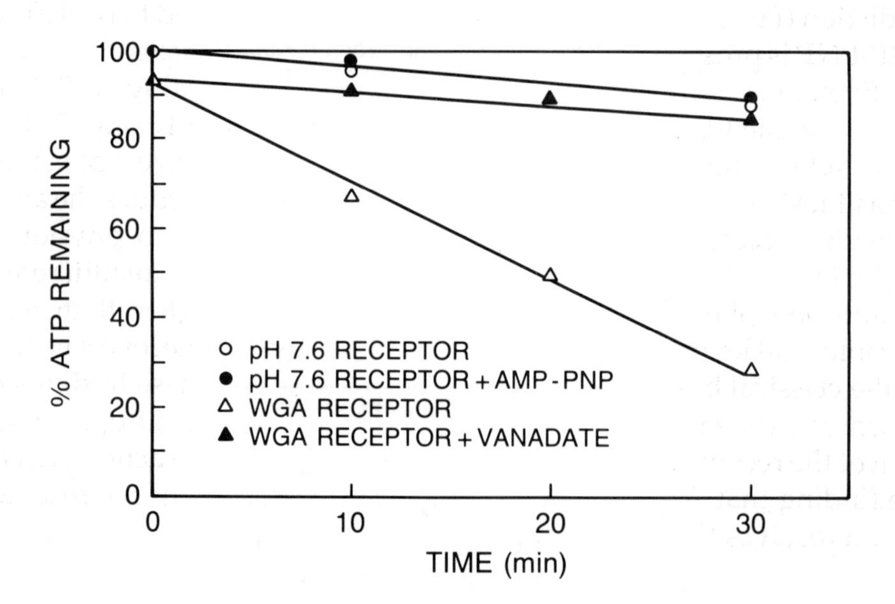

Fig. 10. Comparison of the ATPase activityof wheat germ purified(WGA) or pH 7.6 receptors in the presence or absence of ATPase inhibitors. Results are corrected for the disappearance of ATP in the absence of receptor (3.3% in 30 min).

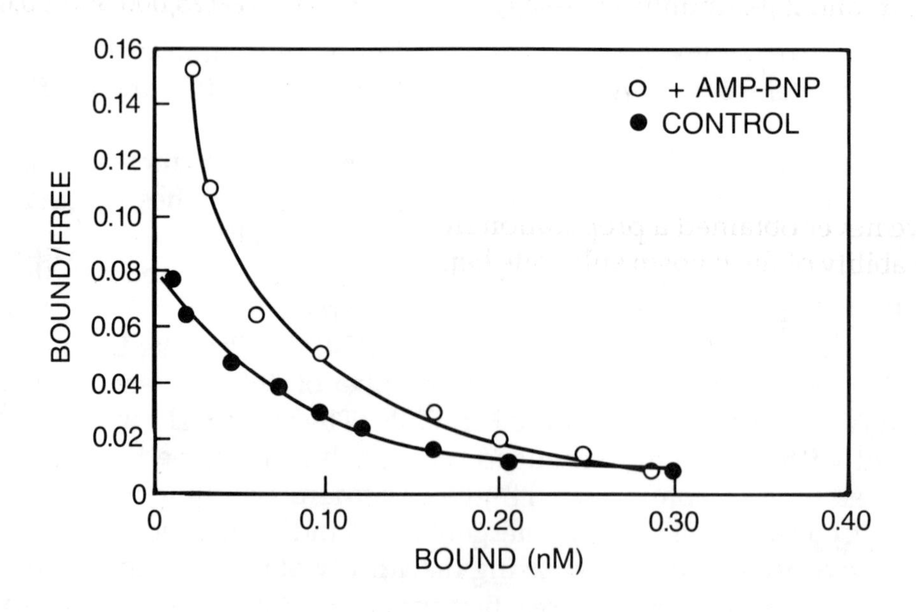

Fig. 11. Scatchard analysis of the binding of ^{125}I-insulin to the detergent exchanged pH 7.6 receptor. AMP-PNP, when present was 250 μM.

prediction (Fig. 5). The curve for pH 7.6 receptor is shifted to the left when AMP-PNP is present. The same is true for the detergent exchanged receptor. Even more significant is the effect produced by adding ATP (1 m*M*). In this case the ED_{50} for insulin is reduced by a factor of 10 (Fig. 5A). As a result, substantial autophosphorylation is observed at levels of insulin as low as 1 n*M*. Thus, the effect of AMP-PNP or ATP is to reduce the amount of insulin necessary to stimulate autophosphorylation to physiological insulin levels (Tasaka et al., 1975). The concentrations of insulin required for autophosphorylation have been considerably higher than normal physiological levels. They have been higher also than the levels necessary for the classical biological actions of insulin. Both of these findings have presented a dilemma for those who have postulated that autophosphorylation of the receptor is an obligatory step in the biological action of insulin. The finding that ATP binding increases the sensitivity of the receptor to insulin resolves this difficulty.

6. Properties of the Rat Liver Receptor

6.1. Binding and Molecular Weight

The rat liver receptor is composed, like the human placental receptor, of 2 α- and 2 β-subunits (Massague et al., 1981) of M_r 125,000 and 90,000, respectively. A major problem encountered in purification of insulin receptors in general, but especially from rat liver, is the proteolytic cleavage of the β-subunit to produce the β' (M_r 45,000) species.

Addition of a "cocktail" of protease inhibitors to the elution buffer affords significant protection against proteolytic destruction, however, we have never obtained a preparation free of the β' fragment. We also tested the ability of the trypsin substrate benzoylarginine ethyl ester (BAEE) and the trypsin inhibitor N^{α}-*p*-tosyl-L-lysyl)chloromethane (TLCK) to protect the β-subunit from proteolysis because of a report by Kathuria et al. (1986) that BAEE suppressed destruction of human placental receptor. Neither the substrate nor the trypsin inhibitor diminished β-subunit cleavage with the rat liver preparation. A comparison of the relative amounts of α, β, and β' subunits can be seen (Fig. 12) on an autoradiogram derived from an SDS/PAGE separation of ^{125}I labeled receptor preparations eluted from the insulin affinity column in the presence of the various inhibitors.

The ^{125}I insulin binding capacity of the purified rat liver receptor is quantitatively similar to that of the placental receptor, i.e., 25–30 μg/mg receptor protein, and the binding isotherms are curvilinear. A comparison

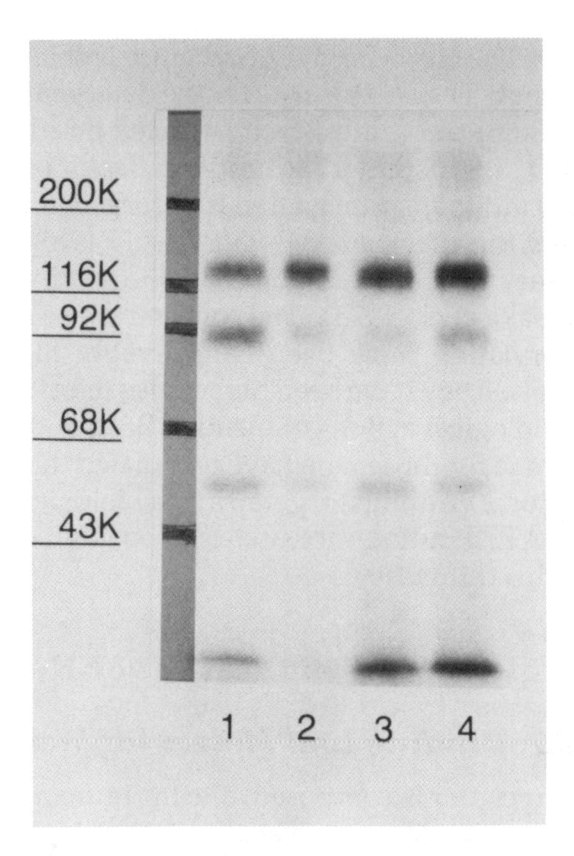

Fig. 12. Autoradiogram of ^{125}I labeled highly purified rat liver insulin receptor separated by SDS/PAGE (reducing conditions). Receptor preparations were eluted from the insulin affinity resin with buffer containing various additives. Relative amounts of α- and β-subunits were measured by scanning the autoradiogram with a densitometer. Numbers on the left refer to mol wt of standards shown in the tracing in lane 1. Additives (lane 2) cocktail β/α ratio 1.03; (lane 3) BAEE, β/α ratio 0 .21; (lane 4) TLCK, β/α ratio 0 .11; (lane 5) none, β/α ratio 0 .21.

of the binding curves for the rat liver and human placental receptors indicates that the rat liver receptor binds less firmly, an observation that supports the adsorption-desorption behavior on the insulin affinity resins.

We have compared the effect on the rat liver and placental receptors of dithiothreitol (DTT) reduction on their insulin binding behavior (Hofmann et al., 1987) and find that although DTT in sufficiently high concentration destroys the binding activity of both receptors, low concentrations of DTT affect the two receptors differently. With the placental receptor, DTT reduces the number of receptor sites without changing the binding affinity. This is consistent with the premise that loss of binding activity is caused by an all or none modification at the binding site; i.e., binding is either present or absent. Reduction of the rat receptor is accompanied by an increase in binding affinity and a gradual loss of receptor sites. This would suggest that reduction is at least a two-step phenomenon involving first an increase in affinity followed by loss of binding. These results are qualitatively similar to those of Massague and Czech (1982) who evaluated the changes in insulin binding of rat liver and human placental receptors in membrane fractions after exposure to DTT.

6.2. Autophosphorylation

Proteolytic cleavage of the β-subunit is accompanied by loss of insulin-dependent autophosphorylation. Addition of the "cocktail" affords a preparation with autophosphorylating activity (Fig. 13) that is significantly better than that obtained when the inhibitors are absent. Furthermore, the rat liver receptor is remarkably stimulated by insulin. A 25–50-fold increase in phosphorylation was elicited with insulin as compared with a 4–6-fold increase for the placental receptor. Maximal activation of rat liver receptor autophosphorylation requires a higher concentration of insulin than is necessary for the placental receptor under the same conditions. This finding, too, is consistent with the weaker insulin binding affinity of the rat liver receptor.

In short, by using three criteria, adsorption-desorption behavior from four insulin affinity columns; ^{125}I-insulin binding; and behavior toward reduction with DTT, we have shown that the α-subunits of the receptors from rat liver and placenta are different. The insulin sensitivity of the autophosphorylation reaction of the two receptors is another distinguishing feature.

Morgan et al. (1986) found that an antibody directed to the α-subunit of the human placental receptor failed to cross-react with the α-subunit of the rat receptor, indicating that the α-subunits are structurally different. The data presented here indicate that the two receptors are functionally different as well.

7. Outlook for Receptor Isolation by Avidin-Biotin Technology

The efficacy of the avidin-biotin method for construction of affinity resins for peptide hormone receptor isolation has been demonstrated by the preparation of highly purified human placental receptor with excellent binding and autophosphorylating activities. This is only one way in which this technology can be applied. There are a number of peptide hormone receptors that have not yet been solubilized with retention of hormone

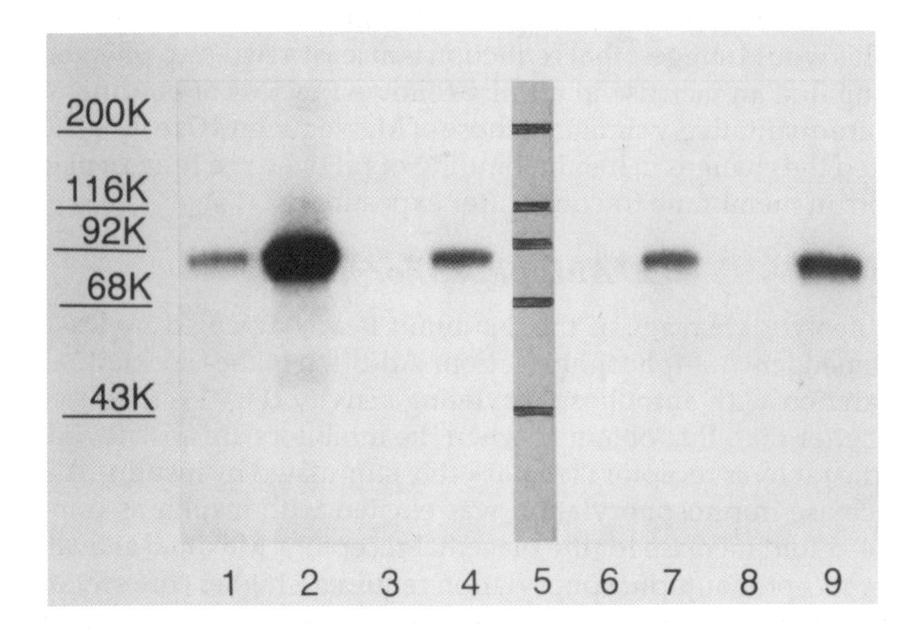

Fig. 13. Autoradiogram of SDS/PAGE separated rat liver insulin receptor following autophosphorylation with ^{32}ATP. The highly purified receptor preparations shown here have been eluted from the insulin affinity column with buffer containing various additives. Lanes 1, 3, 6, and 8 represent basal; lanes 2, 4, 7, and 9 represent insulin-stimulated autophosphorylation. Additives were present as follows: lanes 1 and 2, cocktail; lanes 3 and 4, TLCK; lanes 6 and 7, BAEE; lanes 8 and 9, PMSF.

binding and, thus, present a formidable isolation problem. One such receptor is that for adrenocorticotropic hormone (ACTH). We have recently begun to apply the technique to isolation of a crosslinked hormone: receptor complex. In this case, iminobiotin, an analog of biotin in which a guanido group replaces the ureido group of biotin (for directions on the synthesis and use of this analog for the preparation of iminobiotinyl derivatives of hormones, *see* Finn and Hofmann, 1985). This compound has the unique property of pH dependent (Green, 1966) binding to avidin (or suc-avidin, or streptavidin). Binding is strong at pH 11.0 when the guanido group is uncharged and weak at pH 4.0 when it is charged. The purification scheme in this situation differs from that described for biologically active soluble receptors. In this case, once the iminobiotin is attached to a hormone at a suitable site, the iminobiotinylhormone is crosslinked to its receptor. The crosslinked complex can then be solubilized and chromatographed on a suc-avidin resin at an alkaline pH. When contaminating proteins have been removed the iminobiotinylhormone: receptor complex can be retrieved, in good yield, by lowering the pH of the eluting buffer.

The fact that the affinity column can be prepared by simply mixing the iminobiotinylhormone: receptor complex allows the crosslinked, solubilized complex to be attached to the resin without undergoing chemical modifications. It may also be possible to use reversible crosslinkers so that the iminobiotinylhormone can be separated from the receptor once purification has been achieved. The purified iminobiotinylhormone: receptor complex can then be used for sequence information.

The potential for affinity purification using the avidin-biotin system is only beginning to be realized. The method of column formation and elution is general and should be applicable to a wide variety of hormone receptors. What is essential for the success of this approach, as well as for other methods that rely on affinity chromatography using hormone as a ligand, is a clear understanding of the sites on the hormone where derivatization does not lead to inactivation so that full advantage may be taken of the single property that distinguishes one receptor from another, namely specific binding to its hormone. Once this information is available, it should be possible to construct an effective affinity resin using the principles learned through the development of insulin receptor affinity resins.

Acknowledgment

This work was supported by NIH Grant DK 21292.

References

Brooker, R. J. and Slayman, C. W. (1982) *J. Biol. Chem.* **257,** 12051–12055.

Carpino, L. A. (1957) *J. Am. Chem. Soc.* **79,** 98–101.

Cuatrecasas, P. (1972) *Proc. Natl. Acad. Sci. USA* **69,** 318–322.

DeMeyts, P. (1976) in *Methods in Receptor Research,* part I (Blecher, M., ed.), Dekker, New York, pp. 301–383.

Duckworth, W. C. and Kitabchi, A. E. (1981) *Endocrine Reviews* **2,** 210–233.

Ebina, Y., Ellis, L., Jarnagin, K., Edery, M., Graf, L., Clauser, D., Ou, J.-H., Masiarz, F., Kan, Y. W., Goldfine, I. D., Roth, R. A., and Rutter, W. J. (1985) *Cell* **40,** 747–758.

Finn, F. M. and Hofmann, K. (1985) in *Methods in Enzymology,* vol 109 (Birnbaumer, L. and O'Malley, B. W., eds.), Academic, New York, pp. 418–445.

Finn, F. M., Titus, G., and Hofmann, K. (1984) *Biochemistry* **23,** 2554–2558.

Finn, F. M., Titus, G., Horstman, D., and Hofmann, K. (1984) *Proc. Natl. Acad. Sci. USA* **81,** 7328–7332.

Fujita-Yamaguchi, Y., Choi, S., Sakamoto, Y., and Itakura, K. (1983) *J. Biol. Chem.* **258,** 5045–5049.

Fujita-Yamaguchi, Y. and Kathuria, S. (1985) *Proc. Natl. Acad. Sci. USA* **82,** 6095– 6099.

Geiger, R., Schöne, H.-H., and Pfaff, W. (1971) *Hoppe Seylers Z. Physiol. Chem.* **352,** 1487–1490.

Green, N. M. (1966) *Biochem. J.* **101,** 774–780.
Green, N. M. (1975) *Adv. Protein Chem.* **29,** 85–133.
Hofmann, K., Finn, F. M., Friesen, H.-J., Diaconescu, C., and Zahn, H. (1977) *Proc. Natl. Acad. Sci. USA* **74,** 2697–2700.
Hofmann, K., Romovacek, H., Titus, G., Ridge, K., Raffensperger, J. A., and Finn, F. M. (1987) *Biochemistry* **26,** 7384–7390.
Kasuga, M., Fujita-Yamaguchi, Y., Blithe, D. L., and Kahn, C. R. (1983) *Proc. Natl. Acad. Sci. USA* **80,** 2137–2141.
Kathuria, S., Hertmen, S., Grunfeld, C., Ramachandran, J., and Fujita-Yamaguchi, Y. (1986) *Proc. Natl. Acad. Sci. USA* **83,** 8570–8574.
Lowry, O. H., Rosebrough, N. J., Farr, A. L., and Randall, R. J. (1951) *J. Biol. Chem.* **193,** 265–275.
Massague, J. and Czech, M. P. (1982) *J. Biol. Chem.* **257,** 6729-6738.
Massague, J., Pilch, P. F., and Czech, M. P. (1981) *J. Biol. Chem.* **256,** 3182–3190.
May, J. M., Williams, R. H., and de Haen, C. (1978) *J. Biol. Chem.* **253,** 686–690.
Morgan, D. O., Ho, K., Korn, L. J., and Roth, R. A. (1986) *Proc. Natl. Acad. Sci. USA* **83,** 328–332.
Petruzzelli, L., Herrera, R., and Rosen, O. M. (1984) *Proc. Natl. Acad. Sci. USA* **81,** 3327–3331.
Pullen, R. A., Lindsay, D. G., Wood, S. P., Tickle, I. J., Blundell, T. L., Wollmer, A., Krail, G., Brandenburg, D., Zahn, H., Gliemann, J., and Gammeltoft, S. (1976) *Nature* **259,** 369–373.
Resh, M. D. (1985) in *Molecular Basis of Insulin Action* (Czech, M. P., ed.), Plenum, New York, pp. 451–464.
Ridge, K. D., Hofmann, K., and Finn, F. M. (1988) *Proc. Natl. Acad. Sci. USA* **85,** 9489–9493.
Scatchard, G. (1949) *Ann. NY Acad. Sci.* **51,** 660–672.
Shia, M. A., Rubin, J. B., and Pilch, P. F. (1983) *J. Biol. Chem.* **258,** 14450–14455.
Stekhoven, F. S. and Bonting, S. L. (1981) *Physiol. Revs.* **61,** 1–76.
Tasaka, Y., Sekine, M., Wakatsuki, M., Ohgawara, H., and Shizume, K. (1975) *Horm. Metab. Res.* **7,** 205–206.
Udenfriend, S., Stein, S., Bohlen, P., Dairman, W., Leimgruber, W., and Weigele, M. (1972) *Science* **178,** 871–872.
Ullrich, A., Bell, J. R., Chen, E. Y., Herrera, R., Petruzzelli, L. M., Dull, T. J., Gray, A., Coussens, L., Liao, Y.-C., Tsubokawa, M., Mason, A., Seeburg, P. H., Grunfeld, C., Rosen, O. M., and Ramachandran, J. (1985) *Nature* **313,** 756–761.
White, M. F., Haring, H. U., Kasuga, M., and Kahn, C. R. (1983) *J. Biol. Chem.* **259,** 255–264.
Wilden, P. A. and Pessin, J. E. (1987) *Biochem. J.* **245,** 325–331.
Williams, P. F. and Turtle, J. R. (1979) *Biochim. Biophys. Acta* **579,** 367–374.

Insulin Receptor Purification

Julie D. Newman and Leonard C. Harrison

1. Introduction

The insulin receptor is an integral membrane glycoprotein (M_r 300–350,000) consisting of two α subunits (M_r 130–135,000) and two β subunits (M_r 90–95,000) joined by disulfide bonds (Fig. 1). The α subunit binds insulin and the β subunit contains tyrosine kinase activity and is autophosphorylated after insulin binding. Cloning of the human receptor (Ebina et al., 1985; Ullrich et al., 1985) has demonstrated that it is translated as a single-chain precursor of 1370 (Ullrich et al., 1985) or 1382 (Ebina et al., 1985) amino acids, containing the α and β subunits in tandem. It has been proposed that the subunits are formed by proteolytic cleavage at a tetrapeptide site [arg-lys-arg-arg] after posttranslational folding and inter-subunit disulfide bonding (Ullrich et al., 1985; Ebina et al., 1985).

During the last decade, a number of studies have described purification of insulin receptors from human placenta (Harrison and Itin, 1980; Siegel et al., 1981; Finn et al., 1982,1984b; Fujita-Yamaguchi et al., 1983; Newman and Harrison, 1985), human brain (Roth et al., 1986), rat liver (Cuatrecasas, 1972; Jacobs et al., 1977) and rat skeletal muscle (Wang, 1986). Insulin receptors are also present in a variety of cells that are not classic target cells for insulin action, e.g., human erythrocytes (Im et al., 1983; Ward and Harrison, 1986), human lymphocytes (Gavin et al., 1973) and human lymphoblastoid cell lines (Lang et al., 1980; Harrison et al., 1982;

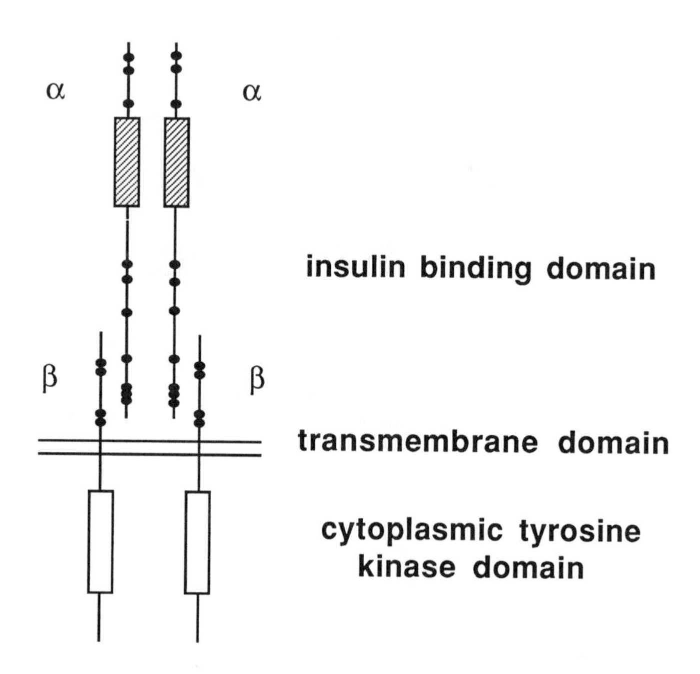

Fig. 1. Schematic representation of the insulin receptor structure. Hatched boxes represent cysteine-rich domains, filled circles represent individual cysteine residues, and open boxes represent the tyrosine kinase domains.

Newman and Harrison, 1989; Newman et al., 1989), human monocytes (McElduff et al., 1985a), a human monocyte cell line, U-937 (McElduff et al., 1985b), a human myeloid erythroleukemia cell line, K562 (Tang et al., 1983), and a rat insulinoma cell line, RIN-m5F (Colman and Harrison, 1984). LeRoith's group has also studied insulin receptors from nonmammalian species, including chicken and lizard (Shemer et al., 1986; Simon and LeRoith, 1986) and have shown that there is strong evolutionary conservation of receptors. Drosophila has also been shown to contain a specific insulin-binding glycoprotein and insulin-dependent tyrosine kinase, and a Drosophila genomic sequence has been isolated that is homologous to the kinase domain of the insulin receptor (Petruzzelli et al., 1986).

Since the insulin receptor cDNA sequence has been available, a number of site-directed mutagenesis experiments have been described (Chou et al., 1987; Ebina et al., 1987; Maegawa et al., 1988; McClain et al., 1988) that have allowed analysis of the sequences of the receptor that are responsible for its biological properties. However, a number of experimental situations still necessitate the use of chromatographic techniques to purify insulin receptors in order to study receptor structure/function. Moreover, the same techniques may be used to purify cloned receptors. In this review, we describe and compare a number of different techniques that can be used to purify insulin receptors from different sources.

2. Chromatography Techniques Used for Purification of Insulin Receptors

Solubilization of insulin receptors from membranes or cells is necessary prior to chromatography. The nonionic detergent Triton X-100 is the most widely used agent for solubilizing receptors that maintain their native properties (Harrison et al., 1978), although octylβ-glucoside has also been used (Gould et al., 1981). We routinely include 0.1 *M* sodium phosphate in the solubilization mixture, since phosphate is also capable of solubilizing many membrane proteins (Dey et al., 1981). For membranes containing ~10 mg protein/mL or for <5 x 10^7 cells/mL, we generally add Triton X-100 to a final concentration of 1% (v/v), and a protease inhibitor cocktail consisting of 2 m*M* phenylmethylsulfonyl fluoride (PMSF), 1000 kallikrein inhibitory U/mL aprotinin, 2 m*M* ethylene diaminetetraacetic acid (EDTA), and 100 U/mL bacitracin.

2.1. Lectins

The insulin receptor contains 15 potential *N*-glycosylation sites in the α subunit and 6 in the β subunit (Ullrich et al., 1985). Immobilized lectins have been successfully used to partially purify insulin receptors from detergent-solubilized membranes or cells, since the binding of glycoconjugates to lectins is reversible and is inhibited by specific sugars. Wheat germ agglutinin was found by Hedo et al. (1981) to be the most efficient lectin for insulin receptor purification with regard to recovery (98%) and purification (22-fold). Although concanavalin A has been used to purify insulin receptors (Williams and Turtle, 1979), it gives poor recovery (20%) and only fivefold purification when eluted with its specific sugar, α-methylmannoside (Hedo et al., 1981).

Immobilized wheat germ lectin can be purchased (Pharmacia, Sweden; Vector Laboratories Inc., Burlingame, CA) or prepared using cyanogen bromide (CNBr)-activated agarose (Pharmacia, Sweden) (Lotan et al., 1977; Adair and Kornfeld, 1974). Wheat germ lectin can also be easily coupled to commercially available activated agaroses such as 1,1'-carbonyldiimidazole (CDI)-activated agarose (Pierce Chemical Company, IL) (Newman and Harrison, unpublished observations) or Affi-Gel 10

(Biorad, CA), as used by Pike et al. (1986). Most preparations contain between 1–5 mg lectin/mL swollen gel.

Solubilized receptor preparations can either be batch-adsorbed for 2–18 h or recycled 3–4 times over immobilized wheat germ lectin columns. Samples applied may contain up to at least 2.5% Triton X-100, since this concentration of detergent does not affect the lectin (Lotan et al., 1977). Approximately 25 mg of crude placental membrane glycoproteins can be bound per mL of lectin-agarose. After washing the columns extensively with 50 m*M* Hepes, pH 7.6, 0.15*M* NaCl, 0.1% Triton X-100, glycoproteins can be eluted using the same buffer containing 0.3*M* *N*-acetylglucosamine (Hedo et al., 1981).

The extent of purification of insulin receptors following wheat germ lectin chromatography varies 4–20-fold between laboratories. This variation could be a result of differences in gel matrices, the quality of lectin bound, or the composition of crude receptor preparations. We have found that wheat-germ lectin-CDI-agarose yields greater recovery of receptors compared to commercial preparations (Newman and Harrison, unpublished observations).

The concentration of sodium dodecyl sulfate (SDS) used to wash lectin columns after desorption of receptors must be carefully chosen. Although Hedo et al. (1981) originally recommended that lectin columns be washed with 50 m*M* Hepes, pH 7.6 containing 0.01% SDS, this concentration of SDS may not be high enough to remove residually bound glycoproteins from all types of lectin columns. For example, some glycoproteins are retained by wheat germ lectin-Sepharose 4B beads (Pharmacia 1976) even in the presence of 0.07% SDS. Kasuga et al. (1985) regenerated wheat-germ lectin agarose by washing with 0.1% SDS; however this wash must be used cautiously, since brief exposure (1 h) to 0.05% SDS has been found to release 10% of lectin subunits from the gel (Lotan et al., 1977). Hedo (1984) recommends regeneration of columns with buffer containing 0.02% SDS, followed by elution buffer containing 0.3*M* *N*-acetyl glucosamine.

Affinity chromatography using immobilized wheat germ lectin is a useful simple technique for first-step partial purification of insulin receptors from crude preparations. It also has the advantage of removing insulinase or insulin degrading activity (Harrison and Itin, 1980; Hedo et al., 1981) that is present in most tissues, including liver, kidney, muscle, brain, and erythrocytes (Bond and Butler, 1987). Receptors are eluted at neutral pH and not subjected to the acidic conditions (pH 5) used for the elution of receptors from insulin affinity columns (*see* Section 2.2.), which may affect their structure or function.

2.2. Insulin Affinity

Insoluble insulin-agarose derivatives were originally shown to mimic the biological effects of native insulin (Cuatrecasas, 1969; Turkington, 1970; Blatt and Kim, 1971; Oka and Topper, 1971) and, subsequently, have been used for affinity purification of insulin receptors.

2.2.1. Insulin-CNBr-Activated-Sepharose

Cuatrecasas (1972) compared the effectiveness of a number of different insulin-CNBr-activated-Sepharose derivatives. Affinity resins were constructed in which insulin was either directly attached to the polymer backbone or spatially separated from the backbone via various spacer arms. One of the most effective adsorbents was one in which the B1 phenylalanine residue of insulin was coupled to the gel support via a spacer of *N*-hydroxysuccinimide ester of 3,3'-diaminodipropylamino-succinyl agarose in 0.1*M* sodium phosphate, pH 6.4, 6 *M* urea. Insulin receptors have been purified from rat liver (Cuatrecasas, 1972; Jacobs et al., 1977) and human placenta (Siegel et al., 1981) by adsorption onto insulin-Sepharose and elution with urea, which dissociates the insulin–receptor complex. However, the use of 4.5*M* urea in the eluant buffer has been shown to cause marked denaturation of the receptor, resulting in a low specific binding activity (*see* Table 1).

The elution buffer for insulin-Sepharose chromatography was later modified by Fujita-Yamaguchi et al. (1983) by combining low pH and high ionic strength (50 m*M* acetate buffer, pH 5.0, 1 *M* NaCl, 0.1% Triton X-100) to enable purification of receptors with full binding activity (5 nmol insulin bound/mg protein) (Table 1). After neutralization of the eluate, receptors retained full insulin binding capacity for up to 6 d. Nevertheless, it is important to elute receptors in the shortest time possible, to minimize their exposure to acid pH. In fact, to circumvent possible receptor denaturation, some investigators have avoided elution from affinity gels and have used receptors purified in the immobilized form in their experiments (e.g., Yu and Czech, 1986).

Preparation of insulin-Sepharose is a moderately lengthy procedure, and uses potentially carcinogenic chemicals such as dioxane. However, the main disadvantage of this affinity resin, in our experience, is that although all preparations are capable of *adsorbing* receptors, functional receptors can only be *eluted* from certain batches. Variability between batches has been observed by others (Jacobs and Cuatrecasas, 1985) but no adequate explanations have been found. For this reason, we attempted to find a simple method for preparing insulin affinity gels using a commer-

Table 1
Comparison of Various Techniques of Insulin Receptor Purification

	Elution buffer	Insulin binding capacity, nmol/mg protein	Purification factor[a]	Reference
Human placenta				
Insulin receptor polyclonal anti-body-agarose	2.5*M* $MgCl_2$, pH 6.5, 0.1% Triton X-100	0.7	333	Harrison and Itin, 1980
Insulin-Sepharose	4.5*M* urea	0.3	123	Siegel et al., 1981
Insulin-Sepharose	low pH, high salt[b]	5.0	2457	Fujita-Yamaguchi et al., 1983
Insulin-biotin-avidin	low pH, high salt[b]	3.1		Finn et al., 1984
Insulin-CDI-Agarose	low pH, high salt[b]	6.0	2000	Newman and Harrison, 1985
Rat liver				
Insulin-Sepharose	4.5*M* urea	2	8000[c]	Cuatrecasas, 1972
Insulin-Sepharose	4.5*M* urea	0.4	2000	Jacobs et al., 1977

[a]Compared to Triton-solubilized membranes.
[b]50 m*M* sodium acetate, pH 5, 1*M* NaCl, 0.1% Triton X-100.
[c]This value may be greatly overestimated because of the very low protein concentration in the insulin-Sepharose eluate.

cially available activated agarose, which allowed reproducible purification of insulin receptors with full binding activity. This method is described in the next section.

2.2.2. *Insulin-CDI-Activated-Agarose*

We have described purification of human placental insulin receptors using insulin coupled directly (without a spacer or leash) to a commercially available activated agarose preparation, 1,1'carbonyldiimidazole (CDI)-activated agarose (Reacti-Gel (6x), Pierce Chemical Company, IL) (Newman and Harrison, 1985). Coupling is performed simply by mixing the washed gel with 2 mg insulin/mL gel in 0.3 *M* sodium phosphate, pH 7.4 for 18 h on a rotator. The amount of insulin bound can be measured by adding a tracer amount of ^{125}I-labeled insulin. Unreactive groups are then blocked by mixing for 3 h with 1*M* ethanolamine, pH 9. Before use, columns are washed with 50 m*M* Tris pH 7.4, 1*M* NaCl; 6*M* urea in 0.1*M* sodium phosphate, pH 7.4; and 50 m*M* Tris pH 7.4, 0.1% Triton X-100, 0.1 m*M* PMSF. Insulin receptors partially purified by wheat germ lectin chromatography (Section 2.1.) are routinely batch adsorbed for 2–18 h at 4°C to insulin-CDI-agarose. Columns are then washed with 50 m*M* Tris pH 7.4, 1*M* NaCl, 0.1% Triton X-100, 0.1 m*M* PMSF and purified receptors eluted using low pH, high ionic strength buffer (50 m*M* acetate buffer, pH 5.0, 1*M* NaCl, 0.1% Triton X-100, 0.1 m*M* PMSF) and neutralized immediately by the addition of 0.25 vol of 1*M* Tris, pH 7.4. Our routine scheme of purification is shown in Fig. 2. A 20 mL insulin-CDI-agarose column is capable of extracting >80% of the receptors present in 40 mg of wheat germ lectin–purified glycoproteins. Recovery of receptors can be improved by reapplying unadsorbed material to fresh columns or by using larger columns.

Human placental receptors were purified 2000-fold (compared to Triton-solubilized membranes) using insulin-CDI-agarose. The insulin binding capacity of the purified receptor was 6 nmol insulin/mg protein (Table 1), similar to that obtained by Fujita-Yamaguchi et al. (1983). Receptors purified using insulin-CDI-agarose retained their insulin binding capacity for >8 wk when stored at either 4°C, or –70°C. Storage at –20°C resulted in rapid decline in insulin binding activity (Newman and Harrison, 1985). We have also identified a subtype of the classical insulin receptor, an atypical insulin receptor, which is copurified by insulin-CDI-agarose chromatography (and also after insulin-Sepharose chromatography). This receptor can be distinguished by its high binding affinity for insulin-like growth factors (Jonas et al., 1986).

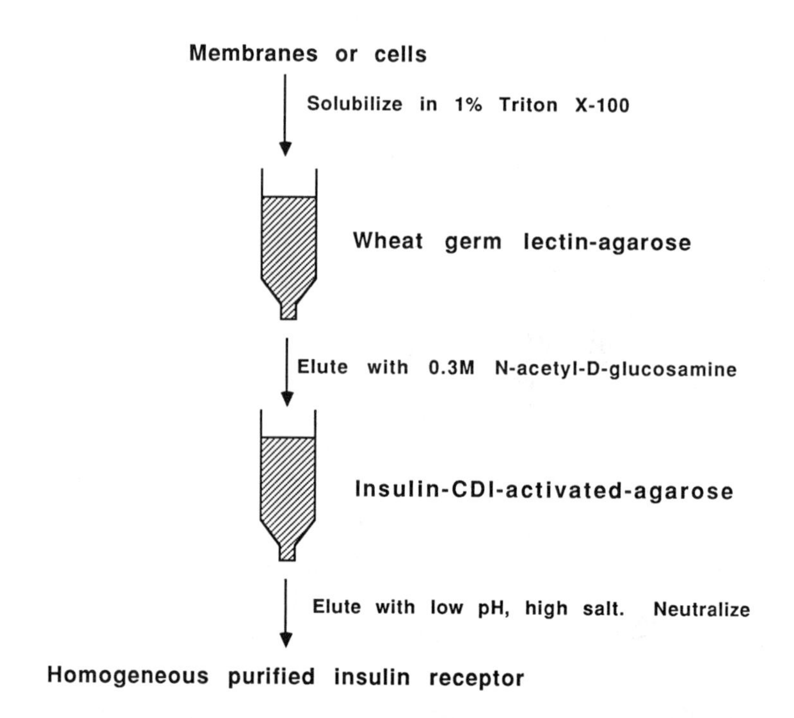

Fig. 2. Standard scheme for insulin receptor purification.

CDI-agarose gels have a number of advantages compared to CNBr-activated gels. They have a stable, uncharged *N*-alkylcarbamate covalent linkage, which is resistant to ligand leakage and minimizes nonspecific adsorption (Bethell et al., 1979,1981). In contrast, ligands coupled to CNBr-activated gels are slowly released into solution as a result of hydrolysis. CDI also produces a higher concentration of activated sites compared to CNBr (Bethell et al., 1981).

We have reproducibly purified receptors with full binding activity using at least six different batches of CDI-agarose. CDI-agarose affinity gels are much easier and faster to prepare than standard CNBr-activated gels (Section 2.2.1.) or avidin-biotin matrices (Section 2.2.3.) and enable purification of receptors with the highest reported insulin-binding activities. Other ligands can also be coupled easily to CDI-agarose (Bethell et al., 1981); we have successfully purified human and pig thyrotropin (TSH) receptors using TSH-CDI-activated-agarose gels (Leedman et al., 1989; Leedman and Harrison, 1990).

2.2.3. Insulin-Biotin-Avidin-Sepharose

A number of different biotinylinsulin-avidin gels have been prepared by Finn and Hofmann (Finn et al., 1982,1984a,b; *reviewed by* Hofmann and Finn, 1985; Finn and Hofmann, 1988). Biotin-coupled insulin is noncovalently attached to immobilized succinoylavidin. Insulin receptors can be desorbed from these gels using a low pH, high ionic strength medium (such as that used for insulin-Sepharose by Fujita-Yamaguchi et al., 1983) or by using biotin. The efficiency of these gels increases with usage, once irreversible binding sites on the gel are occupied by receptors, recovery of receptors subsequently is improved (Finn and Hofmann, 1988).

This affinity technique has been used to prepare homogeneous functional insulin receptors from 3T3-L1 adipocytes (Kohanski and Lane, 1985a,b). Interestingly, rat liver receptors do not have the same binding or elution behavior on different insulin-succinoyl-avidin gels as human placental receptors (Finn and Hofmann, 1988). Eluted insulin receptors are stable for >4 wk at 4°C in the presence of PMSF and sodium azide, but their binding activity is destroyed by freezing in dry ice/acetone (Finn et al., 1984b). Receptors isolated by this technique have a lower specific binding activity (3.1 nmol insulin/mg protein) than those obtained using either insulin-CDI-agarose (Newman and Harrison, 1985) or insulin-Sepharose (Fujita-Yamaguchi et al., 1983) (*see also* Table 1).

2.3. Antibody Affinity

Insulin receptor autoantibodies are present in patients with the type B syndrome of insulin resistance and acanthosis nigricans (Flier et al., 1975; Kahn et al., 1976) and have been shown to mimic the physiological properties of insulin mediated by receptor binding. These polyclonal antibodies were originally thought to selectively precipitate insulin receptors (Harrison et al., 1979), although we have subsequently shown that the sera from three patients (B-2, B-5, and B-8) also precipitate insulin-like growth factor I receptors (Jonas et al., 1982).

Immobilized receptor autoantibodies were used to affinity purify insulin receptors from human placenta (Harrison and Itin, 1980). Wheat germ lectin-purified receptors were adsorbed to antibody-agarose columns prepared by coupling receptor autoantibodies (IgG fraction) to CNBr-activated Sepharose 4B. Receptors were eluted using the chaotropic agent 2.5*M* $MgCl_2$ at pH 6.5 in 0.1% Triton X-100. This purification step resulted in a 20-fold increase in binding capacity compared to wheat germ lectin-agarose eluates (Table 1). However, the purification factor was probably greatly underestimated, since the eluant itself was shown to irreversibly decrease the binding capacity of the receptor by up to 80%.

Insulin receptors have also been purified using a monoclonal antireceptor antibody affinity column (Roth and Cassell, 1983; Roth and Morgan, 1985; Roth et al., 1986). Elution of receptors that retained binding and kinase activities was accomplished using 1.5*M* $MgCl_2$, pH 6.5 or 1 μM insulin.

2.4. Other

A number of other chromatographic techniques have been used for insulin receptor purification, but none have proved to be as effective as the combination of sequential wheat germ lectin and insulin affinity chromatography. Successful purification procedures must exploit highly specific interactions between ligand and receptor as in insulin affinity chromatography, because of the extremely low abundance of the insulin receptor. Other purification methods that have been used are described briefly below.

2.4.1. Ion-Exchange

Rat liver insulin receptors were partially purified (7–17-fold, 65–70% recovery) using DEAE-cellulose (Cuatrecasas, 1972; Jacobs et al., 1977). Insulin receptors bind to this ion-exchange resin in sodium acetate buffer, pH 6.3, and can be eluted with a linear gradient of 0.1–1*M* ammonium acetate, pH 6.3, 0.2% Triton X-100. This step may increase recovery from subsequent insulin affinity chromatography (Jacobs et al., 1977).

2.4.2. Gel Filtration

Sepharose 6B chromatography has been used to purify insulin receptors from human placenta (Williams and Turtle, 1979). Maturo and coworkers (Maturo and Hollenberg, 1978,1985; Maturo et al., 1983) and Koch et al. (1986) used Sepharose 6B and CL-6B gel filtration, respectively, to separate two insulin receptor species that are interconverted by disulfide-sulfhydryl exchange. $\alpha_2\beta_2$ Heterotetrameric and $\alpha\beta$ heterodimeric insulin receptor complexes have also been separated by Bio-Gel A gel filtration (Sweet et al., 1987).

2.4.3. Hydrophobic

The insulin receptor has a large number of hydrophobic residues in the β subunit. Chromatography using phenyl-Superose or phenyl-Sepharose, which purify proteins on the basis of their hydrophobic interactions, has recently been used to purify part of the insulin receptor β subunit expressed in insect cells using the baculovirus system (Herrera et al., 1988).

2.4.4. Sulfhydryl Affinity

The insulin receptor contains a number of disulfide bonds whose disulfide-sulfhydryl exchange may be important in control of receptor function (Clark and Harrison, 1983). Aglio et al. (1985) adsorbed insulin receptors to a sulfhydryl affinity column, an organomercurial-agarose derivative (Affi-Gel 501, Biorad, CA) and eluted receptors with 10 m*M* dithiothreitol. Insulin promoted the binding of receptor to this column perhaps by exposing cryptic sulfhydryl residues.

3. Insulin Receptor Kinase

3.1. Special Precautions Necessary to Preserve Receptor Kinase Activity and Prevent β Subunit Degradation

Although homogeneous insulin receptors with full binding capacity can be purified using insulin-CDI-agarose or insulin-Sepharose as described in Section 2.2., these preparations do not necessarily have intact or functional β subunits (Roth et al., 1983; Hofmann et al., 1987). The insulin receptor β subunit is preferentially degraded at the carboxyl-terminus to form first a M_r 88,000 and then a M_r 50,000 form with a consequent rapid decrease of kinase activity, in purified receptor preparations stored at 4°C (Kathuria et al., 1986). Thus, special precautions are necessary to preserve receptor kinase activity. These include the rapid preparation of membranes and affinity chromatography, maintenance of the temperature at 0–4°C, especially during membrane solubilization, inclusion of 10% glycerol in all buffers to stabilize kinase activity (Pike et al., 1984) and elution of insulin affinity columns with a buffer containing 1 m*M* dithiothreitol, 10% glycerol, sodium acetate pH 5.5 (a slightly higher pH than used by Fujita-Yamaguchi, 1983 or Newman and Harrison, 1985).

Inclusion of cocktails of protease inhibitors have also been used to prevent proteolysis and degradation of the β subunit. McElduff et al. (1985b) disrupted cells in a buffer containing six different protease inhibitors: 4 m*M* PMSF, 1 m*M* EDTA, 1 μg/mL aprotinin, 10 μg/mL leupeptin, 0.5 inhibitor U/mL α_2-macroglobulin and 1 m*M* *N*-ethylmaleimide (NEM). However, it must be noted that 1 m*M* NEM has also been shown to cause a 25% reduction in insulin-stimulated receptor autophosphorylation, and 10 m*M* NEM causes a 70% reduction (Pike et al., 1984). β subunit degradation is more pronounced during purification of the rat liver receptor compared to the human placental receptor. Finn and Hofmann

(1988) have used an anti-protease cocktail containing 1.5 μg/mL aprotinin, pepstatin, antipain and leupeptin, 15 μg/mL benzamidine, and 2 mg/mL bacitracin during insulin-biotin-avidin chromatography of rat liver receptors.

3.2. *o-Phosphotyrosyl-Binding Antibodies*

The catalytically active phosphorylated form of the insulin receptor kinase has been purified using an o-phosphotyrosyl-binding antibody affinity column (Pang et al., 1985). This procedure can be utilized because the intact $\alpha_2\beta_2$ form of the receptor undergoes insulin-stimulated autophosphorylation at tyrosine residues. Human placental receptors purified from wheat germ lectin were chromatographed on *o*-phosphotyrosyl-binding antibodies bound to protein A-Sepharose. After binding of preexistent phosphoreceptors to the antibody column, autophosphorylation of receptors in the flow through was induced by incubation with insulin and ATP, and chromatography was repeated. Intact $\alpha_2\beta_2$ forms (>80% purity) were eluted with 10 m*M* *p*-nitrophenyl phosphate.

o-Phosphotyrosyl-binding antibody affinity chromatography has the advantage of resolving the intact $\alpha_2\beta_2$ form of the receptor from those forms with partially degraded β subunits (*see* Section 3.1.) that do not undergo insulin-stimulated autophosphorylation.

4. Future Perspectives

Although the insulin receptor has a relatively low abundance (between 0.01 and 0.05% of membrane proteins) sequential chromatography using wheat germ lectin and insulin affinity gels has enabled purification of receptors to homogeneity. Molecular cloning techniques may produce sufficient quantities of purified insulin receptor to allow its structural characterization by X-ray crystallography.

Acknowledgments

The authors are grateful to Margaret Thompson for secretarial assistance and to the National Health and Medical Research Council for support.

References

Adair, W. L. and Kornfeld, S. (1974) *J. Cell Biochem.* **249,** 4696–4704.

Aglio, L. S., Maturo, J. M. III, and Hollenberg, M. D. (1985) *J. Cell Biochem.* **28,** 143–157.

Bethell, G. S., Ayers. J. A., Hancock, W. S., and Hearn, M. T. W. (1979) *J. Cell Biochem.* **254,** 2572–2574.

Bethell, G. S., Ayers, J. S., Hearn, M. T. W., and Hancock, W. S. (1981) *J. Chromatography* **219,** 361–372.

Blatt, L. M. and Kim, K. H. (1971) *J. Biol. Chem.* **246,** 4895–4898.

Bond, J. S. and Butler, P. E. (1987) *Ann. Rev. Biochem.* **56,** 333–364.

Chou, C. K., Dull, T. J., Russell, D. S., Gherzi, R., Lebwohl, D., Ullrich, A., and Rosen, O. M. (1987) *J. Biol. Chem.* **262,** 1842–1847.

Clark, S. and Harrison, L. C. (1983) *J. Biol. Chem.* **258,** 11434–11437.

Colman, P. G. and Harrison, L. C. (1984) *Biochem. Biophys. Res. Commun.* **124,** 657–662.

Cuatrecasas, P. (1969) *Proc. Natl. Acad. Sci. USA* **63,** 450-457.

Cuatrecasas, P. (1972) *Proc. Natl. Acad. Sci. USA* **69,** 1277–1281.

Dey, A. C., Sheilagh, R., Rimsay, R. L., and Senciall, I. R. (1981) *Anal. Biochem.* **110,** 373–379.

Ebina, Y., Araki, E., Taira, M., Shimada, F., Mori, M., Craik, C. S., Siddle, K., Pierce, S. B., Roth, R. A., and Rutter, W. J. (1987) *Proc. Natl. Acad. Sci. USA* **84,** 704–708.

Ebina, Y., Ellis, L., Jarnagin, K., Edery, M., Graf, L., Clauser, E., Jing-hsiung, O., Masiarz, F., Kan, Y. W., Goldfine, I. D., Roth, R. A., and Rutter, W. J. (1985) *Cell* **40,** 747–758

Finn, F. M., Titus, G., and Hofmann, K. (1984a) *Biochemistry* **23,** 2554–2558.

Finn, F. M., Titus, G., Horstman, D., and Hofmann, K. (1984b) *Proc. Natl. Acad. Sci. USA* **81,** 7328–7332.

Finn, F. M., Titus, G., Nemoto, H., Noji, T., and Hofmann, K. (1982) *Metabolism* **31,** 691–698.

Finn, F. M. and Hofmann K. (1988) in *Receptor Biochemistry and Methodology* vol 12A (Venter, J. C. and Harrison, L. C. eds.), Arthur R. Liss, New York, pp. 3–14.

Flier, J. S., Kahn, C. R., Roth, J., and Bar, R. S. (1975) *Science* **190,** 63–65.

Fujita-Yamaguchi, Y., Choi, S., Sakamoto, Y., and Hakura, K. (1983) *J. Biol. Chem.* **258,** 5045–5049.

Gavin, J. R., Gorden, P., Roth, J., Archer, J. A., and Buell, D. N. (1973) *J. Biol. Chem.* **248,** 2202–2207.

Gould, R. J., Ginsberg, B. H., and Spector, A. A. (1981) *Biochemistry* **20,** 6776–6781.

Harrison, L. C., Billington, T., East, I. J., Nichols, R. J., and Clark, S. (1978) *Endocrinology* **2,** 1485–1495.

Harrison, L. C., Flier, J. S., Roth, J., Karlsson, F. A., and Kahn, C. R. (1979) *J. Clin. Endocrinol. Metab.* **48,** 59–65.

Harrison, L. C. and Itin, A. (1980) *J. Biol. Chem.* **255,** 12066–12072.

Harrison, L. C., Itin, A., Kasuga, M., and Van Obberghen, E. (1982) *Diabetologia* **22,** 233–238.

Hedo, J. A. (1984) in *Receptor Biochemistry and Methodology,* vol. 2 (Venter, J. C. and Harrison, L. C., eds.), Arthur R. Liss, New York, pp. 45–60.

Hedo, J. A., Harrison, L. C., and Roth, J. (1981) *Biochemistry* **20,** 3385–3393.

Herrera, R., Lebwohl, D., Garcia de Herreros, A., Kallen, R. G., and Rosen, O. M. (1988) *J. Biol. Chem.* **263,** 5560– 5568.

Hofmann, K. and Finn, F. M. (1985) *Ann. N.Y. Acad Sci.* **447,** 359–372.

Hofmann, K. Romovacek, H., Titus, G., Ridge, K., Raffensperger, J. A. and Finn, F. M. (1987) *Biochemistry* **26,** 7384–7390.

Im, J. H., Meezan, E., Rackley, C. E., and Kim, H. D. (1983) *J. Biol. Chem.* **258,** 5021–5026.

Jacobs, S. and Cuatrecasas, P. (1985) *Methods Enzymol.* **109,** 399–405.

Jacobs, S., Shechter, Y., Bissell, K., and Cuatrecasas, P. (1977) *Biochem. Biophys. Res. Commun.* **77,** 981–988.
Jonas, H. A., Baxter, R. C., and Harrison, L. C. (1982) *Biochem. Biophys. Res. Commun.* **109,** 463–470.
Jonas, H. A., Newman, J. D., and Harrison, L. C. (1986) *Proc. Natl. Acad. Sci. USA* **83,** 4124–4128.
Kahn, C. R., Flier, J. S., Bar, R. S., Archer, J. A., Gorden, P., Martin, M. M., and Roth, J. (1976) *N. Engl. J. Med.* **294,** 739–745.
Kasuga, M., White, M. F., and Kahn, C. R. (1985) *Methods Enzymol.* **109,** 609–621.
Kathuria, S., Hartman, S., Grunfeld, C., Ramachandran, J., and Fujita-Yamaguchi, Y. (1986) *Proc. Natl. Acad. Sci. USA* **83,** 8570–8574.
Koch, R., Deger, A., Jäck, H.-M., Klotz, K.-N., Schenzle, D., Krämer, H., Kelm, S., Müller, G., Rapp, R., and Weber, U. (1986) *Eur. J. Biochem.* **154,** 281–287.
Kohanski, R. A. and Lane, M. D. (1985a) *J. Biol. Chem.* **260,** 5014–5025.
Kohanski, R. A. and Lane, M. D. (1985b) *Ann. N.Y. Acad. Sci.* **447,** 373–385.
Lang, U., Kahn, C. R., and Harrison, L. C. (1980) *Biochemistry* **19,** 64-70.
Leedman, P. J. and Harrison, L. C. (1990) in *Receptor Purification* (Litwack, G., ed.), Humana Press, USA.
Leedman, P. J., Newman, J. D., and Harrison, L. C. (1989) *J. Clin. Endocrinol. Metab.* **69,** 134–141.
Lotan, R., Beattie, G., Hubbell, W., and Nicolson, G. L. (1977) *Biochemistry* **16,** 1787–1794.
Maegawa, H., McClain, D. A., Freidenberg, G., Olefsky, J. M., Napier, M., Lipari, T., Dull, T. J., Lee, J., and Ullrich, A. (1988) *J. Biol. Chem.* **263,** 8912–8917.
Maturo, J. M. and Hollenberg, M. D. (1978) *Proc. Natl. Acad. Sci. USA* **75,** 3070– 3074.
Maturo, J. M. and Hollenberg, M. D. (1985) *Can. J. Physiol. Pharmacol.* **63,** 987–993.
Maturo, J. M., Hollenberg, M. D., and Aglio, L. S. (1983) *Biochemistry* **22,** 2579–2586.
McClain, D. A., Maegawa, H., Levy, J., Huechsteadt, T., Dull, T. J., Lee, J., Ullrich, A., and Olefsky, J. M. (1988) *J. Biol. Chem.* **263,** 8904–8911.
McElduff, A., Comi, R. J., and Grunberger, G. (1985a) *Biochem. Biophys. Res. Commun.* **133,** 1175–1180.
McElduff, A., Grunberger, G., and Gorden, P. (1985b) *Diabetes* **34,** 686–690.
Newman, J. D., Campbell, I. L., Maher, F., and Harrison, L. C. (1989) *Mol. Endo.* **3,** 597–602.
Newman, J. D. and Harrison, L. C. (1985) *Biochem. Biophys. Res. Commun.* **132,** 1059–1065.
Newman, J. D. and Harrison, L. C. (1989) *Int. J. Cancer* **44,** 467–473.
Oka, T. and Topper, Y. J. (1971) *Proc. Natl. Acad. Sci. USA* **68,** 2066–2068.
Pang, D. T., Sharma, B. R., and Shafer, J. A. (1985) *Arch. Biochem. Biophys.* **242,** 176–186.
Petruzzelli, L., Herrara, R., Arenas-Garcia, R., Fernandez, R., Birnbaum, M. J., and Rosen, O. M. (1986) *Proc. Natl. Acad. Sci. USA* **83,** 4710– 4714.
Pharmacia (1976) Technical bulletin: Wheat germ lectin-Sepharose 6MB.
Pike, L. J., Eakes, A. T., and Krebs, E. G. (1986) *J. Biol. Chem.* **261,** 3782–3789.
Pike, L. J., Kuenzel, E. A., Casnellie, J. E., and Krebs, E. G. (1984) *J. Biol. Chem.* **259,** 9913–9921.
Roth, R. A. and Cassell, D. J. (1983) *Science* **219,** 299–301.
Roth, R. A., Mesirow, M. L., and Cassell, D. J. (1983) *J. Biol. Chem.* **258,** 14456–14460.
Roth, R. A. and Morgan, D. O. (1985) *Pharmacol. Ther.* **281,** 1–16.
Roth, R. A., Morgan, D. O., Beaudoin, J., and Sara, V. (1986) *J. Biol. Chem.* **261,** 3753–3757.
Shemer, J., Penhos, J. C., and LeRoith, D. (1986) *Diabetologia* **29,** 321–329.
Siegel, T. W., Ganguly, S., Jacobs, S., Rosen, O. M., and Rubin, C. S. (1981) *J. Biol. Chem.* **256,** 9266–9273.
Simon, J. and LeRoith, D. (1986) *Eur. J. Biochem.* **158,** 125–132.

Sweet, L. J., Morrison, B. D., and Pessin, J. E. (1987) *J. Biol. Chem.* **262,** 6939–6942.
Tang, X. Z., Tally, M., Jondal, M., and Hall, K. (1983) *Biochem. Biophys. Res. Commun.* **117,** 823–834.
Turkington, R. W. (1970) *Biochem. Biophys. Res. Commun.* **41,** 1362– 1367.
Ullrich, A., Bell, J. R., Chen, E. Y., Herrera, R., Petruzzelli, L. M., Dull, T. J., Gray, A., Coussens, L., Liao, Y.-C., Tsubokawa, M., Mason, A., Seeburg, P. H., Grunfeld, C., Rosen, O. M., and Ramachandran, J. (1985) *Nature* **313,** 756–761.
Wang, C. (1986) *Biochim. Biophys. Acta.* **888,** 107–115.
Ward, G. M. and Harrison, L. C. (1986) *Diabetes* **35,** 101–105.
Williams, P. F. and Turtle, J. R. (1979) *Biochim. Biophys. Acta.* **579,** 367–374.
Yu, K.-T. and Czech, M. P. (1986) *J. Biol. Chem.* **261,** 4715–4722.

Insulin Receptor Purification

J. M. Maturo III and M. D. Hollenberg

1. Introduction

Early work on the purification of the insulin receptor by affinity chromatographic techniques (Cuatrecasas, 1972a) literally set the stage for the small- and large-scale purification of most plasma membrane receptors for hormones and other biologically active agents. Since the early 1970s, little has changed in the basic approach to the isolation of the insulin receptor. The affinity chromatographic isolation protocol played a major role in isolating sufficient amounts of receptor for the microsequence determination, which ultimately, via cloning of the receptor cDNA, led to the elucidation of the complete sequence of the receptor (Ebina et al., 1985; Ullrich et al., 1985). Important modifications of the original isolation protocol have been introduced over the years, including the use of wheat germ agglutinin affinity chromatography (e.g., *see* Cuatrecasas and Tell, 1973; Harrison and Itin, 1980) and the development of an improved buffer system for the elution of the receptor from the insulin-agarose affinity gel (Fujita-Yamaguchi et al., 1983). The use of wheat germ agglutinin proved superior than the use of concanavalin A; and the omission of urea from the insulin-agarose elution buffer improved the recovery of ligand binding activity and tyrosine kinase activity in the isolated receptor. Apart from insulin- and lectin-agarose affinity chromatography, antireceptor antibodies or anti-insulin antibodies have also proved of use for the affinity-purification of the native or crosslink-labeled receptor (Harrison and Itin, 1980; Siegel et al., 1981; Heinrich et al., 1980; Armstrong et al., 1982). The crosslink-labeling procedures, whereby radiolabeled insulin is bound covalently to the receptor either with a photoaffinity probe (e.g., Yip and

Moule, 1983; Jacobs et al., 1979) or with a chemical crosslinking reagent (e.g., Pilch and Czech, 1980) have been of considerable use for receptor characterization. Studies using the crosslink-labeling approach have complemented work aimed at receptor isolation. This chapter will focus entirely on the methods that have been developed for receptor purification.

2. Receptor Purification with the Use of Insulin Agarose

2.1. Tissue Source and Preparation of Membranes

Although a number of the early studies of the insulin receptor were focused on the adipocyte as a tissue source (Cuatrecasas, 1971), chromatographic isolation procedures required a more convenient receptor-rich tissue source. Both human placenta and rat liver have proved to be attractive sources of insulin receptors. Methods developed for these two tissues have been adapted for use with other less abundant sources of receptor.

2.1.1. Human Placenta

Although insulin receptor can be recovered from fresh-frozen human placental tissue, the biochemical properties of the receptor (e.g., tyrosine kinase activity; subunit composition) recovered from a frozen preparation can be open to question. For our own work, we have preferred to use freshly recovered tissue obtained at Caesarian section and stored on ice (usually 1–2 h) prior to processing. All procedures are done at 4°C after rinsing the tissue thoroughly with chilled 0.9% (w/v) NaCl to remove adhering blood, the umbilical cord, chorion, and amnion are removed. The remaining tissue is rinsed again three times with four volumes of 0.9% (w/v) NaCl. The tissue is then minced to pieces approx 2 cm in diameter and rinsed again thoroughly to remove blood. The minced tissue is then "blenderized" using a rotary blade household blender (full speed for 30 s at 4°C) and the blenderized tissue is rapidly combined with 10 vol of a protease inhibitor-supplemented homogenization buffer at pH 7.5 having the following composition: 50 m*M* Tris HCl, 250 m*M* sucrose, 2 m*M* EDTA, 2 m*M* EGTA, 0.1 m*M* phenylmethylsulfonyl chloride (PMSF: added from a stock 0.1 *M* solution in isopropanol), 0.5 μg/mL leupeptin, 1 μg/mL pepstatin A, 500 kallikrein inhibitor U/mL of aprotinin, 1 m*M* 1,10-phenanthroline, and 25 m*M* benzamidine. The diluted "blenderized" tissue is then homogenized with a large-scale Ross tissue homogenizer (Model LAB-ME) at full speed for 2 min at 4°C. The resulting homogenate is filtered through two layers of cheesecloth to remove debris and is then centrifuged at 10,000*g* for 30 min at 4°C to remove a "low speed" pellet. The supernatant is then supplemented with 0.1*M* NaCl and 0.2 m*M* $MgSO_4$ and recentrifuged at 48,000*g* for 40 min at 4°C. The pelleted membranes are resuspended and washed three times with 50 m*M* Tris HCl buffer, pH 7.5 supplemented with 1 m*M* PMSF. For each wash, the pellets are disrupted at 4°C with a Polytron PT-10 rotary knife homogenizer at setting 4 for 30 s, and are repelleted at 4°C by centrifugation at 48,000*g* for 30 min. This procedure yields pellets with a white to slightly pink color. Except for minor variations (e.g., the inclusion of proteolysis inhibitors) this procedure is essentially the same as the one described some time ago for isolating membranes from either placenta (Hock and Hollenberg, 1980) or liver (*see below*, Cuatrecasas, 1972a). The isolated membranes can be extracted immediately with detergent-containing buffer to yield soluble receptor for further studies (e. g., Maturo et al., 1983).

2.1.2. Rat Liver

Rat liver tissue (frequently from male albino CD-l animals) has long been used as an abundant source of insulin receptor (Cuatrecasas, 1972a; Freychet et al., 1971). Upon sacrifice, rapidly removed livers are minced with scissors and rinsed free from blood with 0.9% (w/v) NaCl. The original procedures (e.g., *see* Cuatrecasas, 1972a,b) used Polytron homogenization in 0.25*M* sucrose-containing buffers followed by differential centrifugation to yield membranes suitable for receptor isolation. One feature of the original isolation procedures relates to the addition of divalent cation (e.g., Mg^{2+}) to the buffers to facilitate the recovery of a "microsomal" membrane fraction. This approach has been retained in a number of modified membrane isolation procedures (e.g., Hock and Hollenberg, 1980) that can be used directly to obtain receptor-rich membranes from liver or placenta (e.g., Cuatrecasas, 1972a; Maturo and Hollenberg, 1978; Hock and Hollenberg, 1980). The main significant modification of the original procedure (e.g., Cuatrecasas, 1972a) relates to the addition of proteolysis inhibitors to the first homogenization buffer and to the buffers subsequently used to resuspend and wash the membrane pellets (*see above*). The details of the several modified procedures based on the original work of Cuatrecasas (1972a) have been adequately recorded elsewhere (Cuatrecasas and Hollenberg, 1976; Maturo and Hollenberg, 1978; Hock and Hollenberg, 1980; Maturo et al., 1983).

2.1.3. Other Sources of Receptor

Many other tissues have been used for the partial purification of the insulin receptor. Of note is the early work with adipose tissue, which has been studied in depth as an insulin-responsive target. Following the work of Rodbell (1964), which made available, via a collagenase digestion procedure, a preparation of isolated adipocytes, it proved possible to isolate the insulin receptor from adipocyte membranes (Cuatrecasas, 1972b). Typically, fat cells harvested by collagenase digestion of epididymal adipose tissue from about 25 male albino CD-1 rats (150–200 g) are suspended in 20 mL of the protease inhibitor-supplemented homogenization buffer described above in Section 2.1.1. The cell suspension is then homogenized at 4°C using a Polytron PT-10 rotary knife homogenizer at setting 4 for 90 s. The membrane fraction is then pelleted by centrifugation at 48,000*g* for 30 min at 4°C. As with other membrane preparations, the initial pellet is resuspended and washed three times in 5 mL of 50 m*M* Tris HCl pH 7.5, containing 1 m*M* PMSF. Resuspension is achieved using a 15-s burst of the Polytron homogenizer; the resuspended membranes are then harvested by centrifugation at 48,000*g* for 30 min at 4°C. The final washed pellet can be extracted immediately for recovery of the insulin receptor (e.g., *see* Maturo and Hollenberg, 1979). Although the long list of tissues from which insulin-receptor bearing membranes have been isolated (ranging from lymphocytes to fibroblasts) is beyond the scope of this chapter, it is of some interest that the recent isolation of the insulin receptor from rat skeletal muscle (Whitson et al., 1988) echos work done over 25 years ago, suggesting the presence of insulin receptor in rat sarcolemma tissue (Edelman et al., 1963).

2.2. Receptor Solubilization

2.2.1. Choice of Detergent

Careful work by Cuatrecasas (1972a,b) documented the conditions under which a nonionic detergent like Triton X-100 could solubilize the insulin receptor from membrane preparations. The rationale for choosing one detergent over another is beyond the scope of this chapter and is discussed in some detail elsewhere (Hollenberg, 1988). The method for detecting soluble receptor represents an important key for monitoring the efficiency of the extraction procedure; the polyethylene glycol (PEG) method (Cuatrecasas, 1972a) for assaying receptor binding is outlined below in Section 4.2. In essence, concentrations of Triton X-100 between 1 and 2% (v/v) release the receptor quantitatively from almost all membrane preparations. Other detergents such as Lubrol-WX (Cuatrecasas, 1972a) and Ammonyx-LO (Maturo et al., 1978) can yield results comparable with those using Triton X-100 (*see* Table 1). In general, the solubilization procedures developed to recover the insulin receptor can be used for isolating a wide variety of membrane-anchored binding proteins (e.g., Kulakowski et al., 1981).

2.2.2. Solubilization Procedure

For solubilization, membrane preparations are suspended in buffer (e.g., 50 m*M* Tris HCl, pH 7.5, supplemented with 1 m*M* PMSF) containing from 1 to 2% v/v detergent (usually, Triton X-100). With 2% v/v detergent, the concentration of membrane protein should be between 6–8 mg/mL; for 1% Triton X-100, the membrane protein should be 5 mg/mL or lower. The suspension is then homogenized (e.g., Polytron PT-10, for 10 s at setting 4) and allowed to incubate. For optimal yield of insulin receptor, membranes are extracted for either 1 h at 24°C or 6 h at 4°C. If necessary, extraction can be done for 17–24 h at 4°C without adverse effects on receptor yield. After extraction, the suspension is clarified by centrifugation for 150,000*g* for 60 min at 4°C. The resulting supernatant contains 90–95% of the insulin receptor in a water soluble state (Table 1).

2.3. Ion Exchange Chromatography

Although ion-exchange chromatography does not provide for a large degree of purification, this step (either batchwise extraction, or column chromatography) provides for either a reduction in the detergent concentration (e.g., to 0.1% Triton X-100) or for an exchange of detergents (e.g., during a wash of the ion exchange column). This step is the first in the sequence used for insulin receptor isolation; the diethylaminoethyl (DEAE) group is preferred. Although any DEAE derivative can be used, DEAE-agarose (0.02 meq/mL; 45 mg of hemoglobin binding capacity/mL) is more easily handled than most preparations of DEAE-cellulose. A 50 mL plastic syringe, with a 45 mL bed volume can be used as a convenient column. This column can adsorb more than 95% of the receptor extracted from membranes resulting from 500 g of a receptor-rich tissue like human placenta. Solubilized receptor (*see above*) is applied (50–60 mL/h) at 24°C to a column preequilibrated with 50 m*M* Tris HCl, pH 7.5 containing 0.1% v/v Triton X-100. The column is then washed with an amount of equilibration buffer that is 10 times that of the applied sample. The receptor is then eluted from the column either batchwise, with 0.1% Triton-supplemented Tris buffer pH 7.4 containing 0.5*M* NaCl or with a sodium chloride gradient (10–200 m*M*) in the same buffer; the receptor elutes between 80–100 m*M* NaCl. The receptor-containing fractions are used directly for the following lectin purification step.

2.4. Lectin-Agarose Affinity Chromatography

As mentioned above, wheat germ agglutinin (WGA) has become the preferred lectin for insulin receptor purification. A commercially available WGA-agarose preparation (Vector Laboratories, 7.0 mg WGA/mL

Table 1
Solubilization of Insulin Binding Material from Human Placenta

Aliquot assayed	Binding of insulin by extracts[a]	
	Ammonyx-LO	Triton X-100
	fmoles bound/mg protein	
Supernatant	55 (65%)	165 (89%)
Pellet	16	2

[a]Membranes from placenta were extracted with either 0.2% Ammonyx-LO or 2% Triton X-100 for 1 h at 24°C. Aliquots of the supernatant and of the pellet (resuspended in the same vol as the supernatant) obtained by centrifugation at 200,000 × g were assayed for insulin binding with 2.5 $\times 10^{-9}M$ ^{125}I-insulin.

packed gel; 1.1 mole glycoprotein binding capacity per mole of immobilized lectin) has reproducibly yielded reliable results. The WGA-agarose is preequilibrated with 50 m*M* Tris HCl buffer, pH 7.5 containing 0.5*M* NaCl and 0.1% Triton X-100. Conveniently, a 20 mL plastic syringe, containing a packed bed vol of 20 mL, serves as a column. The receptor-containing fractions eluted from the DEAE column are applied slowly (10 mL/h at 24°C) to the lectin column. Insulin-binding material that passes unadsorbed through the lectin column at this point (1–5% of total receptor) can be immediately reapplied to the column to improve yield, if desired. After the adsorption step, the column is washed with 10 column vol of equilibration buffer. The receptor is then eluted from the column in buffer supplemented with 10% (w/v) *N*-acetylglucosamine (NAG). The receptor appears in the first few fractions (e.g., 2 mL/fraction) eluted from the column.

2.5. Insulin-Agarose Affinity Chromatography

2.5.1. Preparation of Affinity Matrix

Activated agarose suitable for derivatization can be prepared using gel activation with cyanogen bromide either as outlined originally by Cuatrecasas (1972c) or by a more convenient simplified CN Br procedure (March et al., 1974). The key for receptor isolation is the use of a spacer arm to which insulin is subsequently coupled. The *N*-hydroxysuccinimide ester of succinyldiaminodipropyl-agarose has proved most useful. This derivative can now be obtained commercially (e.g., Sigma) for the coupling of insulin; alternatively, the insulin-succinylaminodipropylamino agarose (insulin-agarose) matrix (1–5 mg insulin/mL settled gel) is also commercially available. Another convenient coupling matrix, containing a spacer arm (Affi-Gel 15, BioRad) has also proved useful in our hands for the synthesis of affinity columns; insulin is coupled to this support exactly as outlined by the manufacturer. For most purposes, small amounts of commercially available insulin agarose derivative will yield adequate results for the purification of analytical amounts of insulin receptor. To prepare larger amounts of affinity matrix, insulin can be coupled to the active ester agarose derivatives essentially, as described previously (Cuatrecasas, 1972c; Maturo and Hollenberg, 1985). In brief, insulin is dissolved (1 mg/mL) in 0.1*M* sodium phosphate buffer, pH 7.5, containing 6*M* freshly deionized urea. Two volumes of the insulin-containing solution are added to one vol of the settled coupling gel, which has been equilibrated with the same urea-containing buffer. The coupling reaction is then allowed to proceed with gentle mixing for 1 h at 4°C. The insulin coupling reaction is terminated by the addition of 2 vol buffer containing 0.5*M* glycine; the incubation is continued for 24 more h to allow the glycine to react with remaining sites on the activated agarose gel. The gel is then washed free of reactants and of urea with large volumes of 50 m*M* Tris HCl, pH 7.5. On average, the coupling procedure yields an affinity matrix containing from 300–400 µg insulin/mL of packed gel (method of Cuatrecasas, 1972c) to 700–800 µg/mL of packed gel (using Affi-Gel 15).

2.5.2. Column Preconditioning

Although a wide range of column sizes can be used (e.g., bed vol ranging from 0.5–25 mL), preparative (as opposed to analytical columns in Pasteur pipets) columns are conveniently poured (10–20 mL bed vol) in a 25 mL plastic syringe (2.7× 5.3 cm).

The poured gel bed is washed in sequence with large volumes (e.g., ten column vol) of the following solutions:

1. 50 m*M* Tris HCl buffer, pH 7.5;
2. 50 m*M* Tris HCl buffer, containing 1*M* NaCl;
3. 0.1*M* sodium phosphate buffer, pH 7.5, containing 6*M* freshly deionized urea;
4. 50 m*M* sodium acetate, pH 5.0, containing 1*M* NaCl.

After this sequence of washes, the column is eluted further with large volumes (10–30 column vol) of 50 m*M* Tris HCl buffer, pH 7.5, containing 1*M* NaCl and 0.1% (v/v) Triton X-100. The receptor-containing solution is then applied directly to this preconditioned column (*see below*).

2.5.3. Sequential Use of Columns for Receptor Isolation

Starting with the solubilized membrane extract, the receptor is added first to the ion exchange column (DEAE) and second (salt-elute from DEAE) to the WGA-agarose column. The material eluted with NAG from the lectin column is passed slowly (6–8 mL/h) over the insulin-agarose column. At this low rate of flow, more than 80% of the soluble receptor is retained by the column. Following sample application, the column is washed at a faster flowrate with 3 column vol of 1 m*M* Tris HCl buffer, pH 7.5. Finally, receptor is eluted from the column with 2 column vol of 50 m*M* sodium acetate, pH 5.0. Fractions (1–5 mL) are collected in tubes containing an equal volume of 1*M* Tris HCl buffer, pH 7.5; the Tris buffer returns the pH of the eluted fractions toward neutrality, so as to preserve the binding and kinase activity of the receptor.

After elution, the receptor-containing fractions can be concentrated (usually to 1 mL) using either Amicon PM-30 membranes or the Pro Di Mem PA-30 membrane system (Bio-molecular Dynamics). In other laboratories, centricon microconcentrators have proved useful for this purpose (Sweet et al., 1987). The result of a typical purification protocol, using the sequential method outlined above, is recorded in Table 2.

3. Receptor Purification Without the Use of Insulin-Agarose

The protocol outlined above is little changed from the procedure first reported by Cuatrecasas (1972b). We have used the procedure successfully in many of our studies of the receptor properties. Other procedures in the literature often omit the ion-exchange step, and some omit even the lectin-agarose step. Most, however, still use the insulin-agarose step. For instance, one published protocol uses gel exclusion chromatography of soluble receptor on Sephacryl S-400 followed by insulin-agarose affinity chromatography; the insulin-agarose step was used to switch the receptor from Triton X-100 into 0.6% *n*-octyl-β,D-glucopyranose (Sweet et al., 1987). As mentioned above, immunoaffinity chromatography using antireceptor antibodies has also yielded purified receptor preparations. The major drawback for this approach relates to the quantity of antireceptor antibody required to yield receptor in the amounts equivalent to the yield of the insulin-agarose columns.

In our own work, we developed a method of receptor preparation that does not expose the receptor to insulin during the procedure (Maturo et al., 1983). The protocol, which employs sequential DEAE, lectin-agarose, and gel filtration chromatography, provides for a 1000–2000-fold purification of the receptor. This receptor preparation displays hydrodynamic properties that are distinct from the receptor that has been exposed to insulin during the course of isolation (Maturo et al., 1983). In brief, the alternative isolation protocol was as follows: The clarified soluble receptor preparation was applied at 4°C at a flowrate of 50 mL/h to a DEAE-Sephadex column (5 × 20 cm) 7.5, containing 0.1% v/v Triton X-100. The column was washed with 2 column vol of buffer and was eluted with a 2 L linear gradient of NaCl (40–400 n*M*) in the same Tris-Triton buffer. The insulin-binding material, detected by the poly(ethylene glycol) (PEG) method (*see below*), was eluted when the NaCl concentration reached approx 100 m*M*. The pooled insulin-binding fractions from the DEAE column (about 300 mL) were applied at a flowrate of 20 mL/h at 4°C to a hydroxylapatite (Fast Flow, Calbiochem) column (1.5 × 10 cm) equilibrated with 10 m*M* Tris-HCl-0.1% Triton X-100. After sample application, the column was washed with 2 column vol of buffer, and a 500 mL linear gradient of sodium phosphate (30–100 m*M*, pH 7.5) in the Tris-Triton buffer was begun. Insulin-binding fractions (eluted at about 60–70 m*M* phosphate) were pooled (about 30 mL) and were adsorbed at a flowrate of about 20 mL/h to a column (0.6 × 9 cm) of Con A-agarose (12 mg of lectin/mL of packed gel) that was preequilibrated with 10 m*M* Tris-HCl buffer, pH 7.5, containing 1 m*M* $CaCl_2$, 0.1 m*M* $MnCl_2$, and 0.1% v/v Triton X-100. The lectin column was washed with 100 mL of buffer, and insulin-binding material was eluted with 25 mL of Tris-Triton buffer, containing 0.5*M* methyl α-D-mannopyranoside (Sigma); 1 mL fractions were collected and assayed for insulin binding. Receptor-containing fractions were pooled (12–15 mL), dialyzed against sugar-free Tris-Triton buffer, and applied to a column (0.6 × 9 cm) of *Ricinus communis* 120-agarose (Vector Laboratories; 6 mg of lectin/mL of packed gel). The lectin column was washed with sugar-free buffer and was eluted with buffer containing 0.5*M* D-galactose; 1 mL fractions were collected and assayed for insulin binding. Receptor-containing fractions were pooled (12–15 mL), and 5 mL aliquots were applied to a column (2.6 × 93 cm) of Sepharose 6B (noncrosslinked) that was preequilibrated with 10 m*M* Tris-HCl buffer, pH 7.5, containing 0.1% Triton X-100. The column was eluted at 4°C with buffer at a flowrate of 27 mL/h, and 3 mL fractions were collected. Fractions (nos. 74–84) containing insulin-binding activity ($K_{av} \cong 0.31$) were pooled and concentrated with a YM-30 filter (Amicon).

Table 2
Purification of Human Placental Insulin Receptor

Step	Volume, mL	Total protein, mg	Insulin binding[a] activity, μg/mg	Total binding[a] activity, μg	Purification, -fold	Yield, %
(1) Solubilized membrane	50	150	0.009	1.35	1	100
(2) DEAE-agarose	40	128	0.010	1.28	1.1	95
(3) WGA-agarose	40	12.2	0.070	0.80	7.3	59
(4) Insulin-agarose	1	0.025	24.4	0.61	2711	45

[a]Insulin activity binding was determined using $2.5 \times 10^{-9}M$ ^{125}I-insulin. At this insulin concentration, the receptor is only partly saturated. Thus, the figures underestimate the total amount of receptor present at each stage of purification. Precipitates, containing insulin bound to the receptor, were collected by centrifugation at 12,000*g*. Radioactivity was determined by a gamma counter. Insulin binding was corrected for nonspecific binding as described in Section 4.2. Protein was determined as described (Maturo et al., 1983), except for step 4, which was determined by amino acid analysis.

4. Analytical Procedures

Although the following procedures are not focused directly on receptor isolation, they are basic methods that are used routinely in connection with soluble receptor detection. For completeness, the following sections outline the abbreviated protocols for radiolabeling insulin and for detecting the recetor by polyethylene glycol (PEG) precipitation of the insulin-receptor complex.

4.1. Iodination of Insulin

A variety of procedures can be found in the literature for radiolabeling insulin with ^{125}Iodine to high specific activity. The method outlined in the following paragraphs has been found to be convenient, reproducibly yielding ^{125}I-labeled insulin (100–200 μCi/μg) suitable for receptor studies.

Stock peptides (about 1 mg/mL) are routinely dissolved in 50 m*M* sodium bicarbonate. The precise concentration of peptide can be measured spectrophotometrically using measurements at 215 and 225 nm for samples diluted in phosphate buffer, according to the formula:

$$(E_{215} - E_{225}) \times 155 = \mu g/mL$$

Carrier-free Na^{125} (or rarely, ^{131}I) (1–3 mCi in a vol of 5–10 μL) is added to 100 μL of 0.25*M* sodium phosphate buffer, pH 7.5 (4.15 g of Na_2HPO_4 + 0.51 g of $NaH_2PO_4 \bullet H_2O$/100 mL) and polypeptide (5 μL of a solution 1 mg/mL in 0.05*M* $NaHCO_3$) is then added with a glass capillary so as to avoid bubbling the solution. Immediately, 20 µL of a freshly prepared solution of chloramine-T (0.5 mg/mL in distilled H_2O) is added with gentle agitation, and the oxidation reaction is allowed to proceed for 20–30 s; then, sodium metabisulfite (20 µL of 1 mg/mL in distilled H_2O) is added, and the consequent reduction step is allowed to proceed for a period of 10–15 s. The reaction mixture is then diluted with 200 µL of 0.1*M* sodium phosphate buffer, pH 7.5, containing 0.1% w/v crystalline bovine albumin. An aliquot (10 µL) is quickly withdrawn so as to quantitate by crystal scintillation counting the amount of radioactive iodine present in the reaction medium. The remaining solution is transferred either to a pre-equilibrated (albumin-containing neutral phosphate buffer) 10 mL column (disposable 10 mL pipet, 18 × 0.9 cm) of Sephadex G-15 or to a heavy-walled 12 mL conical centrifuge tube containing a talc pellet (25 mg, Gold Leaf Pharmacal Co.), which is then crushed and triturated wth a Pasteur pipet so as to adsorb the iodinated peptide. The small amount of solution remaining in the reaction vessel (12 × 75 mm glass test tube) is diluted with approx 0.5 mL of albumin-containing buffer and saved for the measurement of the percentage of ^{131}I or ^{125}I incorporated into peptide. The gel filtration column is eluted with the 0.1% albumin-containing 0.1*M* phosphate buffer pH 7.5 at a flowrate of about 0.5 mL/min; the radiolabeled peptide, collected in 0.5 mL fractions from the column, is usually recovered in fractions nos. 4–8. In order to recover the peptide from talc, the talc-adsorbed peptide is suspended in about 10 mL of albumin-containing buffer and pelleted by centrifugation; the pellet is washed four more times in this manner. The labeled polypeptide is then eluted from the final pellet by suspension in 2–3 mL of a solution comprised of 3 mL of 1 *N* HCl, 2.5 mL of H_2O, and 0.5 mL of 20% crystalline albumin (either in H_2O or in Krebs-Ringer-bicarbonate buffer, pH 7.4). The suspension is clarified by centrifugation (3000 rpm, 40 min) and the ^{125}I-labeled peptide is transferred to a tared vial and the exact vol determined by weight. Several drops of 0.25*M* phosphate buffer, pH 7.4, are added, and the solution is adjusted to near neutrality (pH 4–7, indicator paper) by the dropwise addition of 1 *N* NaOH up to an amount just under half the original vol measured as described above. Should the solution inadvertently become alkaline, a drop of 1 *N* HCl is added, and the pH is then readjusted to near neutrality. Although the gel filtration method of recovery is by far the most convenient, and is suitable for preparing radiolabeled EGF, the talc-adsorption procedure can prove to be an advantage for certain peptides (like insulin) in that the intact radiolabeled peptide can be purified from "damaged" peptide reaction productions that do not adsorb well to the talc. An aliquot (20 µL) of the solution of radiolabeled peptide is withdrawn to measure the precipitability of the preparation by trichloroacetic acid and the specific radioactivity of the peptide. As an alternative to the two procedures outlined above, the entire reaction mixture can be purified by high pressure liquid chromatography. Although this procedure is not necessary for work with radiolabeled insulin and EGF-URO, the separation of mono- and di-iodo derivatives of certain peptides may be essential for the reliable estimate of binding to solubilized receptors.

The incorporation of ^{125}I is estimated in the following manner: An aliquot, e.g., 50 µL, of the residual diluted reaction mixture is mixed into 1.0 mL of 0.1*M* phosphate buffer, pH 7.5, containing 1% w/v bovine albumin. An equal aliquot (50 µL) is withdrawn for measurement of radioactivity. Trichloroacetic acid (0.5 mL of a 10% w/v solution in H_2O) is then added, and the precipitate is chilled in ice and then sedimented in a clinical centrifuge. An aliquot of the supernatant (50 µL) is withdrawn, and the radioactivity is measured. It is assumed that the radioactivity remaining in the supernatant represents nonincorporated ^{125}I. The fraction (f) of ^{125}I incorporated into the peptide is given by the formula

$$f = (\text{RA initial} \times 1.5\ \text{RA final})/\text{RA initial}$$

where RA initial and RA final refer to the radioactivity in aliquots before and after the addition of trichloroacetic acid, respectively. If, in the initial reaction mixture containing 5 µg of peptide, there were CµCi present, the specific activity (SA) of the preparation would be given by the equation

$$SA = (fC/5)\ \mu Ci/\mu g$$

In a procedure identical to the one described in the preceding paragraph, the fraction of ^{125}I-labeled peptide precipitated by trichloroacetic is determined for an aliquot of the purified iodopeptide solution. In practice 90–99% of the radioactivity is precipitated for a preparation of [^{125}I]peptide that is fully biologically active and yields good binding data. The number of radioactive iodoine molecules, on average, which are incorporated can be calculated using a value of 1.62×10^7 mCi/milliatom for ^{131}I and 2.8×10^6 mCi/milliatom for ^{125}I.

4.2. Measurement of Insulin Binding for Soluble Receptor Preparations

A key for all isolation protocols is the ability to measure insulin binding either in the initial membrane extract, or in fractions recovered from the chromatography procedures. The polyethylene glycol (PEG) precipitation procedure, adapted from immunoassay procedures (Desbuquois and Auerbach, 1971) for insulin receptor detection (Cuatrecasas, 1972b), has been used for many years with only minor modification (e.g., *see* Cuatrecasas and Hollenberg, 1976). For measurement of insulin binding, 5–50 μL of detergent-extracted receptor is added to 0.2 mL Krebs-Ringer bicarbonate buffer, 0.1% w/v albumin, pH 7.4, containing [^{125}I]insulin with (control tubes only) or without 25–50 μg of native insulin/mL. Phosphate buffer (0.1*M*, pH 7.4) may also be used for the binding assay provided the pH is maintained between pH 7.0–7.4; buffers containing Tris-HCl, however, appear to interfere with the polyethylene glycol precipitation. Equilibration of binding is achieved in about 20–50 min at 24°C, at which time 0.5 mL of ice-cold 0.1 *M* sodium phosphate buffer (pH 7.4) containing 0.1% bovine gamma-globulin (a carrier for precipitation) is added and the tubes are placed in ice. Cold 25% w/v polyethylene glycol (0.5 mL) is added (final concentration, 10% w/v), and the tubes are thoroughly mixed and placed in ice for 10–15 min. The suspension is then filtered under reduced pressure on cellulose acetate (Millipore EG or EH) filters, and the collected precipitate is washed with 3 mL of 8% w/v polyethylene glycol in 0.1*M* phosphate buffer (pH 7.4) before measurement of trapped radioactivity by crystal scintillation counting. In some instances, it may be possible to collect and wash the ligand-receptor precipitate by centrifugation with a microfuge. As for the measurements of binding with cells and membranes, the specific binding is determined by subtracting from the total radioactivity bound, that which remains bound in the presence of a high concentration (25–50 μg/mL) of native insulin. Under the above conditions, less than 0.5% of the total free [^{125}I]insulin is precipitated or adsorbed nonspecifically and nearly quantitative precipitation of the insulin receptor complex occurs.

Concentrations of polyethylene glycol less than 8% (w/v) incompletely precipitate the complex; concentrations higher than 12% significantly precipitate free insulin. The presence of γ-globulin is essential as a carrier for the precipitation reaction, but concentrations above 0.1% (w/v) cause precipitation of free insulin. If the pH of the buffer containing the γ-globulin is above 8 or below 7, the complex is less effectively precipitated; phosphate buffers (0.1*M*, pH 7.4) can be used effectively in the incubation medium. A final concentration of Triton X-100 in the assay mixture in excess of 0.2% v/v results in decreased insulin binding. The membrane extracts are therefore diluted before assay so that the final concentration of Triton X-100 is usually less than 0.1% and always less than 0.2% (v/v). For the estimation of rate data (association/dissociation), it is assumed that the addition of the cold polyethylene glycol-γ-globulin solutions stop the binding reaction at the timed intervals. It is expected that for each hormone studied, different conditions will be necessary to precipitate the maximum amount of hormone receptor complex, leaving most of the free ligand in solution. The procedure that has been used for the insulin receptor may serve as a prototype for the study of other solubilized receptors.

Summary

With the use of the above procedures, it should be possible to isolate soluble insulin receptor from a variety of tissue sources. It can also be said that virtually the same procedures, using Agarose derivatized with insulinlike growth factor-I (IGF-I) in place of insulin-agarose, will yield purified IGF-I receptor. Nonetheless, it should be pointed out that both the insulin receptor and the IGF-I receptor are able to bind either insulin or IGF-I (albeit with quite different affinities, e.g., *see* Bhaumick et al., 1981, 1982); further, polyclonal antibodies directed against the insulin receptor can also recognize the IGF-I receptor (e.g., *see* Armstrong et al., 1983). Thus, if tissues contain both insulin and IGF-I receptors, affinity methods (either using insulin/IGF-I-agarose or agarose-antibody columns) will not yield purified receptor preparations (e.g., for insulin) that are entirely free from small amounts of the second receptor (e.g., the one for IGF-I; *see* Armstrong et al., 1983). Indeed, monoclonal antibody-agarose columns have proved of importance for the isolation of IGF-I receptor for sequencing studies (Ullrich et al., 1986). As a final comment, one can add that the general scheme of receptor isolation outlined in this chapter should prove of use (and has indeed done so) for the purification of a variety of membrane receptors for hormones and other agents.

Acknowledgments

The work summarized in this chapter owes a great debt to Pedro Cuatrecasas, who pioneered the development of many of the methods used for receptor isolation and characterization.

References

Armstrong, G. D., Hollenberg, M. D., Bhaumick, B., and Bala, R. M. (1982) *J. Cell. Biochem.* **20,** 283–292.

Armstrong, G. D., Hollenberg, M. D., Bhaumick, B., Bala, R. M., and Maturo, J. M. III. (1983) *Can. J. Biochem. Cell Biol.* **61,** 650–656.

Bhaumick, B., Bala, R. M., and Hollenberg, M. D. (1981) *Proc. Natl. Acad. Sci. USA* **78,** 4279–4283.

Bhaumick, B., Armstrong, G. D., Hollenberg, M. D., and Bala, R. M. (1982) *Can J. Biochem.* **60,** 923–932.

Cuatrecasas, P. (1971) *Proc. Natl. Acad. Sci. USA* **68,** 1264–1268.
Cuatrecasas, P. (1972a) *Proc. Natl. Acad. Sci. USA* **69,** 318–322.
Cuatrecasas, P. (1972b) *J. Biol. Chem.* **247,** 1980–1991.
Cuatrecasas, P. (1972c) *Proc. Natl. Acad. Sci. USA* **69,** 1277–1281.
Cuatrecasas, P. and Hollenberg, M. D. (1976) *Adv. Protein Chem.* **30,** 251–451.
Cuatrecasas, P. and Tell, G. P. E. (1973) *Proc. Natl. Acad. Sci. USA* **70,** 485–489.
Desbuquois, B. and Aurbach, G. D. (1971) *J. Clin. Endocrinol.* **33,** 732–738.
Ebina, Y., Ellis, L., Jarnagin, K., Edery, M., Graf, L., Clauser, E., Ou, J., Masiarz, F., Kan, Y. W., Goldfine, I., Rother, R. A., and Rutter, W. J. (1985) *Cell* **40,** 747–758.
Edelman, P. M., Rosenthal, S. L., and Schwartz, I. L. (1963) *Nature* **197,** 878–880.
Freychet, P., Roth, J., and Neville, D. M., Jr. (1971) *Biochem. Biophys. Res. Commun.* **43,** 400–408.
Fujita-Yamaguchi, Y., Choi, S., Sakamoto, T., and Itakura, K. (1983) *J. Biol. Chem.* **258,** 5045–5049.
Harrison, L. C. and Itin, A. (1980) *J. Biol. Chem.* **255,** 12066–12072.
Heinrich, J., Pilch, P. F., and Czech, M. P. (1980) *J. Biol. Chem.* **255,** 1732–1737.
Hock, R. A. and Hollenberg, M. D. (1980) *J. Biol. Chem.* **255,** 10731– 10736.
Hollenberg, M. D. (1988) in *Methods in Neurotransmitter Receptor Analysis* (Yamamura, H. I., Enna, S. J., and Kuhar,M. J., eds.), Raven Press, New York, in press.
Hollenberg, M. D. and Cuatrecasas, P. (1976) in *Methods in Cancer Research,* vol. XII (Busch, H., ed.), pp. 317–366.
Jacobs, S., Hazum, E., Schechter, T., and Cuatrecasas, P. (1979) *Proc. Natl. Acad. Sci. USA* **76,** 4918–4921.
Kasuga, M., Fujita-Yamaguchi, T., Glithe, D., White, M., and Kahn, C. R. (1983) *J. Biol. Chem.* **258,** 10973–10980.
Kulakowski, E., Maturo, J. M. III, and Schaffer, S. (1981) *Arch. Biochem. Biophys.* **210,** 204–209.
March, S. C., Parikh, I., and Cuatrecasas, P. (1974) *Anal. Biochem.* **60,** 149– 152.
Maturo, J. M. III and Hollenberg, M. D. (1978) *Proc. Natl. Acad. Sci. USA* **75,** 3070–3074.
Maturo, J. M. III and Hollenberg, M. D. (1979) *Can. J. Biochem.* **57,** 497–506.
Maturo, J. M. III and Hollenberg, M. D. (1985) *Can. J. Physiol. Pharmacol.* **63,** 987–993.
Maturo, J. M. III, Shackelford, W. H., and Hollenberg, M. D. (1978) *Life Sci.* **23,** 2063–2072.
Maturo, J. M. III, Hollenberg, M. D., and Aglio, L. S. (1983) *Biochemistry* **22,** 2579–2586.
Pilch, P. F. and Czech, M. P. (1980) *J. Biol. Chem.* **255,** 1722–1731.
Rodbell, M. (1964) *J. Biol. Chem.* **239,** 375–380.
Siegel, T. W., Ganguly, S., Jacobs, S., Rosen, O. M., and Rubin, C. S. (1981) *J. Biol. Chem.* **255,** 1722–1731.
Sweet, L., Morrison, B. D., Wilden, P. A., and Pessin, J. E. (1987) *J. Biol. Chem.* **262,** 16730–17638.
Ullrich, A., Bell, J., Cen, E., Herrera, R., Petruzzelli, L. M., Dull, T. J., Gray, A., Coussens, L., Liao, Y., Tsubokawa, M., Mason, A., Seeburg, P., Grunfeld, C., Rosen, O., and Ramachandran, J. (1985) *Nature* **313,** 756–761.
Ullrich, A., Gray, A., Tam, A. W., Yang-Feng, T., Tsubokawa, M., Collins, C., Henzel, W., Le Bon, T., Kathuria, S., Chen, E., Jacobs, S. Francke, U., Ramachandran, J., and Fujita-Yamaguchi, Y. (1986) *Embo J.* **5,** 2503–2512.
Whitson, R. H., Grimditch, G. K., Sternlicht, E., Kaplan, S. A., Barnard, R. J., and Itakura, K. (1988) *J. Biol. Chem.* **263,** 4789–4794.
Yip, C. C. and Moule, M. L. (1983) *Diabetes* **32,** 760–767.

Isolation of Fibronectin Receptors

Kenneth M. Yamada and Susan S. Yamada

1. Fibronectin Cell-Interaction Sites and Integrin Receptors

Fibronectin is an adhesive glycoprotein involved in a variety of cellular functions, including adhesion of cells to extracellular matrices, embryonic cell migration and differentiation, wound healing, and hemostasis (Yamada and Olden, 1978; Hynes, 1985; Ruoslahti and Pierschbacher, 1987; Mosher, 1988). As might be expected for a protein with such varied functions, fibronectin can interact with more than one cell surface receptor. These fibronectin receptors interact with specific peptide recognition sequences in fibronectin, of which there appear to be at least four. Most cell interactions occur with the central fibronectin cell-adhesive domain containing the sequence Gly-Arg-Gly-Asp-Ser (GRGDS) combined with a second crucial site of ≤37 amino acids (Ruoslahti and Pierschbacher, 1987; Obara et al., 1988; Yamada, 1988). The central 3 amino acids of the GRGDS sequence appear to be the most important in this pentapeptide, and receptors that bind to this adhesive recognition sequence are termed RGD receptors. In animal studies, synthetic peptides containing such sequences can block the function of RGD receptors crucial for embryonic cell migration, platelet adhesion and aggregation, and experimental metastasis (*reviewed in* Yamada, 1988). The RGD sequence appears to require a second, synergistic site for effective interactions with nucleated mammalian cells and to bind the presently best-characterized fibronectin receptor termed $\alpha_5\beta_1$ or VLA-5 (Obara et al., 1988,1989).

Two other adhesive recognition sequences in fibronectin are found in regions that are present in only certain fibronectin molecules owing to alternative splicing of its precursor messenger RNA (Hynes, 1985; Mosher, 1988). These sites, termed IIICS or V regions, are recognized in a cell-type specific fashion by neural crest derivatives, such as sympathetic and sensory ganglion cells, as well as by melanoma cells. The key amino acid sequences involved in cell surface recognition of these sites are a 25-residue sequence including an active Glu-Ile-Leu-Asp-Val sequence and the much less active, independently expressed Arg-Glu-Asp-Val sequence (Humphries et al., 1987). A class 1 integrin appears to be involved in binding of the first, major cell-type specific site, but its identity remains to be determined (M. J. Humphries, personal communication).

The receptors used by cells to bind to fibronectin are part of large family of adhesion receptors termed integrins (Hynes, 1987; Buck and Horwitz, 1987; Ruoslahti and Pierschbacher, 1987; Juliano, 1987; Hemler, 1988; Yamada, 1988). A current classification of this family of noncovalently associated heterodimers, many members of which have been cloned and sequenced, is shown in Table 1. There are three major subfamilies at present, each of which is based on a shared beta subunit that is characteristic of that family. The alpha subunits provide the ligand specificity for the different receptors within each family. In addition, at least two alpha subunits can combine with alternative beta subunits to form novel receptors.

The best-characterized fibronectin receptor is $\alpha_5\beta_1$, also known as VLA-5. The morphology of the isolated receptor resembles two other integrins, and is shown in Fig. 1 as determined by a variety of ultrastructural methods by Nermut et al. (1988). It is noteworthy that the receptor is predicted to extend a substantial distance outward from the plasma membrane (≥200 angstroms) on two spindly legs contributed by the alpha and beta subunits. Each of these subunits has a large, glycosylated extracellular domain predicted to contribute to the mushroom-shaped head, legs or stalks formed of repeated beta sheets or repetitive cysteine-rich units, a transmembrane segment, and a short cytoplasmic domain (Fig. 1).

Other class 1 integrins serve as receptors for collagen (e.g., VLA-2) or laminin (e.g., VLA-3, VLA-6, and others); none of these other receptors have yet been shown to recognize the RGD sequence. One emerging current concept is that individual class 1 integrin receptors can often have multiple ligand specificities, which can vary according to the specific alpha subunit involved. Class 2 integrins are characteristic of leukocytes, and mediate cell–cell adhesion of these cells and endothelial cells. This class does not directly recognize the RGD sequence in a short peptide. Class 3

Table 1
The Integrin Family of Receptors[a]

Fibronectin/VLA receptors	Leukocyte receptors
$\alpha_1\beta_1$ (laminin and collagen receptor)	$\alpha_L\beta_2$ (LFA-1; I-CAM-1 and I-CAM-2 receptor)
$\alpha_2\beta_1$ (collagen receptor for types I–VI)	$\alpha_M\beta_2$ (Mac-1; C3bi receptor)
$\alpha_3\beta_1$ (fibronectin, laminin, collagen receptor)	$\alpha_X\beta_2$ (p150,95)
$\alpha_4\beta_1$ (cell–cell adhesion and binds fibronectin)	**Platelet/vitronectin receptors**
$\alpha_5\beta_1$ (fibronectin receptor)	$\alpha_{II}b_3$ (GP IIb-IIIa, fibrinogen, fibronectin, von Willebrand factor, vitronectin receptor)
$\alpha_6\beta_1$ (laminin receptor)	$\alpha_v\beta_3$ (vitronectin, von Willebrand factor, thrombospondin receptor)
$\alpha_0\beta_1$ (avian band 1; could be $\alpha_1\beta_1$ or $\alpha_2\beta_1$)	**Alternative beta subunit receptors**
	$\alpha_6\beta_4$ (laminin and (?) fibronectin receptor)
	$\alpha_v\beta_x$ (fibronectin and vitronectin receptor)

[a]Based on data reviewed in Hynes, 1987; Buck and Horwitz, 1987; and Yamada, 1988; as well as personal communications from M. Hemler, W. Carter, C. Damsky, D. Cheresh, R. Hynes, V. Quaranta, A. Sonnenberg, and T. Springer. Some of these assignments of receptor function are still tentative.

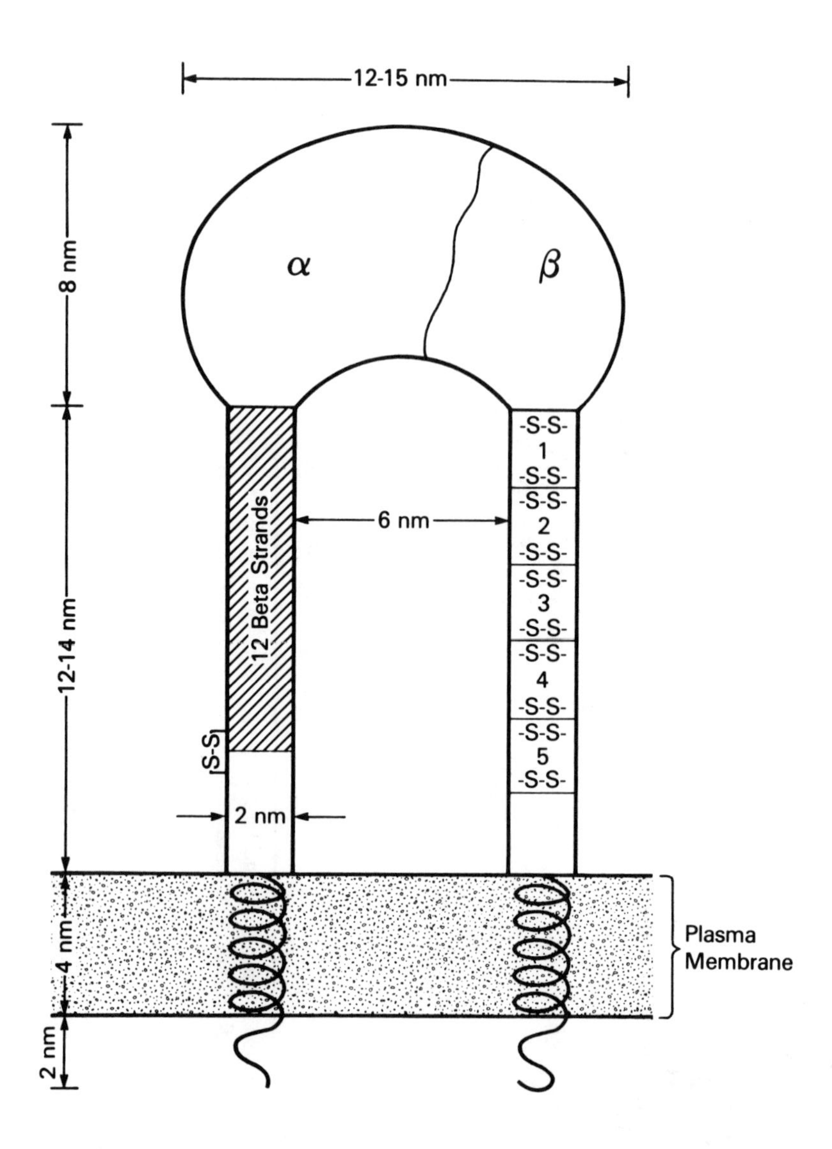

Fig. 1. Morphology of human fibronectin receptor. The VLA-5 fibronectin receptor is shown diagrammatically based on measurements using a variety of electron microscopic methods (Nermut et al., 1988). The receptor is comprised of alpha and beta subunits supported on spindly legs that are tentatively predicted to be highly structured, containing either repeating beta strands or disulfide-crosslinked repeats. The receptor is anchored by transmembrane segments, with short cytoplasmic domains.

integrins include the platelet glycoprotein IIb/IIIa and the vitronectin receptor. The former protein is crucial for platelet aggregation, and both recognize short RGD sequences directly.

2. Fibronectin Receptors

As noted above, there are several integrins that function as fibronectin receptors. In addition, more may be identified, e.g., for the alternatively spliced IIICS cell-adhesion sequences. How cells regulate and use more than one fibronectin receptor will be interesting to unravel in the future.

2.1. Avian Integrins

Monoclonal antibodies that block cell adhesion were used to isolate and partially characterize the first fibronectin receptors described; these receptors were isolated as part of a mixture of what are now often termed class 1 integrins (Neff et al., 1982; Greve and Gottlieb, 1982; Knudsen et al., 1985; Hasegawa et al., 1985). These monoclonal antibodies, termed CSAT or JG22, appear to recognize primarily the class 1 beta subunit of the complex. The most successful isolation protocols use monoclonal antibody affinity chromatography of detergent homogenates of partially purified membrane fractions from chick embryos, with a mild acid precipitation step to remove contaminants. The purified antigen consists of three major bands by SDS polyacrylamide electrophoresis (Fig. 2a). Band 1 provisionally corresponds to the alpha-0 subunit (Table 1), which might actually be equivalent to VLA-l and/or VLA-2; band 2 corresponds to alpha-3, and probably alpha-4 and alpha-5; band 3 corresponds to the shared beta subunit of class 1 integrins. These preparations are probably a complex mixture of all class 1 alpha subunits that bind to beta-1 molecules.

The integrins isolated by these antibodies appear to be heterodimeric complexes of subunits that are approx 140,000 daltons in size after reduction; the integrity of the dimer is necessary for function (Knudsen et al., 1985; Hasegawa et al., 1985; Buck et al., 1986; our unpublished results). This "140K complex" was initially proposed to be a transmembrane linkage between fibronectin and the cytoskeleton on morphological grounds (e.g., *see* Mueller et al., 1988). The complex binds to fibronectin, albeit weakly in vitro under physiological conditions; binding can be demonstrated directly by column retardation assays, as well as biologically by competition with cell adhesion to fibronectin (Horwitz et al., 1985; Akiyama et al., 1986). The binding is readily blocked by the GRGDS(P) peptide. This receptor complex appears necessary for various aspects of cell adhesion, migration, and invasion (e.g., Savagner et al., 1986).

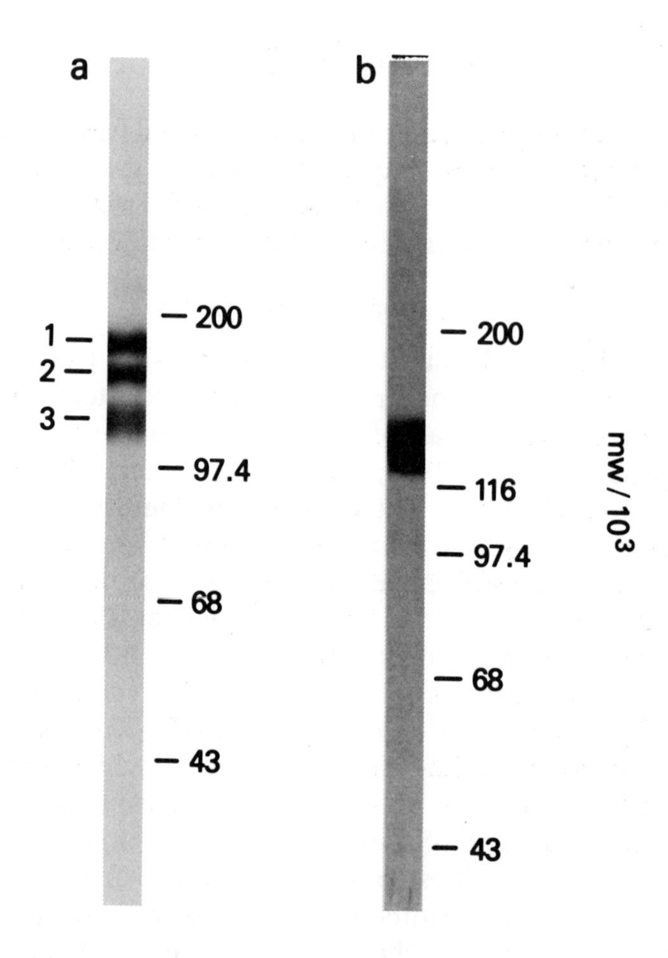

Fig. 2. Purified fibronectin receptors. Lane a shows avian integrin receptors purified as described in the text from chicken embryos using monoclonal antibody affinity chromatography, then analyzed by SDS polyacrylamide gel electrophoresis (without chemical reduction) and stained for protein with Coomassie brilliant blue. Bands 1, 2, and 3 are marked on the left, and molecular weight markers (in kilodaltons) are indicated on the right. Lane b shows human VLA-5 fibronectin receptor purified from human placenta by affinity chromatography as described in the text. The sample was reduced with dithiothreitol, then analyzed by SDS polyacrylamide gel electrophoresis and silver stained.

2.2. Mammalian Fibronectin Receptors

Because the avian integrins bind rather weakly to fibronectin in vitro in assays used to date, the best-characterized fibronectin receptors have been those that can be isolated by fibronectin affinity chromatography, i.e., mammalian fibronectin receptors (*reviewed in* Akiyama and Yamada, 1989; *see also* the chapter in this volume by Brentani for a novel approach to their isolation). The reason for the higher avidity of solubilized mammalian receptors for fibronectin compared with embryonic avian integrins is unexplained at present; however, it has recently been possible to purify avian $\alpha_5\beta_1$ receptors by affinity chromatography in the presence of manganese, which appears to substantially increase binding to affinity columns (R. Chiquet-Ehrismann and R. O. Hynes, personal communications). The affinity of fibronectin binding to intact mammalian cells as estimated by Scatchard analysis is $7 \times 10^{-7}M$, and there are 100–500,000 receptors per fibroblastic cell (e.g., *see* Akiyama et al., 1985; Akiyama and Yamada, 1985).

Isolation protocols for mammalian fibronectin receptors have generally been modeled on the methods of Pytela et al. (1985). The affinity chromatographic ligand is usually a fibronectin proteolytic fragment containing the central cell-adhesive domain, from which have been removed the domains that might bind to other extracellular molecules such as collagen, fibrin, and heparan sulfate. An excess of crude detergent-solubilized tissue homogenate is applied to columns and washed; receptor-ligand binding may be stronger in Triton X-100 than in *N*-octyl-β-D-glucopyranoside (S. K. Akiyama and K. M. Yamada, unpublished results). Thus, the method described below involves binding of the receptor to the affinity support in Triton, followed by washing in this relatively inexpensive detergent, then brief washing and elution in *N*-octyl-β-D-glucopyranoside.

For elution, one approach uses the inactive Gly-Arg-Gly-Glu-Ser-Pro (GRGESP) or GRGES peptide to preelute any nonspecifically bound protein, then the bound receptor is eluted specifically by the active GRGDSP or GRGDS peptides. A much more economical alternative is to elute the receptor with EDTA (Pytela et al., 1985; Pytela et al., 1987; Roberts et al., 1988), which yields preparations of similar purity. Remaining residual materials are eluted with 6*M* urea, although usually surprisingly little additional material is recovered. This type of affinity chromatographic protocol is effective for isolation of fibronectin receptor from both tissue homogenates and in vitro cultivated cells. In all cases to date, the protein isolated at physiological ionic strength is the VLA-5 molecule containing an alpha-5 subunit bound noncovalently to the shared beta-1 subunit. The beta subunit is nicked proteolytically near the bottom of the leg close to the transmembrane domain, but is held together by disulfide bonding (Argraves et al., 1987); the significance of this cleavage is not yet known.

2.3. Broad-Specificity VLA-3 Receptor

The VLA-3 receptor consists of an alpha-3 subunit bound noncovalently to a beta-1 subunit, and is thought to be a multiple-ligand receptor that binds to fibronectin, various collagens, and laminin. Monoclonal antibody inhibition studies document the role of this receptor in cell adhesion, and very recent binding studies support the notion of multiple specificities for this molecule. Although it binds to fibronectin, the VLA-3 receptor appears to do so much less efficiently than VLA-5; however, if the latter is absent, VLA-3 can bind to fibronectin, but apparently does so effectively only in the presence of manganese ion (Wayner and Carter, 1987; M. J. Elices and M. E. Hemler, personal communication). It will be of considerable interest to learn whether this protein recognizes each ligand separately, or whether there is a specific shared determinant that is shared by these seemingly quite distinct proteins.

2.4. Platelet Receptor GP IIb/IIIa

One of the best-characterized integrins is the important platelet receptor GP IIb/IIIa (e.g., *see* review by Phillips et al., 1988). This heterodimer mediates platelet adhesion to fibrinogen, fibronectin, von Willebrand factor, and vitronectin. Inhibition of its function or genetic defects result in faulty platelet aggregation and adhesion to substrates, causing bleeding disorders. This receptor binds directly to the RGD (RGDF, RGDS, or RGDV) sequence in these proteins, and the majority of the binding appears to be to this simple sequence. However, significant binding affinity appears to be contributed by a second site that remains to be characterized (M. H. Ginsberg, personal communication). Unlike the mammalian fibronectin receptor, this receptor is readily dissociated and inactivated by treatment with EDTA.

2.5. Heparan Sulfate Proteoglycan

Fibronectin binds to heparin and heparan sulfate, and thus an additional cell-surface "receptor" for fibronectin could be heparan sulfate proteoglycan intercalated into the plasma membrane. This type of function for cell adhesion has been demonstrated in mammary epithelial cells by Saunders and Bernfield (1988).

3. Isolation of Fibronectin Receptors—General Comments

The following detailed protocols focus on two of the best-characterized classes of fibronectin receptors. A protocol for the isolation of the platelet multifunctional receptor GP IIb/IIIa has been published by Fitzgerald et al. (1985). It is important to stress that the following protocols can be scaled down directly for small-scale preparations, although care to avoid highly dilute receptor preparations is recommended. For discussions of the rationale for specific steps, see the preceeding sections reviewing each receptor as well as the original descriptive studies.

4. Isolation of Avian Fibronectin Receptor

4.1. Preparation of JG22E Monoclonal Antibody Column

The JG22E hybridoma is a subclone of the original JG22 line, and is available as "JG22" from the Developmental Studies Hybridoma Bank, c/o Thomas August, The Johns Hopkins University School of Medicine. The monoclonal antibody can be isolated from ascites or from serum-free medium as follows:

Culture the JG22E hybridoma in medium composed of a 1:1 mixture of Dulbecco's Modified Eagle's Essential Medium and Ham's F12 with 4 m*M* glutamine, 1 m*M* sodium pyruvate, 0.4 mg/mL charcoal-extracted bovine serum albumin (Collaborative Research), 50 U/mL penicillin, 50 μg/mL streptomycin, and insulin-transferrin-selenous acid supplement (ITS, Collaborative Research). After growing 1–2 L of cell suspension, allow the cells to become very dense and to begin to die. When most of the cells are dead, centrifuge out the cells and add to the supernatant solution solid ammonium sulfate to 50% saturation. Adjust the pH to 7.5–8.0 with NaOH, and incubate with gentle stirring overnight at 4°C. The next day, sediment the precipitated antibody at 10,000 × g for 20 min in a refrigerated centrifuge. Resuspend the pellet in 0.1*M* borate buffer pH 8.5, at a ratio of 15 mL/L of original cell suspension. Dialyze vs the borate buffer for 2–3 d, changing the buffer two times per day. In preparation for DEAE chromatography, dialyze the antibody solution against low ionic strength buffer as follows: Dialyze vs 10 m*M* sodium phosphate pH 7.4 and 70 m*M* NaCl overnight. Then dialyze vs 10 m*M* sodium phosphate pH 7.4 and 50 m*M* NaCl for 3 h.

To prepare the DEAE column, weigh out 10 g of preswollen DE 52 (Whatman) per liter of original cell suspension. Suspend well in 300 mL 0.3*M* sodium phosphate pH 7.4. Let the beads sediment, discard the supernatant solution, and repeat. Pack 8 mL beads into a column, and wash extensively with 10 m*M* sodium phosphate pH 7.4 and 50 m*M* NaCl. Centrifuge the dialyzed antibody at 17,000*g* for 15 min, and discard the pellet. At 4°C, apply the antibody solution to the DEAE column, and collect the flow-through fractions. Measure the absorbance of the fractions at 280 nm, and pool the peak fractions. Adjust the NaCl concentration of the peak to 0.15 M, and add sodium azide to 1–3 m*M* for storage. This antibody will not survive freezing, so storage on ice is recommended.

Couple the purified monoclonal antibody to CNBr-activated Sepharose 4B (Pharmacia) at high concentration (at least 25 mg of protein per g of dry beads) following the manufacturer's recommended protocol. The IgG-Sepharose should be prewashed as recommended, then subjected to one round of elution with pH 11.0 diethylamine buffer immediately before the first use to remove residual antibody.

4.2. Purification of Avian Receptor

Remove heads and intestines from 36 13-d-old White Leghorn chicken embryos. Wash the embryos three times in a buffer composed of 0.25*M* sucrose, 20 m*M* Tris-HCl pH 7.5, 2 m*M* EDTA, and 2 m*M* phenylmethylsulfonyl fluoride (PMSF)* at 4°C. Homogenize in a blender in 350 mL of the same buffer for two one-minute periods, cooling the blender container on ice for several min in between. Filter the homogenate through six layers of cheesecloth. Homogenize again, either with 15–20 strokes in a Dounce homogenizer, or preferably for 20 s in a Polytron homogenizer (Brinkman). Centrifuge in a refrigerated centrifuge at 2000*g* for 10 min, and collect the supernatant solution through cheesecloth. Isolate the crude membrane fraction by centrifuging in an ultracentrifuge at 200, 000*g* for 60 min.

Suspend the pellet in 50 mL of a buffer composed of 1.5% Nonidet-P40, 10 m*M* Tris-HCl pH 7.5, 0.1*M* NaCl, 1 m*M* EDTA, and 1 m*M* PMSF using ten strokes of a Dounce homogenizer. Add 10% Nonidet-P40 to make a final concentration of 1.5%. Stir at 4°C for 60 min. Centrifuge in an ultracentrifuge at 200,000*g* for 30 min. Collect the supernatant solution and add 1*M* sodium acetate to make a final concentration of 20 m*M*. Adjust the pH to 5.5–6.0 with 1*M* acetic acid, and incubate at 4°C for 30 min. Centrifuge in a refrigerated centrifuge at 27,000*g* for 20 min. Collect the supernatant

*Immediately before use, prepare a 200 m*M* solution of PMSF in 95% ethanol. Dilute this solution into the buffer with vigorous stirring to make the desired final concentration of PMSF.

solution, add 1*M* Tris-HCl to make a final concentration of 50 m*M*, and adjust the pH to 7.5 with NaOH.

Wash a 10 mL column of JG-22E-Sepharose prepared as described in Section 4.1 in the same buffer in which the pellet was resuspended. Incubate the membrane extract and the JG-22E-Sepharose at 4°C for 5 h or overnight, with constant agitation (e.g., with a rotating mixer; stirring with a stir bar should be avoided). Wash the beads containing bound antigen twice in a buffer composed of 0.5% Nonidet-P40, 10 m*M* Tris-HCl pH 7.5, 0.1*M* NaCl, and 1 m*M* PMSF. Pack the beads in a column and wash with 300–400 mi of the same buffer. Then wash the column with 50 mL of a buffer composed of 1% *N*-octyl-β-D-glucopyranoside (OGP), 10 m*M* Tris-HCl pH 7.5, 0.1*M* NaCl, and 1 m*M* PMSF. Elute the receptor from the column with 50 m*M* diethylamine pH 10.6, 1% OGP, 0.1*M* NaCl, and 1 m*M* PMSF. Collect 60 drop fractions in tubes containing 80 µL 1*M* Tris-HCl pH 7.5. Locate the peak fractions by reading the absorbance at 280 nm. Pool the peak fractions and adjust the pH to 7.5. Dialyze the pooled peak fractions vs 0.75% (w/v) OGP in Dulbecco's phosphate-buffered saline without Ca^{2+} or Mg^{2+} and with 1 m*M* PMSF. Concentrate the receptor preparation by burying the dialysis bag in Ficoll (Pharmacia) for several hours, and centrifuge in a refrigerated centrifuge at 40,000*g* for 30 min. Assay for protein with the BCA assay (Pierce). Expected yields are in the range of 0.5–1 mg receptor. Figure 2a shows the purified avian receptor preparation, consisting of three major bands, termed band 1 (M_r = 155,000), Band 2 (M_r = 135,000), and band 3 (M_r = 120,000).

The JG-22E-Sepharose column can be reused up to about 10 times. After each use, clean the column by post-elution with diethylamine-OGP buffer at pH 11. Wash and store the column in 0.1*M* sodium borate pH 8.0, 0.1*M* NaCl, and 1–3 m*M* sodium azide.

5. Isolation of Mammalian Fibronectin Receptors

5.1. Preparation of Fibronectin Affinity Support and Tissue

Prepare the 110K cell-binding domain of fibronectin exactly according to Zardi et al. (1985) using thermolysin digestion and hydroxyapatite chromatography. Couple the 110K fragment to CNBr-Sepharose 4B (Pharmacia) according to the manufacturer's instructions, using 10 mg 110K fragment per gram of Sepharose. This column can be reused many times. The 110K fragment column should be protected by a guard column of Sepharose 4B to trap molecules that may stick nonspecifically to the column matrix and to prevent the column from becoming clogged with aggregates from the placenta homogenate. The entire receptor purification protocol should be performed at 4°C.

Human placenta should be cut into 0.5–1 cm^2 pieces, avoiding the thicker sheets of connective tissue. After rinsing in Dulbecco's phosphate-buffered saline, freeze the pieces of placenta by dropping them into liquid nitrogen onto a cheesecloth net for collection, then store in or above liquid nitrogen.

5.2. Purification of Human Fibronectin Receptor

Prepare a 50–60 mL column of 110K fragment-Sepharose protected by a 25 mL Sepharose 4B guard column. Wash both columns with 150 mL of column buffer (1% *N*-octyl-β-D-glucopyranoside (OGP), Dulbecco's phosphate-buffered saline with Ca^{2+} and Mg^{2+} (PBS^+), and 1 m*M* phenylmethylsulfonyl fluoride (PMSF). (*See* footnote under avian receptor purification for methods for addition of PMSF to solutions.) Partly defrost 75 g of human placenta. Using a blender with a 350 mL container, homogenize the placenta in 150 mL PBS^+ with 10 μg/mL each of leupeptin and aprotinin plus 1 m*M* PMSF. Measure the volume and add OGP to a concentration of 75 m*M* and PMSF to 3 m*M*. Stir the homogenate for 30 min at 4°C and centrifuge in an ultracentrifuge at 200,000*g* for 30 min.

Load the supernatant solution on the column at a rate of 150 mL/h with the guard column in place. Wash the column with 300 mL of column buffer at a rate of 200 mL/h. When the colored homogenate has cleared the guard column, remove the guard column and discard it. Elute the receptor from the column with 150 mL 10 m*M* EDTA, 1% OGP, Dulbecco's phosphate-buffered saline without Ca^{2+} and Mg^{2+} and 1 m*M* PMSF at a rate of 150 mL/h. To regenerate the column, wash it with 175 mL of 6*M* urea, 1% Triton X-100, PBS^+, and 1 m*M* PMSF, and then with 175 mL of 1% Triton X-100, PBS^+, and 1 m*M* PMSF at a rate of 200 mL/h. Store the column in PBS^+ and 1–3 m*M* sodium azide.

Analyze 40 μL aliquots of the column fractions after reduction by 0.1*M* dithiothreitol by SDS-polyacrylamide gel electrophoresis using a 4% stacking gel and a 7.5% resolving gel (Laemmli, 1970). Stain the gel with a silver stain (BioRad). Pool the peak fractions.

5.3. Further Purification

Incubate the pooled receptor preparation with 0.25 mL of a 50% suspension of agarose-bound wheatgerm agglutinin (Vector Laboratories). Incubate overnight at 4°C with agitation. Centrifuge out the beads, and wash them six times with 1 mL of column buffer. Elute the receptor by incubating the beads for 15 min with agitation in 1 mL 0.5*M* *N*-acetyl glucosamine, 30 m*M* OGP, PBS$^+$, and 1 m*M* PMSF. Repeat the elution two more times. Analyze 5–10 µL from each fraction on an SDS 4/7.5% polyacrylamide gel after chemical reduction of the sample, and silver stain the gel. Pool the fractions containing the receptor and concentrate (Centricon 30, Amicon). Dialyze vs column buffer to remove *N*-acetyl glucosamine. The usual yield from this procedure is approximately 0.5 mg receptor per 75 g of placenta starting material. Figure 2b shows the purified human receptor preparation, which consists of a single band of M_r = 140,000 after chemical reduction.

If necessary, the purified receptor can be subjected to a final high resolution size-fractionation step by chromatography on a Superose 6 FPLC column (HR 10/30, Pharmacia). Inject 0.5 mL of receptor onto a 1 × 30 cm column and elute at 0.3 mL/min in 1% OGP and 1 m*M* PMSF in PBS$^+$ (prefiltered through a 0.2 µm filter). Collect 0.6 mL fractions and determine the peak by analysis of fractions by SDS polyacrylamide gel electrophoresis, as above.

5.4. Automation and Additional Applications

Since the protocol for purification of the human fibronectin receptor involves a number of washes by different solutions at 4°C, it can be simplified by automation. Although other systems can be envisioned, the simplest uses a 6-way valve and a peristaltic pump preceeding the affinity column. The valve is fed by reservoirs containing the homogenate and eluants, and is controlled by a programmable timer. The only required operator intervention is then the removal of the precolumn; manual collection of the EDTA eluate is simple, but this step could also be programmed.

The affinity chromatography and monoclonal antibody receptor isolation protocols can be adapted with minor changes (mainly scaling down) to the isolation of receptors from cultured cells. For affinity chromatography, the receptor can be extracted from monolayers by direct incubation with 100 m*M* octyl glucoside, e.g., in 1–2 mL of solution per roller bottle rotated in a roller bottle apparatus at 4°C. As for intact tissue, the receptor can be purified readily from such crude homogenates, although additional purification steps are desirable (Akiyama and Yamada, 1987).

Receptors isolated according to these protocols are biologically active and are immunogenic. However, pretreatment with glutaraldehyde to increase immunogenicity may help in antibody production. In general, antibodies to the beta subunits appear to be easiest to obtain, but monoclonal antibodies against the alpha subunits can be obtained. The monoclonals can be either biologically active or nonperturbing; the former are obviously valuable for functional studies, but the latter are useful for receptor mobility studies on living cells (Duband et al., 1988).

References

Akiyama, S. K., Hasegawa, E., Hasegawa, T., and Yamada, K. M. (1985) *J. Biol. Chem.* **260,** 13256–13260.

Akiyama, S. K. and Yamada, K. M. (1985) *J. Biol. Chem.* **260,** 4492–4500.

Akiyama, S. K. and Yamada, K. M. (1987) *J. Biol. Chem.* **262,** 17536–17542.

Akiyama, S. K. and Yamada, K. M. (1989) *Biochim. Biophys. Acta,* in press.

Akiyama, S. K., Yamada, S. S., and Yamada, K. M. (1986) *J. Cell Biol.* **102,** 442–448.

Argraves, W. S., Suzuki, S., Arai, H., Thompson, K., Pierschbacher, M. D., and Ruoslahti, E. (1987) *J. Cell Biol.* **105,** 1183–1190.

Buck, C. A. and Horwitz, A. F. (1987) *Ann. Rev. Cell Biol.* **3,** 179–205.

Buck, C. A., Shea, E., Duggan, K., and Horwitz, A. F. (1986) *J. Cell Biol.* **103,** 2421–2428.

Duband, J.-L., Nuckolls, G. H., Ishihara, A., Hasegawa, T., Yamada, K. M., Thiery, J. P., and Jacobson, K. (1988) *J. Cell Biol.* **107,** 1385–1396.

Fitzgerald, L. A., Leung, B., and Phillips, D. R. (1985) *Anal. Biochem.* **151,** 169–177.

Greve, J. M. and Gottlieb, D. I. (1982) *J. Cell. Biochem.* **18,** 221–229.

Hasegawa, T., Hasegawa, E., Chen, W.-T., and Yamada, K. M. (1985) *J. Cell. Biochem.* **28,** 307–318.

Hemler, M. E. (1988) *Immunol. Today* **9,** 109–113.

Horwitz, A., Duggan, K., Greggs, R., Decker, C., and Buck, C. (1985) *J. Cell Biol.* **101,** 2134–2144.

Humphries, M. J., Komoriya, A., Akiyama, S. K., Olden, K., and Yamada, K. M. (1987) *J. Biol. Chem.* **262,** 6886– 6892.

Hynes, R. (1985) *Ann. Rev. Cell Biol.* **1,** 67–90.

Hynes, R. O. (1987) *Cell* **48,** 549–554.

Juliano, R. L. (1987) *Biochim. Biophys. Acta* **907,** 261–278.

Knudsen, K. A., Horwitz, A. F., and Buck, C. A. (1985) *Exp. Cell Res.* **157,** 218–226.

Laemmli, U. K. (1970) *Nature* **283,** 249–256.

Mosher, D. F., ed. (1988) *Fibronectin,* Academic, New York.

Mueller, S. C., Hasegawa, T., Yamada, S. S., Yamada, K. M., and Chen, W.-T. (1988) *J. Histochem. Cytochem.* **36,** 297–306.

Neff, N. T., Lowrey, C., Decker, C., Tovar, A., Damsky, C., Buck, C., and Horwitz, A. F. (1982) *J. Cell Biol.* **95,** 654–666.

Nermut, M. V., Green, N. M., Eason, P., Yamada, S. S., and Yamada, K. M. (1988) *EMBO J.* **7,** 4093– 4099.

Obara, M., Kang, M. S., and Yamada, K. M. (1988) *Cell* **53,** 649–657.

Obara, M., Shinagawa, S., and Yamada, K. M. (1989) in preparation.

Phillips, D. R., Charo, I. R., Parise, L. V., and Fitzgerald, L. A. (1988) *Blood* **71,** 831–843.

Pytela, R., Pierschbacher, M. D., Argraves, W. S., Suzuki, S., and Ruoslahti, E. (1987) *Methods Enzymol.* **144,** 475–489.

Pytela, R., Pierschbacher, M. D., and Ruoslahti, E. (1985) *Cell* **40,** 191–198.

Roberts, C. J., Birkenmeier, T. M., McQuillan, J. J., Akiyama, S. K., Yamada, S. S., Chen, W.-T., Yamada, K. M., and McDonald, J. A. (1988) *J. Biol. Chem.* **263,** 4586–4592.

Ruoslahti, E. and Pierschbacher, M. D. (1987) *Science* **238,** 491–497.

Saunders, S. and Bernfield, M. (1988) *J. Cell Biol.* **106,** 423–430.
Savagner, P., Imhof, B. A., Yamada, K. M., and Thiery, J. P. (1986) *J. Cell Biol.* **10** 2715–2727.
Wayner, E. A. and Carter, W. G. (1987) *J. Cell Biol.* **105,** 1873–1884.
Yamada, K. M. (1988) in *Fibronectin* (Mosher, D. F., ed.) Academic, New York, pp. 47–191.
Yamada, K. M. and Olden, K. (1978) *Nature* **275,** 179–184.
Zardi, L., Carnemolla, B., Balza, E., Borsi, L., Castellani, P., Rocco, M., and Siri, A. (1985) *Eur. J. Biochem.* **146,** 571–579.

Receptor Purification Using the Complementary Hydropathy Approach

Ricardo Brentani and Renata Pasqualini

1. Introduction

Many cellular events, such as migration, adhesion, and cell proliferation, are dependent on the interactions between specific cell surface receptors and cell matrix macromolecular components.

The latter are often very large molecules, frequently engaged in multiple interactions. One can then imagine that full interactions, for instance between adherent cell and their substratum, comprise binding of distinct domains of such macromolecules and perhaps distinct cell receptors differing among themselves by the specific recognition of simple domains by each.

It becomes critical, therefore, in order to obtain fuller comprehension of the molecular mechanisms involved, to devise experimental approaches that enable us to predict, with a reasonable certainty, the chemical nature of such interacting domains.

It has been hypothesized that the complementary DNA sites at the same DNA locus might code for interacting peptides (Biro, 1981). It was further suggested that the chemical basis for such interactions is the complementary hydropathy of the encoded peptides (Blalock and Smith, 1984). Implicit is the assumption that strongly hydrophilic amino acids will interact with strongly hydrophobic one, and vice versa. The chemical plausibility for such interactions as well as the biological relevance for such a genetic arrangement have been recently discussed (Brentani, 1988).

This hypothesis was initially validated by reports on the successful prediction of the binding sites for ACTH (Bost, Smith, and Blalock, 1985); g-endorphin (Carr, Bost, and Blalock, 1986); and luteinizing hormone releasing hormone (LHRH) (Mulchahey et al., 1986). For the applicability of such a concept to peptide interactions in general, it was essential to show that similar predictions could be validated also at the level of protein–protein interactions, where the conformation of the interactants is "frozen" and, therefore, there is less margin for the accommodation of possibly ill-fitting theoretically predicted analogs.

We have shown (Brentani et al., 1988) that one could predict the nature of the binding site of the 140 kDa fibronectin receptor (Pytela, Pierschbacher, and Ruoslahti, 1985) from the nucleotide sequences coding for the RGDS domain present both in human (Oldberg, Linney and Ruoslahti, 1983) and rat (Schwarzbauer et al., 1983) fibronectin known to be its actual binding site to one of its cell specific receptors. The peptides trp-thr-val-pro-thr-ala (WTVPTA) and gly-ala-val-ser-thr-ala (GAVSTA), predicted from the rat and human complements, respectively, inhibit binding of labeled fibronectin to receptor-rich human osteosarcoma cells MG-63; and antibodies raised against one of them (WTVPTA) coupled to keyhole lympet hemocyanin (KLH) can react, in ELISA assays, with either of them to the same extent. Furthermore, such antibodies can recognize, in Western blots, the 140 kDa FN receptor both in blots of whole-cell extracts as well as after affinity purification on a FN-fragment Sepharose column. It was further shown that an affinity column prepared by coupling WTVPTA to CNBr-activated Sepharose could selectively retain FN. The retained FN could be removed by washing the column with RGDS-containing solvents but not by RGES, already known to be an ineffective analog of RGDS, one of the cell binding domains of FN (Pierchbacher and Ruoslahti, 1984). More recently, we have shown that the same antibody that recognizes the 140 kDa FN receptor also recognizes platelet FN receptor GPIIIa in Western blots (Pasqualini, Chamone, and Brentani, 1989).

In Fig. 1 we show that, despite conspicuous differences in the primary structure of the rat and human complements, when their hydropathy plots are compared, they are strikingly similar. Also shown, and of great interest, is the hydropathy plot of a peptide ser-ala-val-gly-thr-len (SAVGTL), a peptide found in the sequence of cloned integrin (Tamkun et al., 1986) and human FN receptor (Argraves et al., 1987).

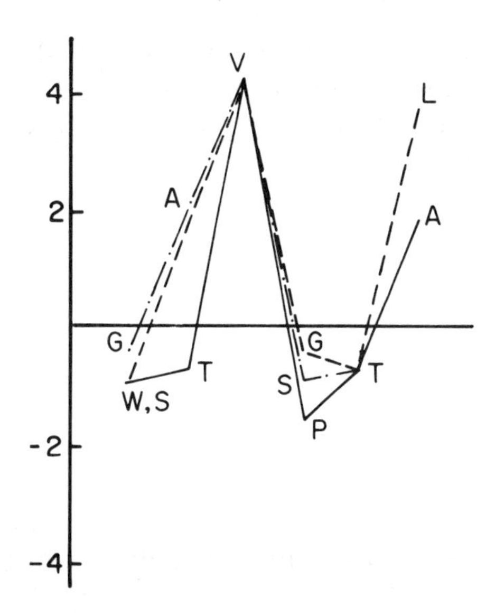

Fig. 1. Hydropathy plot of peptides complementary to the RGDS domain of fibronectin. WTVPTAS, rat complement; GAVSTA, human complement; SAVGTL, a peptide present in both human and chick integrins. For further details, *see text*.

Peptides displayed in Fig. 1 were shown to inhibit binding of labeled affinity purified GPIIIa, whereas the related peptides GAGSTA and GARSTA, which differ from GAVSTA only at a single position, do not. We believe that the reason for their ineffectiveness lies in their considerably different hydropathy (Pasqualini, Chamone, and Brentani, 1989).

The high specificity of the interaction between anti-WTVPTA antibodies and FN receptors has led us to attempt to use such antibodies for the purification of FN receptors.

2. Experimental

In order to ascertain the specificity of the anti-WTVPTA antibody, we performed ELISA inhibition assays. Inhibition of reactivity of the antibody against WTVPTA immobilized in 96-well plastic plates was determined by adding increasing concentrations of synthetic peptide (Fig. 2).

To attempt to purify the FN receptor with this antibody, human osteosarcoma MG-63 cell extracts were prepared by suspending cells in a buffer containing 100 m*M* octylglucoside, 3 m*M* PMSF, $CaCl_2$, and $MgCl_2$ 1 m*M* in PBS. The mixture was incubated at 4°C for 15 min and centrifuged for 15 min in an Eppendorf microfuge.

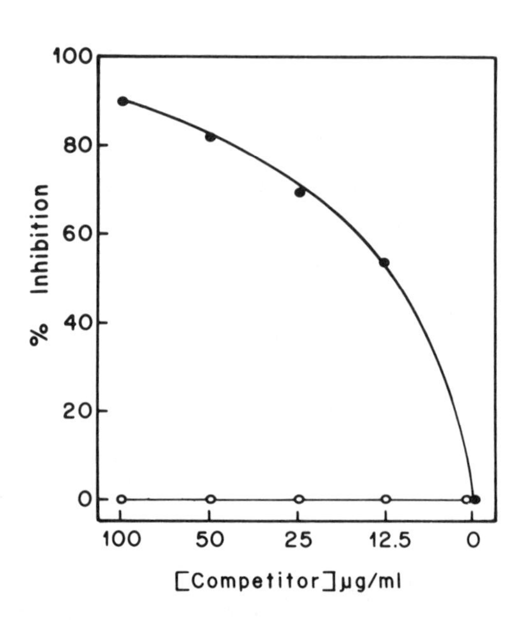

Fig. 2. Inhibition studies. The effect of increasing concentrations of WTVPTA on the recognition of this peptide by anti-WTVPTA antibodies was determined. • WTVPTA; O unrelated peptide.

Immunoprecipitations were performed using 300 µL of the extract, precleared by a 30-min incubation with rabbit normal serum followed by 100 µL of *S. aureus* (10% suspension). After 15 min of centrifugation in an Eppendorf microfuge, the supernatants (cleared extracts, 300 µL) were incubated with 3 µL of antihuman FN receptor antibody (kindly given by E. Ruoslahti, Cancer Res. Center, La Jolla, CA) (as a positive control), anti-WTVPTA antiserum, anti-SAVGTL antiserum, or normal mouse serum, overnight at 4°C. One volume of a 10% *S. aureus* was added, and left for 2 h at room temperature. After a 15-min microcentrifugation (Eppendorf), the pellets were washed 5 times with PBS pH = 7.5, resuspended in reducing sample buffer (240 m*M* Tris HCl, pH = 6.8; 0.8% SDS; 40% glycerol, 200 m*M* 2-mercaptoetanol and 0.2% Bromophenol blue), and analyzed in a 7.5% SDS-PAGE, after heating to 100°C for 5 min.

Figure 3 shows the SDS PAGE profile of the immunoprecipitated material. Anti-WTVPTA and anti-SAVGTL antisera were able to immunoprecipitate the 140 kDa band corresponding to FN receptor under reducing conditions. Exactly the same gel electrophoretic profile was obtained after immunoprecipitation of the same MG-63 cell extract with a polyclonal rabbit antihuman hFN antiserum. To further characterize this material, the

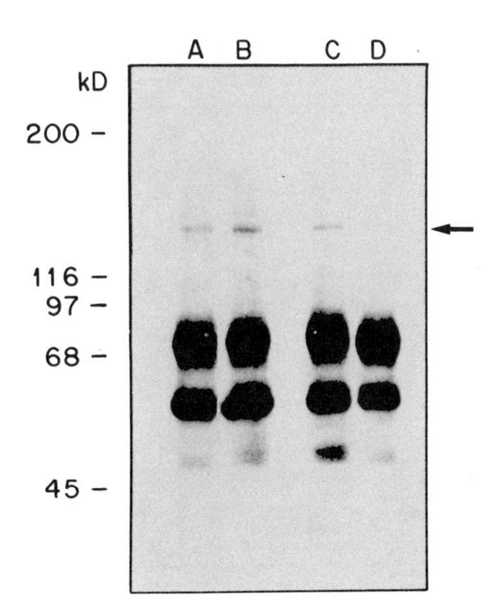

Fig. 3. Immunoblot of proteins immunoprecipitated from whole cell extracts of MG63 human osteosarcoma cells. Lane A, antifibronectin receptor rabbit antiserum; Lane B, anti-WTVPTA mouse antiserum; Lane C, anti-SAVGTL mouse antiserum; Lane D, mouse preimmune serum. *See text* for experimental details.

140kD band was extracted from a silver stained gel (Merril and Pratt, 1986) by incubating the electrophoretic gel slice containing the protein for 24 h at 22°C in 0.4 mL of the following buffer: 50 m*M* Tris-HCl pH = 7.5, 0.1 m*M* EDTA, 0.1% SDS, 5 m*M* DTT, and 100 m*M* NaCl. The supernatant was then mixed with 5 volumes of cold (–20°C) acetone. The mixture was incubated for 30 min on ice and centrifuged for 10 min in an Eppendorf microfuge. The protein pellet was washed with 150 μL of 100% ethanol, vacuum-dried, and resuspended in PBS (Hatamochi et al., 1988). This material was probed in an ELISA assay with the polyclonal rabbit anti-hFN antiserum. Figure 4 shows that recognition by this antiserum is definite, indicating that indeed anti-WTVPTA is effectively immunoprecipitating the hFN present in human osteosarcoma cells.

We have shown, using a microscale method, that it is possible to employ antibodies raised against a short peptide predicted to at least mimic the putative binding site, to purify the human fibronectin receptor, a large protein. Clearly, the availability of monoclonal antibodies could provide enough material to prepare affinity columns for the large scale chromatographical purification of such a receptor.

It must be kept in mind that the synthetic peptides employed for obtaining the antibody may have little resemblance to the actual sequence of the receptor binding site. They simply mimic the corresponding epitope.

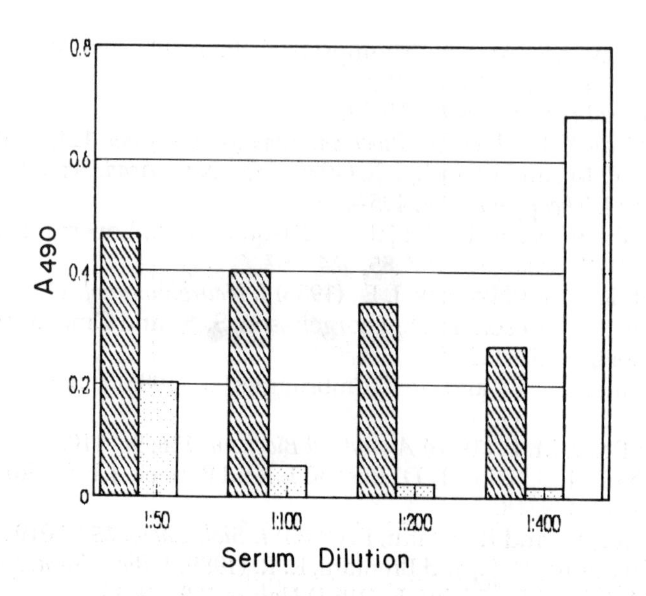

Fig. 4. Probing the identity of the 140 kDa band immunoprecipitated by anti-WTVPTA antibody. After eluting the 140 kDa band from a silver stained polyacrylamide gel, an ELISA assay with purified antifibronectin receptor rabbit antibody was performed. *See text* for further experimental details.

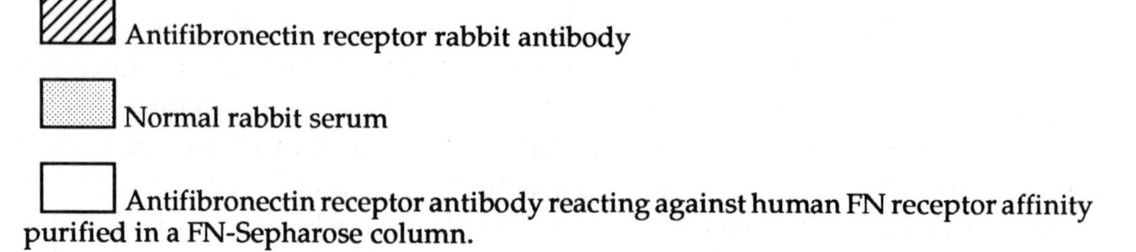

This is why the antibody recognizes the receptor molecule. Peptides displaying hydropathic complementarity to the ATP binding domain of the c-raf protein, deduced either from its complementary mRNA or with a computer program that generates peptides with hydropathic complementarity to a given sequence were synthesized. These peptides were shown to bind specifically to the c-raf protein (Fassina et al., 1989) even though there was a large difference in their amino acid sequence. On the other hand, the above results indicate that the same approach might be of value for the purification of other protein interactants.

References

Argraves, W. S., Suzuki, S., Arai, H., Thompson, K., Pierschbacher, M. D., and Ruoslahti, E. (1987) *J. Cell Biol.* **105,** 1183–1190.

Biro, J. (1981) *Medical Hypotheses* **7,** 981–993.

Blalock, J. E. and Smith, E. M. (1984) *Biochem. Biophys. Res. Com.* **121,** 203–207.

Bost, K. L., Smith, E. M., and Blalock, J. E. (1985) *Proc. Natl. Acad. Sci. USA* **82,** 1372–1375.

Brentani, R. (1988) *J. Theor. Biol.* **135,** 495–499.

Brentani, R. R., Ribeiro, S . F., Potocnjak, P., Pasqualini, R., Lopes, J. D., and Nakai, C. (1988) *Proc. Natl. Acad. Sci. USA* **85,** 364–367.

Carr, D. J. J., Bost, K. L., and Blalock, J. E . (1986) *J. Neuroimmunol.* **12,** 329–337.

Fassina, G., Roller, P. P., Olson, A. D., Thorgeirsson, S. S., and Omichinski, J. G. (1989) *J. Biol. Chem.* **264,** 11252–11257.

Hatamochi, A., Paterson, B., and de Crombrugghe, B. (1986) *J. Biol. Chem* **261,** 11310–11314.

Merril, C. R. and Pratt, M. E. (1986) *Analytical Biochem.* **156,** 96–110.

Mulchahey, J. J., Neill, J. D., Dion, L. D., Bost, K. L., and Blalock, J. E. (1986) *Proc. Natl. Acad. Sci. USA* **83,** 9714–9718.

Oldberg, A., Linney, E., and Ruoslahti, E. (1983) *J. Biol. Chem.* **258,** 10193–10196.

Pasqualini, R., Chamone, D. F., and Brentani, R. R. (1989) *J. Biol. Chem.* **264,** 14566–14570.

Pierschbacher, M. D. and Ruoslahti, E. (1984) *Nature* **309,** 30–33.

Pytela, R., Pierschbacher, M. D., and Ruoslahti, E. (1985) *Cell* **40,** 191–198.

Schwarzbauer, J. E., Tamkun, J. W., Lemischka, I. R., and Hynes, R. O. (1983) *Cell* **35,** 321–431.

Tamkun, J. W., De Simone, D. W., Fonda, D., Patel, R. S., Buck, C., Horwitz, A. F., and Hynes, R. O. (1986) *Cell* **46,** 271–282.

Purification of the Human Interferon-γ Receptor by Ligand Affinity

Daniela Novick, Dina G. Fischer, and Menachem Rubinstein

1. Introduction

Interferon-γ (IFN-γ) is a product of activated T-lymphocytes (Kasahara et al., 1983) and low density lymphocytes (Fischer and Rubinstein, 1983). In addition to its antiviral and growth inhibitory activities, IFN-γ is one of the major immunoregulatory lymphokines. Its immunoregulatory functions include macrophage activation, growth, differentiation, and maturation of various immunocytes and induction of class I and II MHC gene products both in macrophages and in cells of nonhematopoietic origin (for a review *see* Trinchieri and Perrusia, 1985). A survey of its various activities reveals that IFN-γ elicits 50% of its maximal effect at concentrations of 0.5–1 pM. Indeed, the level of IFN-γ in blood or in other tissue fluids rarely exceeds the lower limit of detection, which is about 10 antiviral U/mL (10 pM). Therefore, a highly sensitive and selective system must be present in the cells in order to trace IFN-γ and respond to it. Such a system is the cell surface receptor. In this respect, IFN-γ is similar to other polypeptide hormones, lymphokines, and other cytokines, all acting through specific cell-membrane receptors. This array of polypeptides and their receptors serve as an efficient chemical signaling network that is essential for the existence of multicellular organisms.

The receptor for IFN-γ on various human cells was characterized. High affinity for IFN-γ (Kd ~$1 \times 10^{-10}M$) and a single class of binding sites were demonstrated in these cells. Moreover, receptor mediated endocytosis followed by ligand degradation were consistently observed. However, the receptors on monocytes and on cells of nonhematopoietic origin differed in several aspects, including molecular weight as determined by crosslinking experiments with ^{125}I-IFN-γ (95,000 in HeLa cells; 140,000 in monocytes); down regulation by exposure to excess ligand that was observed in monocytes and not in HeLa cells; and finally, the receptor on monocytes was found to be acid-labile, whereas that on HeLa cells was resistant to acid treatment. These differences indicate that the monocyte receptor is inactivated following internalization, whereas the HeLa receptor retains its structure and recycles back to the cell surface (Fischer et al., 1988). These differences in the properties and fate of these two receptor subtypes may be the molecular basis for the observed differences in biological activity. For example, IFN-γ induces an antiviral state in cells of nonhematopoietic origin but not in monocytes or other cells of hematopoietic origin (Orchansky et al., 1986).

The cumulative data on the mode of action and more recently, on the structure of various receptors, indicate that they share many common features. Structurally, these receptors are membrane anchored proteins made of one or more subunits. The structure of several receptors has already been deduced. These are made of three domains: extracellular, ligand binding domain; transmembrane hydrophobic domain and intracellular domain (e.g., EGF, insulin, and IL-2 receptors, Ullrich et al., 1984,1985; Leonard et al., 1984). The structure of the IFN-γ receptor is not yet known but it was already shown to be a membrane-bound protein. This receptor was recently purified by affinity chromatography on both immobilized interferon-γ and specific monoclonal antibodies (Novick et al., 1987; Aguet and Merlin, 1987; Calderon et al., 1988; Novick et al., 1989). Specific antireceptor antibodies were already used for screening of cDNA expression libraries and receptor related DNA sequences were identified (Aguet and Merlin, 1987). In the following chapters, the purification of the IFN-γ receptor from fibroblasts and placenta and the development of specific monoclonal antibodies to the receptor will be described in detail. These procedures were taken from our recent publications (Novick et al., 1987; Novick et al., 1989).

2. Receptor Isolation Procedures

2.1. Preparation of Immobilized Human IFN-γ

Human interferon-gamma was produced by recombinant Chinese hamster ovary cells (CHO, Mory et al., 1986) and purified to homogeneity by immunoaffinity chromatography on a monoclonal antibody column (Novick et al., 1983). IFN-γ (5 mg in 20 m*M* phosphate buffer pH7, 1 mL) was bound to Affigel-10 (1 mL, BioRad, Richmond CA) overnight at 4°C. The resin was then washed and blocked with glycine (1*M*, 1 mL) for 1 h at room temperature. The resin was then washed briefly with 50 m*M* Na_2CO_3 in 0.5*M* NaCl followed by phosphate-buffered saline (PBS) until neutrality. About 80% of the ligand was immobilized. This resin was kept at 4°C in PBS with 0.02% NaN_3 and was found to be stable for at least 6 mo.

2.2. Preparation of Cell Membranes

Cell membranes were obtained either from cultured foreskin fibroblasts or from term-placentae. All procedures were performed at 0–4°C. Fibroblasts (3×10^{10} cells) were washed with Dulbecco balanced salt solution and suspended in a hypotonic buffer (final concentration: 10 m*M* Tris-HCl pH 7.5, 1 m*M* $MgCl_2$ 1 m*M* $CaCl_2$ 1 m*M* PMSF, 22 TIU/mL aprotinin). The suspension was vigorously mixed on a Vortex shaker and allowed to stand for 5 min before being spun at three different stages: 700xg for 5 min, 3500xg for 10 min, and 40,000xg for 1 h. The 40,000xg precipitate was collected.

Placental membranes were isolated according to the procedure described by Hock and Hollenberg (1980) with some modifications. Fresh placenta was separated from its amniotic sac and cut into pieces of about 20 g. These pieces were minced by a meat grinder into saline and washed thoroughly with saline. Two volumes of Tris-sucrose buffer (20 m*M* Tris-HCl pH 7.4, 0.25*M* sucrose, 1 m*M* PMSF) were added and the pieces were homogenized by an Ultratorax homogenizer (5×30 s, setting 5). The homogenate was filtered through 2 layers of gauze and spun at 600 xg for 10 min to remove the cell debris. The resulting supernatant was spun at 10,000 xg for 30 min. The supernatant was brought to a final concentration of 0.1*M* NaCl and 0.2 m*M* $MgSO_4$ and spun at 48,000 xg for 40 min. The resulting pellet was resuspended in Hepes buffer (50 m*M*, pH 7.6 containing 1 m*M*

PMSF, 20 TIU/mL aprotinin and 0.02% NaN_3) and the suspension was quickly frozen by liquid nitrogen. A preparation of 900 mg membranes (30 mg/mL) was obtained from two placentae (700 g).

2.3. Solubilization and Chromatography of the Receptor

Either intact fibroblasts (30 mL of cell pellet) or cell membranes (30 mg/mL, 15 mL) were suspended in an equal volume of a solubilization buffer (final concentration: 1% Triton X-100, 10 m*M* Hepes, 150 m*M* NaCl, 1 m*M* PMSF, 20 TIU/mL aprotinin, pH 7.5 and left at 4°C for 1 h with occasional shaking. The mixture was then spun (10,000xg, 15 min, 4 °C), the supernatant was collected and spun in an ultracentrifuge (100,00xg, 60 min, 4°C). The supernatant was applied to an immobilized IFN-γ column (1 mL) at a flowrate of 15 mL/h at 4°C. The column was washed with 30 mL phosphate buffered saline (PBS) containing 0.1% Triton X-100 and then the receptor eluted with a buffer consisting of 50 m*M* Na_2CO_3, 500 m*M* NaCl, 0.05% Triton X-100, pH 11. Ten fractions of 1 mL were collected and immediately neutralized to pH 7.5 with 3*M* acetic acid. Column fractions were monitored by IFN-γ binding activity (*see* Section 3.2.).

Further purification of the receptor was obtained by size-exclusion chromatography. A receptor preparation (34 μg) from the ligand affinity column was concentrated by ultrafiltration (Centricon-10, Amicon Corp.) to a volume of 170 μL and was applied to a Sephacryl S-300column (0.6 × 21 cm, Pharmacia). The column was preequilibrated and eluted with phosphate buffer 10 m*M*, pH 7.4 containing 0.05% Triton X-100 and 650 m*M* NaCl. Fractions of 200 μL were collected at a flowrate of 200 μL/min. Each fraction was tested for protein content and IFN-γ-binding activity (*see* Section 3.2.). The Sephacryl S-300 column was calibrated with molecular weight marker proteins (catalase 260,000; Aldolase 160,000; and bovine serum albumin 67,000).

3. Receptor Characterization Procedures

3.1. Preparation of ^{125}I-IFN-γ

Purified IFN-γ (5–50 μg) in sodium borate buffer (40 μL, 0.1*M*, pH 8.5) was labeled with *N*-succinimidyl 3-(4-hydroxy, 5[^{125}I]-iodophenyl) propionate (Bolton and Hunter, 1973, 250 μCi, New England Nuclear, Cambridge, MA). The labeling was performed at 0°C for 1 h followed by 18 h at 4°C. The reaction mixture was then separated on a Sephadex G-25 (medium) column (0.9 × 10 cm), which was pre-equilibrated and was eluted with phosphate-buffered saline (PBS) containing 0.25% gelatin. Eluted fractions (1 mL) were tested for IFN activity and for radioactivity. The recovery of antiviral activity was quantitative, the specific activity was 200–500 cpm/U and the product was stable for at least 6 wk. A sample of ^{125}I-IFN-γ (100,000 cpm) was analyzed by electrophoresis on sodium dodecyl sulfate (SDS)-polyacrylamide (15%) slab gel according to Laemmli (1970). After electrophoresis, the gels were autoradiographed. The typical bands of IFN-γ (25,000, major; 21,000, minor; 17,000, trace) were observed. Recently, we used the chloramine-T method of labeling (Hunter, 1978) at 4°C for 30 s, followed by the same isolation procedure. The recovery of biological activity was quantitative, the specific activity was 800–1000 cpm/U, and the product was stable for at least 3 wk.

3.2. Binding of ^{125}I-IFN-γ to Its Receptor on Cell Surface and in Solution

Adherent cells (HeLa, WISH, or FSll) were seeded in 23 mm wells (12 well plates, Costar) and incubated for 20 h to form a confluent monolayer (2 x 10^6 cells/well). Growth medium was replaced by RPMI-1640 medium supplemented with 10% fetal bovine serum and 20 m*M* Hepes (0.5 mL) together with various amounts of ^{125}I-IFN-γ. Incubation was carried out at 4°C for 2.5 h. The cells were then washed 3 times with cold PBS to remove free IFN and the cell monolayer was detached with trypsin. Cell-associated ^{125}I-IFN-γ was determined by measuring the radioactivity of the detached cells. The affinity of ^{125}I-IFN-γ for the receptor was calculated by a Scatchard analysis of the binding data with the aid of the LIGAND program (Munson and Rodbard, 1980). The affinity of unlabeled IFN-γ for the receptor was determined by binding experiments in which a constant amount of ^{125}I-IFN-γ (10–20% of the saturation level) was added to the cells together with increasing concentrations of unlabeled IFN-γ. Cell-associated radioactivity was determined after 2.5 h at 4°C. Binding data were analyzed by the LIGAND program taking ^{125}I-IFN-γ and unlabeled IFN-γ as different ligands.

Binding of ^{125}I-IFN-γ to the soluble receptor was performed on aliquots (20–30 μL) of the solubilized receptor from various purification steps. These aliquots were mixed with ^{125}I-IFN-γ (250U) either with or without unlabeled IFN-γ (100,000 U) in 20 m*M* Hepes buffer pH 7.5 containing 0.1% BSA (total volume 200 μL). The mixture was incubated for 2 h at 4°C, rabbit

IgG (0.1% in 0.1*M* phosphate buffer pH 7.5, 0.5 mL) was then added, followed by PEG-8000 (22% in 0.1*M* phosphate buffer pH 7.5, 0.5 mL). The mixture was left for 10 min at 4°C and then passed through a 0.45 μ filter (25 mm HAWP, Millipore). The filters were washed with cold PEG-8000 solution (8% in 0.1*M* phosphate buffer, pH 7.5), and then counted. Background counts were determined in the absence of soluble receptors and were subtracted from all readings. The affinity of the solubilized receptor to ^{125}I-IFN-γ was determined similarly by binding a constant amount of the receptor (0.2 μg) with increasing concentrations (0–6000 pM) of ^{125}I-IFN-γ. The dissociation constant was calculated with the aid of a LIGAND program.

3.3. Crosslinking of ^{125}I-IFN-γ to the Receptor on Intact Cells and in Solution

Either HeLa or WISH cells (1×10^8 cells, obtained fram EDTA-treated monolayers) were suspended in 1 mL of cold Dulbecco balanced salt solution. ^{125}I-IFN-γ (20,000 U, 2×10^6 cpm was added with or without a 100-fold excess of unlabeled IFN-γ and the suspension was left for 90 min at 4°C. Disuccinimidyl suberate (DSS, dissolved in DMSO) was then added to a final concentration of 1 m*M* and the suspension was left for 20 min at 4°C. The cells were then washed with Dulbecco balanced salt solution and then solubilized in 0.5 mL Tris buffer pH 6.5 containing Triton X-100 (1.5%), PMSF (1 m*M*), and aprotinin (20 TIU/mL). The suspension was left for 1 h at 4°C and then spun (27,000xg, 20 min, 4°C). The supernatant was immunoprecipitated by incubation with rabbit anti IFN-γ serum (1:50) for 2 h at room temperature followed by addition of protein-A Sepharose. The sepharose beads were washed three times with 10 m*M* phosphate buffer pH 7.5 containing 1% BSA, 1% NP-40 and 2*M* KCl and then twice with PBS. The beads were suspended in a sample buffer containing 2% β-mercaptoethanol and the supernatant analyzed by SDS-PAGE followed by autoradiography.

Crosslinking of ^{125}I-IFN-γ to the solubilized receptor was performed by mixing aliquots (0.5–2.5 μg) with ^{125}I-IFN-γ (250 U, 2×10^5 cpm) for 1 h at room temperature. DSS (dissolved in DMSO) was added to a final concentration of 0.3 m*M*. The crosslinking was stopped after 20 min at 4°C by the addition of Tris-HCl buffer pH 7.5 to a final concentration of 20 m*M*. Aliquots were either analyzed directly by SDS-PAGE followed by autoradiography or immunoprecipitated with rabbit anti-IFN-γ serum as described for crosslinking with intact cells.

4. Monoclonal Antibodies to the Receptor

4.1. Immunization of Mice and Cell Fusion

BALB/c mice were immunized subcutaneously with a preparation of human IFN-γ receptor obtained from placental membranes. This preparation was purified by ligand affinity chromatography followed by size exclusion chromatography on Sephacryl S-300 (*see* Section 2.3.) and finally adsorbed on agarose beads coupled to monoclonal anti-IFN-γ antibodies (Novick et al., 1983). Two injections were given in complete Freund's adjuvant and the other two were given in 1-wk intervals without an adjuvant. Each mouse received ~30 μg of affinity purified receptor per injection. The last boost was given intraperitoneally 4 d before fusion. Sera were checked for their ability to block the binding of ^{125}I-IFN-γ to HeLa cells (*see below*). Spleen cells (200×10^6) from a mouse exhibiting a titer of 1:500 in this assay were fused with 40×10^6 NSO/L myeloma variant (NSO cells were kindly provided by C. Milstein, MRC, Cambridge, UK) (Eshhar et al., 1980). Hybridomas were selected in Dulbecco's modified Eagle's medium, supplemented with 1 m*M* pyruvate, 2 m*M* glutamine, penicillin (10 U/mL), streptomycin 20 μg/mL, fungizone 1 μg/mL, 10% FBS, and containing HAT (Littlefield, 1964). Hybridomas that were found to secrete anti-IFN-γ receptor antibodies were cloned by the limiting dilution method (Zola and Brooks, 1986).

4.2. Screening of Hybridoma Supernatants

Hybridoma supernatants were tested for the presence of anti-IFN-γ receptor antibodies both by competitive inhibition of binding of ^{125}I-IFN-γ to HeLa cells and by neutralization of antiviral activity of IFN-γ on WISH cells.

4.2.1. Inhibition of ^{125}I-IFN-γ Binding to Cells by Antireceptor Antibodies

HeLa cells were seeded in 96 well microtiter plates (50,000 cells/well) in the presence of dexamethasone (10^{-6}*M*) (Finbloom et al., 1985). After 24 h medium was discarded, cells were washed with ice-cold phosphate buffered saline containing Ca^{2+}, Mg^{2+} (PBS) and sodium azide (0.02%). Hybridoma supernatants (50μL/well) were added, and the plates were left for 2 h at 4°C. Following two washings with ice cold PBS containing 2% FBS and 0.02% sodium azide (PBS-2%), ^{125}I-IFN-γ was added to each well (50 μL,

200,000 cpm) and the plates were left for 2 h at 4°C. The plates were then washed 4 times with PBS 2%, harvested with NaOH (0.75*M*, 125 µL) and the content of each well was counted.

4.2.2. Neutralization of Interferon-γ Activity

Hybridoma supernatants (50 µL) were added to cultures of WISH cells in 96 well plates, incubated for 2 h at 37°C and followed by the addition of IFN-γ (20 U/mL, 50 uL). The plates were incubated overnight at 37°C, vesicular stomatitis virus was added, the plates were further incubated overnight and the extent of the cytopathic effect was determined by staining with crystal violet (Rubinstein et al., 1981). For determination of neutralizing titer, hybridoma supernatants (or ascitic fluids) were serially diluted prior to the neutralization assay. One neutralizing unit is defined as the amount of antibody sufficient for neutralizing one unit of IFN-γ.

4.3. Characterization of Antireceptor Antibodies

4.3.1. Titration of Antibody Binding Capacity

A competitive inhibition of binding was performed in 24 well plates (Costar) on HeLa cells (250,000 cells/well without dexamethasone). The assay was done in the same manner as described for the one in 96 well microtiter plates (*see* Section 4.2.1.), except for the volumes (250 µL of serially twofold diluted hybridoma supernatants and 250 µL of ^{125}I-IFN-γ, 200,000 cpm). Cells were harvested with trypsin (250UL), the wells were further washed with PBS (2 × 250 µL) and the combined cells and washings were counted.

4.3.2. Binding of Antireceptor Antibodies to Cells and Competition by IFN-γ

HeLa cells (3 × 10^5/well) were seeded and grown for 24 h at 37°C in 24 well plates. Medium was discarded, various concentrations of antireceptor antibodies (250 µL, in RPMI containing 10% FBS and 0.04% sodium azide) were added together with IFN-γ (0–6 µg/mL) and the plates were incubated for 3 h either at 37°C or at 4°C. Following two washings with PBS-2%, ^{125}I-goat antimouse serum (250 µL, 100,000 cpm) was added. The plates were left for 5 h at room temperature, washed with PBS-2% (3 × 1 mL), harvested with trypsin and counted.

4.3.3. Inhibition of IFN-γ Induced Class II MHC Antigens (HLA-DR) by Antireceptor Antibodies

HeLa cells (5 × 10^4 cells/well) were seeded in 96 well plates and incubated for 3 h at 37°C in RPMI 1640 medium (100 µL, containing 1% FBS, RPMI-1%). Various monoclonal antibodies were added in serial twofold dilutions (in 50 μL RPMI-1%), and the plates were further incubated at 37°C for 3 h. IFN-γ (60 U/mL, 50 μL RPMI-1%) was then added and the plates were further incubated for 40 h at 37°C. The plates were then washed with cold PBS (3 × 100 μL), and fixed with formaldehyde (3.5% in PBS, 100 μL) for 30 min at 0°C. The plates were then rinsed with cold PBS and incubated with a solution of BSA (100 μL, 0.5% in 50 m*M* Tris-HCl, 150 m*M* NaCl, pH7.5) for 30 min at 0°C. The plates were then rinsed with cold PBS and incubated for 1 hat room temperature with monoclonal anti HLA-DR (L-243 ascitic fluid diluted 1:500 in 50 μL RPMI-1640 medium containing 0.1% BSA and 0.1% sodium azide). The plates were then rinsed with PBS and incubated for 30 min at room temperature with ^{125}I-protein A (10^5 cpm/well in 50 μL RPMI-1640 medium containing 0.1% BSA and 0.1% sodium azide). Excess of radioactivity was removed by washing with PBS containing 0.05% Tween 20 (3 × 200 μL). The cells were then solubilized with NaOH (0.75 *N*, 200 μL) and counted.

4.3.4. Western Blotting with Monoclonal Antibodies

Samples of purified receptor (500ng/slot) were analyzed by SDS-PAGE under reducing conditions and electroblotted onto nitrocellulose sheets (BA 85, Schleicher and Schuell) at 60 volt, 250mA in 25 m*M*, Tris-HCl/10 m*M*, glycine buffer (pH 8.5)/20% methanol for 2 h at 4°C. After electroblotting the nitrocellulose sheet was incubated at least 3 h at 37°C with 10% nonfat milk in PBS containing 0.05% Tween-20 and 0.02% sodium azide (blocking buffer). The nitrocellulose was incubated for 2 h at room temperature with a mixture of the three anti IFN-γ receptor monoclonal antibodies (immunoglobulin fraction of ascitic fluids, 10 mg/mL diluted 1:150 in the blocking buffer). Following washings in 0.05% Tween-20 in PBS, the nitrocellulose sheet was incubated for 3 h at room temperature with ^{125}I-goat anti-mouse serum (0.7×10^6 cpm/mL, in the blocking buffer). The sheet was then washed, dried, and autoradiographed.

4.4. Immunoaffinity Chromatography of the Receptor

Immunoglobulin fraction of ascitic fluids (10 mg, 1 mL) obtained by precipitation with ammonium sulfate (final concentration 50% saturation) was coupled to 1 mL of agarose-polyacrylic hydrazide (Wilchek and Miron, 1974). A crude placental membrane preparation was solubilized by 1% Triton X-100 and the 100,000xg supernatant was applied to the antibody column at 4°C at a flowrate of 0.2 mL/min. The column was washed with PBS containing 0.1% Triton X-100 (40 mL) and eluted by citric acid (50 m*M*, pH2) containing 0.05% Triton X-100 and 0.02% sodium azide. Eluted fractions were neutralized by Hepes buffer (1*M*, pH 8.5) and kept at 4°C.

4.5. Other Procedures

4.5.1. SDS-Polyacrylamide Gel Electrophoresis

SDS PAGE was carried out according to Laemmli (1970) using 7.5% polyacrylamide gel for receptor preparations and 15% for ^{125}I-IFN-γ. Following electrophoresis protein bands were visualized by silver staining (Oakley et al., 1980) or by autoradiography.

4.5.2. Protein Determination

Protein was determined by the fluorescamine method (Stein and Moschera, 1981). Crystalline bovine serum albumin was used as a standard.

4.5.3. Interferon Assay

Interferon was assayed by its ability to inhibit the cytopathic effect of vesicular stomatitis virus (VSV) on WISH cells (Rubinstein et al., 1981). A reference standard of interferon-γ Gg23-901-503 was obtained from the NIH and used in all assays.

5. Results

5.1. Ligand Affinity Chromatography of the Receptor

Affinity chromatography on an immobilized IFN-γ column was used as the main step of IFN-γ receptor purification from cultured fibroblasts. A purification profile of the receptor from a Triton X-100 extract of whole cells is shown in Fig. 1. This figure demonstrates the ratio of specific activities in the eluted fractions vs the load fraction, indicating a high degree of purification. When the column effluent was rechromatographed on the IFN-γ column, an additional amount of receptor (equivalent to 50% of the amount obtained in the first chromatography) was bound and eluted. In fact, three cycles of chromatography were required for complete depletion of the binding activity from the crude extracts. SDS-PAGE of aliquots from the various fractions is shown in Fig. 2. Under reducing conditions a major band of apparent M_r = 95,000 was visualized by silver staining of the gel. Additional bands of M_r = 80,000 and 60,000 were seen. Under nonreducing conditions the major protein band was shifted from M_r = 95,000 to M_r > 200,000 (not shown).

Similar patterns of binding activity and protein bands were obtained when a Triton X-100 extract of isolated membranes, rather than the whole cells, was used as a starting material for the purification of the receptor. In this case a purification factor of 300–600 was obtained;however, since the

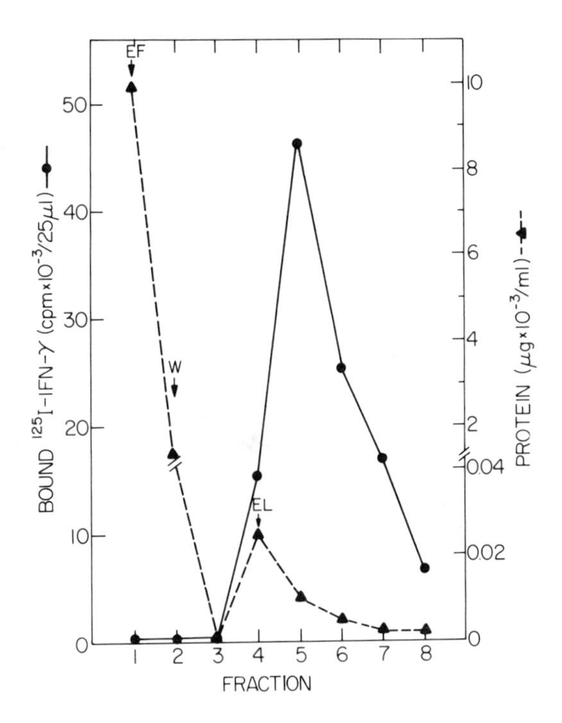

Fig. 1. Ligand-affinity chromatography of the IFN-γ receptor. Triton X-100 extract (100 mL) of whole cells (3×10^{10}) was applied to an IFN-γ agarose column (1 mL) at 4°C and at a flowrate of 15 mL/h. The column was then washed and eluted as described under "Methods." Aliquots (25 μL) of individual fractions [effluent (EF, 100 mL); wash (W, 2 × 15 mL) and elutions (EL, 5 × 1 mL)] were assayed for ^{125}I-IFN-γ binding (●—●) and for protein by the fluorescamine method (▲---▲). Background counts (8500 cpm) were subtracted from all readings.

yields of purified receptor were lower than those obtained with whole cell extracts, the membrane isolation step was omitted. When placenta was used as a starting material instead of cultured fibroblasts, the step of membrane isolation was mandatory. Membranes were then solubilized and applied to the IFN-γ-agarose column.

Chromatography on Sephacryl S-300 was performed in an attempt to further purify the receptor from both fibroblasts and placenta and in order to determine which of the three bands is associated with the receptor. The profiles of protein content and IFN-γ binding activity of the Sephacryl S-300 column using fibroblasts as a source of receptor are shown in Fig. 3. Since the chromatography was performed under nonreducing conditions, the receptor behaved as an aggregate that migrated as a protein of $M_r >$ 95,000. Analysis of the various fractions by SDS-PAGE under reducing conditions revealed a partial separation of the $M_r = 95{,}000$ and $M_r = 80{,}000$

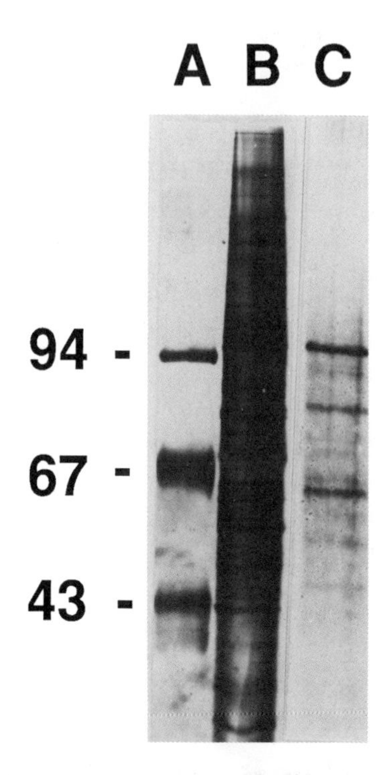

Fig. 2. SDS-PAGE of the purified IFN-γ receptor. Lanes: a. molecular weight markers (from top: phosphorylase 94,000; bovine serum albumin 67,000; oval bumin 43,000); b. Triton X-100 whole cell extract (100,000xg supernatant); c. purified receptor from the IFN-γ agarose column. Proteins were visualized by silver staining.

protein, and the peak of binding activity coincided with the peak of the $M_r = 95,000$ protein. In this case too, the 95,000 band shifted to $M_r > 200,000$ under nonreducing conditions (not shown). The purification of the receptor from fibroblasts and from placenta is summarized in Table 1.

5.2. Characterization of the Purified Receptor

5.2.1. Molecular Weight Determination of the IFN-γ Receptor Complex

Crosslinking experiments of ^{125}I-IFN-γ to intact cells and to the purified receptor preparations were performed. The specificity of the interaction was demonstrated by competition with an excess of unlabeled IFN-γ. SDS-PAGE of the crosslinking reaction products from intact cells followed by autoradiography revealed a broad band of $M_r = 105,000–125,000$. A similar analysis performed on the crosslinking products of ^{125}I-IFN-γ and the

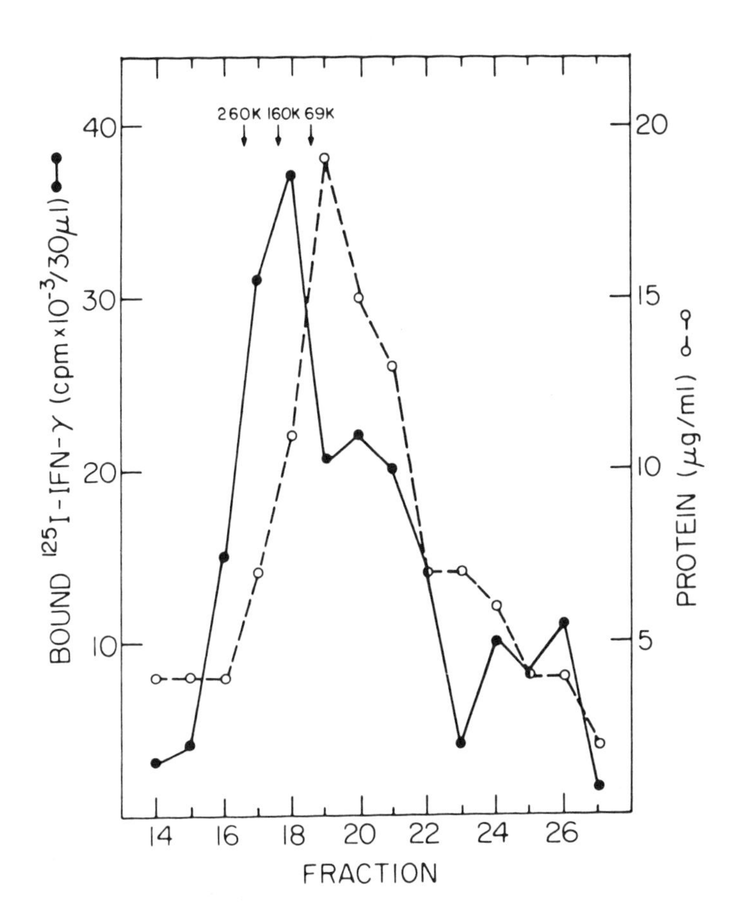

Fig. 3. Steric exclusion chromatography of the receptor. Receptor preparation from the IFN-γ agarose column was concentrated by ultrafiltration and a sample (34 μg in 170 μL) was applied to the Sephacryl S-300 column (0.9 × 21 cm) as described under "Methods." Aliquots (30 μL) from each fraction were assayed for ^{125}I-IFN-γ binding (●——●) and for protein by the fluorescamine method (O - - - O).

purified receptor from a fibroblast membrane preparation gave two bands of M_r = 105,000 and 125,000 as well as a broad band of M_r = 65,000–85,000 (Fig. 4).

5.2.2. Determination of Affinity of ^{125}I-IFN-γ to Intact Cells and to the Solubilized Receptor

Binding experiments with ^{125}I-IFN-γ were performed on a variety of intact cells including foreskin fibroblasts, transformed fibroblasts (HeLa), and epithelial cells. A dissociation constant (K_d) of 3–5 × $10^{-10}M$ and recep-

Table 1
Purification of Human IFN-γ Receptor

Exper.	Step	Protein mg	IFN-γ binding pmol	Specific activity pmol/mg	Purification -fold	Recovery %
1	Solubilized fibroblasts[a]	445	40	0.09	1	100
	IFN-γ agarose eluate	0.025	6	240	2670	15
2	Membrane fraction of fibroblasts	156	48	0.3	1	100
	Solubilized membranes[a]	48	16	0.3	1	33
	IFN-γ agarose eluate	0.08	8	100	325	17
3	IFN-γ agarose eluate of fibroblasts	0.034	2.6	76	1	100
	S-300 fractions 17–20	0.010	4	400	4.5	150
4	Solubilized membranes of placenta	330	8	0.02	1	100
	IFN-γ agarose eluate	0.28	1.6	5.7	286	20

[a]100,000*g* supernatant.

tor density of 10,000–20,000 molecules/cell were calculated in all cases with the aid of the LIGAND program. A typical saturation curve and Scatchard analysis are shown in Fig. 5. The linear Scatchard plot indicates a single type of high affinity binding sites.

Experiments done in order to determine the affinity of ^{125}I-IFN-γ to the purified receptor from fibroblasts yielded a typical saturation curve (Fig. 6, inset). Scatchard analysis performed by the LIGAND program revealed a second order reaction, corresponding to the single class of the high affinity binding sites found in intact cells (Fig. 6). A K_d value of $2.2 \times 10^{-10}M$ was calculated from the binding data.

5.3. Production and Characterization of Monoclonal Antibodies to the Receptor

5.3.1. Screening of the Hybridomas

Hybridoma supernatants were screened for the presence of anti-IFN-γ receptor antibodies by a competitive binding assay and by neutralization of IFN-γ activity. Three out of 468 hybridomas screened were found

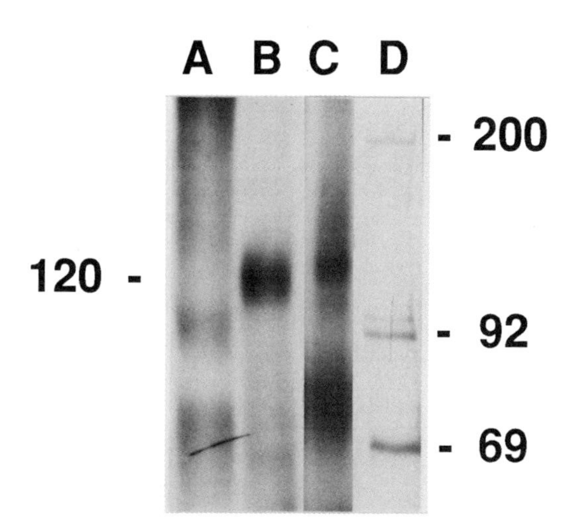

Fig. 4. SDS-PAGE of ^{125}I-IFN-γ crosslinked to the receptor. ^{125}I-IFN-γ was crosslinked to itself (lane a), to WISH cells (lane b). ^{125}I-IFN-γ was also crosslinked to the purified receptor (lanes c). [^{14}C]-methylated molecular weight markers (lanes d) were (from top): myosin 200,000; phosphorylase B 100,000 and 92,000; bovine serum albumin 69,000.

to inhibit the binding of ^{125}I-IFN-γ to HeLa cells, and one out of the three was also positive in the neutralization assay. The positive clones were further grown, subcloned, and the cells were injected into mice for generation of ascitic fluids. The immune response in mice, the screening and the extent of antibody production in tissue culture and in ascitic fluids were all followed both by the binding assay (Table 2) and by the neutralization assay (Table 3).

5.3.2. Characterization of the Positive Hybridomas

All subclones of antibody No. 177 blocked the antiviral activity of IFN-γ. The neutralizing titer of culture fluids of the various subclones ranged from 4,000–30,000 U/mL. No such blocking activity was observed with the other two antibodies (Table 3). In a control study, none of these antibodies were found to block the antiviral activity of IFN-α. None of the antibodies had an intrinsic capacity to elicit an antiviral state in the cells. The anti-receptor monoclonal antibodies were tested for their ability to block the induction of HLA-DR antigens in HeLa cells by IFN-γ. Once again, antibody No. 177-1 exhibited high blocking activity and 50% inhibi-

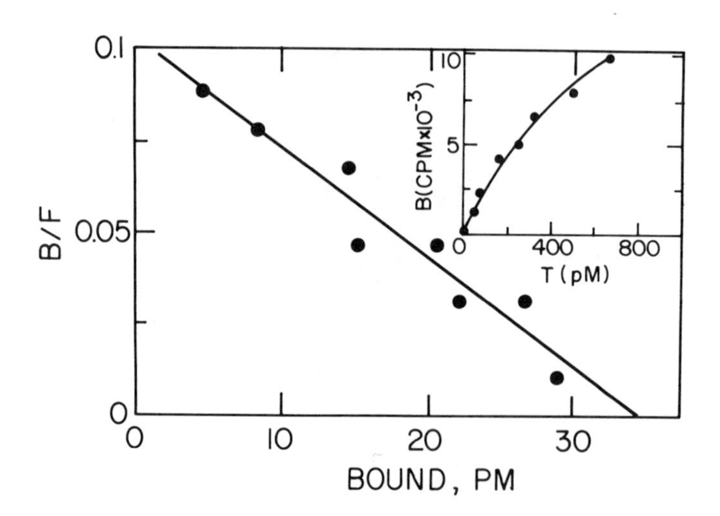

Fig. 5. Equilibrium binding of ^{125}I-IFN-γ to HeLa cells. HeLa cells (2.4 × 10^6) were incubated with various concentrations of ^{125}I-IFN-γ (12.1 µCi/µg) for 2.5 h at 4°C and assayed for binding as described under "Methods." The binding data were analyzed by LIGAND program. Nonspecific binding was 0.02 ± 13% and the K_d was 3.3 × $10^{-10}M$ ± 18%. Inset: Plot of the saturation curve.

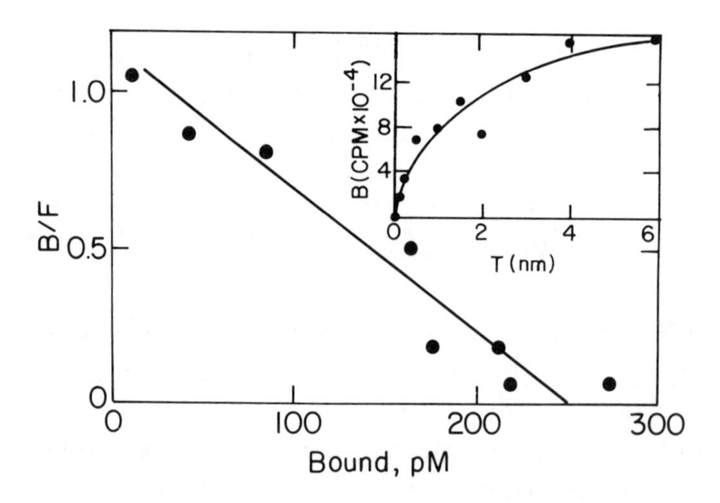

Fig. 6. Equilibrium binding of ^{125}I-IFN-γ to the purified receptor. Aliquots of the receptor from the IFN-γ agarose column (0.2 µg) were incubated together with increasing concentrations of ^{125}I-IFN-γ for 2 h at 4°C as described under "Methods." Binding data were analyzed by the LIGAND program. Nonspecific binding was 0.03 ± 22% and the K_d was 2.2 × $10^{-10}M$ ± 16%. Inset: Plot of the binding curve.

Table 2
Inhibition of ^{125}I-IFN-γ Binding to HeLa Cells by Antireceptor Antibodies

Sample	Inhibition in microplates			Inhibition in 24 well plates		
	antibody dilution	cpm	%	antibody dilution	cpm	%
Immune serum, mouse	1:500	50	83	1:500	550	61
Neg. control serum	1:500	350	0	1:500	1400	0
Hybridoma 37[a]	—	112	68	1:100	250	83
				1:2500	890	40
Hybridoma 177[a]	—	76	78	1:50	900	39
				1:250	1080	27
Hybridoma 183[a]	—	96	73	1:250	370	75
				1:1250	1020	31
Negative Hybridoma	—	350	0	—	1475	0
Ascitic fluid 183-2 immunoglobulins, 16 mg/mL				1:200,000	560	62

[a]Hybridoma supernatants from the first screen were tested in microplates, whereas supernatants of positive clones that were grown in larger amounts were tested in 24 well plates.

tion was observed at ascitic fluid dilution of 1:20,000. Antibodies 37-1 and 183-2 exhibited only a marginal blocking effect. Incubation of HeLa cells with any of the three antibodies in the absence of IFN-γ did not induce HLA-DR antigens. In control experiments a neutralizing monoclonal anti-IFN-γ antibody No. 166-5 (Novick et al., 1983) inhibited the IFN-γ induced HLA-DR, whereas no such inhibition was observed with a monoclonal anti IFN-α antibody No. 7 (Novick et al., 1982).

The three monoclonal antibodies inhibited the binding of ^{125}I-IFN-γ to cells at 4°C (Table 2). However, only antibody 177-1 inhibited the biological activities of IFN-γ. Therefore, comparative binding studies were performed at 37°C in the presence of sodium azide (to prevent internalization). Antibody 177-1 had a significantly higher binding capacity to cells as compared with the other two antibodies. Indeed the addition of IFN-γ at 37°C caused a significant displacement of the bound antibodies No. 37-1 and 183-2 and only minimal displacement of the neutralizing antibody No. 177-1.

5.3.4. Immunoaffinity Chromatography of Placental IFN-g Receptor and Western Blotting

An immunoadsorbent was prepared from an immunoglobulin fraction of ascitic fluid No. 183. A solubilized placental membrane preparation was loaded on the column and the purified IFN-γ receptor was eluted at

Table 3
Neutralization of IFN-γ Activity in WISH Cells by Antireceptor Antibodies

Sample	Titer, U/mL
Immune serum, mouse	35,000
Control serum	<60
Hybridoma 37	<60
Hybridoma 177	2,000
Hybridoma 177-1[a]	4,000
Hybridoma 177-10[a]	30,000
Hybridoma 183	<60
Ascitic fluid 177-10[a]	600,000

[a]A subclone of antibody No. 177.

low pH. The column was monitored by binding of radiolabeled IFN-γ and a purification of 4250-fold was achieved in one step (Table 4). Analysis of the purified receptor preparation by SDS-PAGE under reducing conditions and silver staining revealed the presence of a major band corresponding to a mol wt of 88,000 (Fig. 7). This mol wt band is lower than the one obtained with fibroblast receptors (95,000).

Western blotting of the load fraction, effluent, and eluate was performed. Analysis of an aliquot from the load fraction that consisted of solubilized placental membranes revealed a single band of mol wt 88,000. When this membrane preparation was passed on the immunoaffinity column, the 88,000 band could not be detected in the effluent fraction. Analysis of aliquot from the eluate revealed several bands ranging from mol wt 120,000–50,000. Although a distinct band of mol wt 88,000 was seen, the major band corresponded to a mol wt of 120,000 (Fig. 8).

6. Discussion

In the present study, solubilization and purification of the IFN-γ receptor from human fibroblasts are described. Affinity chromatography on an immobilized IFN-γ column is the main purification step. The purified receptor exhibits a major band of M_r = 95,000 and retains the same affinity to IFN-γ as the receptor on intact cells.

Based on the calculated number of receptors on intact cells and an assumed mol wt of the receptor (95,000) a batch of 3 x 10^{-10} cells should contain at least 50–100 µg of receptor and therefore a 10,000-fold purifica-

Table 4
Immunoaffinity Chromatography of Placental IFN-γ Receptor

Step	Protein, mg	^{125}I-IFN-γ binding, pmol	Specific activity, pmol/mg	Purification, fold
Solubilized membranes	140.6	0.58	0.004	—
Eluate	0.02	0.34	17	4250

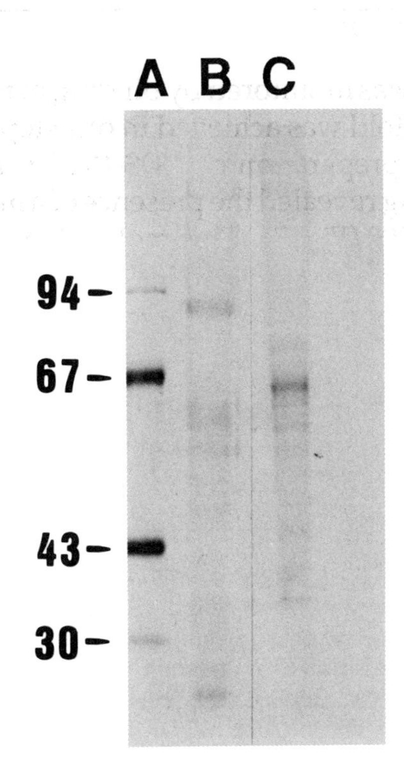

Fig. 7. SDS-PAGE of immumoaffinity-purified IFN-γ receptor. Aliquots of solubilized membrane receptor (lane C, 0.9 μg), immunoaffinity purified receptor (lane B, 0.6 μg), sample mediun alone (lane D), and molecular weight markers (lane A, phosphorylase 94,000; bovine serum albumin 67,000; ovalbumin 43,000 and carbonic anhydrase 30,000) were electrophoresed in the presence of β-mercaptoethanol in polyacrylamide gel. Protein bands were visualized by silver staining.

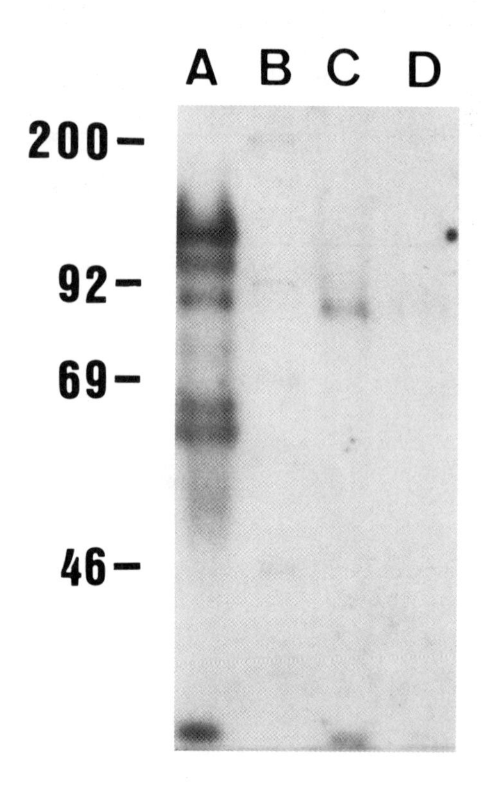

Fig. 8. Western blotting of the IFN-γ receptor. Affinity purified receptor preparation (0.6 μg, lane A), ^{14}C-labeled molecular weight markers (myosin, 200,000; phosphorylase B, 100,000 and 92,000; bovine serum albumin, 69,000 and ovalbumin, 46,000, lane B), solubilized membrane preparation (0.9 μg, lane C), and unbound fraction of the immunoaffinity chromatography (0.8 μg, lane D), were electrophoresed and electroblotted onto nitrocellulose sheets. The nitrocellulose sheet was reacted with antireceptor monoclonal antibodies followed by ^{125}I-goat anti-mouse serum and autoradiographed.

tion (starting with cell extract) is required in order to obtain a pure receptor. Our affinity chromatography yielded 25–90 μg of a partially purified receptor.

The observation that IFN-γ binding activity was found in Sephacryl S-300 fractions that contained a protein of $M_r = 95{,}000$ and was much lower in fractions that did not contain this protein (e.g., Fig. 3 fractions 18 vs 23) indicated that this protein is the receptor for IFN-γ or at least part of it. Crosslinking experiments of ^{125}I-IFN-γ to intact cells yielded a complex of mol wt 105,000–125,000 (Fig. 4). Similarly, bands of M_r = 125,000 and

105,000 were visualized by autoradiography when the purified receptor from fibroblasts was crosslinked to ^{125}I-IFN-γ and the product analyzed by SDS-PAGE. The additional broad band of M_r = 65,000–85,000 could either stem from degradation of the receptor or represent a subunit of the receptor molecule. Assuming a 1:1 molar ratio of ^{125}I-IFN-γ and its receptor, the net mol wt of the receptor is 80,000–95,000. Indeed, these calculated values are in agreement with the bands observed in silver-stained polyacrylamide gels of the purified fibroblast receptor, namely, a major band of M_r = 95,000 and a band of 80,000 (Fig. 2). The M_r = 60,000 band may correspond to the broad band of M_r = 65,000–85,000 that was observed by autoradiography. The purified receptor exhibited the same affinity for ^{125}I-IFN-γ as the receptor on intact cells. Based on the Scatchard analysis, only 2.5% of the protein in the purified receptor preparation represented active receptor molecules. The other proteins in the preparation probably include inactive receptor molecules that migrate as a band of M_r = 95,000 on SDS-PAGE under reducing conditions. The additional bands are probably related to degradation products as well as to nonreceptor proteins that were bound to the IFN-γ-agarose column.

The purified receptor preparation was used to develop monoclonal antibodies and three positive hybridomas were obtained. One antibody (No. 177) was characterized by its ability to inhibit the binding of ^{125}I-IFN-γ to cells (at 4°C); to block the antiviral activity of IFN-γ; to block the induction of HLA-DR by IFN-γ and by its ability to bind specifically to a solubilized IFN-γ receptor. Two other monoclonal antibodies (Nos. 37 and 183) inhibited the binding of ^{125}I-IFN-γ to cells at 4°C but were unable to block the biological activities of IFN-γ. Obviously, all these biological activities are determined at 37°C. Since binding is a prerequisite to biological activity, we tested whether antibody No. 177 had a higher affinity for the receptor as compared with the other two antibodies and whether IFN-γ could displace the antibodies from the receptor at 37°C. Indeed, we found that at 37°C antibody No. 177 had the highest affinity for the receptor. Moreover the two other antibodies could be displaced from cell surface by an excess of IFN-γ, whereas almost no such displacement of antibody No. 177 was observed. It is noteworthy that none of the antibodies exhibited anti- viral activity or HLA-DR inducing activity when incubated with cells in the absence of IFN-γ.

Immunoaffinity chromatography of placental membrane extract yielded a purified IFN-γ receptor that retained its binding capacity. The purified receptor was analyzed by SDS-PAGE followed by silver Staining and by Western blotting. Analysis of this preparation by SDS-PAGE followed by Silver staining revealed a major protein band of mol wt 88,000

(Fig. 7). It is noteworthy that the isolated placental receptor exhibited the same molecular weight as calculated for the receptor of the amniotic cell line WISH, assuming a 1:1 receptor to ligand ratio (Sarker and Gupta, 1984; Rubinstein et al., 1986). However, other values of molecular weight of the receptor were reported as well, based mainly on crosslinking experiments with ^{125}I-IFN-γ. The different values stem from receptor heterogeneity (Rubinstein et al., 1986; Fischer et al., 1988) and from the possible binding to the receptor of either dimers or monomers of IFN-γ.

In the present study, Western blotting analysis of the placental receptor preparation revealed a band of mol wt 88,000, but additional intense bands were observed as well (Fig. 8). These additional bands were not observed in the crude membrane preparation and they probably stem from minor forms of the receptor that were concentrated by the affinity column and were below the limit of detection in the silver-stained gel. The high intensity of the interaction of these minor denatured protein bands with the monoclonal antibody in the Western blot as compared with that of the major 88,000 nondenatured protein might be a result of differences in their affinity for the antibody.

In conclusion, monoclonal antibodies that bind specifically to the receptor and block the binding of IFN-γ were developed. One of the antibodies blocked several biological activities of IFN-γ, thus providing a useful tool in studying the mechanism of action of IFN-γ.

Acknowledgments

We thank Zelig Eshlar for helping us to prepare the monoclonal antibodies. The technical assistance of Rachel Eisenstadt is acknowledged. M. R. has the Edna and Maurice Weiss chair in interferon research.

References

Aguet, M. and Merlin, G. (1987), *J. Interferon Res.* **7,** 683 (Abstract).
Aguet, M. and Merlin, G. (1987), *J. Exp. Med.* **165,** 988–999.
Bolton, A. E. and Hunter, M. W. (1973), *Biochem. J.* **133,** 529–539.
Calderon, J., Sheehan, K. C. F., Chance, C., Thomas, M. L., and Schreiber, R. D. (1988), *Proc. Natl. Acad. Sci. USA* **85,** 4837–4841.
Eshhar, Z., Ofarim, M., and Waks, T. (1980), *J. Immunol.* **124,** 775–780.
Finbloom, D. S., Hoover, D. L., and Wahl, L. M. (1985), *J. Immunol.* **135,** 300–305.
Fischer, D. G. and Rubinstein, M. (1983), *Cellular Immunol.* **81,** 426–434.
Fischer, D. G., Novick, D., Orchansky, P., and Rubinstein, M. (1988), *J. Biol. Chem.* **263,** 2632–2637.
Hock, R. A. and Hollenberg, M. D. (1980), *J. Biol. Chem.* **255,** 10731–10736.

Hunter, M. W. (1978) in *The Handbook of Experimental Immunology.* (Weir, D. M., ed.), Blackwell Press, Oxford, Chapter 14, p. 141.

Kasahara, T., Hooks, J. J., Dougherty, S. F., and Oppenheim, J. J. (1983), *J. Immunol.* **130,** 1784–1789.

Laemmli, U. K. (1970), *Nature* **227,** 680–685.

Leonard, W. J., Depper, J. M., Crabtree, G. R., Rudikoff, S., Pumphery, J., Robb, R. J., Kronke, M., Svetlik, P. B., Peffer, N. J., Waldman, T. A., and Greene, W. C. (1984), *Nature* **311,** 626–635.

Littlefield, T. W. (1964), *Science* **145,** 709.

Mory, Y., Ben-Barak, J., Segev, D., Cohen, B., Novick, D., Fischer, D. G., Rubinstein, M., Kargman, S., Zilberstein, A., Vigneron, M., and Revel, M. (1986), *DNA* **5,** 181–193.

Munson, P. J. and Rodbard, D. (1980), *Anal. Biochem.* **107,** 220–239.

Novick, D., Eshhar, Z., and Rubinstein, M. (1982), *J. Immunol.* **129,** 2244–2247.

Novick, D., Eshhar, Z., Gigi, O., Fischer, D. G., Friedlander, J., and Rubinstein, M. (1983), *EMBO J.* **2,** 1527–1530.

Novick, D., Orchansky, P., Revel, M. ,and Rubinstein, M. (1987), *J. Biol. Chem.* **262,** 8483–8487.

Novick, D., Fischer, D. G., Reiter, Z., Eshhar, Z., and Rubinstein, M. (1989), *J. Interferon Res.* **9,** 315–328.

Oakley, B. R., Kirsh, D. R., and Morris, N. R. (1980), *Anal. Biochem.* **105,** 361–363.

Orchansky, P., Rubinstein, M., and Fischer, D. G. (1986), *J. Immunol.* **136,** 169–173.

Rubinstein, S., Familletti, P. C., and Pestka, S. (1981), *J. Virl.* **37,** 755–758.

Rubinstein, M., Fischer, D. G., and Orchansky, P. (1986) in *Interferons as Cell Growth Inhibitors and Antitumor Factors* (R. M. Friedman, T. Merigan, and T. Sreevalsan, eds.), Alan R. Liss Inc., New York, pp. 269–278.

Sarkar, F. H. and Gutpa, S. L. (1984), *Proc. Natl. Acad. Sci. USA* **81,** 5160–5164.

Stein, S. and Moschera, J. (1981), *Methods Enzymol.* **79,** 7–16.

Trinchieri, G. and Perussia, B. (1985), *Immunol. Today* **6,** 131–136.

Ullrich, A., Coussens, L., Hayflick, J. S., Dull, T. J., Gray, A., Tam, W., Lee, J., Yarden, Y., Liberman, T. A., Schlessinger, J., Downward, J., Mayes, E. L. V., Whittle, N., Waterfield, M. D., and Seeburg, P. H. (1984), *Nature* **309,** 418–425.

Ullrich, A., Bell, J. R., Chen, E. Y., Herrera, R., Petruzzelli, L. M., Dull, T. J., Gray, A., Coussens, L., Liao, Y. C., Tsubokawa, M., Mason, A., Seeburg, P. H., Grunfeld, C., Rosen, O. M., and Ramachandran, A. (1985), *Nature* **313,** 756–761.

Wilchek, M. and Miron, T. (1974), *Methods Enzymol.* **34,** 72–76.

Zola, H. and Brooks, D. (1986) in *Monoclonal Hybridoma Antibodies: Techniques and Applications* (J. G. R. Hurrel, ed.), CRC Press, Inc., Boca Raton, Florida, pp. 34, 35.

Purification of the Cholecystokinin Receptor

Laurence J. Miller

1. Introduction

Cholecystokinin (CCK) is a gastrointestinal and neural peptide hormone with physiological activities at multiple target tissues (Mutt, 1980). These include the classical targets of the gallbladder muscularis smooth muscle and the pancreatic acinar cell (for which this hormone was named "cholecystokinin-pancreozymin") (Ivy and Oldberg, 1928; Jorpes and Mutt, 1966), as well as smooth muscle and some nerves at multiple levels of the digestive tract, including esophagus, stomach, intestine, and colon (Mutt, 1980). In addition, several potential activities have been described for this hormone at the levels of the brain and spinal cord (Dockray, 1982; Yaksh et al., 1982). Thus, there are multiple receptor subtypes from which to choose for CCK receptor purification.

Although debate exists regarding the classification of these receptor subtypes (Dourish and Hill, 1987; Von Schrenck et al., 1988), a reasonable current grouping would include the following.

1. The "peripheral" type receptor on the pancreas and gallbladder (and possibly smooth muscle of the pylorus), which possesses a high degree of

selectivity, requiring the carboxyl-terminal heptapeptide, including sulfation of the tyrosine, for activity (Ondetti et al., 1970; Villanueva et al., 1982).

2. Another type of receptor is found on smooth muscle along the digestive tract other than that in the gallbladder and the pylorus. This receptor subtype is less selective, with better recognition of the desulfated peptide, gastrin, and even the carboxyl-terminal tetrapeptide than the classical "peripheral" type receptor (Bitar and Makhlouf, 1982; Miller, 1984).
3. Central nervous system receptors have been termed "type A" if they have specificities like the "peripheral" type receptor, or
4. "Type B" if they express less selectivity, being less sensitive to the sulfation state of the tyrosine and recognizing the carboxyl-terminal tetrapeptide almost as well as longer fragments of this hormone (Moran et al., 1986). The CCK antagonist L-364,718 (Chang and Lotti, 1986) has been quite useful in distinguishing these receptor subtypes, blocking the Type A, but not the Type B receptors. Type A receptors are present on the area postrema, nucleus of the solitary tract, and interpeduncular nucleus of the brain, and certain regions of the spinal cord (Hill et al., 1987,1988). Type B receptors are the predominant type of receptor found on cerebral cortex (Moran et al., 1986).
5. The classical gastrin receptor found on the gastric parietal cell recognizes gastrin and CCK with similar affinities (Soll et al., 1984). Preliminary observations suggest that these receptor subtypes are all biochemically distinct (Pearson and Miller, 1987; Miller, 1984; Sakamoto et al., 1984; Baldwin et al., 1986; Matsumoto et al., 1987).

Many of these receptor subtypes have been characterized pharmacologically in functional assays with a variety of agonists and antagonists (Mutt, 1980; Maton et al., 1986), and by direct radioligand binding studies (Miller et al., 1981; Shaw et al., 1987; Innis and Snyder, 1980), but biochemical characterization of receptors for CCK has only been performed on pancreas (Rosenzweig et al., 1983; Pearson and Miller, 1987), gallbladder (Shaw et al., 1987; Schjoldager et al., 1988), gastric smooth muscle tumors (Miller, 1984), brain (Sakamoto et al., 1984), and retina (Bone and Rosenzweig, 1988). The overwhelming amount of work has focused on the pancreatic receptor. For this reason, we will focus primarily on this tissue that expresses the "peripheral" type CCK receptor for the remainder of this chapter. Though it is likely that optimal purification schemes for each different subtype of CCK receptor will differ, we anticipate that there will be a great deal of homology in these receptors, and many of the general themes in purification strategies should be consistent among these.

2. Affinity Labeling of the Pancreatic Cholecystokinin Receptor

Cholecystokinin, like many other peptide hormones, occurs as a family of linear peptides that range in length from 58 to 8 amino acid residues, all sharing their carboxyl terminal region (Mutt, 1980; Eysselein et al., 1984). Primary structure-activity studies of the pancreatic CCK receptor have localized the receptor-binding domain of this hormone to be its carboxyl-terminal heptapeptide (Ondetti et al., 1970; Villanueva et al., 1982). In fact, that fragment has been demonstrated to be fully efficacious and as potent as the native peptide in stimulating pancreatic enzyme secretion and gallbladder contraction, the classic activities of this hormone.

Affinity labeling is a powerful method for the biochemical characterization of a receptor that involves the use of a radiolabeled probe to bind specifically to the receptor of interest and the use of a crosslinking technique to establish a covalent bond between the two molecules (Pilch and Czech, 1984). The method of crosslinking can employ a chemically-reactive bifunctional reagent or the use of a photolabile moiety attached to, or incorporated within, the receptor probe (Bayley, 1983). Although each reagent has its own peculiar selectivities and reactivities, the common requirement is the approximation of appropriately-reactive groups on both ligand probe and receptor after binding. For this reason, there is great potential importance in the site of covalent attachment relative to the binding domain for these studies.

The pancreatic CCK receptor has provided a uniquely strong example of this principle. The probe first used for affinity labeling this receptor was Bolton-Hunter labeled CCK-33, in which the sites available for crosslinking, amino groups on lysines in positions one and eleven of the radioligand probe, are far removed from the carboxyl-terminal receptor-binding domain (Rosenzweig et al., 1982,1983). Using that probe with a variety of bifunctional crosslinking reagents (disuccinimidyl suberate, *m*-maleimidobenzoyl-*N*-hydroxysuccinimide ester, ethylene glycol *bis*(succinimidyl succinate), dimethyl suberimidate, and *N*-hydroxysuccinimidyl-4-azidobenzoate), yielded the specific labeling of a M_r = 80,000 pancreatic plasma membrane glycoprotein (variably described to be between M_r = 76,000–95,000) (Rosenzweig et al., 1983; Svoboda et al., 1982; Sakamoto et al., 1983; Madison et al., 1984). In an attempt to confirm that this protein was indeed the hormone-binding site, we developed affinity-labeling probes that were much shorter than CCK-33, with sites of crosslinking adjacent to the carboxyl-terminal receptor-binding domain (Pearson and Miller, 1987; Pearson et al., 1986,1987a,b). In fact, these decapeptides, crosslinked through their alpha amino groups, labeled a M_r = 85,000–95,000 glycoprotein that was distinct from that labeled with the CCK-33-based probe (Pearson and Miller, 1987; Pearson et al., 1987c).

In efforts to further identify the pancreatic CCK-binding protein, we developed probes that incorporated photolabile residues within the theoretical receptor-binding domain (Powers et al., 1988; Klueppelberg et al., 1988a). Photoaffinity labeling through a *p*-nitro-phenylalanine in the position of Phe^{33} at the carboxyl-terminus of this domain also identified the M_r = 85,000–95,000 protein (Powers et al., 1988). Similarly, photoaffinity labeling through a 6-nitro-tryptophan in the position of Trp^{30} in the center of this domain identified the same M_r = 85,000–95,000 protein (Klueppelberg et al., 1988a).

Additional evidence that the M_r = 85,000–95,000 bands labeled by each of these "short" and "intrinsic" probes with sites of covalent attachment that span the receptor-binding domain indeed represented the same binding site was collected in deglycosylation (Pearson et al., 1987c) and protease peptide mapping (Klueppelberg et al., 1988b) studies. The native M_r = 85,000–95,000 bands were each shown to possess core proteins of M_r = 42,000 that were produced on treatment with endo-β-*N*-acetylglucosaminidase F (Pearson et al., 1987c). These deglycosylated core proteins each produced identical peptide maps after treatment with Staphylococcus aureus V8 protease (Klueppelberg et al., 1988b).

This line of evidence suggests that the M_r = 85,000–95,000 pancreatic plasma membrane glycoprotein represents the CCK-binding subunit of the CCK receptor. The remainder of the subunit structure of this receptor is currently unclear. The very common labeling of the distinct M_r = 80,000 protein that has been observed by multiple groups using the "long" CCK-33-based probes under a variety of crosslinking conditions supports the identity of that as another subunit, but it does not appear to be covalently attached to the M_r = 85,000–95,000 subunit. It is not clear whether CCK actually binds to the M_r = 80,000 subunit, or whether that subunit is affinity labeled owing to its proximity to the hormone-binding domain that resides on another subunit. In addition, there is some evidence that a subunit of approximate M_r = 40,000 can be linked to the M_r = 80,000 subunit part of the time via a disulfide bond (Rosenzweig et al., 1983; Sakamoto et al., 1983).

3. Biochemical Characterization of the CCK-Binding Subunit

An understanding of the characteristics of the M_r = 85,000–95,000 protein will certainly be of help in designing a strategy for its purification. Of note, under all electrophoretic conditions employed to date, this band migrates on a SDS-polyacrylamide gel as a poorly-defined broad band. This is fully consistent with its glycoprotein nature (Hames, 1984; Segrest et al., 1971). A series of studies using deglycosylating enzymes with varying specificities and using several lectins has demonstrated that this represents a *N*-linked, complex, sialo-glycoprotein with several carbohydrate chains (Pearson et al., 1987c). We have carefully demonstrated that the highest and lowest portions of the broad band represent differentially-glycosylated forms of the same core protein, that possess similar peptide maps (Klueppelberg et al., 1988b). There is strong precedent for glycoproteins to express charge heterogeneity and to migrate aberrantly on SDS-polyacrylamide gels (Hames, 1984; Segrest et al., 1971).

The number of receptor sites becomes a key factor in the purification of those sites. Unfortunately, no tissue or cell line has been described to possess more than approximately 5000 CCK receptors per cell (Miller et al., 1981; Rosenzweig et al., 1983). This sets the stage for a difficult purification, since the receptor is clearly a very minor membrane protein.

Use of crosslinking reagents with unique reactivities in affinity labeling studies can provide additional biochemical information about the protein being labeled. *m*-Maleimidobenzoyl-*N*-hydroxysuccinimide ester is a bifunctional crosslinking reagent that contains both amino-reactive and sulfhydryl-reactive moieties. Since the peptide probe contains no sulfhydryl groups, any covalent labeling requires a sulfhydryl group on the target protein. Since this has been a very useful reagent for the affinity labeling of the CCK receptor, that molecule must include at least one free sulfhydryl group (Pearson and Miller, 1987; Madison et al., 1984). In addition, brief reduction of this receptor by exposure to dithiothreitol prior to crosslinking with this reagent has been demonstrated to enhance the efficiency of affinity labeling, supporting the presence of additional intramolecular disulfide bonds (Madison et al., 1984).

Further useful information can be suggested by what is known about the CCK stimulus-secretion cascade. Strong evidence exists that this receptor becomes associated with a guanine nucleotide binding protein after ligand binding (Williams and McChesney, 1987; Lambert et al., 1985). Though there is debate whether this cascade actually mediates CCK-induced secretion (Gaisano et al., 1989), the association of these proteins may be utilized in a purification strategy.

4. Fractionation of CCK Receptor-Bearing Tissues

Since CCK is a peptide hormone that cannot diffuse across biological membranes, its receptors should, by definition, be in the plasma membrane with their hormone-binding domain outside the cell. There are autoradiographic (Rosenzweig et al., 1983) and fractionation (Pearson et al., 1987b) data to demonstrate that this is true. A good fractionation procedure, therefore, has the opportunity to enrich for receptor (reported to be up to 150-fold [Pearson and Miller, 1987; Shaw et al., 1987]), at the same time removing many proteolytic enzymes that could potentially damage the receptor. Additionally, incompletely processed forms of the receptor present in intracellular biosynthetic compartments can be eliminated by a fractionation step as well. Various schemes for fractionating the pancreas to produce an enriched plasma membrane fraction have been reported (Miller et al., 1981; Sakamoto et al., 1983). Some schemes focus on high degrees of recovery, whereas others focus on high degrees of enrichment. At this stage, therefore, the sparse number of receptors present per cell must be balanced against the potential damage to those sites by a preparation that is directed toward high recovery with a low degree of purity of the final product.

Fractionation procedures for nonpancreatic tissues bearing CCK receptors are also reported (Miller, 1984; Shaw et al., 1987; Innis and Snyder, 1980). Note that a high degree of enrichment is the key to the ability to affinity label the gallbladder muscularis smooth muscle CCK receptor (Shaw et al., 1987), with a preparation less highly enriched in plasma membrane markers expressing fewer binding sites and not useful for biochemical characterization (Steigerwalt et al., 1984).

5. Solubilization of the CCK Receptor

Once the membrane or cellular source of the CCK receptor has been prepared, the next step in the purification involves the solubilization of this molecule. Like other proteins that are receptors for peptide hormones, the CCK receptor is an integral membrane protein that requires detergents for its removal from the lipid bilayer. An important decision must be made whether the remainder of the purification requires the receptor to be in a nondenatured form that is capable of binding ligand, or whether denaturation is acceptable. Many integral membrane proteins, including receptor molecules, have been successfully solubilized with a variety of detergents (Hjelmeland and Chrambach, 1984). Several of these detergents have been demonstrated to be capable of extracting the CCK receptor from the plasma membrane (digitonin, CHAPS, CHAPSO, Zwittergent 3-14, octyl β-D-glycopyranoside, nonidet P-40, Brij-99, deoxycholate, and sodium dodecyl sulfate) (Szecowka et al., 1985; Lambert et al., 1985; Zahidi et al., 1986), however, solubilization of this receptor in a functional form has proven to be exquisitely difficult (Szecowka et al., 1985).

To date, the best detergent identified for the solubilization of the CCK receptor in a functional form is digitonin (Szecowka et al., 1985). This detergent is a cardiac glycoside that has a sterol structure similar to cholesterol. In fact, its ability to solubilize membrane proteins seems to depend on its binding and the displacement of membrane cholesterol. Optimal conditions for CCK receptor solubilization from pancreatic plasma membranes utilizes one or two percent digitonin. However, using this technique, although the binding activity was no longer sedimentable after digitonin solubilization, the apparent affinity of CCK for solubilized receptor was much lower than that for native receptor, and the number of sites per unit protein was actually lower than that in a highly enriched plasma membrane preparation (Szecowka et al., 1985).

One group has taken advantage of the observation that under solubilization conditions in which binding of native ligand to receptor was quite poor, binding of the CCK receptor antagonist L-364,718 was maintained (Duong and Vlasuk, 1988). They followed the ability of the receptor to bind to this antagonist as a marker of the presence of the receptor through the steps of their purification. There are, however, a much larger apparent number of antagonist-binding sites than CCK-binding sites on intact pancreatic acinar cells (Wank et al., 1988). This has raised concern over the identity of the molecule purified in this type of strategy.

The approach we have taken has relied on the covalent labeling of the CCK receptor using affinity labeling techniques, following the labeled receptor through the subsequent steps of the purification. Because of the theoretical advantages of using a single covalent link to the receptor generated by photolysis of a photolabile probe rather than bifunctional chemical crosslinking, we have used ^{125}I-D-Tyr-Gly-[(Nle^{28,31}, *p*-NO$_2$-Phe^{33}) CCK-26-33] for tracer labeling the CCK receptor for purification. This approach, obviously, has both advantages and disadvantages. The major advantage is that the solubilization and subsequent steps need not require maintenance of the receptor in its native, functional state. However, it is critically important that the purification steps chosen not include those in which labeled molecules might migrate differently from native molecules. Of note, this would probably preclude the use of charge separation techniques such as ion exchange or chromatofocusing after use of an aminoreactive crosslinking reagent. Similarly, the native receptor is very large relative to the labeled probe, but a purification scheme including the separation of receptor fragments may well be a problem because of differences in the migration of fragments that are labeled from those that are not.

Interestingly, it has been demonstrated that the presence of hormone can affect the size and purity of the receptor being solubilized. If the receptor was removed from the bilayer before it came in contact with CCK, it was solubilized independent of a guanine nucleotide binding protein (Williams and McChesney, 1987). If CCK was allowed to bind to its receptor prior to solubilization, the guanine nucleotide binding protein was solubilized within this complex (Williams and McChesney, 1987).

6. Purification of the Solubilized CCK Receptor

6.1. General Considerations

In light of the characteristics of the solubilized CCK receptor, any number of potential purification schemes can be designed. Several general considerations must be kept in mind in choosing a purification scheme. A particularly important component is the decision regarding the scale of the purification. Small scale purifications of this receptor have led to excessive losses of receptor, whereas larger scale purifications have required concentration steps and removal of extraneous proteins so as not to overload columns or gels. Every additional step that must be included in the strategy means additional losses of receptor. This becomes particularly important since this receptor is present in such sparse numbers to begin with. Electroelution of this receptor from SDS-polyacrylamide gels has also been particularly inefficient, leading us to minimize the number of such steps in the purification strategy. A number of potential separation modalities to be considered are listed below.

6.2. Ligand Affinity Chromatography

Affinity steps have been extremely useful in the purification of many types of receptors. The more specific and selective the nature of the affinity, the more useful the step. Among the most useful such purifications are ligand affinity steps. Unfortunately, as noted above, solubilization markedly interferes with the high affinity binding of native ligand to this receptor (Szecowka et al., 1985). Because of this, it is difficult to be certain that the molecule purified on a CCK-affinity column is the high affinity receptor. Since binding of the CCK antagonist L-364,718 was maintained better than agonist binding after solubilization (Duong and Vlasuk, 1988), an antagonist-affinity column might have unique advantages. However, there is no easily accessible site to attach this antagonist to a solid support and to still maintain its binding activity, thus limiting this approach.

6.3. Immunoaffinity Chromatography

An immunoaffinity column would be similarly useful in a purification scheme for this receptor. Certainly, such strategies have been very useful in the purification of many other sparse receptors. Once again, however, no CCK receptor antibodies have been reported to date. Discussing such a step in a purification strategy is actually quite circular, since a successful strategy for generating an antibody might well depend on a successful strategy for purifying the receptor. Multiple groups have unsuccessfully tried to raise CCK receptor antisera using a variety of immunogens, including intact cells, enriched plasma membranes, SDS-polyacrylamide gel slices, and early fractions from purification schemes. Developing a monoclonal antibody to this receptor should theoretically require less purified immunogen, however, this has also not been reported. Although a monoclonal antibody to the parietal cell gastrin receptor has been reported (Mu et al., 1987), we have found no crossreactivity of that antibody with the CCK receptor.

Antibodies are currently available that are directed toward a number of guanine nucleotide binding proteins. Since it is possible to solubilize a complex of CCK receptor with its associated guanine nucleotide binding protein (Williams and McChesney, 1987), an antibody that recognizes that protein in the receptor-associated state might be useful for the immunoaffinity purification of the receptor. Here, too, unfortunately, the guanine nucleotide binding protein that becomes associated with the pancreatic CCK receptor appears to be novel, and not the type to which antisera currently exist (Merritt et al., 1986).

6.4. Lectin Affinity Chromatography

The glycoprotein nature of this receptor makes the use of a lectin affinity step a logical choice. Both the use of wheat germ agglutinin (Pearson et al., 1987c) and Ulex-europaeus agglutinin (Duong and Vlasuk, 1988) have been reported to be useful. This is a very useful step to enrich for plasma membrane glycoproteins if the fractionation chosen produced a high yield, low purity plasma membrane fraction. It can also be used easily in nondenaturing conditions if ligand binding is important. It must be kept in mind that the carbohydrate portion of this glycoprotein is heterogeneous, and a highly selective lectin may well only provide partial yields. We have evidence of this using the fucose-binding lectin Ulex-europaeus agglutinin.

6.5. Organomercurial Affinity Chromatography

Because of the sulfhydryl content of the CCK receptor, an organomercurial affinity column might also be a useful additional step in a purification scheme. We have demonstrated that both the M_r = 80,000 and M_r = 85,000–95,000 proteins bind to such a column (unpublished observations).

6.6. Size Separations

Separations based on size have also been used with this receptor (Duong and Vlasuk, 1988). It is unlikely that such a step would produce a high degree of enrichment since the receptor band is quite broad on a polyacrylamide gel, and since it migrates in a common weight range; however, it is a useful step to reduce the amount of protein or salts to be loaded onto another type of column that might be sensitive to these. An interesting strategy for purification of this receptor might take advantage of the aberrant migration of a heavily glycosylated protein on an SDS-polyacrylamide gel (Segrest et al., 1971). Running the native protein on the first dimension of a gel, cutting out the relevant slice, deglycosylating the material in the slice either enzymatically using endo-β-*N*-acetylglucosaminidase F or chemically using hydrogen fluoride, and rerunning the products on another SDS-polyacrylamide gel should provide dramatic enrichment. Unfortunately, the carbohydrate portion of this receptor is very important in maintaining its solubility, and after deglycosylation it is difficult to keep this receptor in solution without it aggregating. In our hands, this has led to very poor yields of receptor in preparations that have included a deglycosylation step.

6.7. Charge Separations

Charge separations have also been reported for this receptor (Duong and Vlasuk, 1988). S-Sepharose was used with NaCl elution (Duong and Vlasuk, 1988). Such a strategy is probably not useful to produce a high degree of enrichment, since the receptor possesses significant charge heterogeneity owing to its carbohydrate. It is, however, useful as a concentration step. Once again, we should emphasize that certain types of crosslinking (such as that employing reagents that modify amino groups) would likely interfere with this type of purification.

7. Specific Purification Schemes for the CCK Receptor

To date, the most convincing report, though preliminary, of a complete purification of the CCK receptor to homogeneity was that by Duong and Vlasuk (1988). In this, they initially generated total rat pancreatic membranes, and solubilized these with 1% digitonin. They followed the receptor through the purification scheme using its ability to specifically bind the CCK receptor antagonist, ^{3}H-L-364,718. They sequentially used three column chromatography steps, followed by SDS-polyacrylamide gel electrophoresis. The first column was a S-Sepharose ion exchange step. This provided an opportunity to concentrate the receptor while eliminating many proteins. The second column utilized the fucose-binding lectin, Ulex-europaeus agglutinin. The third column utilized size separation on a Sephacryl-S300. Silver stain of the SDS-polyacrylamide gel used to separate the products that eluted from the Sephacryl-S300 column within the activity peak demonstrated only a single protein band at approximately M_r = 90,000, whereas iodination of this material demonstrated that it contained the major M_r = 90,000 protein and minor species of M_r = 200,000, 47,000, and 26,000.

We have followed a similar purification scheme, incorporating highly purified plasma membranes instead of the crude particulate and following photoaffinity labeled receptor rather than the ability to bind antagonist. We started with freshly harvested rat pancreata, that were promptly fractionated using a scheme that included flotation of plasma membranes up into a sucrose gradient (Miller et al., 1981). A small aliquot of these membranes were affinity labeled using the "intrinsic" photoaffinity labeling probe ^{125}I-D-Tyr-Gly-[(Nle^{28,31},*p*-NO_2-Phe^{33}) CCK-26-33] and ultraviolet photolysis (Powers et al., 1988). The labeled membranes were then mixed with the remainder of the membranes and together carried through the purification scheme. The receptors present in these membranes were solubilized using 1% digitonin, and followed through the steps of cation exchange chromatography and lectin affinity chromatography on a fucose-binding column. The radioactivity being followed in our preparation behaved similarly to the protein purified by Duong and Vlasuk (1988) on similar columns. As further confirmation that we were still following the labeled receptor protein, this material was then run on an SDS-polyacrylamide gel. Autoradiography of this gel indeed demonstrated that the radioactivity migrated within the M_r = 85,000–95,000 band. This provides further assurance that the protein purified by Duong and Vlasuk represents the pancreatic CCK receptor.

8. Conclusion

Much is now known about the biochemical characteristics of the pancreatic cholecystokinin receptor. The complete subunit structure of a macromolecular assembly is unclear, but the hormone-binding subunit of this receptor is a M_r = 85,000–95,000 *N*-linked complex sialoglycoprotein with a M_r = 42,000 core protein that contains at least one free sulfhydryl group and an intramolecular disulfide bond. The carbohydrate component of this receptor seems to be important in maintaining its hydrophobic core in solution. This molecule can be purified to homogeneity using a series of steps, including cell fractionation, plasma membrane solubilization with digitonin, and using sequential steps of cation exchange chromatography, lectin affinity chromatography, and SDS-polyacrylamide gel electrophoresis.

References

Baldwin, G. S., Chandler, R., Scanlon, D. B., and Weinstock, J. (1986) *J. Biol. Chem.* **261,** 12252–12257.

Bayley, H. (1983) *Lab Tech. Biochem. Molec. Biol.* **12,** 1–163.

Bitar, K. N. and Makhlouf, G. M. (1982) *Am. J. Physiol.* **242,** G400–G407.

Bone, E. A. and Rosenzweig, S. A. (1988) *Peptides* **9,** 373–381.

Chang, R. S. L. and Lotti, V. J. (1986) *Proc. Natl. Acad. Sci. USA* **83,** 4923–4926.

Dockray, G. J. (1982) *Br. Med. Bull.* **38,** 253–258.

Dourish, C. T. and Hill, D. R. (1987) *Trends Pharmacol. Sci.* **8,** 207–209.

Duong, L. T. and Vlasuk, G. P. (1988) *FASEB J.* **2,** A1796.

Eysselein, V. E., Reeve, J. R., Jr., Shively, J. E., Miller, C., and Walsh, J. H. (1984) *Proc. Natl. Acad. Sci. USA* **81,** G565–G568.

Gaisano, H. Y., Klueppelberg, U. G., Pinon, D., Pfenning, M. A., Powers, S. P., and Miller, L. J. (1989) *J. Clin. Invest.* **83,** 321–325.

Hames, B. D. (1984) in *Gel Electrophoresis of Proteins a Practical Approach* (Hames, B. D. and Rickwood, D., eds.), IRL Press, Washington, DC.

Hill, D. R., Shaw, T. M., and Woodruff, G. N. (1987) *Neurosci. Lett.* **79,** 286–289.

Hill, D. R., Shaw, T. M., and Woodruff, G. N. (1988) *Neurosci. Lett.* **89,** 133–139.

Hjelmeland, L. M. and Chrambach, A. (1984) in *Membranes, Detergents, and Receptor Solubilization,* vol. 1 (Venter, J. C. and Harrison, L. C., eds.), Alan R. Liss, Inc., New York, pp. 35–46.

Innis, R. B. and Snyder, S. H. (1980) *Proc. Natl. Acad. Sci. USA* **77,** 6917–6921.

Ivy, A. C. and Oldberg, E. (1928) *Am. J. Physiol.* **86,** 599–613.

Jorpes, E. and Mutt, V. (1966) *Acta Physiol. Scand.* **66,** 196–202.

Klueppelberg, U. G., Gaisano, H., Powers, S., and Miller, L. J. (1988a) *Gastroenterology* **94,** A408.

Klueppelberg, U. G., Powers, S. P., and Miller, L. J. (1988b) *Gastroenterology* **94,** A230.

Lambert, M., Svoboda, M., Furnelle, J., and Christophe, J. (1985) *Eur. J. Biochem.* **147,** 611–617.

Madison, L. D., Rosenzweig, S. A., and Jamieson, J. D. (1984) *J. Biol. Chem.* **259,** 14818–14823.
Maton, P. N., Jensen, R. T., and Gardner, J. D. (1986) *Horm. Metabol. Res.* **18,** 2–9.
Matsumoto, M., Park, J., and Yamada, T. (1987) *Am. J. Physiol.* **252,** G143–G147.
Merritt, J. E., Taylor, C. W., Rubin, R. P., and Putney, Jr., J. W. (1986) *Biochem. J.* **236,** 337–343.
Miller, L. J. (1984) *Am. J. Physiol.* **247,** G402–G410.
Miller, L. J., Rosenzweig, S. A., and Jamieson, J. D. (1981) *J. Biol. Chem.* **256,** 12417–12423.
Moran, T. H., Robinson, P., Goldrich, M. S., and McHugh, P. (1986) *Brain Res.* **362,** 175–179.
Mu, F.-T., Baldwin, G., Weinstock, J., Stockman, D., and Toh, B. H. (1987) *Proc. Natl. Acad. Sci. USA* **84,** 2698–2702.
Mutt, V. (1980) in *Gastrointestinal Hormones* (Glass, G. B. J., ed.), Raven Press, New York, pp. 169–221.
Ondetti, M. A., Rubin, B., Engel, S. L., Pluscec, J., and Sheehan, J. T. (1970) *Am. J. Dig. Dis.* **15,** 149–156.
Pearson, R. K. and Miller, L. J. (1987) *J. Bio. Chem.* **262 (2),** 869–876.
Pearson, R. K., Hadac, E. M., and Miller, L. J. (1986) *Gastroenterology* **90,** 1985–1991.
Pearson, R. K., Miller, J. J., Powers, S. P., and Hadac, E. M. (1987a) *Pancreas* **2 (1),** 79–84.
Pearson, R. K., Powers, S. P., Hadac, E. M., Gaisano, H., and Miller, L. J. (1987b) *Biochem. Biophys. Res. Commun.* **147,** 346–353.
Pearson, R. K., Miller, L. J., Hadac, E. M., and Powers, S. P. (1987c) *J. Biol. Chem.* **262,** 13850–13856.
Pilch, P. F. and Czech, M. P. (1984) in *Membranes, Detergents, and Receptor Solubilization*, vol. 1 (Venter, J. C. and Harrison, L. C., eds.), Alan R. Liss, Inc., New York, pp. 161–175.
Powers, S. P., Fourmy, D., Gaisano, H., and Miller, L. J. (1988) *J. Biol. Chem.* **263,** 5295–5300.
Rosenzweig, S. A., Miller, L. J., and Jamieson, J. D. (1982) *Fed. Proc.* **41,** 1183.
Rosenzweig, S. A., Miller, L. J., and Jamieson, J. D. (1983) *J. Cell. Biol.* **96,** 1288–1297.
Sakamoto, C., Goldfine, I. D., and Williams, J. A. (1983) *J. Biol. Chem.* **258,** 12707– 12711.
Sakamoto, C., Williams, J. A., and Goldfine, I. D. (1984) *Biochem. Biophys. Res. Commun.* **124,** 497–502.
Schjoldager, B., Powers, S. P., and Miller, L. J. (1988) *Am. J. Physiol.* **255,** G579–G586.
Segrest, J. P., Jackson, R. L., Andrews, E. P., and Marchesi, V. T. (1971) *Biochem. Biophys. Res. Commun.* **44,** 390–395.
Shaw, M. J., Hadac, E. M., and Miller, L. J. (1987) *J. Biol. Chem.* **262,** 14313–14318.
Soll, A. H., Amirian, D. A., Thomas, L. P., Reedy, T. J., and Elashoff, J. D. (1984) *J. Clin. Invest.* **73,** 1434–1447.
Steigerwalt, R. W., Goldfine, I. D., and Williams, J. A. (1984) *Am. J. Physiol.* **247,** G709–714.
Svoboda, M., Lambert, M., Furnelle, J., and Christophe, J. (1982) *Regul. Pept.* **4,** 163–172.
Szecowka, J., Goldfine, I. D., and Williams, J. A. (1985) *Regul. Pept.* **10,** 71–83.
Villanueva, M. L., Collins, S. M., Jensen, R. T., and Gardner, J. D. (1982) *Am. J. Physiol.* **242,** G416–G422.
Von Schrenck, T., Moran, T. H., Heinz-Erian, P., Gardner, J. P., and Jensen, R. T. (1988) *Am. J. Physiol.* **255,** G512–G521.
Wank, S. A., Gardner, J. P., and Jensen, R. T. (1988) *Biomed. Res. Suppl.* **1,** 186.
Williams, J. A. and McChesney, D. J. (1987) *Regul. Pept.* **18,** 109–117.
Yaksh, T. L., Abay, E. O., and Go, V. L. W. (1982) *Brain Res.* **242,** 279–290.
Zahidi, A., Fourmy, D., Darbon, J.-M., Pradayrol, L., Scemama, J.-L., and Ribet, A. (1986) *Regul. Pept.* **15,** 25–36.

Index